International Series in Operations Research & Management Science

Volume 188

For further volumes:
http://www.springer.com/series/6161

Charles S. Tapiero

Engineering Risk and Finance

Springer

Charles S. Tapiero
Department of Finance and Risk Engineering
Polytechnic Institute of New York University
Brooklyn, NY, USA

ISSN 0884-8289
ISBN 978-1-4614-6233-0 ISBN 978-1-4614-6234-7 (eBook)
DOI 10.1007/978-1-4614-6234-7
Springer New York Heidelberg Dordrecht London

Library of Congress Control Number: 2012953261

Printed on acid-free paper

Springer is part of Springer Science+Business Media (www.springer.com)

Preface

Risk and uncertainty are neither topics of recent interest nor a fashion arising due to an increased awareness that uncertainty prevails—fed by information, financial crises, an economy in turmoil, a networked world, and an economic environment increasingly unpredictable. To better mitigate the implications of uncertainty on our life, on our work, on the economy, on our health, and on our environment, we construct risk models. These are models of uncertainty, framing uncertainty in terms of what we know and can predict and provide estimates to their consequences (whether adverse or not). These models are defined using many considerations, predictable factors—some external, some strategic, some based on statistical estimates, some on partial information, some derived from what we actually do, some due to neglect, etc. In these cases, "risk models" seek to construct and define a coherent and practical set of measures, which are analyzed and used to confront objectively and subjectively (based on our values and preferences) the uncertainty we face. These themes underlie also the world of finance.

These elements are common to many disciplines that concern individuals, large and small firms, industries, governments, and societies at large. Industrial risks, strategic (military, economic and competitive) risks, nuclear risks, health and bio-risks, marketing and financial markets risks, environmental risks, contagion risks, etc. are all models of uncertainty with risks defined, measured, assessed, analyzed, and controlled that we seek to value, price, and manage. An interdisciplinary convergence of risk models and their techniques arises due to their common concerns. Various professions have increasingly learned from each other, developing the common means that lead to such a convergence and contributing to the engineering of risks, their management, their valuation, and pricing through contracted exchanges and financial markets. A horizontal risk convergence is prevalent across disparate professions facing similar risk models that contribute to both mutual learning and exchange. For example, statistical controls are applied to control food safety, to health care, to track and audit tax returns, etc. A risk convergence—both horizontal and vertical—has contributed to a greater awareness that risk is no longer a "derivative" or a "consequence" but an integral part of everything we are and we do, what we pay for, and what we seek to profit from.

The commonalities of risks, the need to mitigate, share, transfer, and trade risks, have increasingly contributed to the need for a common valuation of risks, its exchange price, and thereby to the special role of money (and therefore finance) as a common "risk metric." This book recognizes both the specificity of risks in its many manifestations and, at the same time, the special importance that finance assumes with the growth of financial markets and insurance where "risks of all sorts" are being exchanged.

The many definitions of "Uncertainty", "Risk" and money can only be covered partially. There is an extraordinarily large number of publications, academic, practical, philosophical, ethical, religious, social, economic, financial, technical (statistical, stochastic models, etc.) that preclude a truly representative coverage. Every aspect of uncertainty and risk models (whether technical or conceptual) is both specific and general at the same time. Setting even its principle elements is overreaching. For this reason, the intent of this book is to provide a partial coverage of elements that seek to bridge theoretical notions of risks and their uses in economics and finance, as well as use examples and applications to highlight their importance.

The book is both narrative and quantitative, outlining a large variety of uncertainty and risk related issues, with examples that emphasize their useful applications. Quantitative techniques particularly based on probability and statistical techniques are both essential to construct risk models and tools to analyze and control risks. Elements of these techniques are presented in this text in three quantitative chapters reviewing basic notions of probability, statistics, and stochastic process modeling. An additional chapter (Chap. 12) is also used to provide an intuitive outline of game theory. These chapters are kept at an introductory level, although some sections require prior studies in applied probability and statistics.

A quantitative formulation is required to both anchor the definition and provide a frame of reference for risk models. The need for quantitative tools in risk analysis and convergence does not negate or reduce the importance for a greater understanding of what is uncertainty, what are risk models, and what are the principles that can reconcile their conceptual meaning and uses in finance. This book, in an attempt to do so, albeit only in a limited sense, focused on many applications and problems. In particular, the book emphasizes the irrevocable interdependence of defining risks, measuring them, and the techniques to assess, to value, to price, and to control financial risks. In some chapters, new approaches to pricing and controlling risks are introduced. These span the development of multi-agents expanded CCAPM pricing model (Consumption Capital Assets Pricing Model) and strategic (game like) statistical controls in the regulation of financial firms.

Although the book emphasizes primarily economic–financial and management problems, other issues and application problems are discussed. In particular, legal issues, health care, and extreme risks are used to emphasize that risk models and techniques, albeit often used in different ways, are in fact quite similar. Chapters 6, 7, 8, 9, and 10 in particular are devoted to the economics, the valuation, and the price of risk and their models, while Chap. 11 is devoted to risk and strategic risks controls and regulation.

To complement some of the topics covered in the text, an extensive list of references is included in a special section at the end of each chapter, directing the reader to specific references for further applications and study.

In writing this book, I surveyed an extremely large number of papers on fundamental risk theories, some on quantitative risk measurement, valuation, and pricing and some derived risks and papers easily accessible through the Internet. In particular, these papers are accessible through academic services and Web sites such as sciencedirect.com, the SSRN (Social Science research Network), GLORIA (for financial credit risks and derivatives) and Web sites with a special focus in risks of all sorts. I soon realized that there is little one may innovate or add to the extraordinary and accessible explosion of currently published and working papers or to an endless list of econometric and statistical studies outlining educated viewpoints and diffused freely. Yet, I also realized that such an explosion of knowledge is also confusing, difficult to digest, and contains fundamental ideas drowned by information excess. In fact, most of the fundamental theories and applications of risk related papers we mostly refer to are in fact pre-Internet research papers or fundamental theories. This may explain the selection of references used in this book that may seem outdated. It also reinforced my belief that writing books to integrate diffused knowledge is probably more important today than it ever was before. Thus, while I do not believe that this book will add any particular or specific knowledge (except hopefully for some particular and selected problems in risk valuation and control in chapters 8–11), I hope that it will provide an overview of risk in its multiple manifestations, risk models, and uncertainty and thus lead to a better understanding of what is risk and how we may be able to value, price, and confront its consequences.

"Engineering Risk and Finance" is structured as follows. The first two chapters provide a cursory overview of basic concepts such as risk and uncertainty, risk manifestations across numerous areas. A broad overview of conceptual approaches to risk management is also outlined. There is an extensive literature on risk management in all professions that the reader may wish to consult as well. These two chapters are nontechnical providing some motivation for subsequent and technical chapters. The second part, consisting of Chaps. 3, 4, and 5, are essentially technical, reviewing well-known risk and probability models applied to a variety of risk problems to highlight their usefulness. Probability and statistics are an inherent part of risk models, their analysis, and their control. Further, often "everything we do or wish to do" is defined in terms of probability and statistical notions. An appreciation of what these probabilities mean, how they are defined and used is necessarily important for any text on risk. Chapter 3 covers basic probability models, moments, distributions, and their use in selected risk models. Chapter 4 is concerned with multivariate models emphasizing the fact that many risks occur due to the dependence of multiple factors. Chapter 5 is concerned with stochastic models and risk modeling in an inter-temporal perspective. Quantitative models are always based on the specific assumptions made that underlie the definition of risk events, their probability, their causal processes, and their consequences.

Appreciating these assumptions, both for their usefulness and their implications is an important part of risk engineering.

For some students and readers, these quantitative notions are well known and can therefore be skipped (although examples are used to highlight their usefulness) while for others, these may be a bit difficult, and therefore, some sections are starred to indicate their difficulty.

Chapters 6, 7, 8, and 9 introduce principles and methods for risk measurement (Chap. 6), valuation (Chap. 7), risk economics (Chaps. 8 and 9), and uncertainty economics (Chap. 10 by Oren Tapiero). In Chap. 6, we distinguish between statistical measurements, measurements of value and deviations underlying a great number of risk measures. For example, techniques such as risk detection, using a standard deviation as a proxy to manage risks, etc. are outlined and illustrated through numerous examples. In Chap. 7, we emphasize risk valuation using a plethora of techniques as well as utility theory in setting a foundation to risk economics. At the same time, the basic concepts of complete markets for (risk) assets pricing is introduced. Chapter 8 pursues these developments to value the risk of more complex problems. In particular, the concept of (utility based) CCAPM to price certain assets is extended to include a variety of other situations. The development of this framework (in particular the multi-agents Extended CCAPM, which I have pursued in a number of academic papers) is somewhat new and provides an opportunity to study a great many situations and problems to price risks assets in terms of real policy variables as well as a function of macroeconomic factors. Applications to a variety of problems, are then used to delineate both the usefulness and the limits of such approaches. For example, pricing the exchange between a debtor and a lender, the risk and price of economic inequality, the price of rationing, the price other regulation, and so on. Chapter 9 provides additional applications extending Chap. 8. Chapter 10 introduces an approach to "Uncertainty Economics". It is based on the Doctoral Dissertation of Oren Tapiero (no coincidence, he is my son). It emphasizes an approach to the incomplete Arrow–Debreu theory of pricing using non-extensiveness, Tsallis (and Boltzmann–Gibbs) Entropy, and Quantum Physics. This chapter may be viewed as providing a quantified approach to "behavioral finance." Chapter 11 provides an overview of risk and strategic control techniques for regulation. Given the profusion of texts in this area, the chapter merely outlines its principles and focuses on strategic control problems (based on Game Theory models). Some of the examples used are based on an outgrowth of my past papers and books published. In addition, given the practical importance of management approaches such as 6 Sigma in industrial risk management, robust decision making and experimental design, queue network risks, and their control, these problems are also introduced because of their importance for risk management. The essential contribution of this chapter, however, is in its formulating and solving several problems of regulation statistical control. Particular cases are developed providing a theoretical framework to assess the efficiency and the implications of regulation controls, on both the regulated and the regulator. Again, references are added in the text for the motivated reader.

Chapter 12 provides finally a partial overview of risk games or strategic risks. Such games are important when consequences depend as well on parties' decisions reflecting their information, their preferences, and agendas. Such risks occur in environmental problems, in supply chains, in competitive economic and financial markets, in contracts negotiations, in cyber-risks, etc. In fact, increasingly, risks have become strategic. It is, therefore, essential that techniques that conceptualize these special characteristics be addressed. In this sense, Chap. 12 is partly an appendix to strategic issues considered in a number of chapters.

This book is intended as a background text for undergraduate and graduate courses in Risk Finance, in Risk Engineering and Management, as well as a book intended for professionals that are both concerned and experienced in some aspect of risk assessment and management techniques. Given the book's finance and interdisciplinary approach, it differs from functional books in these areas in its attempt to view risk as representing common issues faced by many disciplines. As a result, an appreciation of uncertainty and risk, what it means, how they differ, their manifestations, and how to value and manage both uncertainty and risk models are perceived as generic problems relevant to industry, to business, to health care, to finance, etc. Professional readers, aside from financial managers, and financial and risk engineers may, therefore, (hopefully) find some elements in this book to be useful or find another approach to risk and uncertainty which is based on "money valuation" which they may have not been aware of.

Of course, experience and approaches to risks and their management have been devised by numerous professions, resulting from risk technology transfers between these professions and finance. The intent of this book is to capitalize on this "technology transfer." All disciplines concerned by risks and how they define and confront it have contributed an enormous and overbearing number of books, academic papers, and general publications. While the number of papers and books I consulted was extremely large, it is possible that some ideas and some results were reproduced by neglect or due to my being unaware of the appropriate reference. I apologize if this is the case. I have borrowed heavily from articles I have published over the past years as well as new results resulting from my own research and my many collaborative papers. Of course, I would like to express my gratitude to all the collaborators I had over the years and from whom and from each I have learned much.

Finally, I have profited from discussions, comments, and help from many students, colleagues, and friends. Although they are many, I wish to thank my colleagues, Nassim Taleb, Alain Bensoussan, Elizabeth Pathe-Cornell, Pierre Vallois, Raphael Douady, Mirela Ivan, Konstantin Kogan, Oren Tapiero, Mina Teicher, Bertrand Munier, Agnes Tourin, Fred Novomestky, my children Daniel, Dafna, and Oren—all of whom are concerned with risks, financial and global, my students, Jin Qiuzzi, Yijia Long, Ge Yan, and so many others from whom I have learned much. I also wish to thank the Sloan Foundation, and in particular Prof. Dan Goroff for the support and encouragement they have provided.

Not least, I am also thanking my partner Carole, who had the patience to tolerate the endless frustrations to have this book finished.

Finally, I wish to dedicate this book to my mother, Violette Budestchu Tapiero, whose love and care while alive nourished me and all my family.

Brooklyn, NY, USA Charles S. Tapiero

Contents

1 Engineering Risk 1
1.1 Risks and Uncertainty Everywhere 1
1.2 Many Risks 4
1.2.1 Globalization and Risk 4
1.2.2 Space and Risk 5
1.2.3 Catastrophic Risks 5
1.2.4 Debt, Credit and Counter-Party Risk 8
1.3 Industry and Other Risks: Deviant or Money 11
1.3.1 Technology and Risks 11
1.3.2 Technology and Networking 12
1.3.3 Technology and Cyber Risks 13
1.3.4 Example: Technology Risks, Simplicity and Complexity Risk Mitigation 13
1.4 Quality, Statistical Controls and the Management of Quality 14
1.5 Health and Safety Risks 16
1.6 Finance and Risk 18
1.6.1 The Risks of Certainty 18
1.6.2 The Risks of Complexity 19
1.6.3 The Risks of Regulation (and Non Regulation) 19
1.6.4 Micro-Macro Mismatch Risks and Politics 19
1.6.5 Risk and Incomplete Markets 21
1.6.6 Risk Models and Uncertainty 22
1.7 Corporate Risks 23
1.8 Risk and Networked Firms 26
1.8.1 Information Asymmetry 27
1.9 Risks—Many Reasons, Many Origins 28

2 Risk Management Everywhere 33
2.1 Elements of Applied Risk Management: A Summary 33
2.2 Risk Management, Value and Money 34

2.2.1 Insurance Actuarial Risk . . . 36
2.2.2 Finance and Risk . . . 37
2.3 Industry Processes and Risk Management . . . 40
2.4 Marketing and Risk Management . . . 44
2.4.1 Reputation Risks . . . 44
2.4.2 Advertising Claims and Branding Risks . . . 45
2.4.3 IPO, Reputation and Risks . . . 46
2.5 Externalities and Risks Management . . . 49
2.6 Networks and Risks . . . 50

3 Probability Elements: An Applied Refresher . . . 57
3.1 Introduction . . . 57
3.2 Risk and Probability Moments . . . 58
3.2.1 Expectations, Variance and Other Moments . . . 58
3.2.2 The Expectation . . . 58
3.2.3 The Variance/Volatility: A measure of "Deviation" . . . 59
3.2.4 Skewness, Kurtosis and Filtration . . . 59
3.2.5 Range and Extreme Statistics . . . 60
3.3 Applications . . . 61
3.3.1 Skewness in Standardized Stocks Rates of Returns . . . 61
3.3.2 Reliability, Probability Risk Constraints and Deviations' Risks . . . 62
3.3.3 The Hazard Rate and Finance . . . 64
3.3.4 Risk Variance and Valuation . . . 65
3.3.5 VaR or Value at Risk . . . 67
3.3.6 Chance Constraints . . . 68
3.3.7 Type I and Type II Statistical Risks . . . 69
3.3.8 Quality Assurance and Chance Constraints Risks . . . 70
3.3.9 Credit and Credit Granting and Estimation of Default Probabilities . . . 71
3.3.10 Chance Constrained Programming . . . 73
3.3.11 Chance Constraint Moments Approximations . . . 75
3.3.12 Transformation of Random Variables into Normally Distributed Random Variables . . . 75
3.4 Generating Functions . . . 76
3.4.1 The Convolution Theorem for Moment and Probability Functions . . . 77
3.4.2 The Probability Generating Function of the Bernoulli Experiment . . . 79
3.4.3 Additional Examples . . . 81
3.4.4 The PGF of the Compound Poisson Process . . . 82
3.5 Probability Distributions . . . 83
3.5.1 The Bernoulli Family . . . 84

3.5.2 The Binomial and Other Distributions 85
3.5.3 The Poisson Distribution 90
3.5.4 The Conditional Sum Poisson and the Binomial Distribution 91
3.5.5 Super and Hyper Poisson Distributions 92
3.5.6 The Negative Binomial Distribution (NBD) 92
3.6 The Normal Probability Distribution 93
3.6.1 The Lognormal Probability Distribution 94
3.6.2 The Exponential Distribution 95
3.6.3 The Gamma Probability Distribution 95
3.6.4 The Beta Probability Distribution 96
3.6.5 Binomial Default with Learning 97
3.6.6 The Logistic Distribution 97
3.6.7 The Linear Exponential Family of Distribution 98
3.7 Extreme Distributions and Tail Risks 98
3.7.1 Approximation by a Generalized Pareto Distribution . . 100
3.7.2 The Weibull Distribution 100
3.7.3 The Burr Distribution 101
3.8 Simulation 104

4 Multivariate Probability Distributions: Applications and Risk Models 109
4.1 Introduction 109
4.2 Measures of Co-variation and Dependence 110
4.2.1 Statistical and Causal Dependence: An Oil Example 110
4.2.2 Statistical Measures of Co-dependence 112
4.3 Multivariate Discrete Distributions 117
4.3.1 Estimating the Bi-variate Bernoulli Parameters 124
4.3.2 The Bivariate Binomial Distribution 126
4.3.3 The Multivariate Poisson Probability Distribution 127
4.4 The Multivariate Normal Probability Distribution 128
4.5 Other Multivariate Probability Distributions (*Statistics and Probability Letters*, 62, 203, 47–412) 128
4.6 Dependence and Copulas 130
4.6.1 Copulas and Dependence Measures 135
4.6.2 Copulas and Conditional Dependence 136

5 Temporal Risk Processes 139
5.1 Time, Memory and Causal Dependence 139
5.2 Time and Change: Modeling (Markov) Random Walk 141
5.2.1 Modeling Random Walks 142
5.2.2 Stochastic and Independent Processes 143
5.2.3 The Bernoulli-Random Walk: A Technical Definition 143

5.2.4 The Trinomial Random Walk . . . 145
5.2.5 Random Walk as a Difference Equation . . . 145
5.2.6 The Random-Poisson Continuous Time Walk . . . 146
5.2.7 The Continuous Time Continuous State Approximation . . . 148
5.2.8 The Poisson-Jump Process and its Approximation as a Brownian Model . . . 149
5.2.9 The Multiplicative Bernoulli-Random Walk Model . . . 150
5.2.10 The BD Model in Continuous Time with Distributed Times Between Jumps . . . 151
5.3 Inter-Event Times and Run Time Stochastic Models . . . 153
5.4 Randomized Random Walks and Related Processes . . . 154
5.4.1 The Randomized Random Walk Distribution . . . 154
5.4.2 Binomial-Lognormal Process . . . 155
5.5 Markov Chains . . . 156
5.6 Applications . . . 159
5.6.1 The Sums of Poisson Distributed Events Is Also Poisson . . . 159
5.6.2 Collective Risk and the Compound Poisson Process . . . 159
5.6.3 Time VaR . . . 161
5.6.4 A Portfolio Trinomial Process . . . 163
5.7 Risk Uncertainty, Rare Events and Extreme Risk Processes . . . 166
5.7.1 Hurst Index, Fractals and the Range Process . . . 169
5.7.2 R/S and Outliers Risks . . . 172
5.7.3 RVaR, TRVaR and Volatility at Risk . . . 173
5.7.4 The Generalized Pareto Distribution (GPD) . . . 178
5.7.5 The Normal Distribution and Pareto Levy Stable Distributions . . . 180
5.8 Short Term Memory, Persistence, Anti-persistence and Contagion . . . 182
5.8.1 Mathematical Calculations . . . 183
5.8.2 Persistence and the Probability of Losses in a Contagion . . . 188

6 Risk Measurement . . . 195
6.1 Introduction . . . 195
6.2 Big Data and Risk Measurement . . . 199
6.3 Decision and Risk Objective Measurements . . . 201
6.4 Risk Measurement in Various Fields . . . 204
6.4.1 Medical Risk Measurement . . . 204
6.4.2 RAM as Performance and Risk Measures . . . 207
6.4.3 Quality and Statistical Tracking . . . 208
6.4.4 Operations and Services and Risk Measurements . . . 208

6.5 Bayesian Decision Making: EMV and Information 209
6.6 Multi Criteria and Ad-Hoc Objectives 211
6.6.1 Perron-Froebenius Theorem and AHP 212
6.6.2 The Data Envelopment Analysis and Benchmarking . . 212
6.7 Risk Measurement Models: Axiomatic Foundations 213
6.7.1 Coherent Risk Measures 213
6.7.2 Axiomatic Models for Deviation Risk Measurements 215
6.7.3 Absolute Deviations 215
6.7.4 Inequality Measures 216
6.7.5 The Variance and the VaR 216
6.7.6 Entropy and Divergence (Distance) Metrics 216
6.8 Functional and Generalized Risk Measurement Models 218
6.9 Examples and Expectations 219
6.9.1 Models Based on Ordered Distributions' Measurement 220

7 Risk Valuation 223
7.1 Value and Price 223
7.2 Rational Expectations, Martingales and the Arrow-Debreu Complete States Preferences 224
7.2.1 Rational Expectations Models: A Simple Quantitative Definition 227
7.2.2 The Inverse Kernel Problem and Risk Pricing 229
7.3 Utility Models and Valuation 232
7.3.1 Critique of Expected Utility Theory in Measuring Preferences 234
7.3.2 Examples and Problems 235
7.4 Risk Prudence and Background Risk 242
7.4.1 Risk, Uncertainty and Insurance 244
7.5 Expected Utility Bounds 246
7.6 VaR Valuation 246
7.7 Valuation of Operations by Lagrange Multipliers 248

8 Risk Economics and Multi-Agent CCAPM 251
8.1 Introduction 251
8.2 Economic Valuation and Pricing: Supply, Demand and Scarcity 255
8.2.1 Valuation, Risk, and Utility Pricing: One Period Models 256
8.2.2 Aggregate and Competing Consumption and Pricing Risks 259
8.2.3 Two Products and Derived Consumption 260
8.3 The CAPM and the CCAPM 265
8.3.1 The CCAPM Model 266

8.3.2 The Beta Model and Inflation Risk 269
8.4 The Multi-Agent CCAPM Model: A Two Periods Model 270
8.4.1 The CCAPM with Independent Prices 270
8.4.2 Endogenous-Aggregate Consumption and the CCAPM 272
8.4.3 The General Case with Independent Rates of Returns 273

9 Risk Pricing Models: Applications 283
9.1 Debt and Risk Models 283
9.1.1 Market Risk Pricing Models for Credit Risk and Collaterals 284
9.1.2 The Structural-Endogenous Model and the Price of Credit Relative to its Collateral 285
9.1.3 Credit Risk and Swaps: A Reduced Form or Exogenous Models 287
9.1.4 Pricing by Replication: Credit Default Spread 289
9.2 A Debt Multi-Agent CCAPM Model 290
9.3 Global Finance and Risks 297
9.3.1 Pricing International Assets and Foreign Exchange Risk 300
9.3.2 International Credit, Debt Leverage and the Investment Portfolio 310
9.3.3 FX Rates Risk, Bonds and Equity 317
9.4 Additional Applications 323
9.4.1 Finance and Insurance: Pricing Contrasts and Similarities 323
9.4.2 Insurance and Finance: Pricing Examples 325
9.4.3 Contrasts of Actuarial and the Financial Approaches 325
9.4.4 Franchises 326
9.4.5 Outsourcing and Risks 327
9.5 Subjective Kernel Distributions 328
9.5.1 The HARA Utility 329

10 Uncertainty Economics 333
10.1 Introduction 334
10.2 Risk and Uncertainty, Time and Pricing 334
10.3 Assets Pricing with Countable and Non-countable States 336
10.4 Maximization of Boltzmann Entropy 338
10.5 The Subjective, the Q Distributions and BG Entropy 342
10.6 The Tsallis Maximum Entropy and Incomplete States Preferences 344

10.6.1 Tsallis Entropy and the Power Law 345
10.6.2 A Mathematical Note: (Abe 1997; Borges and Roditi 1998) 346
10.6.3 The Maximum Tsallis Entropy and the Power Law Distribution 348
10.6.4 The Tsallis Entropy and Subjective Estimate of the M-Distribution 349
10.6.5 Maximum Tsallis Entropy with Escort Probabilities 351
10.7 Choice, Rationality, Bounded Rationality and Making Decision Under Uncertainty 356
10.7.1 Models Sensitivity and Robustness 357
10.7.2 Ex-Post Decisions and Recovery 362
10.8 Uncertainty Economics, Risk Externalities and Regulation 364
10.8.1 Risk Externalities, Industry and the Environmental: Energy and Pollution 366
10.8.2 Networks and Externalities 368
10.8.3 Infrastructure and Externalities 369
10.8.4 Economics and Externalities: Pigou and Coase 370

11 Strategic Risk Control and Regulation 375
11.1 Introduction 375
11.2 Statistical Risk Control: Inspection and Acceptance Sampling 377
11.2.1 Elements Statistical Sampling 378
11.2.2 Bayesian Controls—A Medical Care Case 382
11.2.3 Temporal Bayesian Controls 385
11.3 Risk Control with Control Charts 387
11.3.1 Interpreting Charts 389
11.3.2 6 Sigma and Process Capability 392
11.4 Queue Control 394
11.4.1 The Simple M/M/1 Queue 395
11.4.2 The Simple M/M/1 Queue and Non-compliance . . . 396
11.4.3 The Continuous CSP-1 Control of Queues and Banking 398
11.4.4 Networks and Queues 400
11.5 Strategic Inspections and Controls (See Also Chap. 12 for a Review of Game Theory) 403
11.5.1 Yield and Control in a Supplier–Customer Relationship 404
11.6 Financial Regulation and Controls 412
11.6.1 Financial Regulation in a Post Crisis World 412

11.6.2 Statistical Controls and Regulation 414
11.6.3 Private Information, Type I and II Risks and Externality Risks 428

12 Games, Risk and Uncertainty 437
12.1 Introduction 437
12.1.1 Games, Risk and Uncertainty 439
12.2 Concepts of Games and Risk 439
12.3 Two-Persons Zero-Sum and Non-zero Sum Games 443
12.3.1 Terms and Solution Concepts 443
12.3.2 The Nash Conjecture 444
12.3.3 The Numerical Solution of Two Persons-Games: The Lemke-Howson Algorithm . . . 449
12.3.4 Negotiated Solution and the Nash Equilibrium 450
12.4 The Stackelberg Strategy 451
12.5 Random Payoff and Strategic Risk Games 452
12.5.1 A Risk Constrained Random Payoff Games: A Heuristic Interior Solution 454
12.6 Bayesian Theory and Bayesian Games 456
12.6.1 Bayes Decision Making 458
12.6.2 Examples: Bayesian Calculus 458
12.7 Mean Field Games and Finance 461

References 465

Index 503

Chapter 1
Engineering Risk

Abstract This chapter provides an overview of the many manifestations of risk and uncertainty, their applications and the factors that contribute to a convergence of multiple approaches to risk engineering. In subsequent chapters non-quantitative and quantitative applications and problems are considered.

1.1 Risks and Uncertainty Everywhere

Risks are to be found "everywhere". They can be large, small or TBTB (Too Big to Bear), they can be predictable or not, they may arise due to conflicts or due to some adverse party, they may be due to a lack or partial information, they may affect us or others (or both) etc. For example, insurance and finance, quality and consultancies, industrial management, logistics, marketing, technology and engineering, health care and delivery, food regulation and control, safety and policing, politics, infrastructures, supply chains etc. are all beset by risks and the many factors, whether controllable or not, that cause such risks (See Fig. 1.1):

> Driving a car; A terrorist attack; Your associate stole your money; Property loss; Supply chains delays; Product recall; Theft; D&O Liability; Emerging and global Markets risks; Nuclear risks; Industry Risks such as Workers compensation costs; Plant security; Unreliability; Breakdowns; Downtime; Health Risks; Diseases and contagion; Health care mistreatment; Pharmaceutical lab. Errors; Misdiagnosis and wrong medicine administered; Financial Loss Risks; Returns risks; Volatility risks; Trading risks, Mergers and Acquisitions Risks; IPO risks; Carbon caps trade risks; Interest rates changes risks; Investment risks; Reputation risks; Options losses as well as vulnerable options risks; Environmental risks; Weather risks; Tsunami's risks; Climate change; Pollution risks etc.;
>
> Supply Chains risks; Contractual risks; Technology risks; Cyber risks; Normal risks (mostly predictable of relatively un-consequential); Catastrophes risks (mostly rare but consequential) such as earthquakes in Japan, in New Zealand, floods in Thailand, and Australia, tornadoes and Hurricanes in the Americas; Man-made risks such as the MBS crisis, Man-Made wars, sovereign debt meltdown, Process and Man-Made systemic risks etc.

C.S. Tapiero, *Engineering Risk and Finance*, International Series in Operations Research & Management Science 188, DOI 10.1007/978-1-4614-6234-7_1,

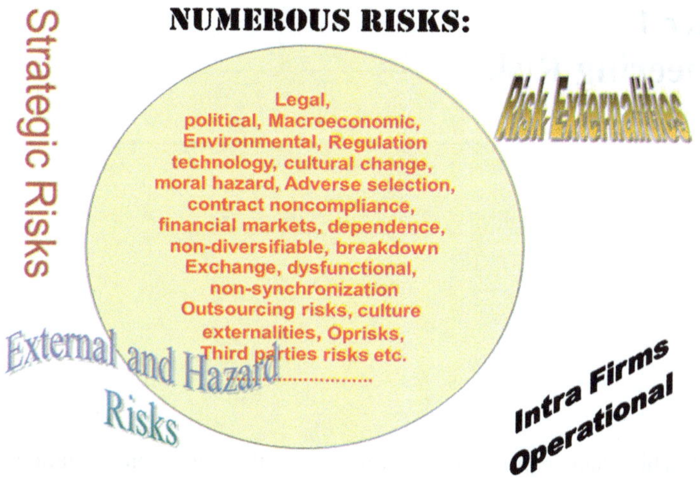

Fig. 1.1 Numerous risks

Risks can have direct, derived and indirect adverse consequences, or outcomes that were not accounted for, that we were ill prepared for or are unaware of. They may affect individuals, firms or the society at large. They result from causes, internally and strategically induced, or occurring externally. Some risks are the result of what we do such as failures, misjudgment or conflictual (strategic) situations, while others result from uncontrollable and unpredictable events or events we cannot prevent. Risk models seek to model uncertainty based on what is known and can be predicted. Risk and uncertainty thus differ appreciably by the countability and accountability of their potential future occurrences (states) and consequences. A definition of risk models involves as a result, a number of factors:

1. Countable and accountable events and their measurements.
2. Probabilities and their distributions defined in terms of countable events and the elaboration of statistical data and its analysis.
3. Risk Consequences, assumed individually and/or collectively or assumed by other parties.
4. Risk Attitudes of individuals, firms, markets, societies or governments
5. Risk valuation, whether subjective or objective with prices defined by the terms of an exchange
6. Risk mitigation and management ex-ante and ex-post, including risk sharing and transfer risk design and generally a multitude of approaches and means set to detect, to control, to prevent and to recover from risk events—once they have occurred.

These are relevant to a broad number of professions, each providing a different approach to the detection, measurement, valuation, pricing and the management of risk which *is motivated by real, economic, financial and psychological needs and the need to deal individually and collectively with problems that result from uncertainty, risk models and their adverse consequences. These may be sustained unequally by individuals and society at large.* For these reasons, risk and uncertainty, their consequences and their management are applicable to many fields where risks and uncertainty prime.

Recurrent crises, the growth and awareness of complexity have reaffirmed the limits of risk models that account for calculated risks and the importance of framing uncertainty into a mold we can better comprehend and manage. A distinction between risk and uncertainty was pointed out originally by Knight (1921) emphasizing that risk is mostly associated to the predictability of future events, while uncertainty is associated to their lack of predictability and thus to consequences that were not accounted for (or are unpredictable). When events are predictable, they can be counted and their consequences assessed to better forecast their propensity to occur. A distinction between what we mean by "predictability" or a lack of it is still a debated question however. Is unpredictability embedded in randomness? Is unpredictability embedded in our lack of understanding, in an over-simplification of intricate relationships, their complexity and dependencies that beset us? Is unpredictability embedded in rare events? Is unpredictability embedded in the strategic encounters of parties with broadly varying agendas, information and power and their asymmetries? While in fact risk models are based on predictability, uncertainty is defined by those risks that are not accounted for. For example, insurance firms mostly agree to sign contracts with all future states accounted for while remaining states are left to the insured who assumes their residual uncertainty. Financial practitioners, some successful such as George Soros (2008), have repeatedly questioned fundamental financial economic theories pricing assets based on discounting future outcomes (see also Chaps. 7 and 8) by pointing out that markets are dotted with "reflexive feedback". Namely, markets "redefine their fundamentals"—the same fundamental they are supposed to imply. Such concepts underlie markets nonlinearities, bifurcations, and complex and chaotic processes leading to new dynamic evolutions (or say financial regimes). These properties of markets are both difficult to predict and thus are sources of uncertainty. Theorists, such as Minsky (1993), hypothesized that financial markets are regime-unstable (presented as an interpretation of the elements of Keynes' Theory of general equilibrium). In this framework, markets have financing regimes under which they are stable and others where they are not. In Minsky's theory there is a natural tendency for the economy to transit from a stable to an unstable system which providing a rationale for the booms and the busts (i.e. a dynamic equilibrium) that we often observe but have difficulty to explain. This inherent instability of financial markets is also difficult to reconcile with measurable risks and predictable outcomes.

"Countability" and "accountability" of specific future states combined with their measurements underlies therefore the many activities that fall under what we call "Risk Management". These are used to assess their causes and mitigate their risks. In this sense:

- *Risk models do not manage uncertainty*
- *Risk management is applied mostly to risk models based on a bounded rationality that uses what we know (our cognitive framework) with what we need (our wants or preferences)*

"Managing uncertainty" is thus defined by the residual set of events and their probabilities that are not framed by risk models. For this reason, management of uncertainty requires mostly ex-post and contingent means to respond to adverse and initially unpredictable events. In some cases, robust management models (Chap. 11) may be used to augment the insensitivity of a risk model to parametric errors, thus expanding their usefulness. This usefulness comes at a price however. Risk models can thus be assessed, valued and managed "rationally" while uncertainty, belongs to the domains of "mystics", based on apprehending facts, if at all, that may exist in our "unconscious states of mind" or confronting ex-post consequences. Similarly, a distinction between risk and uncertainty is expressed by what "we know", by what "we do not know" and our ability to react and recover from events that were not or could not be predicted. These elements are common to a broad number of domains, each defining and confronting uncertainty and framing it into a risk based on ones' own knowledge, based on one's experience, based on one's professional language and based on one's needs and experience in confronting uncertainty. When risk is defined in a common quantitative language such as probabilities, consequences (loss of lives, loss of money, etc.), risk management is also based on common principles and techniques. When risks are valued using money, these become economic and financial problems. Below, we shall consider a number of particular cases and applications that are formalized in subsequent chapters.

1.2 Many Risks

1.2.1 Globalization and Risk

"Globalization" is an economic and political opportunity that has also fostered the growth of many internal and external threats that have previously been kept at bay. It opens markets and removes social and other barriers but increases competition and a global openness on the other. Both, have wanted and adverse consequences. Global risks and their assessment differ from place to place and from situation to situation due to societies' values, traditions and environment. Risks models are thus relative, culture- sensitive and multifaceted, framed in partial beliefs and information, based on nations culture, political environment and agenda's etc. Definition of risk, its measurement and mitigation in such cases ought then to recognize local habits, cultures, their micro-economic and macro-economic effects as well as their latent opportunities and threats. The extensive number of issues that globalization entails and their micro-prudential and macro-prudential implications precludes their full treatment. Instead we outline a series of questions to highlight some risks and/or their causes (for an outline of explicit models in the economics of global finance, see Chap. 9):

- Different laws from country to country with different penalties.
- Regulation differences for industrial standards and for financial regulation.
- Taxation applied differently to local and foreign investors and agents.
- Local foreign inflation versus domestic and global inflation.
- Potential expropriation, nationalization, foreign control, foreign exchange controls.
- Trade restrictions (both symmetric and asymmetric).
- Devaluations of the currency and its convertibility (foreign exchange risks).
- Contracts repudiation, their legal foundations and their enforcement.
- Embargoes.
- Sovereign Default.
- Religions, their beliefs and their certainty.
- Kidnappings, extortions and ransom.
- Political risks, etc.

1.2.2 Space and Risk

The GEO (Global Earth Observation) of the United Nations Center in Geneva has become an important data gatherer and information system center to observe the evolution of the earth's ecosystem using satellite systems. For example, climate and weather shift patterns across the globe, desertification, migration, etc. are using "space" as an observatory of global risks (see Fig. 1.2).

Such systems and the size of "big data" information systems it builds to assess dependence risks at a global scale is based on techniques developed also for "big data" financial systems seeking to track the evolution of commodity and financial assets globally. These systems provide also a set of techniques that are used by emerging firms that propose to use internet data to assess various risks and opportunities.

1.2.3 Catastrophic Risks

Catastrophic Risks are defined by their consequences, some predictable and hopefully rare and some not. In the US, the 9/11/2001 man-made destruction of the twin tower has still lasting effects whose toll is incalculable (wars, the transformation of societies, a global religious conflict, etc.). The Hurricane Katrina in 2005 had human and financial costs that have impacted both the US economy and its

Fig. 1.2 The GEO information system

resilience. Figure 1.3 below, is a reminder of these two events. Below a small sample of such events is outlined:

Water and Tsunamy catatrophies: 2011: Japan Tsunami 2011 and the nuclear station meltdown); 2004: The South East Asia Tsunami December 26: 226 408 deaths; 1931 : Yangtsekiang-Wuhan 400,000 deaths (july–september); 1954, 1959 and 1998 : Dongting 40,000 deaths and 100,000 victims, the Yangtze 3,500 victims

Desertification and Heat: *in Africa 2003*; *Europe* August Temperatures of 50°C with *14,802* deaths in France and *25,000* in Italy),

Extreme Winds: Katrina in 29/08/2005; Bangladesh in 1970 with *400,000* deaths; India-Pradesh 1977 with *10,000* victims

Earth Quakes: Pakistan 2005 with *73,338* deaths; Japan –Kobe with *6,424* deaths, *43,700* wounded and *250,000* homes destroyed ; Yokohama in 1923 with *143,000* victims , China 19736, with *290,000* victims; in 1920 with *180,000* deaths, in 1932 with *80,000* deaths) ,

Volcanic Eruptions: Pompei erased in 79, Martinique (Saint Pierre erased) in 1902 with *30,000* deaths; Colombia 1985 *24,000* deaths

Technological Catastrophic Risks: September 1921, Oppau, Rhénanie, Germany—a mine explosion, 450 deaths and 700 homes destroyed ; April 1942, Tessenderlo, Belgium—hundreds killed ; January 1961, National Reactor Testing Station at Idaho Falls, Idaho—The first nuclear accident in the US; December 1984, Union Carbide accident at Bhopal in India—with 8,000 deaths initially and 20,000 subsequently over 20 years; April 1986, Tchernobyl's—Nuclear accident in the USSR with 5 million persons exposed to excess radiation.

Fig. 1.3 Catastrophic risks

Other types of catastrophic events includes Transport (Trains, Planes, Maritime and others), Oil transport, Environmental (the BP Oil well blow up in 2011–2012), Military, Crimes against Humanity (The Jewish Holocaust in World War II by the German, the Armenian Genocide by the Ottoman Empire, the Genocide of Tutsis in Rwanda by Hutues etc.). These are crime defined by international courts as crimes against humanity including all attacks against a people for a precise reason (ethnic, social, religion, language, etc.)

Additional catastrophic events such as floods; extreme temperatures; extreme winds; earth quakes; volcanoes; earth movements; forest fires; etc. are extreme risks that can degenerate into catastrophic disasters. Their extraordinary consequences that defy predictability have led to the belief that they are extremely rare and therefore not accounted for—when in fact, they recur more often than we would like.

A survey of disasters in websites such as http://www.em-dat.net clearly points out to their growth, breadth and consequences and to costs associated to settlements in risk prone areas. Great efforts are applied to map and predict these risks. For example, some human dense habitats compared to dispersed habitats may be prone to such (predictable) disasters (such as cities constructed at current sea levels). For example, a hurricane striking an empty space is less likely to be catastrophic than its striking a large city.

The number of disastrous "rare events" has grown over time for many reasons. Improved accounting of such events, a growth of the density of human settlements concentrated in specific and risk prone parts of the world, technology, the sophistication of military weapons, etc. The financial cost of disasters is hardly accounted for, although Swiss-Re, a very large Reinsurance firm monitors these risks and use them as a basis to determine the risk premium they extract to insure against such risks. Commensurably, Swiss-Re revenues have increased as well, with losses stimulating a growth in risk premiums for insurance and reinsurance. The variety of these risks has also grown. For example, focusing on natural disasters, causes between 1994 and 2010 have pointed out to the predominance of floods, drought, epidemics and Tsunamis.

Various countries have recognized the importance of these risks and are dealing with them by providing additional protection through regulation, insurance (healthcare, social security), risk incentives and infrastructure investments. In some cases, insurance is used to protect individual parties. For example, insurance firms sell conditional claims and partial insurance contracts for certain catastrophic risks, such as earth quakes, floods, lightning etc. In France, the Government assumes the coverage of all forms of catastrophic events (The Law of July 13, 1982), although in August 1, 1990, the Law has been amended not to include Storms where insured are required to take a special insurance contract whose cost is 12% of the asset coverage, 0.5% for cars, a self-initial participation of 380 Euros for individuals and 1150 Euros for firms and 1530 and 3060 Euros in case of drought. Denmark assumes flood disasters by creating special funds financed by insurance against fire, etc. The study of how a society deals with its problems and their derived consequences is clearly an important topic, revealing its underlying value system and political culture. Such problems are prevalent in health care as well. For example, should universal health care be applied and letting "all insure all" or used conditional defined health care benefits for segments of a society and the remaining assuming their own risks? These are some of the issues that confront and pitch political agendas one against the other, with a political course focused on risk, money and its allocation.

1.2.4 Debt, Credit and Counter-Party Risk

"Credit is a disposition of one man to trust another", (Walter Bagehot, nineteenth century). It is a trust that one bestows on another to meet prior commitments. Credit risk arises then when there is a lack of trust and-or when there are potential adverse consequences and when one of the parties cannot meet its commitments. This lack of trust may arise due to external hazard as well as due to what parties do to one another (for example, the misuse of trust by parties privy to the negotiated agreement, or counterparty risks). In such cases, a financial loss, a "disappointment" and a misuse of trust may occur. These motivate parties to be better informed, do due diligence and seek risk reports. Whether one grants credit or is granted credit, there are risks for the one or the other. The exchange between say a lending bank and a borrower involves therefore a number of considerations and risks each party assumes that define their transaction. A traditional definition to credit risk covers a set of multiple risks, essentially including:

- Default risk (defined in terms of default sources, their legal and operational definitions, and how they are measured—explicitly, or implicitly by the price of the financial transaction between the parties or valued by a financial market where such a contract may be traded.
- Recovery risk (defined by the potential recovery of losses. Such risks vary broadly from country to country and are defined by the obligations the holder of the credit risk owns)
- Collateral risks, or risks associated to the assets used to protect the underlying credit transaction. For example, a home used as collateral may have a value which fluctuates over time.
- Third Parties guarantee risk. For example, insurance firms (such as AIG) insuring a transaction of two parties they have no part in.
- Legal risk are risks associated to a broad set issues that pertain to the contract negotiated between parties with one of the parties violating part or all its negotiated parts with recovery depending on a legal resolution.
- Risk exposure or risk of loss (which underlies Capital Adequacy Regulation—CAR in financial institutions).
- Macroeconomic and external real risks involve factors that affect globally the credit granted and its price. For example risks associated with the cyclical behavior of the economy, sudden jumps in economic variables (such as the financial meltdown of 2008). These risks are particularly important because their effects underlie complete portfolios (for example the Mortgages, the loans, the price of collaterals etc. held by a bank).
- Counterparty risk or the risk inherent in the conflict, information, power and the attitudes to the parties to a credit contract.

In many instances, definitions of credit risk depend on their sources, who the client may be and who uses it. When credit risk is associated to an individual borrower, techniques for credit scoring are applied. Banks in particular are devoting a considerable amount of time and thoughts to define and manage their credit risks.

Two essential risk sources quantified by risk models includes: default by a party to a financial contract and a change in the present value (PV) of future cash flows (which results from changes in financial markets conditions, a change in the economic environment, interest rates etc). This could assume the form of money lent that is not returned.

The terms of credit are expressed by a financial borrowing and lending agreement and the pre-posterior steps taken to assure that the parties meet the terms of their contract. The number of approaches to define these terms of credit is extremely varied. For example, to protect themselves, firms and individuals turn to rating agencies such as Standard and Poor's, Moody's as well as others (such as Fitch Investor Service, Nippon Investor Service, Duff and Phelps, Thomson Bank Watch etc.) to obtain a certification of the risk assumed by financial products (theirs or others they have an interest in). Furthermore, even after careful reading of these ratings, investors, banks and financial institutions proceed to reduce these risks by risk management tools. The number of such tools is also varied. For example, limiting the level of obligations, seeking collaterals, netting, re-couponing, insurance, syndication, securitization, diversification, swaps and so on, are some of the tools financial services firms or banks might use.

Regulatory distortions in credit markets are also a persistent risk theme. Over-regulation hampers economic activity and thereby the creation of wealth while "under regulation" (in particular in emerging markets with cartels and few economic firms managing the economy) can lead to speculative markets and financial distortions. The economic profession has been marred with such problems. For example:

> "One of today's follies, says a leading banker, is that the Basle capital adequacy regime provides greater incentives for banks to lend to unregulated hedge funds than to such companies as IBM. The lack of transparency among hedge funds may then disguise the bank's ultimate exposure to riskier markets. Another problem with the Basle regime is that it forces banks to reinforce the economic cycle—on the way down as well as up. During a recovery, the expansion of bank profits and capital inevitably spurs higher lending, while capital shrinkage in the downturn causes credit to contract when it is most needed to business." *Financial Times*, October 20, 1998, p.17

Some banks cannot meet international CAR standards. For example, Daiwa Bank, one of Japan's largest commercial banks, has withdrawn from all overseas business partly to avoid having to meet international capital adequacy standards. For Daiwa as well as other Japanese Banks, capital reserves have been eroded by a growing pressure to write off bad loans and by the falling value of shares they hold in other companies. These have undermined their ability to meet these capital adequacy standards.

The modern era of finance has further expanded the means to "produce money and credit". Financial products such as financial options, Mortgage Backed Securities (and various securitized assets), home mortgages, personal and business loans, credit cards, bonds (corporate, government, municipalities) etc. have created additional means currently to trade and exchange in financial markets. These means are bonds of all sorts, insurance, securitization, Credit default Swaps (CDS), Multi Names Credit Derivatives (such as CDO's and their variants), that have made it possible for US consumers to consume at an unprecedented pace in history. A lack of credit during the 2007–2009 financial crisis and its aftermath arose due to a "process default" in these instruments resulting in the credit crisis, an immense fall in liquidity and to financial markets meltdown. Banks hoarding cash and investors "refusing" to buy securitized loans as they have done so gingerly in the past, are evidence that the year 2008 was the beginning of a substantial decline in credit liquidity. Even after receiving billions of dollars from the government, banks were reluctant to lend money! The relative ineffectiveness of government intervention, is a testament to the fact that the new economic and financial environment, defined in terms of new Financial and Information Technologies, Globalization, Dependence, and virtual financial transactions has created a family of credit risks hitherto presumed to be non-important.

1.3 Industry and Other Risks: Deviant or Money

Industrial revolutions have consistently transformed products, work procedures, organizations and management—increasing efficiency and redefining risks in terms of industrial needs. When industrial technologies matured and supply competition increased, risk was defined as well in terms of its demand side. In this sense, the definition of risks and risk management have changed hand in hand with the needs of the supply and demand sides. On the supply side, industrial and technological revolutions have transformed the work place, manufactured products, their organization and the definition of their associated risks. In this process industry has become:

- Atomized (with specialized job assignments and standardized products parts) to assure that all components or parts are interchangeable and conforming to specific standards of manufacture.
- Robotized (reduced to elementary functions that require no expertise or judgment, all of which operating in tandem and replacing men by machines)
- Workers de-responsibilized within a complex and interconnected system controlled externally by men and machines
- Work default became more difficult to detect and trace, demanding more extensive and complex control systems.
- Networked processes with products and dependent systems.
- Outsourcing and globalization have contributed further to transform industries into an assembly of intermediaries and "stake holders", each with its own preferences, agenda and information and power asymmetries in a world of increasing complexity with strategic risks, controls and regulation.

These changes, have introduced a persistent process of change by creating new needs and a new environment where to detect and control, to design reliable and risk-proof systems, become important. As a result, while in the risk management process was in-process (such as risk of failures, risks of malfunctions, poor performance, sabotage and other risks etc.), it grew into complex upstream and downstream processes (with supply risks as well as demand, post sales, service and consumption risks translated into money to assess their financial consequences). To these ends statistical and risk control tools were devised including for example work sampling, control charts, quality assurance and more recently, the development of techniques to manage supply and consumption and complex, networking risks.

1.3.1 Technology and Risks

Technology has fed, is fueled, challenged and is feeding economic, industrial and business needs, including among others:

- In industry, mass customization based on automatic and robotized systems, responding to consumers demand for variety and low costs. By adapting traditional industrial processes to be more flexible, with smaller production lots and reduced production delays it developed an efficient economy of scope, respond to consumers demand for greater variety at a better price.

- Global connectivity, both for the opportunities it provides for instant communication, and support global outsourcing and the rise of A-national, stealth and virtual firms and organizations that operate independently of their location and evading regulation and Sovereign and public controls.
- A power asymmetry, providing the few technology savvy, to conduct their affairs, conduct conflicts, cyber-theft, spread terror etc. at an extraordinary and more efficient pace. Technology has thus, in some cases, augmented the risks of tyranny by minorities, by terror operating from a global base.

The insatiable needs of firms to be here and there "stealthly" and safely at all times, the need to communicate and to sustain a state of instant and mobile communication without geography, the need to keep pace and paces away from a future and futuristic transformations falling upon us faster and faster than the present can handle have fueled new inventions and perpetrated new opportunities and risks—social, industrial, political, financial and personal. In such processes, technology and risks have grown to be ever more complex, networked, dependent, more difficult to control and thereby far less predictable. In many respects, technology has become an autonomous process, evolving through a process of "mass innovation" no longer planned or concentrated but diffused globally. Opportunities and risks have thus grown hand in hand with a technological transformation in the hands of "atomic innovators", dispersed globally and able to affect its course and distract its intents. In this sense, technology is both a strategic opportunity and a systemic risk based on an inherent *disequilibrium that defines technology and innovation operating in a global, uncertain and self-sustaining networks* of agents all of whom are in the pursuit of an agenda (Tapiero 1989, 1994b, 1995a).

1.3.2 Technology and Networking

Social and political upheavals originating in major changes in networking are inducing a far greater awareness of global wants and their inequalities, inducing a process of "global equalization" (namely a growth of global "entropy"). These evolutions are based on an exponential growth in exchanges and information, spanning internet systems, IT social media with the capacities to trigger contagions and revolutions. In this process, change and dealing with change have replaced entrenched approaches to local political concerns, to financial and business opportunities. In these processes, global and "stealth firms" have grown, eluding regulation and controls. These have perpetrated again, a new family of risks, more complex and requiring increased controls and sophistication. Some firms, faced with an accelerating change in technologies, are losing their abilities to manage their own technologies and thus outsource functions that require sustaining in-house a technological intelligence. Rather, a technological dependence has increased with agents and intermediaries used to acquire and operate critical functions in banks, hospitals, the military, the police, the government, etc. These trends have induced new risks.

1.3.3 Technology and Cyber Risks

Internet and communications networks (Facebook, Twitter, and the spread of websites, virtual firms, etc.) have also contributed to cyber theft, to identity theft, to loss of storage, to networked computer safety, to cyber wars, to threats to reputation etc. For example, five men, hiding in St Petersburg, Russia are believed to be responsible for spreading a notorious computer worm on "Facebook" and other social networks—and to have pocketed several millions of dollars from online schemes. Persons, businesses and countries are currently challenged at an unprecedented scale—some in their pockets, some in their security and intelligence, and some in confronting new means of destruction. These are precursors of a growth of uncertainty. For example insurance of banking fraud (in credit cards, in financial transactions), cannot be guaranteed fully. Regulation limits (insurance) losses for victims of a cyberstrike to $500, forcing banks to cover the balance. New malware innovations, reaching the internet in their millions each day, are also designed to get around the security fixes of recent years. Qakbot, has been infecting computers since 2009, downloaded from infected Web sites it piggy backs on legitimate online transactions to evade the security provided by changing passwords. The Zeus Trojan that propagates through spam is estimated to have infected 3.6 million computers in the United States and simply waits for users to log on to their bank account and steal their information as they type it. It can even replace the bank's Web pages with its own on the victim's browser and entice the user to divulge even more information (*The NYT*, January 17, 2012, Editorial Page, A 20).

1.3.4 Example: Technology Risks, Simplicity and Complexity Risk Mitigation

Technology expands our capacities to do more, to do better and construct self-organizing systems. It also induces a spiraling growth of "complexity" and the need to control its potential risk consequences. This is the banking system, increasingly relevant to the growth of complexity of financial regulation. For example, the Dodd-Frank act as well as tweaking the Volcker rule by regulators has created complex rules that are both difficult to maintain and control (see Chap. 11). Controlling complexity, is then based on two essential approaches: More advanced technologies inducing higher levels of risk and complexity, thereby leading to a dynamic forward growth of complexity and risks. Such a process underlies the second law of cybernetics, (the law of requisite variety of Ashby 1956). Ashby's law essentially states that complexity induces a need for controls which are at least as complex as the system to be controlled. This induces in turn the growth of more complex systems. In other words, a positive feedback loop process is set in motion, where "complexity begets more complexity". Automation, in general, has created a need for more information needed to control "complexity and its many consequences, evolving into extremely complex an integrated systems that operate, treat and analyze the systems and the information it has produced. Alternatively, complexity may be controlled by simplification. Similar problems arise in taxation, regulation, health care and in the banking sector. For example, can a complex tax and regulation policy attains its purpose or merely provide employment to accountants and lawyers to entangle the many problems and risks that wage and investment earners have? By the same token, would overly complex organizations implode when they can no longer be controlled (for example, implosion of the banking sector in 2008, the lack of internal controls at JP Morgan Chase that led to a loss of 6 Billion Dollars in 2012)? Can the complex regulation mechanisms planned to control systemic financial risks following the 2008–2009 financial crisis overcome the complexity it is creating?

Problem:

Say that a firm (or government) has the right to use a technology (for example, assembling data on the internet to define individual preferences). How does it improve your welfare and the services you may access to and how can it harm you? What are the moral implications and obligations of the firm (government)? How can one mitigate the risks you may have pointed out. In particular, discuss the case of a firm right to drilling for gas in farm next to yours, the case of a chemical firm building a plant near a small city with the right to spill some of its waste, etc. (See also Hansson 1996, 2001; Hansson and Peterson 2001 for a definition and the implication of moral rights and risks as a discourse on the philosophy of risk).

1.4 Quality, Statistical Controls and the Management of Quality

Prior to the first Industrial Revolution, production was an art while quality was the measurement of this art. Each unit produced was handled and signed by an artisan, acting simultaneously as a "designer" and "producer" assuming all the responsibilities of his wares. With the industrial revolution, quality was measured by a process conformance to industrial standards. Subsequently, with the rise of consumers' dominant societies, it has been refined in terms of consumer expectations.

During WWII, a need to control weapons and munitions shipments led to the use of SQC (Statistical Quality Control) and SPC (Statistical Process Control) to inspect and control shipments to Europe and the Pacific. From the mid 1950s, it evolved to TQC (Total Quality Control), TQM (Total Quality Management) and continuous improvement and into comprehensive 6 sigma approaches (see also Chap. 11). Quality Gurus led by prominent Risk Statisticians Deming, Juran and Ishikawa provided a firm statistical foundation to an approach to production and quality-risk management based on, process simplicity (to counter the effects of production complexity), control, design, organization and collaborative management. Subsequently, standardization of processes and their rating rather than just a standardization of products was launched under an ISO (International Standardization Organization) certification, initially implemented to create a barrier to entry to European Industrial market and subsequently used to rate the quality-risk efficiency of industrial and business processes and firms globally.

Subsequently, technology intensive platforms have increased further the complexity, the dependence and fragility of industrial processes that led to greater needs for risk control and preventive measures. However, incoherent technologies grafted on one another resulted in integration of risks creating "technological Towers of Babel" (with components unable to operate as part of a whole system). New risk management approaches were thus needed requiring the "Re-engineering" and "Concurrent Engineering" of industrial systems to be far more coherent and coordinated and operating as a whole (with new Gurus stepping in to highlight the needs of the time).

Concurrence, Robustness and Robust Design were then introduced in the 1970 and 1980s. Such concepts have been extended further to "managerial" approaches embedded in the widely practiced 6 sigma (zero default to be developed in Chap. 11). Such an approach can be summarized simply by the integration of tolerant standards and stringent industrial controls. Explicitly, broader tolerance for manufactured products combined with the stringent tests and controls of manufacturing systems led to manufactured products to be almost always accepted (and thus the zero default 6 sigma industrial process management).

Outsourcing supplies and manufacture started in the 1970s has led to a transformation of the industrial process to be far more focused on the one hand and strategic on the other, to be networked and based on supply chains that consist of suppliers/industrial stakeholders distributed over the globe and assisted by IT networking.

Rather than using just statistical tools, control and risk management have been focused on organizations' design, human and technology resources, on self-regulated systems, the capacity to focus and at the same time, integrate, communicate, interface, monitor and control in an uncertain and competing environment. Risks and their management have in this process been challenged by becoming more important and more difficult to control.

A similar process is currently occurring in the financial world where financial systems are more complex, global, dependent regulated and for more difficult to control. These require therefore the evolutions and the tools needed to sustain their viability.

1.5 Health and Safety Risks

Health risks, Doctors' risks, diagnosis risks, Laboratories risks, contagious disease risks, drug risks, outbreak risks, accidents' risks, experimental risks, liability risks, IT health hospital systems related risks, health services risks, conflict of interest risks etc. are risks we may associate to both health care risks and insurance. For example, what if a primary care doctor misdiagnoses a patient and recommends a treatment which does more harm than good. Such risks may arise because many diseases (some cancers, asthma, etc.) can be misdiagnosed and lead to an incorrect treatment. It may also arise due to conflict of interest with Drug firms stealth payments to Doctors who recommend their drugs or their medical devices (at times in the hundreds of thousands of dollars). While risks causes can vary from an inconvenience to the fatal, the financial consequences—both directly and indirectly can be substantial. For example, while doctors' treatment, incomplete testing of some drugs and their like can incur very large liability costs, there are also reputation risks (see Chap. 2) that can linger over a hospital, a doctor, a drug manufacturer or a financial service. Health risks are laden with partial information, many stakeholders (patients, doctors, nurses, health administrators, health care service and insurers etc.) as well as complex risks arising from multiple and dependent factors. For these reasons, often a unique opinion, a unique test or a unique experiment, may not be sufficient. Health statistics, adapting and applying statistical concepts (commonly used in industrial statistical tests) as well as integrating health care to a greater awareness of costs and the alternative treatment of patients is in fact contributing to a convergence with both risk statistics (based on an extensive experience) and money based on both insurance and financial pricing.

The US Institute of Medicine for example, suggests a risk-based definition of health care quality: Quality of care is the degree to which health services or individuals and populations increase the likelihood ("probability") of desired health outcomes and are consistent with current professional knowledge (and therefore based on current "expectations"). The concern for health care delivery is indeed one of the greater challenges this decade. The growth of health care delivery, health risks and health maintenance costs, render it an essential item in GNP composites. It is also an important part of most Nations' social agendas. The transformation of health care, from a back door cottage industry to a complete and massive "industrial activity" is also an added motivation to alter the traditional means of management and controls of health care delivery risks.

Health care's major players are placing health risks and management at the top of their priorities, each for different reasons, and for each, different consequences.

- For hospitals, ambulatory surgical centers and other patient care sites, their goal is to maintain a competitive advantage in patient care that will differentiate them and provide profits opportunity.
- For Physicians, nurses, and other professionals, good care is the goal of medical practice and the standard by which they will be measured by peers, patients, regulators, and malpractice attorneys. A deviation from recognized and safe

practices induces risks such as legal cost, reputation cost etc., that health care delivery personnel will avoid.

- For major employers, insurance companies and managed health care networks, costs and risks of health treatments are primary considerations when selecting doctors and hospitals when price is not a factor.
- For politicians, governments, regulators and health advocacy groups, risk means the protection of the public welfare and responding to voters and consumer groups. The political debate on health care is assuming a growing importance due to both its costs and the social inequalities it may imply due to the cost of access to quality treatments.

The multiplicity of parties (Hospital Administrators, Government, Doctors, Patients), each clinging to a definition of an acceptable and desirable Health Care performance and its risks introduces some confusion of what ought to be an appropriate level of health care delivery, how to measure it, how to manage its performance and its risks in a real and practical sense. These are extremely important problems that involve personal, strategic, external, and externality risks.

Uncertainty in health care arises due to the extraordinary complexity of the human body and information asymmetries that render the definition of risks, risk models and their management extremely complex. Patients, Doctors and health insurers (or HMO's-Health maintenance organizations, hospitals etc.) have all an agenda and information that need not be shared. Lacking any other possibility, patients, buyers of health care, have relatively little information and choice to manage their health and must rely on the "health care sellers" for advice. In addition, Doctors are paid according to the numbers of procedures they perform, regardless of their results (although numerous attempts are made to construct performance incentives rules). The price paid for this procedure often has no relationship to the cost borne by the patient or the value (or risk) of health to the patient, who might be insured directly or indirectly through an employer. This cost is, in any case, paid up front, and thus there is both a double moral hazard (see Chaps. 7–9) on the seller and on the buyer side. It is not surprising therefore that the operational solution for health care systems is, to say the least, elusive. Patient's assurance, whose purpose is to provide information regarding the health care system and let this information be public (thereby informing patients and letting them make their own decisions) is a provocative idea which may moderate the market perversities of information asymmetry in health care. It may not be implemented simply however.

These types of problems recur in many areas in business when contracts are signed or agreed on between parties that are unequally informed. Banks for example are notorious information asymmetry warehouses, acting at times counter to their clients' interests. The American Medical Association, for example will accept only Doctors that have been properly trained to assure patients. In most cases, Health Insurance firms provide a list of recommended Doctors etc. These are a mere sample of what the health care sector attempts to institute to better mitigate the risks it confronts. Further, while medical and technological advances alter the profession's ability to predict and prevent diseases, it has also increased the risk of patients' medical information and misinformation.

1.6 Finance and Risk

In finance, risks emanates from the definition of financial risk and risk models (namely from model where all potential future events are both counted and accounted for) and, uncertainty (namely from events that were not counted, are unaccounted for and generally, defining all events and consequences not considered by risk models—namely the neglected and the unknown). In the first case, risk are defined by potential financial losses (or a risk exposure) and volatility. In the second (uncertainty) case, risk arises from situations and states that are not accounted for by risk models. Financial risks are many, such as investors' risk of losses, banks' risks, financial systems risks, derived risks to sectors other than financial services, etc. Risk models are themselves "warehouses of risks" as they may induce decision makers into errors with financial consequences. These include essentially:

- The risks of certainty (Sect. 1.6.1)
- The risks of complexity (Sect. 1.6.2)
- The risks of regulation (Sect. 1.6.3)
- The mismatch of micro and macro policies as well as the macro risks not accounted for in micro-financial risk models (Sect. 1.6.4)
- Risk and Incomplete Markets (Sect. 1.6.5)
- Risk Models and Uncertainty (Sect. 1.6.6)
- The risks of unpredictable and neglected or rare events (Chap. 10)
- Political, globalization and strategic risks (Chaps. 8, 9 and 11)

Each of these risks will be considered technically and in greater details in Chaps. 6–10.

1.6.1 The Risks of Certainty

Finance theory (both economic and mathematical) "has done away with uncertainty and risk" by developing risk models based on both countable and accountable future states priced implicitly by current information. In this sense, the "traded future" is priced and exchanged now and therefore, it has neither risk nor uncertainty. Based on fundamental competitive economic models Adam Smith (1776), Walras (1874/1877), Keynes (1935), Arrow (1951a, b, 1963a), and Debreu (1953, 1959) expanded the global Walrasian economic equilibrium to future markets. For that theory to hold a number of assumptions are made leading to a risk model where future states are not defined by "uncertainty" but by financial exchanges that economic agents pursue at a present time. This theory, validated in complete markets (as will be considered in Chaps. 7 and 8) can however mislead investors and traders when its underlying assumptions are violated in fact. For example, the presumption that a theoretical perfect hedge can remove financial risks can be misleading and lead to extremely large losses. Risks of certainty can have dire consequences in the many other facets of our life. For example, decision or policy makers over-confident in the absolute beliefs they hold can lead to disastrous decisions.

1.6.2 The Risks of Complexity

"Complexity risk matters":

> One reason is that the more complex the rules are, the greater the likelihood that smart bankers will find ways to game them. Another is that contradicting regulations, however well meaning, simply do not make the system sager. But the most important reason is that complexity risks is having an effect on business (*NYT*, Joe Nocera, Editorial Page, A21, January 17, 2012).

1.6.3 The Risks of Regulation (and Non Regulation)

Regulation is a two edge sword. On the one hand it mitigates public and systemic financial risks (at a cost to the regulator and to regulated firms), on the other, it can alter economic and financial systems and lead to developments that "were not intended" or to consequences they have not integrated in their risk analyses. These risks may emanate (as indicated above) from responses of regulated who migrate to other countries to evade their compliance to States' regulation. They may also occur because regulation may be extremely complex and hamper firms' profitability and in some cases their ability to function. For example, requiring from banks excessive capital to be set aside, may lead banks to lend less. This often occurs when banks are called to tighten their credit when the economy requires a credit expansion. Such situations are often revealed when the rate of business closures augments due to a lack of credit, increased unemployment and a slowing economy. For example, what if because of regulation a firm has to downsize its activities resulting in a loss of 1,000 jobs. These are persons that would have earned and paid taxes, rather than being a social charge. A regulated firm is of course at risk of not meeting the regulatory requirements and when caught it will be financially penalized. It may also be blackmailed by the regulatory agency into paying fines for it to demonstrate its effectiveness (evidently, this happens when the cost of litigation is smaller than the blackmail price). Such blackmails may increase if the future public agencies will finance themselves through the penalties they can extract.

1.6.4 Micro-Macro Mismatch Risks and Politics

Economic policies and politics are based on microeconomic and both macroeconomic considerations. Often politics trumps both by catering to individual wants and needs without assessing their macro-economic consequences. Further, if all of us reduce at one time our consumption, then the economy will not expand, wealth will not be created, firms may close their doors and a recession will ensue. By the same token if a government's policies are to increase employment by increasing the number of public employees and finance this growth by increasing taxes (to say 75% as it is the case in France, in 2012), there may be consequences to contend with (such as capital flight, a decline in investment, a decline in jobs and a slower economic growth). If one taxes a person's total capital wealth over 1,000,000 Euros yearly at a 3% rate, and if a person capital wealth is his home inherited worth today 3,000,000 Euros (quite a usual price for a Parisian in many parts of the city or in some city centers) it means a yearly tax payment of 90,000 Euros in addition to other income and related taxes. To pay this tax, the fortunate owner of such a home would then have to earn yearly 360,000 Euros since his tax rate will be 75%. Of course, while a few may be hurt, there will simultaneously be a migration of the tax paying rich and an excess supply of expensive homes by owners who cannot afford to own such homes, a flight of capital and of course a financial crisis of the local community that depends on revenues from their valued real estates. In this sense, economic, risks and politics are closely intertwined. In Chap. 8 such problems are considered and priced within the confines of an extended CCPAM (Capital Consumption Asset Pricing Model) risk pricing model that seeks to reconcile some micro & macro considerations.

Micro-Macro risks result from a mismatch between the two, such mismatch leads to financial markets to be "incoherent", on the one hand seizing opportunities that result from subsidies, bubbles and sectorial investments (for example, infrastructure investments that are jobs oriented rather than needed, social programs with no consideration for their costs etc.). Thus, policies that negate the combined macro-economic effects of globalization, technological change, the demographic profile of a country, the migration of industrial capacity and jobs from one country to another, the effects of economic inequalities on national policies etc. are also causes to potential risks. For example, the MBS (Mortgage Backed Securities) crisis of 2008, was such a mismatch—on the one hand setting the condition of "a home for every one" based on low initiation costs and interest, and their long run (and unsustainable) consequences. A blatant additional example, is the EURO crisis of 2012, with some countries' seeking to have "their cake and eat it too", namely, committed to unsustainable expenditures without increasing debt (already deemed too large), with limited taxation, increasing salaries and related benefits while at the same time losing jobs to competitors with lower production costs, reduce inequalities by financial transfers from one population segment to another (without accounting for the ensuing flight of capital), pursuing a technology intensive program without support to the development of its needs for education and manpower retraining etc. In other words, macro-economic policies can trump micro-economic policies instituted by governments, leading to important (an often unintended) risks.

Fig. 1.4 Economic risks

1.6.5 *Risk and Incomplete Markets*

When any of the assumptions made by fundamental finance theory is violated, markets are said to be incomplete. In other words, markets are no longer "risk prone and priced" but "uncertainty prone and mispriced" (Fig. 1.4).

Finance and assets pricing is theoretically based on the exchange between buyers and sellers substituting risks for returns and vice versa. Risks are then shifted for a price (the risk premium) from one party to another, from place to place and from one time to another. Arrow (1951b), Lucas (1978), Black and Scholes (1973) and a legion of financial engineers have devised a theoretical framework based on a one price for all equivalent risk-returns assets and no arbitrage. When markets are not complete (in a theoretical and practical sense), uncertainty primes and theories falters.

Lacking models we can agree on or justify, alternative approaches are based on some rationality, personal and empirical experience, behavioral-psychological constructs to reconcile observed states with theoretical predictions. These models augment our risk comfort, but do not necessarily imply that risks are tamed.

Technically, financial theory falters when:

- States and their consequences are not completely counted or accounted for (either due to a lack of knowledge or to remote events that are extremely unlikely or ignored).
- Markets are neither efficient (i.e. the theoretical assumptions that lead to efficiency are violated) nor are they rational due to asymmetries of power and information of its participants. Or due to a mismatch between micro-financial processes and the structural macro-economic factors that have an important

effects on exchanges in financial markets. Sovereign States interventions in foreign exchange markets, strategic behaviors between contentious Sovereign States acting to support their financial markets, regulatory and support policies, etc. are such examples.

In these conditions, a "state of incompleteness and uncertainty prevail" affecting finance at two levels: "individual-personal" and "financial markets". For individuals, risk is managed based on their potential to bear risks, based on their wealth, their information and risk attitudes. To do so, they construct portfolios to fit their wants. To maintain and sustain the efficient functioning of financial markets, coherent international and national financial policies are set in place to sustain their viability and mitigate the systemic risks that might ensue.

Incompleteness of markets is far more prevalent than presumed. Insurance firms for example, distinguish between what is insurable and what is not in the same manner that we may distinguish between "complete and incomplete". Insurable risks are well defined and experience based and thereby defined, measured, valued and priced. "Un-insurables", however, are contractually avoided by removing all unpredictable events from insurance coverage. Of course, uncertainty is never removed but remains the privy of the insured who assumes it (mostly without being aware of it). In this sense, insurance prices have been somewhat oblivious to markets incompleteness (or have done away with uncertainty) by selecting what to insure and what not based on predictions of their portfolios. By pricing their insured risks and "selling these risks" to financial investors, they have however profited from markets when these are "presumed complete" and thus, they too have removed their risks at the expense of a fraction of the risk premium they have extracted from insured.

1.6.6 Risk Models and Uncertainty

Models are a partial representation of reality, neither claiming, nor fully representing the unknown. It attempts to reconcile what we know by what we assume we do not know. In this sense, risk models, are bounded by our own rationality and knowledge what we know. For this reason it can neither be expected to be always predictive nor accused of predicting the "unknown". It cannot either be assumed to be always objective. Rather, risk models are useful to provide a perspective which has both subjective and objective intents—on the one hand a state of mind to "tolerable uncertainty" and on the other an operational frame of reference to manage risks. They are thus both needed, and are subject to continuous contentions. In a world of global outreach where exchange between agents are beset by partial and information asymmetry, insiders trading, etc., such models are difficult to construct.

Rational risk decision making processes have always been considered partial theories, proved and disproved in a broad set of circumstances and providing Nobel Prizes to those who have been able to expand our horizons in confronting "uncertainty". Psychologists such as Herbert Simon (1955, 1979, 1982), and Kahneman and Tversky (1979, 2010), Kahneman and Lovallo (1993), (with Nobel Prizes granted to Herbert Simon and Daniel Kahneman) have sought to better understand "Models of Man". Kahneman and Tversky suggested an approach coined "Prospect Theory" that questions the underlying assumptions of a universal "rational behavior" framed in a utility function (Von Neumann and Morgenstern 1944, see also Chaps. 7 and 8). Subsequently, numerous studies have advanced the importance of "regrets" and other judgmental aspects of decision making (for example Allais 1953, 1979a, b; Loomes and Sugden 1982, 1987; Gul 1991; Quiggin 1982, 1993) to provide an ex-post perspective to risk assessment and management (see also Chap. 10 for a quantitative approach to such problems).

Such approaches have ushered an "interpretation" of events based on their likelihood and valued in terms of subjective accounting of both their probabilities and their consequences. This assume many forms such as re-normalizing probabilities and renormalizing their consequences to account for the important-versus the less important, the known versus the unknown, the likely versus the less likely, etc. to reflect behavioral and psychological states emanating from theoretical and empirical studies in the social and human sciences. These re-normalizations can result in deformed probability distribution functions with parameters estimated based on an empirical observation of decisions made by decision makers, their risk attitudes and their behavior. These models provide the means to verify theories defined in terms of unconfirmed hypotheses—whether simple or complex.

To frame uncertainty and normalize it as a risk, many approaches and techniques are used. These may include, parametric rationalities embedded in utility functions or future probabilities (see Chaps. 7–9), information constrained principles based on the known and its complement randomized (for example using concepts of Maximum Entropy etc. as outlined in Chaps. 9 and 10) etc. These elements will be presented and used in various situations in this text. Throughout, we will seek to remain in a risk or in an implied risk framework based on some approach to risk "re-normalization" of uncertainty (see also Chaps. 8 and 10 and Oren Tapiero's 2012c thesis.).

1.7 Corporate Risks

Corporate risks have traditionally been used as a panacea for the many ills that Corporate Management deals with. The strategic growth and realignment of corporate strategies of typical Industrial firms (focused on their intra-corporate risks) reinvented as global and market sensitive firms with a broader social and ethical agenda has gradually altered the conception and the management of Corporate Risk. Corporate concerns have expanded corporate risks to be far more in tune with "the new corporate global environment". Many of these risks are difficult to define, to value and to manage. These are strategic-competitive, political-social, sustainable-strategic, external and internal risks. They expanded hand in hand with the growth of:

- Technological Change
- Globalization
- Financial Markets
- Strategic Economic and Competitive Risks
- Sustainability and Risk Externalities

These are today pre-eminent in corporate strategies. Below we outline essential elements of these risks.

Corporate Technology and Risks. Corporations are today consuming far more fast-changing technologies (in particular networking and IT with recurrent security threats), while increasingly losing the controls they have over the use of these technologies through outsourcing and by a lack of competent expertise to face their fast changing pace and their derived risks. IT outsourcing for example implies helplessness in managing technology, embedded in a strategic rationale with long run consequences when they become dependent on external know how rather than proprietary capacities.

Globalization and Corporate Risks. Globalization is both an opportunity and a threat. It is an opening to economic markets while at the same time it raises competitive challenges. It is an opportunity to expand corporate outreach, but it also allows foreign and competing firms to invade what may have been corporate traditional markets. Further, operating in a global environment might differ due to local cultures, legal frameworks, values, traditions etc. Thus, the presumption that globalization will provide forever expanding markets can be presumptuous (e.g. the KOF index of globalization has indicated for 2012 a contraction of global markets).

Corporate Firms and Financial Markets. For corporate firms, financial markets are the means to "liquefy their worth" based on their financial performance, their future prospects and numerous intangible factors we may choose to call "reputation" or "goodwill". Financial markets are creating immense possibilities for raising capital, for risk transfer, for increasing liquidity, to expand and contracts. At the same time corporate real prospects may be raided, its processes and technologies becoming obsolete, global competition challenging its "reputation" (and punished financially if it is tarnished). Warren Buffett, a leading Corporate Leader is credited to saying that a reputation may be built over 25 years and lost in a day! Reputation however can be affected by numerous factors such as: word of mouth, regulatory compliance, meeting expectations, long term (and persistent) financial performance, corporate governance and leadership, corporate social responsibility, talent and corporate culture as well as communication (advertising) and managing crises. Failing in any of these elements has repercussions on reputations risks.

Corporate and Strategic Risks. These risks arise when corporate firms with different agendas compete. Extreme cases may be based on information and power asymmetries that lead to the few to threaten and control firms and in some case markets. Risks are then a consequence of corporate or Sovereign States agendas and their application based on the power and the private information they have. Corporate realignment along Supply Chains, integrated and networked business entities, have augmented the dependence of corporate firms and therefore their risk exposure to both their stakeholders and their competitors. The implication of such risks are that "risk neutrality" and a "conflict free" environment applied to the many problems—operational and otherwise, that corporate firms face, are no longer realistic. Rather, a strategic approach that recognizes both risks and conflict to risk assessment, exposure and its management are to be accounted for. *Sustainability, Risk Externality and Corporate Risks.* Corporate responsibility, ethics and environmental concerns are a sample of issues corporations confront that are related to risk externalities. An externality is a cost or benefit that is experienced by someone who is not a party to the transaction that produced it. Externalities (in particular negative externalities) create a risk for corporate firms—the risk that a divergence between private costs and social costs will entail appreciable reputation, regulation and penalties costs. A sober view and study of corporate risk externalities and its sustainability are therefore essential and their implications internalized in intra-corporate risk prevention and in recovery from particularly disastrous events (for example, the BP oil blow up in 2011).

Risk Externalities are important because they can create incentives to engage in too much or too little of an activity, from an efficiency viewpoint. When the costs and benefits of a transaction are internal, meaning that all costs and benefits are experienced by someone directly involved, we expect the transaction to take place only if its benefits are greater than its costs. Say that you want to buy a car. You may buy it if its value is greater than its price. Car companies also will only produce and sell if the price it expects to get is greater than what it costs to produce. There may be in such cases some economic grounds to negotiate a price that the car company can agree on. In such a process, wealth is created—in the sense that at least one of the parties will pay a price less than the value of the car. On the other hand, if it is worth less than its cost then there will be no agreement on a price and therefore it will not be produced. But externalities throw a kink in this logic. What if the production process or the product-car leads to dumping some of the costs of the car on a "neighbor" (for example, polluting inordinately the air, or spilling wastes and chemicals used in the production process in the river "next door") without the buyer or the car seller sharing or assuming the (externalities) costs and consequences? Thus, while both the buyer and the seller may profit, no wealth will be created as it ignores the externalities emanating from such a transaction. Such problems are notoriously difficult to assess. Risk externality, when recognized, requires that regulation agencies impose penalties on violators to prevent and reduce their costs (see in Chaps. 9 and 11).

1.8 Risk and Networked Firms

Integrated and networked firms (whether loosely or not) are stakeholders or partnerships with a common interest. Yet each may be partly independent and dependent and maintaining a private interest. Supply chain risks for example, are therefore specific, operational, strategic-relational and organizational firms, collaborating in their pursuit of profits (Agrawal and Seshadri 2000; Zsidisin et al. 2001; Kogan and Tapiero 2007). A search for the term "Supply Chains Risks" on the internet will reveal a large number of interviews with practitioners, individual and academic contributions, consulting firms and papers that seek to bring attention to what managers are calling attention to—network derived risks. For example, Chris D. Mahoney (UPS, October 2004, http://www.ism.ws/Pubs/ISMMag/100406.cfm) points out that networked chains that run like clockwork, reducing costs and bolstering customer service have a downside risk. These risks transcend traditional concerns for operational and external (hazard) risks. Risks previously neglected arise because of these entities dependencies, information and power asymmetries, political, strategic and risks externalities.

Marsh Consulting for example, advises corporate and supply chain firms: (in Risk-Adjusted Supply Chain Practice):

- Do you fully understand the dependencies within your supply chain?
- Have you identified the weak links within your supply chain?
- Do you understand the risk that has been inadvertently built into your supply chain?
- Have you identified the supply chain risks that might mitigated, eliminated, or passed on to another supply chain member?
- Did you incorporate the element of risk when making strategic or tactical decisions about your supply chain?
- Is your supply chain nimble and flexible so that you can take advantage of both supply chain risks and opportunities?
- Have you fully captured your enterprise-wide risk profile?
- Do you know which supply chain risks may cause an adverse event, causing significant disruptions?
- Do you have the necessary tools, skills, and resources to model your supply chain, including its risks and vulnerabilities, in order to understand the financial impact that various events and scenarios will have on your supply chain?
- Did you benchmark the activities that make up your supply chain?
- Have you identified and do you monitor key upstream and downstream risk indicators or activities that might result in a disruption of your supply chain?
- Have you fully integrated business contingency plans and emergency (recovery) response plans into supply chain management initiatives?

1.8.1 Information Asymmetry

Information asymmetry risk is a source of uncertainty, providing one of the parties and informational advantage that the other does not have. In such cases, risks depend on parties information, their behaviour, their incentives and their preferences and the manners in which they use their information to their advantage. There are then two types of risks resulting from such situations: Adverse selection risks and Moral hazard risks. These are considered below.

"*Adverse Selection and The Lemon Phenomenon*". Akerlof (1970), points out that goods of different qualities may be uniformly priced when buyers cannot realize that there are quality differences. For example, one may buy a used car, not knowing its true state, and therefore the risk of such a decision can induce a customer to pay a price which does not truly reflect the value of the car. Such situations can explain the need of consumers to buy warranties to protect themselves against post-sales failures or to favor firms who are trustworthy and possess service organizations (in particular when products are complex or involve some up-to-date technologies).

"*The Moral Hazard Problem*". Its risk is that a property that cannot be observed by one of the parties to an exchange, has a risk. For example, a supplier may use that fact to its advantage and not deliver default prone parts. Similarly, banks selling financial products as "safe, secure and highly rated" when they are not are knowingly unloading risks they do not want to bear on unsuspecting investors. We may deal with such problems by regulation for a greater transparency, punishing penalties for purposeful violators who use knowingly such risks to their profit; by using various sorts of (risk-statistical) controls combined with contracts providing incentives to be honest, transparent and reveal what they know. For example, some restaurants might open their kitchen to their patrons to convey a message of truthfulness in so far as cleanliness is concerned. A supplier would let the buyer visit the manufacturing facilities as well as reveal procedures relating to their process controls, machining controls and the production process in general. A drug manufacturer may publish publicly the experimental results of drug tests, etc.

Insurers have recognized the importance of moral hazard risks by building risk prevention incentives in insurance contracts. For example, a car driver with a history of no accidents may have to pay a reduced premium. Similarly, United Health Group Insurance has proposed to Doctors and Medical Providers new incentive payments consisting in switching from "quantitative treatments" to "qualitative benchmarked payments". In this "new system", Doctors that avoid sending patients to hospitals for readmission, etc. and based on treatment outcomes would be paid incentive bonuses or payments. The intent of course, is to reduce costs and institute some form of "quality control" over excessive use of health care facilities.

Examples of hazard risks are numerous. We outline an additional few: An over-insured warehouse may be burned or looted by its owner to collect insurance, or at least result in careless handling of materials which may have risk consequences;

A transporter may not feel sufficiently responsible for the goods shipped by a company to a demand point. The "de-responsabilization" of workers also induces a moral hazard. It is for this reason that incentives, performance indexation to performance and "on-the-job-responsibility" are often used.

1.9 Risks—Many Reasons, Many Origins

Risk models are defined in terms of specific risks delineating their origin. We have referred to these risks earlier in this chapter. Below we summarize some of these origins. These include essentially, external, endogenous, strategic and externality risks. "*External risks*" are derived from events over which we have no control. These can arise due to Natural causes or causes that outstrip our ability to control them. *Endogenous risks* are both man-made, resulting from "what we do" or result from the internal malfunction. In the first case, it can be defined by predictable events with predictable probabilities or be a rare event, ignored or unknown. In some cases these risks are "unintended", resulting from a misunderstanding and a lack of knowledge about the consequences of our own actions. External risks tend to have extensive records and therefore extensive data bases to be defined by probabilities, while endogenous risks are mostly the outcomes of case studies, experimental and simulation studies (Fig. 1.5).

"*Strategic risks*" are motivated endogenous risks. These are based on the intents of interacting parties, such as competing firms, the gaming of regulation and regulators by banks (see Chap. 11), economic exchanges financial and other contracts etc. Chapter 12 provides an outline of gaming models and in particular their risks while Chaps. 8 and 11 devote a particular attention to pricing and control approaches when strategic relationships are dominant (namely in multi-agent systems).

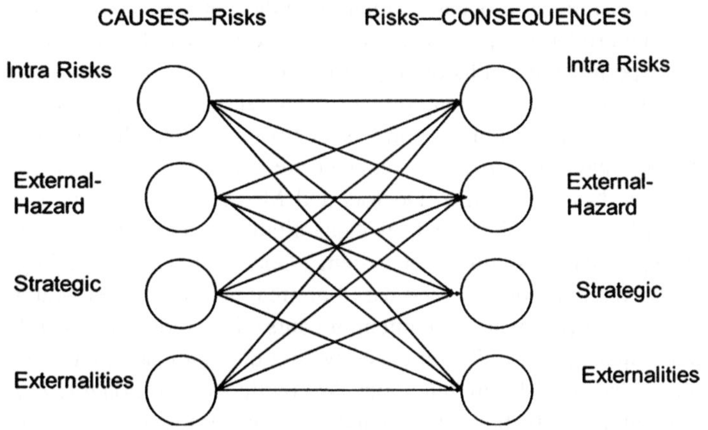

Fig. 1.5 Risk and consequences

"*Risk Externalities*": These risks are associated to situations where parties at fault do not assume the consequences of their actions. These may be "pollution risks", "financial systemic risks" etc. where the public at large must bear these risks. Most business activities have derived external effects not accounted for (in a traditional accounting sense). A driver emits gases when he drives and it is us that are taxed by breathing a deteriorating air quality and be confronted with increased Medicare costs. The same may be said about a smoker; an industrial firm emitting gases and contributing to air pollution etc. In other words, "almost everything we do" has a derived effect which may be assumed by another party, whether consciously or not. When this is the case, there is an externality. Namely, the action of the one affects the other. It is a risk externality when its consequences are negative. It can be a water fracking firm (in search of gas) destroying the water supplies of farmers; it can be a polluting ship; it can be a manufacturer generating irresponsibly waste and pollution in a public area or in another's person property. In such cases, regulation is used to balance between the costs of pollution prevention and its effects on the pollution level. A risk externality can result in the breakdown of the market mechanism in a number of situations. These include when one of the parties has a monopoly power, when there is an information asymmetry and when there is no exchange market (and therefore consequences have no price). For example, there are attempts to "create Co2 markets" in order to use it to price pollution in the hope that investors-speculators as well as pollution producers would act to prevent pollution (since they will be required to buy Co2 emission rights to pursue their pollution creating activity). That intent too has a risk externality as it can lead one to neglect other pollutants that are far more important.

Statistics, Games and Risk: Statistics has, for the most part, presumed that risk and uncertainty are neutral! In other words, they are not motivated. It need not be the case. For example, multiple parties interacting with broadly varying motivations; information asymmetries; power and asymmetric relationships may generate both predictable risks and an unpredictable uncertainty. In the former case, parties may have full knowledge of alternative courses of actions each of the parties have while in the latter case, they may not. Recognizing the importance of these problems and how we may confront strategic decisions and risks has been universally recognized by the economic profession with Nobel Prizes granted to researchers in Game Theory to Nash, Selten and Aumann (see also Chap. 12 for a review of elements of game theory and risks).

Statistic's failure to deal with strategic risks is based on the presumption that risks are mostly defined in terms of random events. Interpreting risk and uncertainty and reducing its effects assume then that our measurements and our acts are independent of strategic factors (such as multiple decision makers with conflicting agendas, counter-parties, etc.). In fact, information asymmetry and strategic conflict lead to strategic risks. For example, insurance contracts with binding clauses may be designed to avoid both unpredictable (uncertain) events and to induce a post contractual behaviour to be compatible with the contract's intentions and to reduce the propensity to claim by insured. Similarly, strategic audits have always a number of messages they convey; a control, a signal to the audited on the firm's intentions and of course to collect information which is needed to reach an economic decision. For example, numerous research papers have pointed out that in a bilateral monopoly (i.e. when only two parties are involved in a decision making, information asymmetry can lead to opportunistic behaviour, or simply said—cheating). The control of exchanges between such parties should therefore account for parties' intentionality imbedded in their preferences, the exchange terms as well as the information each will use in respecting or not the intended terms of their exchange. In some papers (Reyniers and Tapiero 1995a, b), these factors are considered in producer-supplier relationships, in the design of contracts for the control of quality and more recently on the effects of gaming in environmental problems (Tapiero 2004a, 2005b, d).

References and Additional Reading

Risk, uncertainty and their many manifestations and applications have been and promise to remain a continuous source of publications, books, academic papers, theories and an endless number of applications in hand of all professions. The following references are a sample of books and papers that have been consulted in the writing of this chapter and this book in general. These references provide an initial guideline for the reader. Some relevant papers I published and used in this chapter include Tapiero (2010a) on the future of financial risk management, Tapiero (2012b) on the price of safety, Tapiero (2012a) on the convergence of insurance and finance on my paper with Nassim Taleb, (2010) on TBTF and risk externalities as well as my recent book in risk finance (Tapiero 2010a on Risk Finance and Assets Pricing). Beck (1992), provides a background on Risk and Society, Bernstein (1998) is a bestseller on risk, highlighting the many facets and manifestations of risk.

Books in Financial and Insurance Risk Management include among others Alexander (1998) (with subsequent revisions and re-publications), Laffont (1989), Borge (2001), Borch (1974), Beard et al. (1979), Seal (1969, 1978), Doherty (2000), Taleb (2007), Tapiero (2004b) on risk management. Other books related to industry problems may include Tapiero (1996) On the Management and the Control of Quality (see also Daudin and Tapiero 1996); Tapiero 2004a, b, 2006, 2008; Kogan and Tapiero 2007; Tapiero and Kogan 2008 as well as books on reliability and statistical-quality controls (Barlow and Proschan 1965; Montgomery 1991; Tietenberg 1998, McKerlie 1986, Tapiero 2004a, b, 2006, 2008). Classical and important books with important contributions to the discourse of risk and uncertainty include Adam Smith's The Wealth of Nations, (1776), Walras (1874/1877) treatise as well as seminal research and publications by Arrow (1951a, b, 1963a, b, 1982), Debreu (1952, 1953, 1959), Laplace (1774), Keynes (1921, 1935), Hacking Ian (2006) (as well as additional books by the same author), Savage (1954), Ariel Rubinstein (1998), Norbert Wiener (1950), Taleb (2007).

The research literature on risk and uncertainty, their definitions and distinctions, include contributions that have underlies the fundamental of risk finance, decision making and their application. Some outstanding references include Frank Knight (1921), Allais (1953, 1979a, b) and Ramsey (1931), Muth (1961), Arrow (1951b), Rabin (1998), Prigogine and Stengers (1988), Samuelson (1963, 1965), Simon (1955, 1976, 1979, 1982), Thaler et al. (1997), Kahneman and Tversky (1979, 2010), Tversky and Kahneman (1974, 1992). Other references include, Machina (1987), on problems solved and unsolved of choice under uncertainty, Fellner (1961), on the distortion of subjective probabilities as a reaction to uncertainty, Slovic (1972), psychological study of human judgment and their implications for investment decision making, Hansson (1996, 2001), Abdellaoui and Munier (1997, 1999), Abdellaoui et al. (2007), Bauer (1960), Economides (on network risks), (1996), Kutler (2009), McCarthy (1997), Munier (1995), Munier and Tapiero (2008), Tapiero (1994a, b), Diebold et al. (2010), an edited book on papers on the known, the unknown and the unknowable.

References on safety and related problems include, Ale et al. (2000), Ale (2002), Bahr (1997), Bakker and Blom (1993), Barnett (2000), Blom et al. (1998, 2003, 2005, 2006), Dimukes (2005), Gain (2003), Hale (2002), Janic (2000). In addition there are numerous professional surveys and papers including Alliance for Nuclear Safety, (http://a4nr.org/articles/fatalFlaws/2.security.pdf/download. Federal Aviation Administration (FAA) papers and reports (2004, 2005, and others), Boeing (2006), etc.

We shall refer subsequently to numerous books and papers on industry, quality and supply chains. Some include Taguchi (1985), Taguchi and Clausing (1990), Box (1993), Ritchken and Tapiero on option pricing of inventory policies, (1986).

References in Insurance and Finance are naturally extensive including Muth (1961), Lucas (1978), the fundamental papers by Walras, Arrow and Debreu. Other papers include Slutsky (1937), Bohn and Choi (1996), Benston (1996), The Basel Committee (for risk regulation) (2004, 2008), Golub and Crum 2010, Black and Scholes (1973) and a multitude of articles published in the Journal of Finance, Journal of Banking and Finance, Risk and many others about fundamental and practical finance and Financial Engineering. The increase disenchantment with the financial theories has also led to the profusion of a finance based on observed on financial behaviors. Some relevant papers include De Bondt et al. (2008) on behavioral finance, De Bondt (2005) on the psychology of world equity markets, Sheffrin (2008a, b) on a behavioral approach to assets pricing as well as Kahneman and Riepe (1998), on aspects of investors' psychology. Not the least, numerous studies have suggested that unpredictability and chaos define risks (for example see Peter Edgar (1995) book on *Chaos and Order in Capital Markets*). Some of these approaches will be discussed throughout the book.

In Health care, a number of articles consulted in the writing of this book has included Coile (1990), Nasr (2007) and Comaro (2003).

In environmental and risk externalities problems, papers by Angell and Klassen (1999), The Argonne National labs, (2005), Greaker (2003a, b, 2006), Srivastava (1997). In Chap. 10, an extensive list of papers on environmental problems and risk externalities will be suggested.

Chapter 2
Risk Management Everywhere

Abstract The purpose of this chapter is to expand Chap. 1 with additional applications and problems. Elements of risk management are summarized and application areas considered. These include managing risks and money, industrial, and marketing risks, environmental management, as well as network risk models. This chapter is introductory and non-quantitative.

2.1 Elements of Applied Risk Management: A Summary

Risk management is multi-faceted. It is based on both theory and practice. It is conceptual and technical, blending engineering, statistics, behavioral psychology, financial economics into a comprehensive and coherent approach to make, manage and justify risk decisions, to account for their consequences and confront their residual uncertainty. It consists of a broad number of tools based on mixtures of Active-Preventive-Reactive and Passive approaches to risks management. These are:

- Ex-ante risk management tools that consist of risk intelligence and detection, data-information collection; assessing risk exposure; risk analysis and design; risk valuation and pricing; risk sharing; contractual, prevention and control measures; strategic and endogenous risks management.
- Ex-post risk management consists of risks of recovery; regret minimization; robust design, anti-fragility and contingent actions that allows us to mitigate adverse events when they occur.

Under uncertainty, when risks are not predictable, the contingent ability to recover ex-post from risk events is essential. For example, recovering from a Tsunami, recovering from an unplanned bankruptcy, recovering from false accusations etc. are based on contingent means that are set for events with no knowledge whether and when these events occur. The intent of risk management is then to alter desirably their likelihood and mitigate the magnitude of undesirable

C.S. Tapiero, *Engineering Risk and Finance*, International Series in Operations Research & Management Science 188, DOI 10.1007/978-1-4614-6234-7_2,

outcomes if they occur, share or transfer their consequences. *Risk Management uses many techniques including*:

- Risk definition and detection
- Data management and risk analysis
- Risk measurement
- Risk design and optimization
- Risk exposure assessment
- Risk sharing
- Risk valuation and Risk pricing
- Risk management: Ex-Ante, Ex-Post Robust Design and Recovery

Ex-Post risks pertain to observed risk events rather than an estimate of their prospects. Risk management principles are thus based on minimizing (ex-post) regrets, on contingent recovery plans as well as robust designs which are used to maintain system, performance within a broad act of potential conditions. For major ex-post events, FEMA in the United States plans and prepare contingencies for disastrous events, investing in the means to confront emergencies.

This chapter emphasizes a number of approaches to risk management. We consider first the valuation of risk and money with insurance and finance applications, emphasizing both their differences and similarities. The following approaches are presented: "Subjective valuation" and "Rational valuation" consisting of axiomatic approaches and risk pricing. The problems we consider cover areas of interest that cannot be covered exhaustively. For this reason, numerous problems and other applications are referred to for further study and research at the end of the chapter A number of fundamental problems are treated quantitatively in subsequent chapters.

2.2 Risk Management, Value and Money

Finance and Insurance are concerned by both risk pricing and valuation and their management—each adopting approaches that are specific to their needs and to the problems they confront. Finance for example, manage risks by hedging through a trade or a series of trades using financial products (such as options, investment portfolios appropriately assembled, securitization, etc.) while insurance manage risks through insurance contracts with various clauses to limit losses and induce insured to prevent losses. It seeks first to avoid as much as possible unpredictable claims and share risk with reinsurance firms to share and mitigate excessive claims.

For finance (and increasingly insurance) risks are priced and managed by using financial instruments (such as options, swaps, etc.). Both finance and insurance do not eliminate risk but only transfer it from one party to another at a price which we call the risk premium. When financial products are so complex that buyers are unable to calculate the risk they bear (as this was the case with structured financial products such as Mortgage Backed Securities and credit derivatives (see Chap. 9), stealth financial risks are held by the structured products holders. These products have transformed both bankers and insurers as intermediaries "re-packaging" elementary financial and insurance products into complex portfolios that they are able to sell at a profit without assuming any of their risks. Both finance and insurance are essential to the "real economy" however, without which there can be no exchange and no business (except by extremely inefficient means). Lacking finance and insurance is therefore immensely costly for an economy or any institution or enterprise confronted with risks. These risks justify the claim that some finance and insurance institutions are "Too Important To Fail" (or "Too Big To Fail") and therefore they ought to be regulated for the risks that their failure can inflict on "society". These firms, provide the means to sustain an industrial activity, to finance a health care system, to provide money for technological innovations and investments, to finance warranties, to insure exports (and therefore create jobs) etc. Finance and insurance and the real economy are thus complementary—one needing the other, one not functioning without the other. Similarly, managing and regulating risks in the one is essential to the other. In this sense, all economic and real activities are all locked in a complex set of relationships that defines and amplifies their dependency, and thereby their risks.

The valuation and pricing of risk is based on how and what risk we actually measure, how valuable they are to us and whether they can be exchanged, or traded (see Chaps. 8–10). Measurements of risks can however be valued financially when they are accepted as a means of exchange or trade (see also Chaps. 6 and 7).

Insurance firms define and value and predict risk in terms of actuarial (statistical) terms and manage the value of their risks through risk aggregation and risk sharing with specific partners (rather than financial markets—although they are increasingly using financial markets). Insurance profits are then made by the spread that an insured is willing to pay (the retail price) for a contractual risk protection (or required by law to seek such a risk protection), and the cost (the warehouse price) of an aggregate portfolio of such contracts. The greater "warehouse-retail" spread price, the greater the insurance firm profit. To manage this spread, insurance firms, construct portfolios with contracts that are statistically independent (i.e. consisting of diversified risks), that are not subject to latent or rare risks and clearly delineated. As a result, insurance firms seek basically to increase their spread, maximize their returns on their capital and minimize risks.

For finance, unlike insurance, the risk spread is derived from standardized "financial and traded commodities" (such as securities, bonds, options etc. that are globally accepted as exchange instruments for returns and risks). For example, a "security" whose price provides a rate of return of 10% while an equivalent riskless bond provides a rate of return of 6%, has a rate of return of 10-6 = 4% which is the price that the buyer of a security receives when it assume the security risks. As a result, in theory and in practice risk is measured and valued based on two essential model-intensive approaches consisting of the following:

1. Asymmetric (individual, firms, investor, or economic agents) preferences that are motivated to exchange. Risk, measured and valued differently by say two parties may then agree to an exchange price that is consistent with their preferences. Such a price is then specific to the parties and to their transaction.
2. The market pricing of risk, based on economic equilibrium models, summarizing the demands of buyers and sellers for "financial commodities" assumed (and therefore believed and accepted) to include and reflect implicitly the price of risk.

The first approach is based on the presumption that "persons" are not indifferent to the size of gains and losses. A risk-averse person for example, would weigh losses more than their equivalent gains! To express such subjective preferences, a number of approaches (based and often derived from expected utility arguments as we shall see in Chap. 7) pretend to represent persons' preferences. In this context, the premium a person is willing to pay to do away altogether or partly with risk indicates his risk aversion. This approach turns out not to be practical because the underlying utility of decision makers is not usually known (although it is often expressed in terms of a number of parameters that one seeks to estimate based on observed prices and trades). For this reason, models that are tested or based on economic experience are used to determine the risk premium an insured ought to be charged. These models are important as they provide a theoretical framework for assessing future risks based on implied prices. For example, insurance firms assess insured claims risks using proprietary actuarial information and formulas on the basis of which a risk premium is calculated (see Chap. 7). As indicated above a gradual shift to marketing standardized insurance products is leading the insurance business to be financial intermediaries. A marketing apparatus for aggregate portfolios consisting of life insurance and other pensions' related products are then "commoditized" and sold in financial markets as "securities". Such products are essentially insurance products dressed in a financial tunic. The importance of insurance liquidity and the ability of insurance firms to meet insured claims require that they conform to certain rules set by insurance regulators. Basel III regulation rules for example, require that they set aside a certain amount of risk free financial capital to meet potential liquidity needs (also called Capital Adequacy Ratio-CAR, Value at Risk-VaR as well as Conditional Value at Risk-CVaR, as outlined in Chap. 6). As a result, insurance firms have become complex "warehouses of risk". In some cases, elements of the insurance business has been assumed by banks while element of the banking business have been assumed by insurance firms—contributing again to an insurance/finance competitive risk convergence.

2.2.1 Insurance Actuarial Risk

An insurance contract substitutes payments now (the risk premium) for potential future losses. The size of current payments and future losses are based on the actuarial assessment and valuation for the probabilities of future losses and their size (Cramer 1930, 1955; Buhlmann 1970, 1980; Gerber 1979; Seal 1969, 1978). The risk of loss is then priced at the insured risk premium, managing assets of the insurance firms, meeting insurance regulation and building the capacity to meet losses when thy occur. In most cases, insurance firms negotiate with insured preventive measures they ought to assume in exchange for a reduced premium. When there is an information asymmetry, with one or the other parties taking advantage of such a situation, risk shifting is counter-productive, providing a gain to the one party at the expense of the other (Arrow 1963a; Akerlof 1970; Hirschleifer and Riley 1979; Holmstrom 1979; Spence 1977). To mitigate these risks, contracts are stated explicitly in terms of contractual and obligations and premium incentives (such as bonus-malus premium payments based on past claims of the insured). At times it is difficult to control what each of the parties does and therefore both inspections and controls as well as recourse to litigation provide means to control and meet the terms of the contract as agreed on and as expected (Reyniers 1992; Reyniers and Tapiero 1995a, b).

Preventive means by insured are many. For example, driving carefully, locking one's own home effectively, installing fire alarm for the plant, etc. are all forms of loss prevention. Car insurance rates tend, for example, to be linked to a person's past driving record, leading to the design of (incentive) bonus-malus insurance premiums. Certain clients (or geographical areas) might be classified as "high risk clients", required to pay higher insurance fees. Inequities in insurance rates will occur, however, because of an imperfect knowledge of the probabilities of claims and because of an imperfect distribution of information between insured and insurers (Borch 1968, 1974).

Traditionally, actuarial science was concerned with quantitative risk measurement and assessment. Tetens and Barrois, as early as 1786 and 1834 respectively, attempted to characterize the "risk" of life annuities and fire insurance so that they may be properly assessed and valued financially. It is due to Lundberg (1909) and a group of Scandinavian actuaries that we owe much of current actuarial theories. Actuarial statistics have initially focused on claims risks arising from bankruptcy, survival (for life insurance), accidents proneness etc. Based on economic and social considerations, "Risk Premium Principles" were suggested to "price fairly insurance contracts". Most of these approaches are based on a willingness to pay of insured, their actuarial propensity to claim and their claim history. In the last decades, risk premiums have increasingly been engineered in terms of the market price for risk.

2.2.2 *Finance and Risk*

Finance is motivated by three basic purposes:

- To price the multiplicity of claims, accounting for risks and deal with the adverse effects of risks and uncertainty;

- To explain and account for investor's behavior. To counteract the effects of regulation and taxes by firms and individual investors (who use a wide variety of financial instruments to bypass regulations and increase the amount of money investors can make while reducing the risk they sustain).
- To provide a rational framework for individuals and firms, financial decisions and to suit investors needs in terms of the risks they are willing to assume and pay for.

Financial risk instruments are used for risk management, allowing both an exchange of financial and real assets, a risk transfer both horizontally (for example across different products, different countries) and vertically (for example across different financial and dependent products such as options and their underlying). Markowitz (1952, 1959) suggested a risk/reward framework based on the mean and the variance of a portfolio. In this case, a portfolio's returns/variance emphasizes their substitution, with one valued in terms of the other. Risk management consists then in a risk minimization (variance) for a preferred return and vice versa, maximizing expected returns subject to a tolerable risk. Markowitz' solution provides then an efficiency curve that outlines investors' portfolio preferences. By including a risks free asset (bonds) in Markowitz' portfolio, relative risk premiums were defined and embedded in the celebrated CAPM (Capital Asset Pricing model, see Chaps. 7 and 8). Subsequent studies have suggested an inter-temporal shift and transfer of risks (see Bismut 1975; Bismut 1978 for initial and important studies in this matter as well as elements of the CCAPM pricing models, Cochrane 2000 and Chap. 8). This has allowed risk managers to better manage the *time phasing of risk and returns*.

Financial liquidity expresses the capacity to meet demands for money. It is also an appreciable source of risk when the demand for liquidity cannot be met. Following the financial crisis of 2007, liquidity was proved to be extremely important, providing immense risks to firms or persons that did not and could not access the liquidity they needed and great profits (or at least no loss) to those who possessed or could access liquidity. Further, the financial crisis has pointed out that liquidity affected significantly equity and debt market prices. The need for liquidity as a fuel of economic development and expansion has led to the stratospheric growth of credit risk and credit derivatives leading both expanding mortgage markets, insurance liquidity but at the same time to its uncontrolled and poorly regulated use leading some to call such financial products "weapons of mass destruction", presumable a dominant factor in the financial crisis and its aftermath.

Example: Managing Credit Risk To manage credit risks banks use a plethora of techniques and financial products they sell and trade in financial markets (in order to share and sell the risks they have assumed). Among the many tools they turn to for risk management are the following:

- Loan underwriting standards (i.e. procedures applied to safeguard loan reimbursements).

- Insurance of credit transactions and insurance of credit portfolios. In this case, either one or both parties, or a third party will act as guarantor to some or the risk implied in the credit transaction.
- Construct credit portfolios and their collaterals priced to be sold in financial markets (called securitization).
- Managing collaterals and hedging their values to maintain the collateral effective use as a guarantor of the loan given to a borrower.
- Diversify the credit risk across different borrowers.
- Managing the credit portfolios in light of macroeconomic developments.
- Manage the counterparty risk, both when credit transactions involve the two parties engaged in the transaction and when a third party assumes its insurance. These problems assume many forms—such as a third party (a rating firm) defining the credit quality of a portfolio, an insurance firm insuring the portfolio, etc. In such cases, issues such as, "is the rating firm providing a high rate to a bank's portfolio because they profit greatly from providing such a rating or engage in extensive other businesses with the bank? Is the insurance firm a traditional insurance firm or a financial and speculating enterprise branded as an insurance and financially safe firm?

Example: Too Big To Fail: Externality Risks and the Fallacy of Large Numbers "The lure for size" embedded in "economies of scale" and the "monopoly power" it provides have led to firms that may be TBTF. When these firms fail they have important risk consequences that are Too Big To Bear for the public at large. This is the case for industrial giants such as GM that has grown into a complex and diversified global enterprises that has accumulated a default risk too large to bear. This is also the case for banks that bear risks that are often ignored and end up also Too Big To Bear. Banks, unlike industrial enterprises, draw their legal rights from a common trust, to manage the supply and the management of money for their own and the common good. Consequences of their failure, overflow into the commons, and contribute to systemic risks, that far outstrip their direct losses. When banks are too big to fail they may use it to both assume excess risks and seek protection when their financial bets sour. Further, they may also price their services to be unrelated to their costs or their quality and exercise unduly their market power. TBTF firms create therefore risk externalities that justify their regulation and tight controls. For example, Frank Rich (The *New York Times*, Goldman Can Spare You a Dime, October 18, 2009) has called attention to the fact that "Wall Street, not Main Street, still rules Washington". Similarly, Rolfe Winkler (Reuters) pointed out that "Main Street still owns much of the risk while Wall Street gets all the profits".

Banks are not the only economic institutions that are TBTF (Too Big To Fail) and can produce risk externalities. In the energy sector, a study by the National Academy of Sciences has pointed out to extremely large hidden costs that are not accounted for by the energy industry, but assumed by the public at large. Fujiwara (2004), using an exhaustive list of Japanese bankruptcy data in 1997, has pointed out that firms fail regardless of their size. Further, since the growth of firms has been fed by debt, the risk borne by large firms seems to have increased significantly, threatening both creditors and borrowers alike. In fact, the growth of size through a growth of indebtedness combined with ''TBTF'' risk attitudes has contributed to moral hazard risks, with firms assuming non-sustainable growth strategies with stealth and important risk externalities.

When size is based on networked firms (such as large global networked financial institutions, "supply chains", health care networks, electrical power grids), their dependence increases and failure can be contagious (see also Mandelbrot and Hudson 2006, Serafɩn Martɩnez-Jaramillo et al. 2010; Allen and Gale 2000). Saito et al. (2007), while examining inter-firm networks noted for example that larger firms tend to have more inter-firms relationships than smaller ones and are therefore more dependent. In particular, Saito points out that Toyota purchases intermediate products and raw materials from a large number of firms; maintaining close relationships with numerous commercial and investment banks; with a concurrent organization based on a large number of affiliated firms. Such networks have augmented both Toyota's dependence and its supply chains risks. Thus, when one supplier controls a critical element needed for the proper function of the whole firm, lacking this element can lead to the whole supply chain to be immobilized. Renault, in France, experienced such a state of affairs when a small plant in Normandie with no more than a hundred employees could strike out the whole Renault complex. By the same token, a small number of traders at AIG could bring its "Too Big To Fail" firm to bankruptcy. Simulation experiments on a network of cooperating banks was conducted by Aleksiejuk and Holyst (2001) to assess the propensity of a systemic breakdown using percolation theory and its contagious effects. Their simulation have shown that sudden withdrawals from a bank can have dramatic effects on the bank stability and may force it into bankruptcy in a short time if it does not receive assistance from other banks. More importantly however, the bankruptcy of a simple bank can start a contagious failure of banks that can lead to a systemic financial failure.

TBTF Risks raise an essential question: Can economies of scale savings compensate their risks. Such an issue has been implicitly recognized by Obama's administration proposal in Congressional committees calling for banks to hold more capital with which to absorb losses. The bigger the bank, the higher the capital requirement should be (*New York Times*, July, 27, 2009, Editorial). However such regulation does not protect the ''commons'' from the risk externalities that banks create and the common sustains. Further, augmenting inexorably the capital banks must hold will necessarily reduce the ability of banks to lend and therefore their profitability.

2.3 Industry Processes and Risk Management

The transformation of industry following the first industrial revolution, has redefined risks to be embedded in parts "uniformity", in "product assurance", in a consistent and a repetitive process. These have led to risks to be measured and managed by the "control of variations". Non-performance or a deviation from specifications as well as departing from a process specs' identified as essential risks. These resulted in an industrial approach to risk management which is adopted as a standard in many other professions. It is based on:

- Control and process design for systems with no default, no errors, no defects and consistency
- The propensity to maintain the manufacturing process in control, i.e. operating to meet its conformance to pre-sets standards through statistical sampling and controls and assure its supplies and delivery
- Zero errors and faults tolerance through prevention, robustness and comprehensive managerial approaches to risk management.
- Manage Variations in all ways feasible and economical. Meet expectations or operate to meet agreed on standards of supply or manufacture.
- Maintain a process propensity to be repetitive with variations within defined bounds.

Risks of default, of breakdowns, of non-conformance, supply risks, quality risks etc. were thus defined and industrial risk management based on maintenance, reliable designs, statistical controls, inventories etc. For example, Shewart (1931) introduced principles of statistical (risk) controls to control parts and products uniformity. The standard deviation (a statistical variation) was associated to risk and used to construct control charts that are used since.

Statistical approaches were applied to all sorts of problems (in agriculture, in insurance, in health care, etc.). Classical tools coined under the names SPC (Statistical Process Control), SQC (Statistical Quality Control) as well as Experimental Design and Robust Control (se Chap. 11). In the 1980s, the number of parties involved in a process (whether industrial, service, health care, development, management, etc.) became so large, interdependent and complex that their management became overwhelming. For this reason more systematic and comprehensive approaches were needed. These led to defining a series of acceptable and easy to understand and apply tools for the many parties involved in a process (Fig. 2.1).

For example, TQM (Total Quality Management) seem to standardize the use of simple statistical tools to allow both their acceptance and simplify the technical exchanges required in quality and risk management. These tools express the need to systematize and quantify: Sources of risks; Measure and accumulate data; Find relationships; Summarize variations; Seek risk causes; Seek what are the important contributing factors to risks. The Table below Fig. 2.1 summarizes both the techniques used and their intent. These approaches are both common and easy to understand and can be used to communicate among and across the parties that are involved in an industrial process. Other approaches and tools were also suggested. They are essentially based on the following: Risk definition (what is most important and what to measure that will help prevent risks and responds to your needs); Measure what is essential and revealing to track, to test, to predict and to

	Brainstorming (to define a relevant and complete states to refer to)
Events **Day 1** **Day 2** **Total** **A** IIIII IIIIII 13 **B** III II 5 **C** IIII III 7	Tally sheets (to accumulate data and reach data supported decisions)
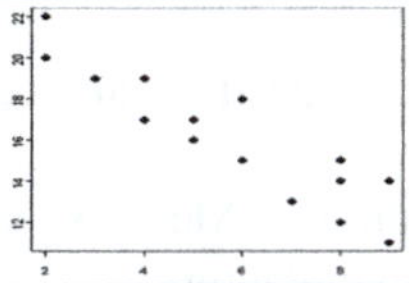	Scatter Plots to demonstrate elemental relationships or a lack of such relationships
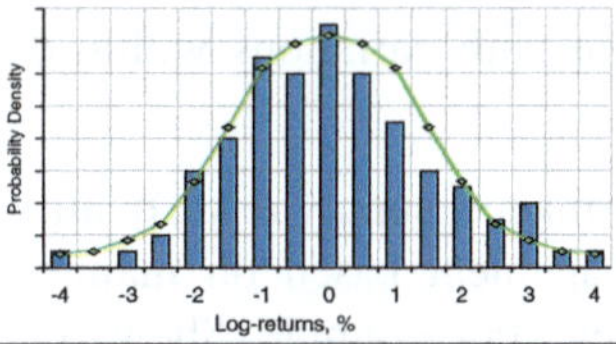	Histograms as an effective way to organize and summarize data into meaningful statistics
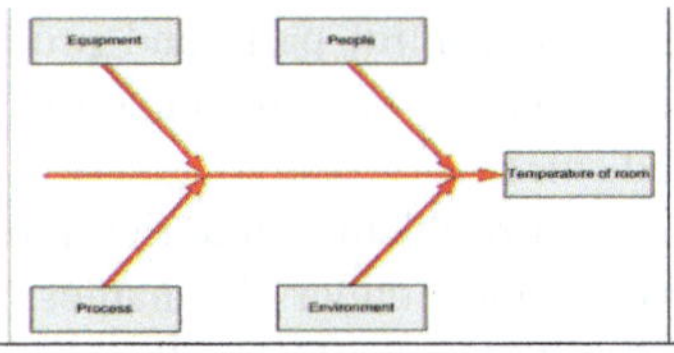	Cause-Effects diagrams to distinguish causes from effects
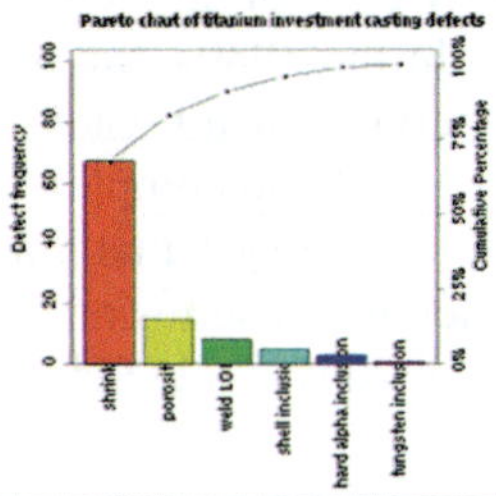	Pareto charts to provide an ordering of what is more important and what is less so.

Fig. 2.1 Tools for quality and risk management

control and finally, improve performance and risk, value and price performance and its risk.

As industry processes became more networked and integrated, risk dependence grew as well, becoming far less manageable. Further, growth in productivity reduced market prices and the need of industries, having mastered the economics of production, dominant needs became needs to market, sale and profits.

A Summary: Industrial Process and Risk Management

Figures 2.2 and 2.3 above summarize some elements defining a process of risk management. We note in particular, “Risk Definition”, “Risk Data Assembly”,

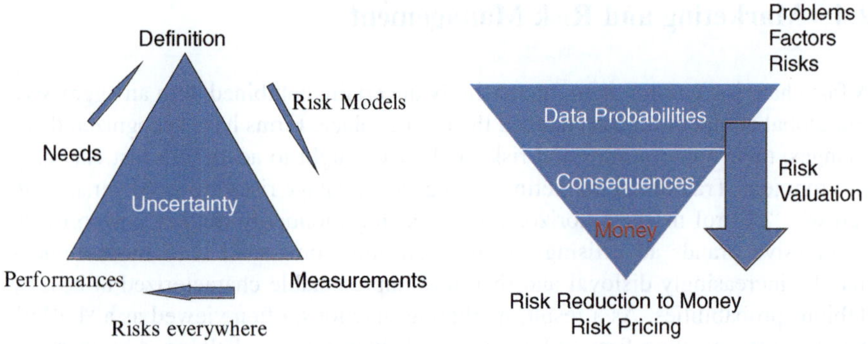

Fig. 2.2 Industrial management and risk everywhere: definition, measurement, performance

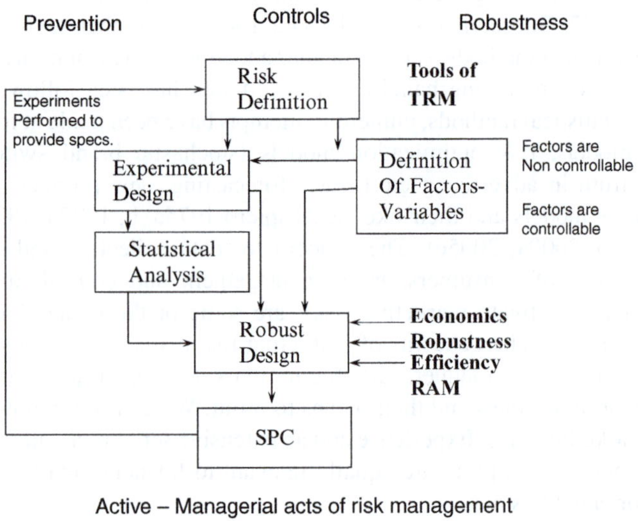

Fig. 2.3 Statistical tools and the management of risk

"Statistical and Experimental Design" and Statistical Control. Examples to these tools to manage risks in other fields, more similar than different, abound. For example, auditing traders, control of shipping lanes and ships entering a port, controlling a portfolio's performance over time, controlling experimental results in a medical lab, are such cases. Over time, firms have grown from industrial to service, to retail, to supply and networked and stealth firms loosely controlled and A-national—belonging to no one in their pursuit of profits and global positioning. Risk management has in this process become far more complex, far more elusive to regulation and controls.

2.4 Marketing and Risk Management

A fast changing environment, increasingly uncertain, combined with an aggressive and global competition has changed the market place. Firms have recognized these changes, their opportunities and risks and have sought to adapt to a fast changing environment. Traditional marketing strategies to manage risks are to "diversify", to "grow", "control markets horizontally" (seeking monopoly market powers) and, "aggressive brands advertising". Consumers too, confronted with more choices may be increasingly disloyal and therefore unpredictable characterized at best by dubious probabilities. As a result, marketing functions, often viewed as a "buffer" between the operating firm and its markets is increasingly challenged by markets, consumers and post sales and complementary services.

Risk and uncertainty have always beset marketing. The beginning of marketing research (and risk) activities appeared in the late 1920s and early 1930s. The first well known text Marketing Research and Analysis by Lyndon O. Brown, did not appear until 1937. Subsequently, (in the late 1940s) a number texts introduced descriptive statistical methods, such as tabular and graphic presentations. Although the lion's share of marketing handling of uncertainty has been followed by data handling and statistical methods, numerous attempts have been made at formulating probability models, risk optimization models (stochastic brand switching and advertising, truth in advertising etc.), sales forecasting, new products prospects, repeat purchase models and their like (see Tapiero 1975a, b, 1977a, 1978b, 1979, 1982, 1982a, b, 2000a, 2005e). The concern for data collection and validation, statistical surveys of consumers, brand management and control, manage the marketing mix to mitigate marketing risks, are some of the issues that confront marketing managers. The concerns of marketing risk management consist then in altering in a desirable manner, the future market states (opportunities), the probabilities of these states and their means to profit. We consider below a sample of specific marketing risks. Experience and an extensive set of retail and consumers studies and their consumption are equally relevant to financial retail firms, banks and clients oriented businesses.

2.4.1 Reputation Risks

A cruise ship, belonging to a large corporate entity in the shipping business, departs from safe sea lanes, ventures dangerously close to shore put its passengers at risk and sink the ship with some passengers drowning and others traumatized. Of course, aside from the loss of property, the potential legal litigation that the Corporate entity, the ship owner, will face, compensation to the 4,000 passengers of the ship etc. will most likely contribute to an immense financial consequence (whether the firm is insured or not). No less important however is the reputation risk—resulting from a number of factors that affect the perceived future safety of the firm performance and thereby its attractiveness. It affects further the current financial performance of such firms with investors and traders factoring future financial consequences into current prices. In this particular case, it may imply a falling stock price, a contagious sales of securities, future consumer behaviors, disappointment-regrets risks, future reservations to be cancelled, a lower demand for the firms services and on. Reputation risks is therefore important and can strike any firm or person, due to word-of-mouth and social media contagion, due to down-rating of a firm, due to incompetence and errors of one or more employee, or firms strategic decisions (such Bank of Americas requesting unreasonable payments for using an ATM machine), due to strategic actions by counter parties or due to external and revealing events (pointing out to an incongruence between a firm's stated reputation), to a firm lies revealed, etc. Reputation risks are also gaining an increased attention. A survey by the Federation of European Risk Management Association (FERMA) with the Institute of Risk Management (RIM) found that reputation risk from social media was cited a "material risk" by nearly 50% of European companies—making it one of the greatest cyber threats that organizations face. Further, reputation risks are not rare. A 2010 study of the world's 1,000 largest companies found that 80% of those firms have a major "reputational event" every five years that causes them to lose a fifth of their value (*Wall Street Journal*, January 17, 2012, p. B5). For these reasons, particular attention is given by corporate firms to engineering their reputation and mitigating its risks by advertising and "false" advertising as the example below highlights.

2.4.2 Advertising Claims and Branding Risks

Advertising claims and consumers' experiences define claims reliabilities, namely the probability that an advertised claim is confirmed or not by the experienced purchaser. Reliable claims are important as they contribute to branding of products and to their re-purchase while unreliable claims can lead to a loss of clients and in some extreme cases (as it is the case in drug advertising claims) to extremely large litigation costs. Since true products characteristics are necessarily random (due to the production process, use and misuse of the product) advertising claims truthfulness is inherently random as well. There is therefore always a probability that an advertising claim is not met. Advertising claims that underestimate product characteristics might be reliable", namely be mostly true, but then they might not entice first time purchasers, while overly optimistic advertising claims might entice first time purchasers but be perceived as unreliable by repeat purchasers who might switch to other competing products. In this sense, the decision to advertise or to claim returns or advantages one does not always have, is necessarily concurrent to the decision to "what to advertise". Such decisions are compounded by the fact that in prevalent marketing philosophies, a consumer is also a consumer of services (such as warranties, product servicing, an investor, an insurance client etc.) and a firm profits not only from the revenues generated at the time of sale but also in derived revenues maintained as long as the customer remains a client of the firm.

2.4.3 IPO, Reputation and Risks

Investopedia's definition of an IPO ('Initial Public Offering—IPO') is "The first sale of stock by a private company to the public" (http://www.investopedia.com/terms/i/ipo.asp#ixzz1kysbF3L5). Typically, IPOs are issued by smaller, younger companies seeking capital to expand as well as "cash in" on their prior achievements and an expectation of future returns. However, IPOs are also used by large privately owned companies seeking to be publicly traded. This is an inverse process where firms are privatized for both profit and to avoid the regulation required by publicly traded companies. IPO are however gambles for the firm, its employee and its investors. For employees, the IPO is both an opportunity and a risk. On the one hand it "liquefies" their holdings, or the options they received instead of, or as a complement to their salary. On the other it is also a risk to their shares' potentially diluted and arising from the legal transformation of the firm that can change appreciably the corporate structure, the work environment and employees' pension holdings.

IPO's are performed in a number of ways, summarized by directly turning to the public (as it was the case with Google's IPO) and IPO's through financial intermediaries. In this latter case, the IPO issuer obtains the assistance of one or several underwriting firms, providing professional advice, access to capital investment and become more credible to prospective market investors. For example, advising on the type of security to issue, whether common or preferred or some financial products denoted in local or foreign currency, their voting powers and other characteristics. Advisors then seek to define a market sensitive price based on their markets analysis with a favorable chance for the IPO to succeed and where to issue the IPO (for example, in Wall Street, in London, in Shanghai, Hong Kong and other countries). The IPO is therefore part of an elaborate process resulting in a "public offering" summarized by the initiated firm "seeking to go to market" or "induced to go to market". The end process is a public offering.

IPOs are not without risk however: Future Expectations and Partial Information providing the IPO issuer an informational advantage. It can also punish the IPO issuer if ex-post, it turns out to have been misleading (whether intended or not). For example, Groupon, "the daily deals" internet firm founded in 2007–2008 that went public in 2012 had an astonishing growth rate going public in 2011 with a market valuation of 20 Billion Dollars. However, once public, Groupon was open to scrutiny and ever since has met a harsh criticism essentially due to its accounting (used to project publicly its financial performance and based on "certain assumptions and forecasts" that the company used). This resulted in the company reporting losses, and a reputation that led the market to reduce its price dramatically (on March 30, 2012, the stock price went down by 6%). For these reasons, the firm, its shareholders and employees and prospective buyers may on the one hand be at risk and on the other be enriched.

For firms, potential risk exposures arise due to regulatory risks (including stock Exchange requirements to comply to a set of rules, increased responsibilities of directors and officers of the firm, etc.); An enhanced corporate governance responsibility including an increased exposure to potential personal liability; An Increased regulatory scrutiny; New regulation in relation to the publication of financials (for example, late publication of financials would lead to regulators imposing fines and/or penalties); In addition, alleged errors, omissions, misstatements in the listing offer document etc. may lead to potential lawsuits inherent in the offering document itself. Shareholders rely on statements made in the document and an alleged misstatement, or breach of the offering, could lead substantial defense costs and the potential settlement cost. Finally, insider trading can result in substantial fines and a legal cost once the IPO has been issued (whether successfully or not).

For shareholders, there are additional risks. Some may result from the dwindling of their hold on the firm, while other result from an increased number of shareholders augmenting the probabilities of potential future lawsuits. Further, a growth in stock market volatility following the IPO, may contribute to a decline in the shareholders' stock prices as well as potential claims occurring to changed and additional geographical locations where the firm was not listed. Finally, while a positive growth in earnings is rarely challenged by shareholders, a disappointing performance of the firm after an IPO and required to divulge internal procedure, accounts and projections, can lead to additional and troubling legal claims and reputation risks.

Client (stock purchasers) risks abound. The firm may provide partial or incomplete information and thereby increases stock purchasers' counter-party risk. The IPO may be a ploy by its management and a number of angel investors seeking a quick turnaround profit rather than a capital expansion for future prospects. Finally, external bubble markets condition or a unique reputation fed by intensive and purposeful advertising to brand the firm as an extraordinary opportunity may also be a source of risk for prospective clients. For the firm, counter party risk may incur potential indemnities to underwriters in the event of being sued as a result of misstatements: The offering document usually allows for indemnities to be provided to the underwriters and/or sponsors of an IPO. If the listing company is sued for misstatement in its prospectus, not only will the directors and officers be potentially liable, but further indemnities may have to be provided to counterparties.

Additional elements that point out to an IPO success are based on stock market valuations and aggregate market and international conditions (such as liquidity, relative returns on stock investments, regulation and the cost of capital etc.). In a global world, IPO will also follow markets where opportunities are largest. Recent statistics have pointed out that over the last 5 years the number of IPO's in the US has been declining dramatically while IPO's in Europe and Asia have been increasing appreciably. These elements point out to both global trends in the competition of financial markets for IPOs. Macroeconomic factors such as market liquidity can also lead to IPOs bubbles or "hot IPO markets" (Chemmanur and He 2011; Ritter 1991; Beatty and Ritter 1986). Market liquidity combined with investors and managers' real options, both contributing to the potential to cash in at advantageous market condition, can usher an "IPOs contagion" of successful IPOs increasing the market hunger for IPO's investments. These elements point out to both opportunities and to IPO risk factors, plagued by uncertainty (and thus gambles).

Pricing an IPO (i.e. a valuation of the IPO initiative when it is going to market) is therefore a challenging and a risk problem with information asymmetries with future events unknown and strategic risks (i.e. game like among an IPO's parties, see Chap. 12). IPO pricing therefore more of an art than a financial problem which is resolved technically. For example, empirical studies report that investment banks and commercial banks that are involved as financial advisors with lucrative fees to IPOs have no particular advantage one over the other when pricing IPOs. Nevertheless, rules were made not to allow banks private research with an interest in an IPO to be published before the IPO. This is currently overturned.

Example: IPO underpricing The risks of over pricing an IPO and face litigation costs, reputation risks and a downfall in a firm market price leads (for risk averse firms) IPOs to be underpriced. There are a number of theories that explain such observations. A number of research papers have indicated that IPO are underpriced as a form of insurance also coined the insurance effect and to lower expected litigation costs (the deterrence effect). Evidence for both these aspects is embedded in three factors that define the initial IPO price:

- Signaling. A lower underpriced IPO signals a quality that allows firms to raise more capital in the future (Welch 1989).
- Information asymmetries. On the one hand it provides an initial and larger demand for the IPO issue (and therefore increases its returns and price) and a reputation risk on the other, if the IPO firm does not meet its projected expectations or turns out to have misled investors.
- Litigation risks whose probability and magnitude are a function of the disparities in actual and reported information.

Under-pricing may also be a compensation for uncertainty that particular and uninformed investors bear compared to informed (insiders) investors. Thus, ex-ante uncertainty is an important determinant of an IPO price that may justify underpricing. To mitigate this uncertainty, some IPO firms turn to financial institutions with a high reputation to manage their IPO (for example, Facebook turning to Goldman Sachs and to leading banks, international IPOs of Chinese firms are also turning to local banks to help the launching of their stock in foreign local markets, etc.).

Example: IPO and Facebook The Facebook IPO has, as the public media explained (May–June, 2012), been disastrous resulting both from the mismanagement of information and claims regarding Facebook potential earning capacity (on what bases these estimates were created?) and on error in trading in Nasdaq that have resulted in both a lack of trust by individual investors as well as discriminatory actions taken by leading banks to favor themselves and their clients. In the weeks that followed the IPO, Facebook lost close to 16% of its valuation. In addition, the prospect of court cases lingering over Facebook and its associated institutions is contributing to risks that will be latent for some time. This has been reflected by an extraordinary decline in its market valuation after its first two weeks.

2.5 Externalities and Risks Management

Environmental pollution that spill over into the public domain with no consequences to its perpetrator are risk externalities. Their consequences can be health, financial, international, political that can linger over long periods of time. Managing these risks involves a broad set of questions and potential means that are prevalent in economic, environmental, social and health sectors. These include for example: regulation; preventive efforts to be implemented by both firms and the public; contingent recovery and emergency preparedness; legal pursuits that perpetrators will have to face; who pays for it and in what proportion; are the responsible parties penalized; etc.

Say that a Chemical firm employing 1,000 persons has an environmental accident due to some malfunction resulting in a public damage. If such a firm is to be punished harshly and its employees joining the lines of the unemployed will their costs be sustained by the public domain? Similarly, rare earth—those materials needed for many and important parts in technology intensive products are known to produce extensive environmental damage. This has led to the production of such rare earth only in a country willing to assume this damage for the returns it brings. However, when there is only one country producing rare earths and it exercises its monopoly power over all other countries, are the political risks and financial costs of not producing rare earths appropriately assessed at their right value? Similarly, car drivers enjoy their car while at the same time they do not assume the environmental consequences of the car's pollution which could have been avoided had they selected to walk or use an environmental efficient public transportation. For these reasons, environmental risks are prevalent in everything we do. On the one hand, they are harmful, on the other they derive from what we need, we want and do. A similar argument can be applied to banks' risk externalities.

Concepts of risks applicable to sustainable development are both interdisciplinary and omnipresent in many disciplines. Each discipline contributes its particular wants and characteristics. For example:

- How to assess the risks of an industrial policy (pollution, costs and efficiency, etc.)
- What are environmental risks and what are their multiple consequences (financial, reputation, regulation, etc.)
- How can we recover from environmental disasters
- What are production risks derivatives and what are their costs and their environmental impact.

- How are environmental risks subcontracted and transferred (for example disposing of nuclear and others wastes in willing and poorer countries)
- What are the strategic (gaming) factors in environmental regulation and what are their consequences—economic and otherwise
- What are potential pollution abatement technologies and how can they be values and priced.

2.6 Networks and Risks

Networks abound and have multiplied in all kinds and all sizes. They also compound risk and uncertainties resulting in greater interdependence and a far greater complexity to contend with. Network risks are prevalent in industrial networks; in supply chains; in the banking systems; in financial markets; in consumers associations; in integrated retail commercial networks with stores, warehouses and supplier organized as franchises (of various sorts and contracts); in health supply networks; in a transportation system; in electrical grids; to model contagions, to model an internet system, etc. A network model is defined by a set of inter connected entities (nodes) defined by lines that connect their nodes (explicitly, this is a mapping of nodes onto themselves. These lines may be uni-directional or be directional, with connections potentially random or quantitatively defined. Networks may also be cyclic or be acyclic (namely, with potential feedback loops or none). A network may be *arborecent*, in which case, each node has a single predecessor but any number of successors (which is the opposite of the assembly process); *Acyclic*, with nodes of connected by any number of predecessors and successors, with no return; *Cyclic* or general, with no restriction on the flows connectedness. Networks and Graphs, whether modeled as a set of deterministic connections or as a set of random and interacting elements (nodes or lines) are in general complex systems that can be analyzed sometimes analytically but mostly using simulation techniques (see Chap. 3). Numerous books and papers in Operations Research, Queuing Theory, Transportation and Management Science, Biomathematics, Engineering Systems, etc. are published referring to network and graph models. For example, Economides (1996) considers economic issues that pertain to networks, FMS are modeled as randomly connected networks of queues while Capobianco (1973), Tapiero et al. (1975), Boots and Tapiero (1973), provided a sampling statistical approach to estimate networks and graphs characteristics (for example, to estimate the connectedness of underground water flows connectivity). Networks may also be used to models to model the interconnectivity of supply chain, international banks as well the operational flow associated to financial transactions. Some of these networks are not easily observable. For example, networks of underground rivers (Boots and Tapiero 1973) but also of banking relationships which can be extensive some of which may in certain cases be unknown even to bank executives. For example, with the demise of Lehman Brothers, it turns out that there were thousands of financial and legal

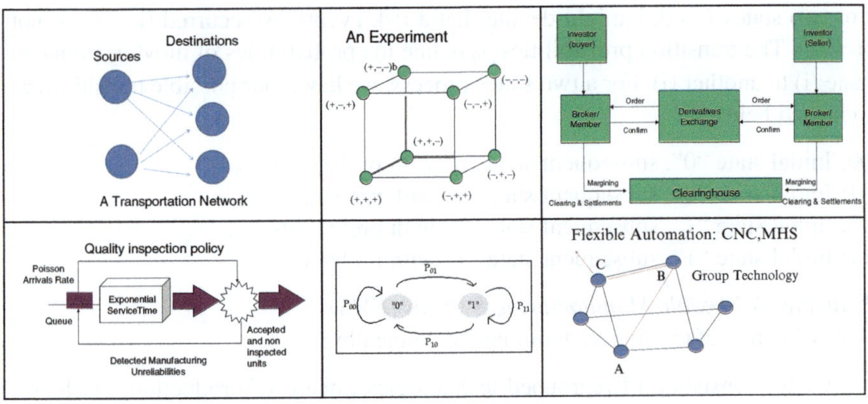

Fig. 2.4 A sample of networks and their use

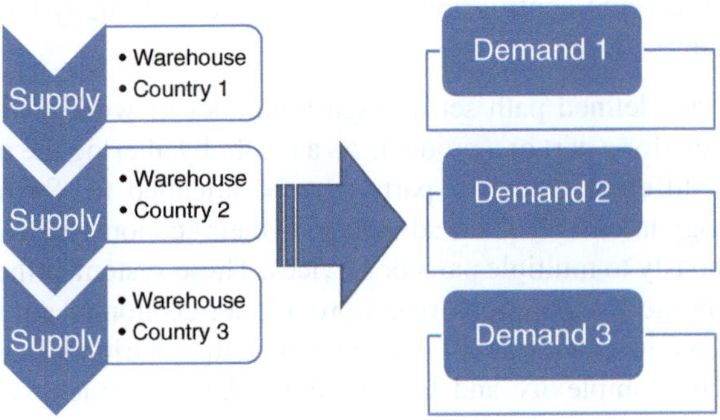

Fig. 2.5 The transportation model

entities in Lehman Brothers that were in fact unknown to the corporate center of Lehman Brothers!

A network can be used to model a system of components parts, each defined by a reliability function; a transportation network defining the flow from a given set of sources to a set of destinations (with flows or times to travel subject to breakdowns or delays); an industrial automatic process such as flexible manufacturing Systems; a network of queues (see Fig. 2.4); a sequence of events leading to a polluting event or used to highlight a causal or a dependent structure; a model of contagion; a supply chain, a set of networked stake-holding firms, communications and telecommunication networks; a cellular pattern; the internet network, social media networks, etc. Networks risk models are abundant, are complex, occur for many reasons and proliferate in all professions. Below specific network models are considered—both to express quantitative relationships and risks.

Example: Networks and Markov Chains Markov chains (see Chap. 5) are used to define the transition probabilities between a set of notes (called sates) as shown in Fig. 2.4. They are modeled as follows. Let there be n states, $i = 1, 2, \ldots n$.

For two states $i = 0, 1$ might denote that a risk event has occurred ($i = 0$) or not ($i = 1$). The transition probabilities p_{ij} define the probabilities of moving from one state (i) to another (j). For a two states process, we have four possible transitions as indicated below:

00: Initial state "0", subsequent state "0" with probability p_{00}
01: Initial state "0", subsequent state "1" with probability p_{01}
10: Initial state "1", subsequent state "0" with probability p_{10}
11: Initial state "1", subsequent state "1" with probability p_{11}

Example: A Flexible Manufacturing Systems (FMS)
A flexible manufacturing system may be defined by:

- "Cells" consisting of regrouped technologies (means of production, machines, expertise, a combination of resources)
- A material handling system (MHS) connecting the "cells", either automatically using a robotized system or not
- Systems to direct and control flows from cell to cell. Usually an integrated IT control system

In a FMS a pre-defined path set through a number of work stations defines a production process for a part or a product. As a result, by altering the path to follow, numerous (and different types) of parts may be attended to. Such systems are usually technology intensive. Their advantage is their economy of scope in attending to simultaneously to multiple parts or services. These system, although complex (and therefore prone to operations risks) provide an economic advantage due to their ability to produce with small lots, allowing the sharing of equipment and technologies. The complexity and the interdependence of such systems require however far more information, far more computer aids, far more automation, greater reliability and risk controls for their operations.

Planning and constructing such systems are an engineering challenge however. The planning of cells is also a complex process based on using a GFT (Group Technology). GT is essentially an approach that considers all the variables relevant to process and its scope. It is in principle applicable to organizational design. In manufacturing it consists in grouping multi-components (technologies, grouping of people, process functions) into cells with similar or complementary characteristics to improve the network ability to respond to needs and various demands. A cell or a group is thus the basic unit in a process that seeks to "more" with "less". The principles of GT is due to the Russian S.P. Mitranov suggested a formal proposal for GT in 1940. It is however in Germany where an Opitz classification system was accepted for GT and widely implemented. Its underlying assumption is that "many problems" have common characteristics and can therefore have a similar solution.

Network risks, whether in a FMS, in transportation, in a banking system or other networks can be classified into the following categories:

- Nodes integrity risks
- Flows-Communications-linkages-line risks
- Coordination risks

- Complexity risks
- Contagion risks

Nodes integrity risks relate to cell malfunctions, breakdown and misprocessing. When a cell malfunctions it is critical if the network as a whole malfunctions (for example, it hinders a needed flow to all other nodes). Node integrity may be directly observable or unobservable (but revealed through either statistical controls or through observables elements.

Flows or line risks assume various forms such as a loss (of money, or a transaction cost, an information imbued with noise, a misinformation), an unexpected time delay when moving from one node to another is random.

Network risk dependence—statistical and contagious are expressed by the statistical relationships that exists explicitly or in a latent and stealth manner between the nodes and the flows from node to node. When malfunction in one node induces a malfunction in another, a contagion risk materializes. For example, say that a network consists of n connected banks and say that one firm defaults. Given its connectedness, to what extent are neighboring banks affected? Similar situations arise in electric network-grid, in supply lines, in health contagions in dense cities etc.

Are certain risks critical and causing a network failure? For example, if the failure of a particular service can cause a global networked failure, then that service may be considered too important to fail (and therefore requiring special attention), while other services may not.

To manage network risks, we design the "network structure": (for example, a transportation system in Fig. 2.5), "manage its content and relationships" and "control the random and dependent factors" that define the flows and the exchanges that occur in the network. In this sense, network risks are generic risk models that may be applied to a wide variety of problems and contexts.

Example: Queues and Networks The mathematical theory of queues has its origin in the study of telephony systems (that are necessarily highly interconnected networks) initiated by Erlang, a mathematician with the phone company in Copenhagen in 1917. In the 1950s numerous applications of queuing models and networks have been used to model an extremely large number of practical problems. The mathematical theory of queues is a subfield of discrete events stochastic processes (for example, see Gross and Harris 1985 and Chap. 5) in its elemental quantitative form. It is based on interacting models of independent stochastic processes.

An extremely large diversity of such models can be constructed, both based on realistic assumptions regarding the "randomness" of events (arrival, queuing, screening/processing/service, etc.) that define a queue's statistical operational characteristics. For example, queues can be considered as cells that are connected and together form a network with queuing systems connected randomly or following a pre-defined path. Such models are used to model power grids, patterns of migrations, communication systems, inter banks flows of money etc. A typical

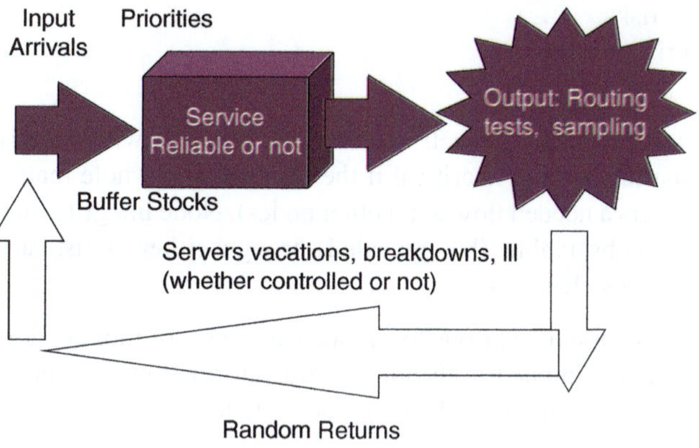

Fig. 2.6 A feedback queue

example includes the FMS (Flexible Manufacturing Systems) modelled as a network of queues where nodes are cells performing specific functions and final products (or services) defined by the path set for the product and its exit from the network. The generic mathematical definition of queues is thus defined in terms of the following:

1. An input or arrival (planned, random, etc.)
2. A waiting line (into a waiting space whether bounded or not)
3. A service (by one or a number of parallel "service stations, with a deterministic or a random service time).
4. Disciplines and priorities in seizing a waiting space, in the queues, in service, etc.

Figure 2.6 above represents a specific queue model where service may be reliable or not operating or not (in which case, it is "on vacation"), with a buffer defining its waiting capacity and routed to some other queue as a function of a pre-determined schedule that defines its requirements. An arrival may be well intended, seeking a "service" they may pay for (for example, entering a cinema) or not (such as going to a free beach). Some may be ill-intended, seeking to harm the arrival process, the service or pass through undetected to create great harm. Such an input might be persons arriving to a security line prior to entering an airport and ought to be prevented. Arrivals are usually random while preventive efforts may consist of inspections and controls as well as external efforts based on the detection of such elements prior to their arrival. Arrivals can of course be non-threatening, or "normal", in which case the time spent waiting for the desired service may be either null or due to controls imposed by threats to the queue system. For example, access to a beach might or might not be controlled, access to a concert hall might require controls at the entrance or not, etc. Awareness of threats is inducing a state of "uncertainty or un-safeness" that controls seek to reduce by their own existence, by their own actions, by their threat to ill-intended parties (even if the probabilities of such threats are very small). In this later case, queue controls are strategic.

References and Additional Reading

Additional references on risk management consulted in this chapter have included Cheng et al. (2004), Hale (2002), Hallikas et al. (2004) as well as Tapiero (1977b) (on dynamical problems), (1978a, 1996). Industrial risk processes references include numerous articles published by Hsu and Tapiero 1987a, b, 1988a, b, 1992, 1994, Hsu et al. 1993, Lee and Tapiero (1986, 1989) as well as Tapiero (1987, 1989, 1995a, 1996, 2007), Zsidisin (2003), Zsidisin et al. (2001) as well as Kogan and Tapiero (2007), Gattorna (1988), on Strategic Supply Chain Alignment, La Londe and Cooper (1989), on Partnership in providing customer service and Marvel (1982), on exclusive dealing. Additional references will be used in subsequent chapters. In particular, Chap. 11 on risk and strategic controls.

Risk processes and safety have extended this theme outlined in Chap. 1 including Blom et al. (2005, 2006), Bohn (2000), Bohn and Choi (1996), CAS (2001), FSA (2005, 2006a, b) and IMO (The International Maritime Organization) (2002).

References on Marketing applications include the paper by Ingene and Hughes (1985), Munier et al. (1999), Spence (on economic search), (1977), Nerlove and Arrow (1962) on advertising problems as well as numerous extensions to the stochastic domain reviewed in 1977 by Sethi and my own papers on these problems in 1975a, b, 1977a, 1978b, 1979, 1982a, b, 2005.

Application to risks externalities in this chapter were based on my own lecture notes and on papers by Greaker (2003a, b, 2006). Additional application on IPO, includes Culp (2006) on Structured Finance and Insurance, Yip (2007), Port Traffic Risks, Cox (2009) on Risk Analysis and Complex Uncertain Systems with applications to health care, Kogan and Tapiero (2007), on supply chains and games, my own book (Tapiero 1996) and the Management and the Control of Quality, Coles (1993) on the the price of quality and Yaniv (1991) on absenteeism and risk.

Chapter 3
Probability Elements: An Applied Refresher

Abstract The quantitative construction of risk models, risk measurement and analysis are essential to obtain a better appreciation of the risks we confront and to mitigate their effects. To do so, probability theory and statistics are necessary. The purpose of this chapter is to provide a cursory and intuitive introduction to basic elements of probability distributions.

3.1 Introduction

Risk analysis and engineering are based on a quantitative definition of risk and its modeling, its measurements, developing the means to detect, control and finally manage risks—their probabilities and their consequences. It is used for many purposes with various applications, depending on their primary concerns. For example, finance is concerned with risk pricing and risk management while industrial production is concerned with standards, RAM (Reliability, Availability, Maintainability) as well as SPC/SQC (Statistical Process and Quality Control), Control Charts, 6 Sigma Process control and quality management etc. (see also Chap. 11). These applications often migrate to other concerns that are defined in similar terms. In health care, such techniques are applied in a laboratory setting as well as in testing and controlling the development of drugs. In other cases, when "health" or a lack of it are defined by a number of states, whether dependent or not, then risks are defined as a function of the probabilities of reaching (or not) these states. Similar approaches are used in insurance as we shall see subsequently. This chapter will introduce elements of probability theory and models. A more formal review of basic probability notions can be found in the references indicated at the end of this chapter.

C.S. Tapiero, *Engineering Risk and Finance*, International Series in Operations Research & Management Science 188, DOI 10.1007/978-1-4614-6234-7_3,

3.2 Risk and Probability Moments

Risk is defined quantitatively by:

- Probability distributions and their moments, including essentially the mean, the variance, the skewness, the kurtosis, the range as well as ordered moments (to be defined below).
- Random events, independent or not, expressed in terms of their statistical covariations (see Chap. 4).
- The valuation of risk and its price

The statistical and applied probability literature abounds with many models used to predict future events assess their consequences (risks) and value them. These models are both single-variate and multi-variate. For simplicity, we shall consider in this chapter single variate models and multi variate models in Chap. 4.

3.2.1 *Expectations, Variance and Other Moments*

Let $(p_i, x_i),\ i = 1, 2, ..n$ define the probability of an event x_i with $0 \leq p_i \leq 1$ and $\sum_{i=1}^{n} p_i = 1$. A *cumulative probability distribution* defines a partial sum of the distribution probabilities:

$$P_k = \sum_{i=1}^{k} p_i \text{ and } 1 - P_k = \sum_{i=k+1}^{n} p_i$$

For a continuous probability distribution $f(z)$, we have instead:

$$F(k) = \int_{-\infty}^{k} f(z)dz, \quad 1 - F(k) = \int_{k}^{\infty} f(z)dz, \quad 0 \leq f(z) \leq 1, \ z \in (-\infty. + \infty)$$

3.2.2 *The Expectation*

The mean of a probability distribution defines its expectation given below in both discrete and continuous states, or:

$$\mu = \sum_{i=1}^{n} p_i x_i \text{ and } \mu = \int_{z \in Z} z f(z) dz$$

Such an expectation can be an expected loss, an expected profit, can be the number of expected treatments needed to overcome a health situation, the expected number of persons travelling in the NY subway at certain times of the day etc. If only some events are deemed risky, say events $i+k, i+k+1, \ldots n$, then the partial moment expectation in discrete and continuous states are:

$$\mu_k = \sum_{i=k}^{n} p_i x_i \quad \text{and} \quad \mu_k = \int_{z \in Z_k \subset Z} z f(z) dz$$

These denote the expected consequence associated to selected tail events. Such measurements are often used to define "tail risks". The variance, skewness and other moments are higher order expectations, whether partial or not. For example, for a given probability distribution $f(z), \ z \subset Z$, the ith moment of a distribution is:

$$\mu_i = E(z^i) = \int_{z \in Z} z^i f(z) dz \text{ and } \mu_k^i = E_k(z^i) = \int_{z \in Z_k \subset Z} z^i f(z) dz$$

3.2.3 *The Variance/Volatility: A measure of "Deviation"*

The variance is a quadratic measure of deviation from expectations. It is often used as a proxi for a risk measure based on the assumption that deviations from expectations are consequential. The variance is then:

$$\sigma^2 = \sum_{i=1}^{n} p_i (x_i - \mu)^2 \quad \text{and} \quad \sigma^2 = \int_{z \in Z} f(z)(x(z) - \mu)^2 dz$$

while the standard deviation is σ, the variance root square.

3.2.4 *Skewness, Kurtosis and Filtration*

Let the random variable $\tilde{R}$ whose probability distribution is conditional on a data set $\Im_n$, implying that estimates of the probability distribution denoted by $f(\tilde{R}|\Im_n)$ is based on the information formally denoted by $\Im_n$. An expectation and its variance are estimated conditional to a filtration (i.e. the data set and all the knowledge we possess at his time), or:

$$\mu_n = E(\tilde{R}|\Im_n) = \int_{\Re} Rf(R|\Im_n)dR,\ \ \sigma_n^2 = E\left((\tilde{R}-\mu_n)^2|\Im_n\right) = \int_{\Re}(R-\mu_n)^2 f(R|\Im_n)dR$$

If the probability distribution is normal (as we shall see later on), its third moment is null. If this is not the case, then the supporting data set would indicate a non-zero third moment, "parametrized" to an index of skewness. Such a moment measures the non-symmetry of the probability distribution defined by:

$$\xi_n = E\left((\tilde{R}-\mu_n)^3|\Im_n\right) = \int_{\Re}(r-\mu_n)^3 f(r|\Im_n)dr \text{ and its Index of skewness} = \frac{\xi_n}{\sigma_n^3}$$

Similarly, the fourth moment indicates the "thickness" of the distribution tail. It is "parametrized" to an index of kurtosis, given by:

$$\psi_n = E\left((\tilde{R}-\mu_n)^4|\Im_n\right) = \int_{\Re}(r-\mu_n)^4 f(r|\Im_n)dr \text{ and its Index of kurtosis} = \frac{\psi_n}{\sigma_n^4}$$

For example, say that a systems' skewness is null. This means that the underlying distribution is symmetric. If in addition, the kurtosis is small, it will reinforce the belief that the underlying distribution is less likely to define events that are "far" from its mean. For example, let ε be a standard normal probability distribution then (as we shall see later on): $E(\varepsilon) = \mu = 0,\ E(\varepsilon^2) = \sigma^2 = 1$ and $E(\varepsilon^3) = 0, E(\varepsilon^4) = 3$. Thus for a standard normal probability distribution, its index of skewness is null while its index of kurtosis is $\frac{\psi}{\sigma^4} = \frac{4}{\sqrt{\pi}}\Gamma\left(\frac{5}{2}\right) = 3$. These moments can be important as they express a deviation from normality and therefore attract a greater attention to factors that contribute to the non-normality of an observed data set. The skewness and the kurtosis of a data set are often used as indicators of departure from the normal distribution. For example, say that standardized rates of returns using a given data set indicates skewness and a kurtosis greater than 3, it would then indicate that the assumption that rates of returns are normally distributed is not supported by the data.

3.2.5 *Range and Extreme Statistics*

Let $H_t^{(n)}$ and $L_t^{(n)}$ be the highs and the lows of a sampled data set $S_t, S_{t-1}, \ldots, S_{t-n}$. Removing the time index t, we have: $H_n = \text{Max}\{S_1, \ldots, S_n\}$, $L_n = \text{Min}\{S_1, \ldots, S_n\}$. The range within a data sample is defined by the difference between the sample highest and the sample lowest value, or:

$$\Re_n = H_n - L_n = \text{Max}\{S_1, \ldots, S_n\} - \text{Min}\{S_1, \ldots, S_n\}$$

It differs from the variance as it is derived from extremes observations in the sample data set. Its importance arises from the following: (1) It can be used to detect variations in volatility (variance). In particular, Irwin 1925 (Barnett and Lewis 1994) used such a statistic to detect outliers, observations that may apparently be out-of sample or jumps in a data set. (2) When a series of range statistics seem to increase, it may imply an increased "disorderliness" (or chaos) and potentially a trend to variance growth. In some cases, the range may be an "early" detector of volatility growth since an unexpected growth in the range points out to volatility growth while a constant range (or with little change) may indicate a mean reversion process. In other words, once the range has increased due to an outlier, mean reversion (a counter movement process, reducing a performance after it has increased over its mean) will be stabilized about the mean. In some cases, the "range/standard deviation" statistic of a security time series measured over a given sample size has also been used to point out to departures from the "normality" of the sample series (see Chap. 5). The analysis of such series using the range have also led to models coined by Mandelbrot as "long memory models" characterized essentially by processes where the variance of the underlying process increases nonlinearly over time (see Chap. 5 for such models as well extensive references at the end of the book). The applications below highlight some practical elements of probability and risk.

3.3 Applications

3.3.1 Skewness in Standardized Stocks Rates of Returns

Consider a sample data set of standardized rates of returns for a stock (in fact, the IBM stock price) given below in Table 3.1 with mean = 0.04 % and volatility = 0.016411. A calculation of other moments for normalized data over a long period of time has indicated that there is almost no skewness but a strong kurtosis. The statistical parameters calculated have pointed out to a kurtosis of 9.91 while the kurtosis of a standard normal probability distribution is 3. Similarly, we see that the data exhibits skewness of 0.043 (while it equals zero for a normal probability distribution). These statistics imply that rates of returns statistics (even for a "stable" and a large firm such as IBM) are not normal (Table 3.2).

A cursory observation of the standardized data has indeed a mean of zero and variance of 1 the skewness althought small is not equal to 0. The data sets exhibit however a kurtosis of 9.91 which is far greater than 3, had the data been distributed as a normal probability distribution. One may then use this observation to question the validity of a risk model which would be based on a normal probability distribution of the data set.

Table 3.1 The data samples

			Mean	STD	Variance	
Date	Adj close	Returns (%)	0.04 %	0.016411	0.000269	Normalized returns
1/2/1962	2.69					
1/3/1962	2.71	0.74				0.426022876
1/4/1962	2.69	−0.74				−0.476719534
1/5/1962	2.63	−2.23				−1.386148925
1/8/1962	2.59	−1.52				−0.953777136
...	...	...				...
1/9/1962	2.62	1.16				0.67878236
6/11/2009	109.4	0.97				0.563482839
6/12/2009	108.21	−1.09				−0.689833506
6/15/2009	107.62	−0.55				−0.359255758
6/16/2009	107.32	−0.28				−0.196879608

Table 3.2 A sample of calculations

Mean	0
Standard error	0.009150084
Median	−0.027020074
Mode	−0.027020074
Standard deviation	1.000041861
Sample variance	1.000083724
Kurtosis	9.919066518
Skewness	0.043580821

3.3.2 *Reliability, Probability Risk Constraints and Deviations' Risks*

Let $f(t)$ be the probability of a component defaulting at time t (the probability of a machine breaking down, the probability of catching the flu at a certain time t). The cumulative probability distribution $F(t) = \int_0^t f(z)dz$ denotes the (cumulative) probability that a default will occur in the time interval $(0, t)$. Thus, the probability of no default in this time interval is its reliability, or:

$$R(t) = 1 - F(t) = \int_t^\infty f(z)dz$$

Reliability may similarly be defined by the probability that a system, an assembly, a medical treatment, a service, etc. performs without default over a given period of time. Statistical properties of reliability are defined in terms of the probability of operating over a useful life time of the "system". As a result, reliability relates to:

- Definition of non-performance (or default) and their statistical characteristics
- An operational environment
- Temporal behavior

Reliability is commonly used to define particular risk characteristic of risk models. For example, it may be used to assess the life time probabilities of patients, to measure a system reliability (how long a system performs as expected), safety, security, consumers satisfaction, mishaps in a lab, etc.. Given the reliability, the hazard rate (the point conditional default probability) is:

$$h(t) = \frac{f(t)}{1 - F(t)} = \frac{f(t)}{R(t)} = -\frac{dR(t)/dt}{R(t)} \quad \text{or} \quad R(t) = 1 - F(t) = e^{-\int_0^t h(x)dx}$$

The residual lifetime at time, $\mu(t)$, is defined by the conditional expected time at t till the event occurs. Setting for simplicity, t = 0. We have:

$$\mu(0) = E(t) = \int_0^\infty tf(t)dx = -\int_0^\infty t\frac{dR(t)}{dt} = -tR(t)\Big|_{t=0}^{t=+\infty} + \int_0^\infty R(t)dt = \int_0^\infty R(t)dt$$

with $\mu(0)$ defining the mean lifetime, or the amount of time between events. It can be used to calculate the mean time between successive "events" (or their frequency).

Empirically, the time performance (reliability) of systems can be statistically modeled and its parameters estimated. Empirical observations in industry for example, have pointed out to three phases is a system lifetime, summarized by a BATHTUB model. These phases consist of:

- A "youth" phase where default occur due to lack of "experience"
- A "tested" phase where default are "stable", recurring with stable probabilities
- An "aging" phase where "deterioration" has set in, augmenting default.

Reliability is a corner stone for the analysis of statistically independent networked (assembled) elements. For example, say that two components in series have reliabilities $R_i(t),\ i = 1, 2$. The system reliability is thus $R(t) = \prod_{i=1}^{2} R_i(t)$ with its hazard rate calculated above. Assume that this system consists of n and m components in parallel. As a result, for parallel components we have the following reliability $R_i(t) = 1 - \prod_{k=1}^{n} (1 - R_{ik}(t),\ i = 1, 2$ and therefore, for the system as a whole, the reliability is:

$$R(t) = R_1(t)R_2(t) = \left(1 - \prod_{k=1}^{n} (1 - R_{1,k}(t)\right)\left(1 - \prod_{j=1}^{m} (1 - R_{2,j}(t)\right).$$

When components are dependent, computing the system reliability is more difficult. Yet, numerical and simulation techniques can be used to estimate (based on specific assumptions) the performance reliability of complex systems (see in Chap. 4 elements of dependent probability models).

Reliability, although derived from a distribution can be specified by a Bernoulli random variable that assumes two values, $R_i(t) = 1, 0$ with $R_i(t) = 1, 0$ denoting the Bernoulli probability. Such a probability is time variant. Further, the reliability of separate components may be statistically dependent (see Chap. 4 for dependent Bernoulli models) either due to a common factor (such as aging) or due to some other statistical factors underlying their function. Reliability is then "some multivariate function" $R(t) = \Phi(R_1(t), R_2(t))$ that expresses their functional and statistical dependence and given by $E(R_1(t)R_2(t)) = \rho(t)R_1(t)R_2(t)$ (with $\rho(t)$ a measure of dependence, see Chap. 4).

Default can of course vary, categorized by its consequences. For example, critical failure may be a complete bankruptcy (rather than say, just a default in payment), a system default leading to the system out of operation, a systemic default, etc. Systemic default for example may be due to a single bank defaulting and leading to financial markets failure or to dependence of some external random variables (such as macroeconomic factors), just as a critical components default in an airplane can lead to its grounding. In all cases, probabilities, causes, risk consequences and risk events detection (i.e. our ability to observe and apprehend a failing element) are needed to manage systems efficiency and economic performance. For example, say that components are a function of a common risk factor $\tilde{\xi}(t)$ then: $(R_1(t), R_2(t)) \equiv \left(R_1(t|\tilde{\xi}(t)), R_2(t|\tilde{\xi}(t))\right)$ then is a dependent reliability system.

3.3.3 *The Hazard Rate and Finance*

In finance, the probability of debt non-payment (say of a loan) at maturity is $F(T)$ and therefore, the probability of payment is $\left(1 - F(T) = e^{-\int_0^T h(x)dx}\right)$. For example, if debt L is to be repaid at time T, its current price at say time "0" ought to be discounted to reflect the risk that it will not be repaid (i.e. the loan unreliability). Since no discount rate is available for such a risk, its price (paid at time T) is expressed in terms of the current and observed price $L(0)$. A simple model based on a "financial discount rate" quantitatively modeled by the hazard rate is then:

$$L(0) = L(T)(1 - F(T)) = L(T)e^{-\int_0^T h(x)dx}$$

This model implies that what traders and investors are willing to pay to finance future debt has an implied probability equal to $L(0)/L(T) = 1 - F(T)$ or a discount rate $-\ln(L(0)/L(T)) = \int_0^T h(x)dx$. For such a model both buyers and sellers of debt must agree to such an exchange. For a fixed discount rate (implying that the default probability distribution is exponential (as we shall see later on), we set $r_0 = h(.)$ and therefore, $L(0) = L(T)e^{-r_0 T}$. The price of the debt at time t is equals the discount payment at time T whose discount rate is r_0.

It is important to distinguish between "a real reliability" and the "the implied reliability" of a financial asset. While the former is an intrinsic real property of an equipment, the latter reflects the choices that are made between buyers and sellers of the debt—namely those that lend and those needy of the loan and both agreeing to "exchange". Reliability is then defined by the choices made by buyers and sellers exchanging ownership of the debt. In Chaps. 7, 8 and 10 such an approach will be shown to be a quantitative cornerstone of assets pricing in finance.

3.3.4 *Risk Variance and Valuation*

The variance or its root, the standard deviation (also called volatility) provides a measure of deviation about an expectation. For example, excess volatility in the stock market may be a signal to market uncertainty (and thus leads to falling prices). Similarly in industry, excess variance in manufacturing production points out to a lack of control and the production of parts deviating from their manufacturing standard. There are important risk pricing and management approaches focused on estimating and controlling the variance. Some outstanding examples include control charts, applied as well to control operations in hospital, in transportation, in logistics, in monitoring one's health etc. In economics and finance, the standard deviation (volatility) is a fundamental variable for pricing options. In fact, it is far more precise to say that the premium paid for an option is rather a premium for a future stock price volatility as options pricing models are a function volatility. In this sense, or current option price implies future volatility of the underlying stock. This volatility is implied (or virtual) in finance while in production, volatility is real, observed and estimated statistically as a property of a process (see Chaps. 6–8).

To highlight the effects of volatility, consider two engineered design (financial assets, or a risk model) each of which is a function of both an expected performance and its standard deviation denoted at time t by $S_i(\mu_i, \sigma_i, t),\ i = 1, 2$ where μ_i and σ_i is the expected performance and its standard deviation (see Fig. 3.1). Assume that performance is advertised to be μ_i (and therefore not deviating from μ_i, corresponding to $\sigma_i = 0$). A 3 terms Taylor series approximation of $S(\mu_i, \sigma_i, t),\ i = 1, 2$ with respect to a null standard deviation is:

$$S(\mu_i, \sigma_i, t) \approx S(\mu_i, 0, t) + \left(\frac{\partial S(\mu_i, 0, t)}{\partial \sigma_i} \sigma_i + \frac{1}{2} \frac{\partial^2 S(\mu_i, 0, t)}{\partial \sigma_i{}^2} \sigma_i{}^2 \right).$$

If $S^*(\mu_i, 0, t)$ is an advertised performance, then its:

$$\tilde{\varepsilon}(\sigma) = \left(\frac{\partial S(\mu_i, 0, t)}{\partial \sigma_i}\sigma_i + \frac{1}{2}\frac{\partial^2 S(\mu_i, 0, t)}{\partial {\sigma_i}^2}{\sigma_i}^2\right)$$

If $\tilde{\varepsilon}(\sigma)$ and its properties can be ascertained, then $S(\mu_i, \sigma_i, t) = S(\mu_i, 0, t) + \tilde{\varepsilon}(\sigma)$ and risk (defined as an expected deviation from an average performance) can be used to educate prospective buyers (or users) of the expected deviations. Similarly, comparing two designs, each with an equal expected performance but a different "standard deviation risk" can be evaluated by:

$$0 < S(\mu, \sigma_1, t) - S(\mu, \sigma_2, t) = \frac{\partial S(\mu, 0, t)}{\partial \sigma_1}\sigma_1 - \frac{\partial S(\mu, 0, t)}{\partial \sigma_2}\sigma_2 + \frac{1}{2}\frac{\partial^2 S(\mu, 0, t)}{\partial {\sigma_1}^2}{\sigma_1}^2 - \frac{1}{2}\frac{\partial^2 S(\mu, 0, t)}{\partial {\sigma_2}^2}{\sigma_2}^2$$

and

$$\frac{\partial S(\mu, 0, t)}{\partial \sigma_1} = \frac{\partial S(\mu, 0, t)}{\partial \sigma_2} < 0 \text{ then :}$$

$$S(\mu, \sigma_1, t) - S(\mu, \sigma_2, t) = \frac{\partial S(\mu, 0, t)}{\partial \sigma_1}(\sigma_1 - \sigma_2) < 0, \sigma_1 > \sigma_2$$

This simple exercise provides a first order and linear approximation in the standard deviation and its effects on alternative designs (Fig. 3.1). In industrial risk management such principles are used to control manufacturing processes, while in finance, similar calculations (although embedded in financial theories) are based on prices rather than performance and provide models to calculate prices. Practically however, it is important to determine the effects of the standard deviations on performance but also its causes (Fig. 3.1).

In industrial and quality management, process improvement is sought to reduce production costs by reducing manufacturing variability (through the use of control charts and related instruments). In finance, however, a risk premium is used, defined as a function of the asset's price volatility—that contributes (among many other factors) to a decline in the asset price. In both cases, greater volatility is not preferred. Economic analyses have pointed out that volatility risks (i.e. asset prices variations as well as random variations in the volatility) are due to common risks such as macroeconomic variables (National Statistics of GNP, GDP, Interest rates, Unemployment etc.), that in turn underlie the performance of corporate securities and all financial assets. In effect, assets' prices variations are continuously fed by information—both objective and subjective based on latent sentiments that contribute to security prices to vary from moment to moment. For this reason, a greater understanding of environmental and common risk factors and their effects on prices and risks are important to better forecast volatility and its consequences. While in

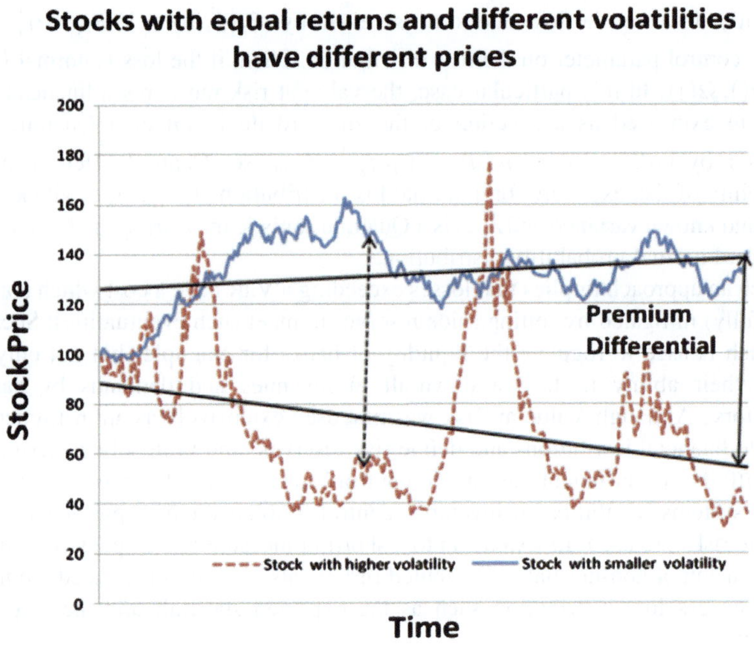

Fig. 3.1 The effects of volatility on a price

both industry and finance volatility is used differently and may express entirely different purposes, its use as a technical and essential element of risk definition is mostly similar.

3.3.5 VaR or Value at Risk

An approach commonly used by banks and financial institutions' risk regulators is an "expectation and threshold" approach commonly coined Value at Risk (VaR) (Artzner 1999; Artzner, et al. 1997, 1999, 2000, 2001; Embrechts 2000a, b; Embrechts et al. 1997, 2001, 2002, 2003). It is defined by the probability that a loss may be greater than a threshold, K or $P\{\tilde{L} > K\} \leq 1 - \varsigma$ where $\tilde{L}$ is a tail random loss, with K a regulated capital deemed adequate to meet critical losses) with a probability at most $1 - \varsigma$. When the tail risks of a loss distribution is appreciable or it has fat tails (i.e. the loss distribution has an appreciable skewness and kurtosis) an excess function defined by a partial expectation of a loss is used to define a more conservative risk measure given by the partial expectation: $e(k) = E\{\tilde{L} - K|\tilde{L} > K\}$. Such measures are also called Tail VaR (TVaR), Conditional VaR (or CVaR), etc. If it is "money based", the threshold defines the amount of money (capital) to set aside in case of a future need to compensate for potential losses (and thereby assure in probability that needs for money can be met).

Explicitly if we set the loss due to a risk event by: $P_t\{\tilde{L}_t(a) > VaR_t(a,\alpha)\} = \alpha$, with a control parameter on the loss distribution. Then, if the loss is normal $\tilde{L}_t(a) \sim \mathbb{N}(\mu(), \Omega())$. In this particular case, the value at risk measures a financial risk exposure expressed as a function of the standard deviation or a Quantile risk specified, by $VaR_t(a,\alpha) = -\mu_t(a) + (\Omega_t(a))^{1/2} Z_{1-\alpha}$ (see Chap. 7). Here, α is the probability of risk exposure for a normal loss distribution $\mathbb{N}(\mu(), \Omega())$ with known mean and known variance and $Z_{1-\alpha}$ is a Quantile statistic measuring the $1-\alpha$ tail of a standard normal probability distribution.

Such an approach implies that losses exceeding a VaR have a cost which may be (hopefully) mitigated by setting aside reserves to meet such eventualities. Such an approach is used to regulate the liquidity of banks for example (but not only) to assure their ability to face a downfall in revenues and demands by banks' depositors.. Although Value at Risk was first used extensively as an instrument to regulate financial institutions and define risk reserves (and profusely criticized for its shortfalls as a sufficient measure to control banks), it has also been proposed to design systems reliability, to inventory control problems and to problems where quantile risks are used. Its extension to a shortfall measure has emphasized further the profusion of non-normally distributed risk events as well as truncated moments of the excess loss distribution such as the expected shortfall and the stop loss measure:

$$E(-\xi|\xi \leq -VaR) = \int_{-\infty}^{-VaR} \xi P_t(\xi)d\xi : \text{ Expected Short Fall}$$

$$E(\text{Max}(\xi - SL, 0)) = \int_{SL}^{\infty} (\xi - SL)f(\xi)d\xi : \text{ Stop Loss Measure}$$

These probability moments are used operationally both as measures and objectives conforming to pre-defined risk design tolerance.

3.3.6 *Chance Constraints*

Chance constraints appear in many statistical and risk models, providing a simple expression for risk exposure. It is simply stated as follows:

$$1 - F(x) = \int_{x}^{\infty} f(z)dz \ \ (\leq \alpha \text{ or } \geq \alpha)$$

For example, let $f(z|u)$ be the probability model for a default, loss, a risk consequence, which is a function of an effort one would apply to mitigate the loss, or $\partial f(z|u)/\partial u < 0$ with u defining the effort (a cost, an activity conducted to prevent the loss, an information acquired, a policy to reduce risk causes) while α is a constraint probability for the loss. In this case, the chance constraint is:

$$1 - F(x|u) = \int_x^\infty f(z|u)dz \le \alpha$$

Say that the effort has a penalty $C(u)$, $\partial C/\partial u > 0$ on the one hand, while the chance constraint must be met, thus, a typical (and extremely simplified) problem consists in defining the effort u that minimizes its penalty while meeting the chance constraint, or $\underset{u \in U}{\text{Min}}\, C(u)$ Subject to : $F(x|u) \ge 1 - \alpha$. Such problems are called chance constrained programming, profusely applied in many situations where probability constraints are used. Below we extend such models to other applications commonly used in risk measurement, tolerance and management.

3.3.7 *Type I and Type II Statistical Risks*

A type I risk is defined when an outcome is rejected because it is "undesirable" when in fact it is not the case. By the same token a type II risk is defined when an outcome is accepted as "desirable" when in fact it is not. Such risks are used to construct statistical tests (applied in Statistics under the Neyman-Pearson Hypothesis Testing). Explicitly, assume that a system (a machine, a process, a patient, etc.) can be in two states, each state defined by a probability distribution:

$$f_i(x|\mu_i, \sigma_i), \ i = 1, 2 \text{ with parameters } \mu_i, \sigma_i$$

Next, say that an observation x is obtained on the basis of which we construct a chance constraint "decision rule" to accept or reject the observation as belonging to one or the other states. In particular, say that: If $x < c$ we accept the hypothesis that this observation is drawn from the distribution $i = 1$ If $x > c$ we accept the hypothesis that this observation is drawn from the distribution $i = 2$. Under such states, we can make two errors, a Type I and II errors, defined below: Type I and II errors $P_I = \int_c^\infty f_1(x|\mu_1, \sigma_1)dx$ and $P_{II} = \int_{-\infty}^c f_2(x|\mu_1, \sigma_1)dx$

The first error consists in rejecting the observation as it is greater than its critical test when in fact it belongs to a "good" state. Let α be the chance constraint of such an error. The second test hypothesis does not reject the observation even though it is drawn from a "bad" state. We let the chance constrained probability of this decision to be β. As a result, the "chance-constraints test" based on specifying the risks we are willing to assume consist in selecting the critical decision parameter c as a function of the type I and II risks we are willing to assume, or:

$$P_I(c,\alpha) = \int_c^{\infty} f_1(x|\mu_1,\sigma_1)dx \le \alpha \text{ or } F_1(c|\mu_1,\sigma_1) \ge 1-\alpha$$

$$P_{II}(c,\beta) = \int_{-\infty}^{c} f_2(x|\mu_2,\sigma_2)dx \le \beta \text{ or } F_2(x|\mu_2,\sigma_2) \le \beta$$

Type I and II risks are used as constraints in the design of statistical tests and control problems for Laboratory tests, in Quality Assurance in industrial processes, in logistics etc. The decision the select both the risks to assume and the critical parameters we use to control these risks depends of course on numerous factors such as the reliability of the inspection procedure set in place, the statistical strength of the test, the costs and risk consequences of failing to detect sub-standards items or products.

Problem: Let $f_i(x|\mu_1,\sigma_1),\ i=1,2$ be two probability distributions (one with a low volatility and therefore "good" and the other with a high volatility, and therefore "bad". In other words,

$$f_i(x|\mu_i,\sigma_i) = \frac{1}{\sqrt{2\pi\sigma_i}} \exp\left(-\frac{1}{2}\left(\frac{x-\mu_i}{\sigma_i}\right)^2\right)$$

How would you construct a critical test parameter c as a function of the type I and II risks that we are willing to assume (see below for the definition of the normal probability distribution)

3.3.8 *Quality Assurance and Chance Constraints Risks*

We consider next two populations, one "good", the other "bad", characterized by Bernoulli probabilities p_G and p_B. Next assume that a set of n statistically independent experiments are performed with r experiments considered "positive" and the remaining ones considered "negative". Again let c be the critical decision parameter. Note that the probability distribution of obtaining r positive outcomes drawn from a "good" population is binomial with parameters (n,p_G) while the probability of obtaining such outcomes from a "bad" population is also binomial with parameters (n,p_B) (see below for a definition of the Binomial probability distribution). Type I and II risks are:

$$P_I(c,\alpha|n,p_G) = \sum_{r=0}^{c}\binom{n}{r}p_G^r(1-p_G)^{n-r} \ge 1-\alpha;$$

$$P_{II}(c,\beta|n,p_B) = \sum_{r=0}^{c}\binom{n}{r}p_B^r(1-p_B)^{n-r} \le \beta$$

There are two test parameters: n and c. These can be selected based on statistical criteria or on the basis of simple economic criteria. For example, say that the cost of each experiment is C and let the probability of making a type I error have a cost π while the cost of a type II error is appreciably larger, say $\pi(1+\delta)$. Thus, a simplistic cost function can be constructed which in expectation is given by:

$$\underset{n,c}{\mathrm{Min}}\,\Phi = nC + n(1-p_G)\pi\alpha + (N-n)p_G\pi(1+\delta)\beta \quad \text{Subject to :}$$

$$P_I(c,\alpha|n,p_G) = \sum_{r=0}^{c}\binom{n}{r}p_G^r(1-p_G)^{n-r} \geq 1-\alpha \text{ and}$$

$$P_{II}(c,\beta|n,p_B) = \sum_{r=0}^{c}\binom{n}{r}p_B^r(1-p_B)^{n-r} \leq \beta$$

where $N-n$ is the total number of accepted parts due to a failed test based on a sample of size n. This problem as well as its generalization to multiple types of tests can be tabulated and the optimal test selected. In fact tests for double test samples, conditional test samples, run samples, sequential samples, rectifying samples, attributes samples and many others, underlie a very rich statistical foundation developing more or less stringent and cost sensitive control techniques. Some references on these problems includes, Tapiero 1996; Daudin and Tapiero 1996; Kenett and Zacks 2004 and a rich literature in Statistical Quality and process Control. A number of selected problems will be considered in Chap. 11 on Strategic and Risk Control. In particular, we shall consider problems of environmental and statistical quality control and the statistical control of regulated banks.

Statistical assurance is practiced in almost all fields. In manufacturing one tests whether a production lot conforms (in probability) to acceptable manufacturing specifications. This is done prior to delivery of products. Samples are selected and tested using stringent statistical tests to mitigate excessive post-sales losses due to default risks. Similarly, medical clinics and laboratories administering numerous (and complex) tests seek to eliminate errors that can have dire human and financial consequences. The errors encountered are numerous including: Tests variations; Physiological variation; Sampling errors; Errors due to complexity of the tests (in particular when samples are tested for a number of attributes that are statistically dependent and relationships difficult to assess), and their like.

3.3.9 Credit and Credit Granting and Estimation of Default Probabilities

In practice, numerous techniques are used to estimate the default probability of a borrower. A common approach used by Banks is the statistical LOGIT model which is cursorily defined as follows. Let x_{ij} be the specific score of a person i on a specific item j. Such information is assembled through questionnaires that borrowers fill when they apply to credit. Let a_j be a set of weights to be applied to each of the items recorded. Given these weights (estimated using a LOGIT or other related models), an aggregate score for a person i is then found to be $S_i = \sum_{j=1}^{p} a_j x_{ij}$. To determine these weights, the LOGIT model assumes a logistic probability distribution to predict the probability of default of the borrower as a function of his score. Explicitly, it assumes a number of observable characteristics (input factors) presumed to predict the probability of default. Say that the probability of the odds of default of a borrower is of the form:

$$\frac{PD_i}{1 - PD_i} = e^{S_i} \text{ or the probability of default is } PD_i = F(S_i) = \frac{1}{1 + e^{-S_i}}:$$

where S_i is the score of the individual borrower which points out to the ln of the odds of default. As a result:

$$\ln\left(\frac{PD_i}{1 - PD_i}\right) = S_i = \sum_{j=1}^{p} a_j x_{ij}$$

A linear predictive model based on this data is then constructed which we can write by:

$$\ln\left(\frac{PD_i}{1 - PD_i}\right) = \alpha + \sum_{j=1}^{k} \beta_j X_{ij} + \varepsilon_i = \tilde{S}_i \text{ or}$$

$$P\{D_i|S_i\} = F(S_i) = \frac{1}{1 + e^{-S_i}} = 1 \bigg/ \left(1 + e^{-\left(\alpha + \sum_{j=1}^{k} \beta_j X_{ij} + \varepsilon_i\right)}\right)$$

As a result the probability distribution for a score is:

$$\frac{dF(\tilde{S}_i)}{dS_i} = f(\tilde{S}_i) = \frac{e^{-\tilde{S}_i}}{\left(1 + e^{-\tilde{S}_i}\right)^2} = e^{-\left(\alpha + \sum_{j=1}^{k} \beta_j X_{ij} + \varepsilon_i\right)} \bigg/ \left(1 + e^{-\left(\alpha + \sum_{j=1}^{k} \beta_j X_{ij} + \varepsilon_i\right)}\right)^2$$

where ε_i is a random error and β_j are parameters to be estimated by the linear regression using the data set banks have regarding their credit applications. In this case, if the LOGIT model is deemed statistically acceptable and if the parameter estimates are also meaningful, we have then a predictive default probability model with the following estimates:

$$\ln\left(\frac{PD_i}{1 - PD_i}\right) = \hat{S}_i = \hat{\alpha} + \sum_{j=1}^{k} \hat{\beta}_j X_{ij}$$

A new borrower, with a characteristic profile X_{ij}, will then have the expected score $\hat{S}_i$. This score is used by the lender to decide to grant or not the loan on the basis of a decisive score S^*. Explicitly, if the score above is the decisive score, a loan is granted or refused by:

$$\begin{cases} \text{Grant a loan if} & \hat{S}_i \geq S^* \\ \text{Do not grant a loan if} & \hat{S}_i < S^* \end{cases}$$

The risk probabilities are then defined by:

1. Type I risk $= P\left(\tilde{S}_i | \hat{S}_i \leq S^*\right)$ The probability of granting a loan and the borrower defaults.
2. Type II risk $=$ 1-$P\left(\tilde{S}_i | \hat{S}_i \geq S^*\right)$ The probability of not granting the loan and borrower not defaulting

For all scores, note that by using the conditional probabilitie $P(D,S) = P(D|S)P(S) = P(S|D)P(D)$and therefore,

$$P(D|S) = \frac{P(S|D)}{P(S)}P(D) = \frac{P(S|D)P(D)}{P(S|D)P(D) + P(S|ND)P(ND)}$$

In other words, given a prior probability of default (expressing general economic conditions for loans failure), based on prior data we have the conditional probabilities $P(S|D)$ and $P(S|ND)$, and obtain a score dependent estimate for the probability of default. Applying the LOGIT model we have then: $f(S_i) = e^{-S_i}/(1 + e^{-S_i})^2$ and therefore, type I and II risks can be calculated. Other models can be constructed in this spirit.

3.3.10 *Chance Constrained Programming*

Chance constrained programming is based on techniques devised to solve optimization problems when either the constraints of a problem and/or its objectives are defined in terms of random elements. Consider the following linear programming problem:

$$\underset{\mathbf{x} \geq 0}{\text{Max}}\, Z = \mathbf{c}'\mathbf{x} \text{ Subject to} : \mathbf{Ax} \leq \mathbf{b}, \mathbf{x} \geq 0 \text{ and explicitly,}$$

$$\text{Max } Z = c_1x_1 + c_2x_2 + \ldots + c_nx_n$$

Subject to:

$$a_{11}x_1 + a_{12}x_2 + \ldots + a_{1n}x_n \leq b_1$$
$$a_{21}x_1 + a_{22}x_2 + \ldots + a_{2n}x_n \leq b_2$$
$$\ldots$$
$$a_{m1}x_1 + a_{m2}x_2 + \ldots + a_{mn}x_n \leq b_m$$
$$x_1 \geq 0, x_2 \geq 0, \ldots, x_n \geq 0$$

The parameters $\{A, b, c\}$ may be random, expressed as a function of a vector of potentially random parameters, or $\{A(\theta), b(\theta), c(\theta)\}$. Such problems are common to many risk related problems where questions such as the following may be raised: What are the random factors of the model; what are their probability distributions, and how do they affect the solution of the problem we are attempting to solve. Further, do we have the information to estimate these parameters? How does such parametric randomness affect the selection of our optimization objective and the problem constraints? What are the numerical techniques we may use to find a solution? There are of course many approaches to handle such problems. We shall consider a few by means of examples.

Say that the vectors **A** and **b** are known and assume that parameters **c** are a function of the random parameter vector θ, or $\mathbf{c}(\theta)$. If the objective is reduced to its expectation (a deterministic linear optimization problem) then, $\mathbf{Z}^* = \mathbf{c}^T\mathbf{x} = E(\mathbf{c}^T(\theta)\mathbf{x}) = \mathbf{c}_1^T\mathbf{x}$ where $\mathbf{c}^\mathbf{T}$ is the transpose of vector **c**. Assume at present that $\{\mathbf{A}, \mathbf{c}\}$ are known but $\tilde{\mathbf{b}}(\theta)$ is random. In this case, $\mathbf{Ax} \leq \tilde{\mathbf{b}}(\theta)$ is a random constraint which we reduce to chance-constraints. To do so, we introduce a vector β of chance constraints probabilities on each of these constraints and reduce the system inequalities to $\mathrm{P}\left[\mathbf{Ax} \leq \tilde{\mathbf{b}}(\theta)\right] \geq \beta$. A chance constrained linear programming problem results, given by:

$$\text{Max} \sum_{j=1}^{k} E(c_j)\, x_j \text{ Subject to : Prob}\left[\sum_{j=1}^{k} a_{ij}x_j \leq \tilde{b}_i\right] \geq \beta_i, i = 1, \ldots, m,$$
$$j = 1, \ldots, k,\ x_j \geq 0$$

Algorithms to the solution of these problems can be found in Tintner and Sengupta (1975) as well as standard operations research books such as Hillier and Lieberman (2010).

Assume at present that the parameters $[\tilde{a}_{ij}]$ have a normal probability distribution and consider again the chance constraints:

$$\text{Prob}\left[\sum_{j=1}^{k} \tilde{a}_{ij}x_j \leq b_i\right] \geq \beta_i, i = 1, 2, \ldots, m \text{ with } E\left(\tilde{a}_{ij}\right) = \alpha_{ij},$$
$$E\left(\tilde{a}_{ij}\,\tilde{a}_{ih}\right) = \sigma_{ijh}, j, h = 1, 2, \ldots k, j$$

In this particular case, the constraint is equivalent to a deterministic quadratic constraint defined by : $\sum_{j=1}^{k} \alpha_{ij}x_j + B_i\left[\sum_{j=1}^{k}\sum_{h=1}^{k} \sigma_{ijh}x_ix_j\right]^{\frac{1}{2}} \geq b_i,\ \mathrm{i} = 1, \ldots,\ \mathrm{m}$ where B_i is the fractile $1 - \beta_i$ of a standard normal probability distribution. If $\beta_i \geq 1/2$, then, $B_i \leq 0$ and the constraint is convex. This property reduces the linear programming problem to a convex programming problem for which efficient numerical algorithms are available.

3.3.11 *Chance Constraint Moments Approximations*

In many cases moments information can be used to establish probability approximations. Below we summarize a few:

1. Tchebycheff's Inequality: Let X be a random variable of a consequential loss, then:

   If $\mu = \text{EX}$ then $P(X \geq \lambda\mu) \leq 1/\lambda$ or $P(X \geq \lambda) \leq \mu/\lambda$, and

   $$P\left(\sum_{i=1} x_i \geq \lambda \sum_{i=1} \mu_i\right) \leq 1/\lambda$$

2. Bienaime-Tchebycheff: $P(|X-\mu| \geq \lambda\sigma) \leq 1/\lambda^2$ or $P(|X-\mu| \geq \lambda) \leq \sigma^2/\lambda^2$

3. Kolmogorov

   $$P(T \geq \lambda\sigma) \leq 1/\lambda^2,\ S_i = \sum_{j=1}^{i} (x_j - \mu_j), i = 1,2,\ldots,n,$$
   $$T = \max_{1 \leq i \leq n} [|S_1|,\ldots,|S_i|,\ldots|S_n|]$$

4. Cantelly: Applicable when the Bienaimee-Tchebycheff inequality is applicable

   $$P(X-\mu \leq \lambda) \leq \sigma^2/(\sigma^2+\lambda^2) \qquad \text{for } \lambda < 0$$
   $$P(X-\mu \leq \lambda) \geq 1-\sigma^2/(\sigma^2+\lambda^2) \quad \text{for } \lambda > 0$$

5. Bernstein: $P(|X-\mu| \geq \lambda) \leq 2\exp\left[\frac{-\lambda^2}{2\sigma^2+2C\lambda}\right]/\lambda^2$ or for independent

   $X = \sum x_i;\ x_i,\ x_j\ \text{E}|x_i - \mu_i|^s \leq \sigma_i^2 s! C^{s-2}/2$ for all integers C>0

3.3.12 *Transformation of Random Variables into Normally Distributed Random Variables*

In many situations, it is useful to transform a given probability distribution into a normally distributed random variable (Malvergne and Sornette 2002). To do so, let $f(x)$ be a probability distribution and let $f(y)$ be a standard normal probability distribution. Set:

$$f(x)dx = \frac{1}{\sqrt{2\pi}} e^{-\frac{y^2}{2}} dy$$

An integration of these equations, yield:

$$F(x) = \int_{-\infty}^{x} f(x)dx = \int_{-\infty}^{y(x)} \frac{1}{\sqrt{2\pi}} e^{-\frac{y^2}{2}} dy = \frac{1}{2}\left[1 + erf\left(\frac{y(x)}{\sqrt{2}}\right)\right]$$

and therefore

$$y(x) = \sqrt{2}erf^{-1}(2F(x) - 1)$$

For example, if a risk model is not normally distributed, a transformation can be found to transform it into a normal distribution. For example, assume that $f(x)$ has a Weibull distribution, whose cumulative density function is given by (see below for a definition of extreme distributions):

$$f(x) = c\mu((x-a)\mu)^c \exp\{-((x-a)\mu)^c\}, x \text{ and } F(x) = 1 - e^{-((x-a)\mu)^c}$$

Then (heuristically):

$$y(x) = \sqrt{2}erf^{-1}\left(1 - 2e^{-((x-a)\mu)^c}F(x)\right)$$

The implication of this approach is that a great many distributions can be transformed in some way to have an underlying normal distribution and interpreted as derived from underlying normal risk probability distributions.

3.4 Generating Functions

Generating functions, as their names implies, generate the properties of their underlying distributions. There include the Probability Generating Function (PGF), the Laplace Transform (LT), the Moment Generating Function (MGF), The Characteristic Function (CF) and well as others (such as Mellin Transforms, Laplace-Stiljes Transforms-LST, Fourier and Fast Fourier Transforms-FT) (see for example Colombo and Lavoine 1972). We consider below a number of such generating functions including:

$$\text{MGF}(\lambda) = E(e^{\lambda z}) = \int_{z \in Z} e^{\lambda z} f(z)dz : \text{ The Moment Generating Function}$$

$$PGF(\nu) = \sum_{i=1}^{n} \nu^{x_i} p_i \ : \ \text{The Probability Generating Function}$$

$$LT(s) = E(e^{-sz}) = \int_{z \in Z} e^{-sz} f(z)dz : \text{ The Laplace Transform}$$

$$\text{CF}(\lambda) = E(e^{i\lambda z}) = \int_{z \in Z} e^{i\lambda z} f(z)dz : \text{ The Charateristic Function, } i = \sqrt{-1}$$

These generating functions are extremely helpful when events are defined and are statistically independent.

The Moment Generating Functions (*MGF*): MGF are tabulated for a large number of probability distributions. Generally assuming that it exists, then it is evident that $\mathrm{MGF}(0) = 1$. Further, the ith moment of the distribution is given by:

$$\frac{\partial^i \mathrm{MGF}(\lambda)}{\partial \lambda^i} = \int_{z \in Z} z^i e^{\lambda z} f(z) dz \text{ and } \frac{\partial^i \mathrm{MGF}(0)}{\partial \lambda^i} = \int_{z \in Z} z^i f(z) dz = E(z^i)$$

For example, i=1 corresponds to the distribution mean, i=2 corresponds to the 2nd moment, and so on.

The Probability Generating Function: It is used mostly for discrete distributions and is defined by:

$$PGF(\nu) = \sum_{i=1}^{n} \nu^{x_i} p_i$$

Assuming that it exists, note that $PGF(1) = 1$ while higher order moments are found by the PGF derivatives, with:

$$\frac{\partial PGF(\nu)}{\partial \nu} = \sum_{i=1}^{n} x_i \nu^{x_i - 1} p_i \text{ and at } \nu = 1, \ \frac{\partial PGF(1)}{\partial \nu} = E(x_i)$$

$$\frac{\partial^2 PGF(\nu)}{\partial \nu^2} = \sum_{i=1}^{n} x_i (x_i - 1) \nu^{x_i - 2} p_i \text{ and at } \nu = 1 \ \frac{\partial^2 PGF(1)}{\partial \nu^2} = E(x_i^2) - E(x_i)$$

Additional derivatives provide higher order moments. Generating functions are in fact a transformation of a distribution into another domain.

3.4.1 *The Convolution Theorem for Moment and Probability Functions*

The sum of two independent random variable each with a moment generating function $(M_X(z), M_Y(z))$ has a moment generating function which is given by:

$$M_{U=X+Y}(z) = M_X(z) M_Y(z)$$

This is proved by a transformation variables as follows. Let $f_{X,Y}(x, y)$ be the joint probability distributions of x and y. Consider their sum and let $u = x + y$ and

$v = x$ be a variables transformation with $x = v$ and $y = u - v$. The Jacobian of this transformation is then given by:

$$J(u,v) = \det\begin{pmatrix} \partial x/\partial u & \partial x/\partial v \\ \partial y/\partial u & \partial y/\partial v \end{pmatrix} = \det\begin{pmatrix} 0 & 1 \\ 1 & -1 \end{pmatrix} = -1$$

and therefore, the joint probability distribution of u and v is given by:

$$f_{U,V}(u,v) = f_{X,Y}(x(u,v), y(u,v))\left|\det\begin{pmatrix} \partial x/\partial u & \partial x/\partial v \\ \partial y/\partial u & \partial y/\partial v \end{pmatrix}\right|$$

or

$$f_{U,V}(u,v) = f_{X,Y}(v, u-v)$$

Therefore, the probability distribution of the sum is:

$$f_U(u) = \int_{-\infty}^{\infty} f_{U,V}(u,v)dv = \int_{-\infty}^{\infty} f_{X,Y}(v, u-v)dv$$

However if X and Y are independent, we have:

$$f_U(u) = \int_{-\infty}^{\infty} f_X(v) f_Y(u-v)dv$$

As a result, for distributions defined in the interval $u \in [0,\infty)$,the MGF of the sum is:

$$\begin{aligned} MGF_U(z) &= \int_0^{\infty} e^{uz} f_U(u)du = \int_0^{\infty}\int_0^{\infty} e^{(x+y)z} f_X(x) f_Y(y)dvdu \\ &= \left(\int_0^{\infty} e^{xz} f_X(x)dx\right)\left(\int_0^{\infty} e^{yz} f_Y(y)dy\right) \end{aligned}$$

and therefore,

$$MGF_U(z) = MGF_X(z) MGF_Y(z)$$

Similarly, for n independent random variables $S = \sum_{i=1}^{n} x_i$, the MGF is $M_S(z) = \prod_{i=1}^{n} M_{X_i}(z)$. It is easy to prove then that the first two moments of the sum distribution is:

$$E(S) = E\sum_{i=1}^{n} x_i = \mu E(n);\quad E(S^2) = \mu^2 E(n^2) + \sigma^2 E(n)$$

Similarly, say that a probability distribution whose moment generating function $M_{X_i}(u|\xi_n)$ is a function of another random variable, ξ_n, assuming some value with probability P_n, $n = 0, 1, 2, 3, \ldots, N$. The generating function of m independent random variable has then a generating function given by: $\prod_{i=1}^{n} M_{X_i}(u|\xi_n)$ and therefore, its expectation is $M_S(u) = \sum_{n=0}^{N}\left(\prod_{i=1}^{n} M_{X_i}(u|\xi_n)\right)P_n$.

Note that this is the case only if the random variables are independent. In other words, let X be the consumption of a consumer and Y be the consumption of another and say that both are statistically dependent, then the MGF of their consumption $U = X + Y$ does not have an MGF which can be calculated by the convolution theorem.

When we use a *Probability Generating Function* (PGF) the convolution theorem is stated and proved as follows: Let $x_i, f_i(.), i = 1, 2$ and their PGF $F_1(z) = \sum_{i=0}^{\infty} z^i f_1(i)$, $F_2(z) = \sum_{i=0}^{\infty} z^i f_2(i)$,. Define $y = x_1 + x_2$ whose probability distribution is: $f_3(y)$ then, by the convolution theorem $f_3(y) = \sum_{i=0}^{y} f_2(y - i) f_1(i)$ with a PGF $F_3(z) = \sum_{i=0}^{\infty} z^i f_3(i)$. As a result,

$$F_3(z) = \sum_{i=0}^{\infty} z^i \left[\sum_{j=0}^{i} f_2(i - j) f_1(j),\right] = F_3(z) = \sum_{i=0}^{\infty} z^i f_1(i)\left[\sum_{j=0}^{\infty} z^j f_2(j)\right]$$

and therefore, $F_3(z) = F_1(z)F_2(z)$. Below we consider one example using the bernoulli probability distribution. Subsequently, other distributions and other generating functions are considered.

3.4.2 The Probability Generating Function of the Bernoulli Experiment

A Bernoulli probability distribution is defined by a random variable $x \in (1, 0)$ with the probability p if x=1 the probability of $1-p$ if x=0. The probability generating function of this distribution is thus:

$$P^*(z) = E(z^x) = pz + 1 - p$$

A binomial probability distribution (to be considered below), consists however in the sum of n statistically independent Bernoulli experiments (random variables) and thus, its probability generating function is:

$$P^*(z) = E(z^x) = (pz + 1 - p)^n$$

whose mean and variance can be verified on the basis of this formula to be: np and $np(1-p)$ respectively.

Say that the bernoulli probability is a random probability with probability distribution $f(p|a,b)$, with parameters (a,b), then, the PGF of the mixture distribution is:

$$P^*(z) = E(z^x) = \int_0^1 (pz + 1 - p)^n f(p|a,b)dp$$

which is the PGF of a family of distributions called Lexis distributions. Note that the moments of this distribution are found by:

$$\frac{\partial P^*(z=1)}{\partial z} = \int_0^1 np(pz + 1 - p)^{n-1} f(p|a,b)dp = n\int_0^1 pf(p|a,b)dp = nE(p)$$

$$\frac{\partial^2 P^*(z=1)}{\partial z^2} = \int_0^1 n(n-1)p^2(pz + 1 - p)^{n-2} f(p|a,b)dp = n(n-1)E(p^2)$$

Since at z=1 $\frac{\partial^2 P^*(z=1)}{\partial z^2} = E(x(x-1)) = E(x^2) - E(x) = \mathrm{var}(x) + [E(x)]^2 - E(x)$ we obtain:

$$\mathrm{var}(x) = n(n-1)\left(\mathrm{var}(p) + [E(p)]^2\right) + \left(nE(p) - n^2[E(p)]^2\right)$$

which reduces to the Lexis formula for the variance:

$$\mathrm{var}(x) = nE(p)[1 - E(p)] + n(n-1)\mathrm{var}(p)$$

An alternative proof applicable to other distributions can be obtained by the Law of Total Variance. In this case, we have:

$$\begin{aligned}\mathrm{var}(x) &= E(x^2) - (E(x))^2 = E[E(x^2|p)] - E\left[(E(x|p))^2\right] \\ &= E\left[\mathrm{var}(x|p) + \left[E(x|p)^2\right]\right] - E\left[E(x|p)^2\right] \\ &= E[\mathrm{var}(x|p)] + \mathrm{var}[E(x|p)]\end{aligned}$$

Applying this formula to the randomized binomial model, we have:

$$\begin{aligned}\operatorname{var}(x) &= E[np(1-p)] + \operatorname{var}[np] = nE(p) - nE(p^2) + n^2\operatorname{var}(p)\\ &= nE(p) - n\left(\operatorname{var}(p) + [E(p)]^2\right) + n^2\operatorname{var}(p)\\ &= nE(p)[1-E(p)] + n(n-1)\operatorname{var}(p)\end{aligned}$$

In this model we see that if p is deterministic then the variance is reduced to that of the binomial distribution. However if p is random and its variance appreciable note that it increases the (randomized) binomial distribution variance at a quadratic rate. This elementary example, highlights therefore the importance parameters uncertainty can have on the moments of the presumed distribution.

3.4.3 Additional Examples

Example: The MGF of a Normal Probability Distribution For an underlying distribution is Normal (Gaussian). Then:

$$f(x) = \frac{1}{\sigma\sqrt{2\pi}}e^{-x^2/2\sigma^2},\ C(\lambda) = e^{-\sigma^2\lambda^2/2}\quad C_{n+1}(\lambda) = C_n(\lambda)C(\lambda),\ C_0(\lambda) = 1$$

or

$$C_{n+1}(\lambda) = e^{-n\sigma^2\lambda^2/2}e^{-\sigma^2\lambda^2/2} = e^{-(n+1)\sigma^2\lambda^2/2}$$

Now assume that the volatility has a probability distribution $g(\sigma)$, then the randomized MGF of the normal distribution is:

$$\int_0^\infty C(\lambda|\sigma)g(\sigma)d\sigma = \int_0^\infty e^{-\sigma^2\lambda^2/2}g(\sigma)d\sigma$$

which can be used to calculate its moments by differentiation of the resulting function.

Problem: The MGF of the Gamma Probability Distribution Let the moment generating function of a Gamma Probability Distribution be:

$$E(e^{\mu x}) = \left(\frac{1}{1-\theta\mu}\right)^\alpha$$

What is the mean, the variance, the skewness and the distribution kurtosis. What is the MGF of a sum of n identically and independently distributed random variables with the same distribution.

Example: The Generating Function of a Sum of Independently Random Variables Define the sum of N independent random events by $\sum_{i=1}^{N} x_i$, each of which has a known PGF. The PGF of the sum and its random sum are respectively: $P_N(z) = \prod_{i=1}^{N} P_{x_i}(z)$ and $P(z) = \sum_{j=1}^{N} \left(q_j \prod_{i=1}^{N} P_{x_i}(z) \right)$

The mean sum and its variance are therefore calculated by the first two derivatives of the Probability Generating Function estimated at z = 1:

$$\frac{\partial P(z)}{\partial z} = \frac{\partial}{\partial z}\left(\sum_{j=1}^{N}\left(q_j \prod_{i=1}^{N} P_{x_i}(z)\right)\right) \text{ and } \frac{\partial^2 P(z)}{\partial z^2} = \frac{\partial^2}{\partial z^2}\left(\sum_{j=1}^{N}\left(q_j \prod_{i=1}^{N} P_{x_i}(z)\right)\right)$$

In particular, say that $P_{x_i}(z) = P_x(z)$ and $q_j = 1$ then: $P(z|N) = (P_x(z))^N$. However, if N is a random sum with a known probability generating function, then the probability generating function of the random sum is $E_N P(z|N) = E(P_x(z))^N = P_N(P_x(z))$.

3.4.4 *The PGF of the Compound Poisson Process*

Let $\{X(t), t \geq 0\}$ be a random variable defined by $X(t) = \sum_{i=1}^{N(t)} y_i$ where y_i are iid (identically and independently distributed) random variable with known probability generating function and $N(t)$ be a Poisson distributed random variable with known mean λt. Then its probability generating function is:

$$N^*(z) = \sum_{i=0}^{\infty} \frac{(\lambda t)^i e^{-\lambda t}}{i!} z^i = e^{-\lambda t} \sum_{i=0}^{\infty} \frac{(\lambda z t)^i}{i!} = e^{-\lambda t} e^{\lambda z t} = e^{-\lambda t(1-z)}$$

while the probability generating function of the CPP (Compound Poisson Process) is:

$$X^*(z) = \sum_{i=0}^{\infty} \prod_{j=1}^{i} y_j^*(z) \frac{(\lambda t)^i e^{-\lambda t}}{i!} = \sum_{i=0}^{\infty} [y^*(z)]^i \frac{(\lambda t)^i e^{-\lambda t}}{i!} = \sum_{i=0}^{\infty} \frac{(\lambda t y^*(z))^i e^{-\lambda t}}{i!} = e^{-\lambda t(1-y^*(z))}$$

Using this last result, we obtain the mean and the variance of the of CPP, or:

$$\begin{aligned} E(X(t)) &= E(N(t))E(y) = \lambda t E(y) \\ \mathrm{Var}(X(t)) &= \mathrm{Var}(N(t))E^2(y) + \mathrm{Var}(y)E(N(t)) = \lambda t\left[\mathrm{Var}(y) + E^2(y)\right] = \lambda t E(y^2) \end{aligned}$$

The proof is as follows. First note that:

$$\Phi(z) = \log(X^*(z)) = -\lambda t(1 - y^*(z)) \text{ and } \frac{\partial \Phi(z)}{\partial z} = \lambda t \partial y^*(z)/\partial z \text{ at z} = 1$$

and therefore, $E(X(t)) = \lambda t E(y)$. Similarly for the variance,

$$\mathrm{Var}(X(t)) - E(X(t)) = \frac{\partial^2 \Phi(z)}{\partial z^2}\bigg|_{z=1}. \text{ since, } \frac{\partial^2 \Phi(z)}{\partial z^2} = \lambda t \frac{\partial^2 y^*(z)}{\partial z^2}$$

Example: Approximation of an Unknown Distribution with Known Moments Let a probability distribution have a mean, variance and skewness. Define a two states probability model x with PGF $P^*(z)$:

$$x = \begin{cases} a & w.p \quad p \\ -b & w.p. \quad 1-p \end{cases} \text{ and } P^*(z) = E(z^x) = \left(pz^a + (1-p)z^{-b}\right)$$

Note that the PGF has three parameters *a,p,b*. Letting n=1, the PGF can be used to calculate its first three moments and equate their mean, variance and skewness to the unknown distribution and thus obtain the approximate bernoulli process whose parameters are parameters *a,p,b*. Note that:

$$\begin{aligned} E(x) &= ap - b(1-p) \\ \mathrm{var}(x) &= E(x^2) - [E(x)]^2 = (a-b)^2(1-p)p \\ E(x^3) &= a^3 p + b^3(1-p) \end{aligned}$$

3.5 Probability Distributions

Probability distributions are models that summarize a data set and the randomness they reveal into recognizable and parametrized forms. Figure 3.2 below summarizes such an approach.

Probability distributions are numerous and varied, seeking to explain the broad variety of statistical data sets one encounters or based on hypotheses (models) of chances events—future or current. These are models with specific and underlying

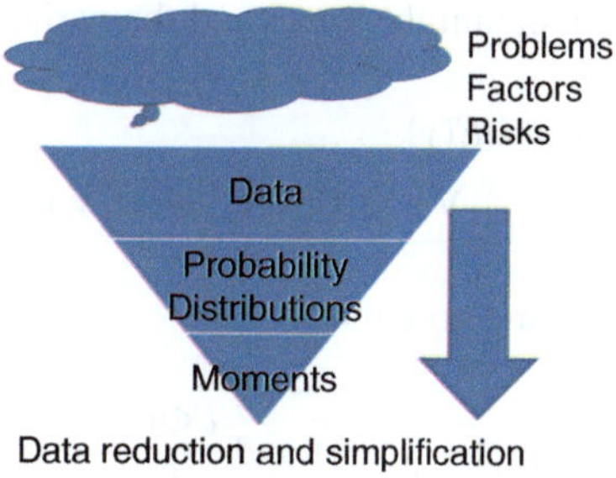

Fig. 3.2 Data reduction

assumptions, including discrete and continuous distributions, single-variate and multi-variate (exhibiting dependence which is of particular relevance to risk), mixture distributions etc.

Probability distributions can also originate in simple scientific or "behavioral hypotheses" to be tested against an unfolding evidence—whether empirical or not. There are families of probability distributions, some related to one another in initially surprising ways, revealed when a common thread to all these families is found. We shall consider below a particularly important family of distributions based on the Bernoulli model.

3.5.1 *The Bernoulli Family*

The Bernoulli family is based on an experiment whose outcome is 1 or 0 and whose probability is given by a fixed number p. Repeating and counting outcomes of the same experiment provides a family of probability distributions, partially represented in the Fig. 3.3 above. In particular:

- The Binomial Distribution results from counting the number of 1's or 0's in n repeated Bernoulli experiments
- The Geometric distribution results from counting the number of experiment to repeat to obtain one outcome 1 (or one 0).
- The Negative Binomial Distribution (NBD) consists in counting the number of experiments to repeat to obtain r outcomes 1 (or 0). It is also a generalized Geometric distribution.
- The exponential distribution results from a continuous limit of the geometric distribution
- The Poisson distribution results from counting the number of 1 and 0 when the Bernoulli probability is small but the number of repeated experiments is large. In this case, it is a special case of the Binomial Distribution. At the same time, the Poisson Distribution results from an Exponential Distribution, expressing a memory-less property when calculating the number of repeated Bernoulli experiments between two 1 events (or 0).

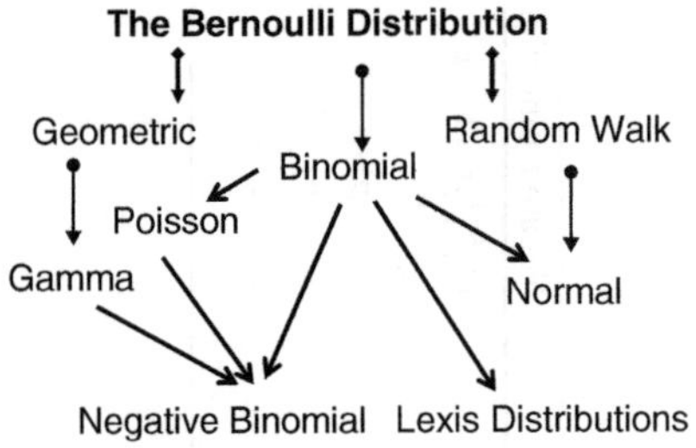

Fig. 3.3 Distributions emanating from the Bernoulli distribution

- The Bernoulli experiment defines the elementary movement (+1 or -1) on a line, also called a random walk.
- The Normal Probability Distribution results from a number of reasons, acting as an attraction pole to numerous probabilities (of the exponential type). At the limit when the number of repeated experiments in a Binomial Models is large, the Normal Probability Distribution provides a good approximation. Similarly, the distance covered on a random walk model over many repeated experiments has at its limit a Normal Probability Distribution.
- Lexis Probability Distributions are mixture, namely a randomization of an underlying Bernoulli experiment in a Binomial Model. The Negative Binomial Distribution can also be obtained by a mixture of the Poisson Distribution when its parameter (mean) has a Gamma Distribution. Mixtures may then be used to model the complexity of the model, combined with a number of parameters that will define the distribution.

These distributions have different functional forms different properties, some of which are summarized in the Table 3.3. However, they can be interpreted as originating from the Bernoulli probability model. Their essential properties includes: The Cumulative Distribution, their mean, their variance, skewness, kurtosis, probability generating function (PGF), defined below.

3.5.2 *The Binomial and Other Distributions*

The Binomial Distribution results from counting the number of 1's or 0's in n repeated and independent Bernoulli experiments. It is given by:

$$P(R = r) = \binom{n}{r} p^r (1-p)^{n-r},\ E(r) = np,\ \mathrm{var(r)} = np(1-p), r = 0, 1, \ldots n$$

This distribution assumes that each Bernoulli event is both identically and independently distributed. Thus, if $x_i,\ i = 1, 2, \ldots n$ denotes the outcome in n Bernoulli experiments, then $\sum_{i=1}^{n} x_i$ denotes a binomial probability distribution with

Table 3.3 Discrete probability distributions

Bernoulli	Binomial	Poisson	Negative binomial
Probability model	$\binom{n}{r} = p^r(1-p)^{n-r}$	$e^{-\lambda}\lambda^r/r!$	$\binom{n+k-1}{n-1}p^n(1-p)^k$
Parameter p	*n,p*	λ	*n,p*
Mean p	*np*	λ	$n(1-p)/p$
Variance p(1–p)	*Np(1–p)*	λ	$n(1-p)/p^2$
PGF $P^*(z) = p + (1-p)z$	$P^*(z) = (p + (1-p)z)^n$	$P^*(z) = e^{\lambda(z-1)}$	$P^*(z) = \frac{(1-p)z}{1-pz}$

$E\left(\sum_{i=1}^{n} x_i\right) = np$ and $\text{var}\left(\sum_{i=1}^{n} x_i\right) = np(1-p)$ as stated above. When the probability p in Bernoulli experiments is not deterministic (for example due to the complexity of the subject test, due to "background risks", the heterogeneity of the tested population, etc.) then assuming that it is random, with a known mean and known variance, we obtain a family of Lexian distribution given by:

$$P(R = r) = \int_0^1 \binom{n}{r} \theta^r (1-\theta)^{n-r} f(\theta) d\theta;$$

$$E(R) = nE(\theta); \text{Var}(R) = nE(\theta)[1 - E(\theta)] + n(n-1)\text{var}(\theta)$$

The implications are that randomness of the "population parameter θ increases the variance of the results at a quadratic rate as a function of the number of experiments. For example, say that a medical survey of 10,000 patients is conducted, on the basis of which the efficiency of a medical treatment is tested, measured by a proportion of successful treatments. Namely, observing r successful treatments. Since patients are (mostly) heterogeneous, the parameter θ is random, the variance of the observation will be extremely large since:

$$\frac{\text{var}(R)}{10,000} = E(\theta)[1 - E(\theta)] + (9,999)\text{var}(\theta)$$

As a result, if $\text{var}(\theta)$ is not infinitely small, its effect can be substantial.

Finally, while the binomial distribution counts the number of outcomes in Bernoulli experiments, the geometric distribution counts the number of events until one first event of one type is obtained. In this case it is given by:

$$g(n) = p(1-p)^{n-1}, n = 1, 2, 3, \ldots.. \text{ with } E(n) = \frac{1}{p}, \text{var}(n) = \frac{1-p}{p^2}$$

This distribution is therefore an inter-event distribution. For example, if a process is prone to default with a probability p, then the probability of default at time n is given by the geometric distribution.

Approximations of the binomial distribution when its parameter p is small and the number of experiments is large, leads to a Poisson distribution while for n sufficiently large, it leads to the Normal probability distribution.

$$\text{Lim}_{p\to 0, np\to\lambda} b(r|n,p) = e^{-\lambda}\lambda^r/r!, \ r = 0, 1, 2, 3, \ldots.., \ \lambda = E(r) = \text{var}(r)$$

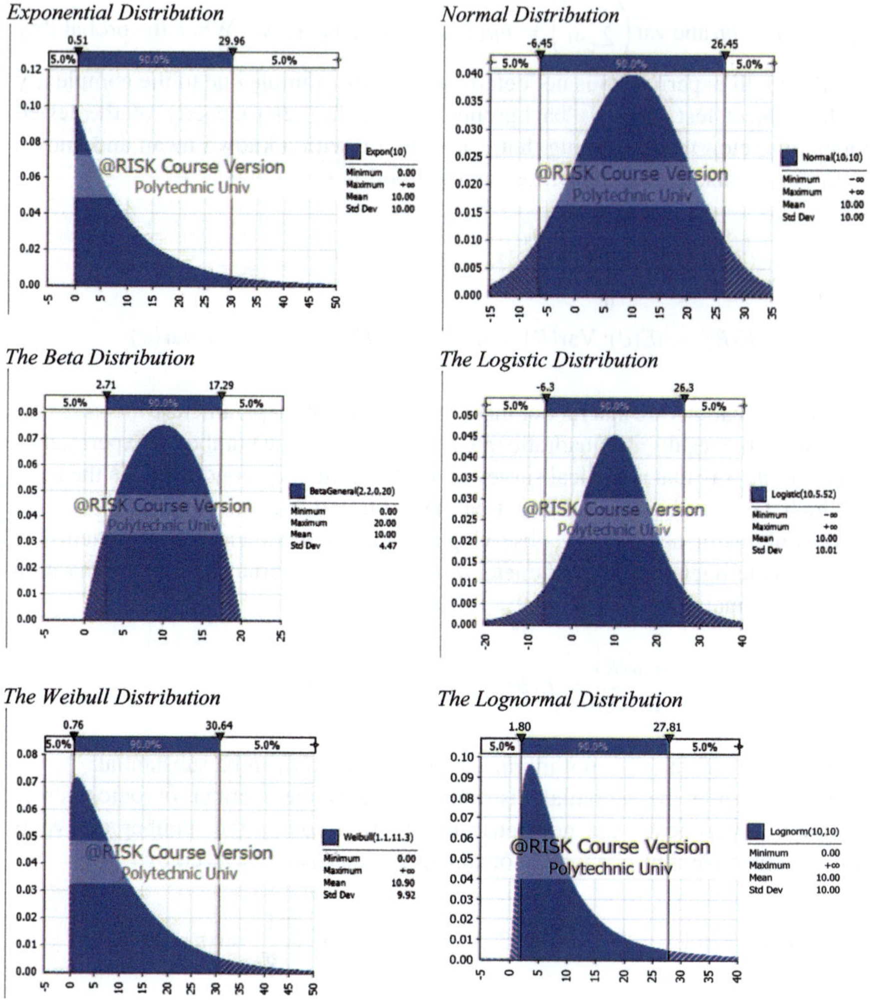

Fig. 3.4 Continuous distributions

and

$$\lim_{n \to \infty} b(r|n,p) = \frac{1}{\sqrt{2\pi}\sigma} \exp\left[-\frac{1}{2}\frac{(x-\mu)^2}{\sigma^2}\right]; \; x \in (-\infty, +\infty), \; \mu = np,$$
$$\sigma^2 = np(1-np)$$

A number of continuous distributions are indicated in Fig. 3.4. These distributions will be defined below.

Example: A Reliability System. Say that a system consists of n independent components. If k units are defective, then the system is in a critical (failing) state.

The probability of any individual component failing is fixed and has a probability p (and therefore a reliability 1-R). What is the probability of a system being in an operating state? An individual's component is clearly a Bernoulli event. Therefore the probability of n events in an operating state has a binomial probability distribution. As a result, the probability of an operating state is:

$$P(X=1)=\sum_{i=1}^{k}\binom{n}{i}p^{i}(1-p)^{n-i} \text{ and } P(X=0)=1-P(X=1)$$

Next assume that the components are set in series, which requires that all components be individually operating and let each component reliability be R_i, $i=1,2,\ldots,n$. The probability of the system being operational is thus equal to the probability that all components have not failed, or:

$$P(X=1)=R_1R_2\ldots R_n=(1-p_1)(1-p_2)\ldots(1-p_n)$$

Networks of independent components can be constructed and their operational state calculated. In fact, many components have time varying reliabilities that may be statistically dependent. For example, the time reliability of an independent series system is:

$$R(t)=R_1(t)R_2(t)\ldots R_n(t)$$

or

$$R(t)=(1-p_1(t))(1-p_2(t))\ldots(1-p_n(t))$$

Note that:

$$\frac{dR(t)}{dt}=\frac{d}{dt}((1-p_1(t))(1-p_2(t))\ldots(1-p_n(t)))=\frac{d}{dt}(1-F(t))=-f(t)$$

where $F(t)$ is the probability of default in the time interval (0,t) while $f(t)$ is the probability of default at a given time t. As a result, the hazard rate of such a series (networked) system is:

$$h(t)=\frac{f(t)}{1-F(t)}=-\frac{\frac{d}{dt}((1-p_1(t))(1-p_2(t))\ldots(1-p_n(t)))}{(1-p_1(t))(1-p_2(t))\ldots(1-p_n(t))}$$

when the components probability are all a function of an external event ξ with probability distribution, say $f(\xi)$, denoting a common cause, then the reliability is random and given by:

$$R(t|\xi)=R_1(t|\xi)R_2(t|\xi)\ldots R_n(t|\xi) \text{ and } R(t)=\int_{\xi\in\Re}R(t|\xi)f(\xi)d\xi$$

If the probability distribution of a common factor $f(\xi)$ is known then reliability has a Bernoulli distribution with:

$$z(t|\xi) = \begin{cases} 1 & \text{with probability } R(t|\xi) \\ 0 & \text{with probability 1-}R(t|\xi) \end{cases}$$

and therefore the joint probability of the Bernoulli and the external event is:

$$\Phi(z, \xi; t) = z(t|\xi)f(\xi)$$

where $\Phi(z, \xi; t)$ is the joint probability distribution at time t with its expected reliability $R(t)$ given above. Other moments and cross moments that are particularly important in risk assessment can be defined on the basis of this distribution as we shall see in Chap. 4. Such problems indicate that while a "strong model assumption" provides greater tractability of the problem at hand, it may in fact hide the risks we are trying to manage.

3.5.3 *The Poisson Distribution*

The Poisson distribution is a "counting" distribution. It is, derived from the Binomial probability distribution when the probability of an individual event occurring is small, their number is large and their mean finite and given by λ. The assumptions underlying the Poisson distribution are:

- The events mean equals λ (the variance as well)
- Independence or memory-less properties: two subsequent events are independent
- Inter-event times are independently and identically distributed as an exponential probability distribution whose mean is $1/\lambda$

In addition, the Poisson distribution has the following properties:

- The sums of Poisson generated events is also Poisson.
- "Fractionating" a Poisson process whose mean is λ by a Bernoulli distribution with parameter P leads also to a Poisson process with parameter λp.

This probability distribution is given by:

$$p(x) = \frac{\lambda^x e^{\lambda}}{x!}, \text{with } E(x) = \text{var}(x) = \lambda$$

with

$$1 - F(k) = 1 - \sum_{i=0}^{k} \frac{\lambda^i e^{-\lambda}}{i!} = \int_0^{\lambda} \left(t^k e^{-t}/k!\right) dt$$

In addition, its Probability Generating Function is given by:

$$P_X(z) = \sum_{x=0}^{\infty} z^x \frac{\lambda^x e^{\lambda}}{x!} = e^{\lambda(z-1)}$$

A Poisson model that counts events leads to a class of models called Compound Poisson. These results in the Poisson sum defined by: $S = \sum_{i=1}^{\tilde{n}} \tilde{z}_i$ where $\tilde{z}_i$, consequences, are statistically independent and $\tilde{n}$ has a Poisson probability distribution. In this case, using the Probability Generating Function is, as calculated earlier:

$$E(S) = E\sum_{i=1}^{\tilde{n}} \tilde{z}_i = \mu\lambda \text{ where } E(\tilde{n}) = \lambda, \text{var}(\tilde{n}) = \lambda$$

$$E(S^2) = \lambda\mu^2(1+\lambda) + \lambda\sigma^2 \text{ and } \text{var}(S) = \lambda(\mu^2 + \sigma^2) = \lambda E(\tilde{z}_i{}^2)$$

Such calculations are used to assess for example, collective risks, summing for example the potential claims in large portfolios where the number of claims at any one time has a Poisson Probability Distribution.

3.5.4 *The Conditional Sum Poisson and the Binomial Distribution*

Say that at time t, the probability of distribution of a demand of a party as a function of time has a Poisson distribution with means $\Lambda_i(t)$, that is:

$$P(x_i(t)) = \frac{e^{-\Lambda_i(t)}(\Lambda_i(t))^{x_i(t)}}{[x_i(t)]!}$$

Now say that there are n parties, each with its own demand, while supplies are limited to the function $D(t)$. In other words, $\sum_{i=1}^{n} x_i(t) = D(t)$ and therefore, if the supply to each party is given by some rule stating that each party's supply equals $\left(x_i(t) \middle| \sum_{i=1}^{n} x_i(t) = D(t)\right)$. Then the supply probability distribution to party i is given by a binomial probability distribution:

$$P\left(x_i(t) \middle| \sum_{i=1}^{n} x_i(t) = D(t)\right) = \binom{D(t)}{x_i(t)} [p_i(t)]^{x_i(t)} [1 - p_i(t)]^{D(t)-x_i(t)}, p_i(t) = \frac{\Lambda_i(t)}{\sum_{i=1}^{n} \Lambda_i(t)}$$

In this case, the supply shortage supply is given by the difference of the two (dependent) random variables $x_i(t) - y_i(t)$ with expected shortage:

$$E(.) = \Lambda_i(t)\left(1 - \frac{D(t)}{\sum_{i=1}^{n}\Lambda_i(t)}\right) > 0$$

3.5.5 *Super and Hyper Poisson Distributions*

The Poisson distribution as noted above is a discrete distribution whose mean and whose variance are equal. Super and Hyper Poisson distribution are functional generalizations of the Poisson distribution that has a variance greater (Super) or smaller (Hyper) than the mean. Such distributions are given by:

$$P(x) = \frac{C_{\theta,\Lambda}\theta^x}{\Lambda^{[x]}}, x = 0, 1, \ldots \text{ and } C_{\theta,\Lambda} = \frac{1}{{}_1F_1(1,\Lambda,\theta)}$$

where ${}_1F_1(1,\Lambda,\theta) = 1 + \frac{\theta}{\Lambda} + \frac{\theta}{\Lambda(\Lambda+1)} + \ldots\ldots, \Lambda > 0, \theta > 0$ and $\Lambda^{[x]} = \Lambda(\Lambda+1)\ldots\ldots(\Lambda + x - 1)$ while the mean and the variance are:

$$E(x) = \theta + (1-\Lambda)\left(1 - \frac{1}{{}_1F_1(1,\Lambda,\theta)}\right), \text{var}(x) = \theta(1 + E(x)) + E(x)(1 - E(x) - \Lambda)$$

Note that when $\Lambda = 1$ we obtain $E(x) = \theta$ and $\text{var}(x) = \theta$. In risk models, such a distribution is useful if events (or data) are assumed to be Poisson distributions but its variance is nor necessarily equal to its mean.

3.5.6 *The Negative Binomial Distribution (NBD)*

The NBD probability distribution, unlike the binomial, counts the number of events that engender k "successes" (i.e. winnings, in a Bernoulli lottery, the number of defaults in a test, the number of periods for k violations, etc.) generated by Bernoulli experiments. The probability distribution is given by:

$$P(x = n + k) = \binom{n+k-1}{k-1} p^n (1-p)^k, \text{n} = 0, 1, 2, \ldots\ldots,$$
$$E(x) = k(1-p)/p, \text{var}(x) = k(1-p)/p^2$$

when k = 1, we then have a geometric probability distribution as indicated earlier $P(x = n) = (1-p)p^{n-1}$. The NBD arises also from a mixture of the Poisson distribution when its mean is randomized by a Gamma probability distribution. In this case, let the Poisson mean distribution be:

$$f(\lambda) = \frac{\lambda^{\alpha-1} e^{-\lambda/\beta}}{\beta^{\alpha}\Gamma(\alpha)}, \ \lambda > 0, \ \alpha > 0, \ \beta > 0$$

then,

$$\int_0^{\infty} p(x|\lambda) f(\lambda) d\lambda = \int_0^{\infty} \frac{\lambda^x e^{\lambda}}{x!} \frac{\lambda^{\alpha-1} e^{-\lambda/\beta}}{\beta^{\alpha}\Gamma(\alpha)} d\lambda = \binom{\alpha+k-1}{\alpha-1} \left(\frac{\beta}{1+\beta}\right)^k \left(\frac{1}{1+\beta}\right)^{\alpha}$$

In general, the moments of a mixture distribution can be calculated as follows:

$$E(x^k) = E_{\lambda}\left(E(x^k|\lambda)\right) \text{ and therefore } \mathrm{var(x)} = E_{\lambda}(\mathrm{var}(x|\lambda)) + \mathrm{var}(E(x|\lambda))$$

3.6 The Normal Probability Distribution

The normal probability distribution is obtained from several vantage points. As noted above, it is obtained as a limit of the binomial in which case:

$$\lim_{n\to\infty} b(r|n,p) = \frac{1}{\sqrt{2\pi\sigma}} \exp\left[-\frac{1}{2}\frac{(x-\mu)^2}{\sigma^2}\right]; x \in (-\infty, +\infty), \mu = np, \sigma^2 = np(1-np)$$

It is also obtained from the Central Limit Theorem that indicates that a sum of random variables tends at the limit to a normal probability distribution. In this sense, this distribution has an attractive pole which we consider subsequently under Pareto Stable distributions, for which the normal distribution is a special case (there are therefore other attractive probabilities). In its standard form, namely applying the transformation,

$$z = \frac{x-\mu}{\sigma}$$

The standard normal probability distribution whose mean is null and its variance 1 is obtained:

$$N(0,1) \sim \frac{1}{\sqrt{2\pi}} \exp\left[-\frac{1}{2}z^2\right]; z \in (-\infty, +\infty)$$

In addition, the normal distribution is symmetric and therefore its third moment is null while its fourth moment (its kurtosis) is 3. This property is often used to verify the applicability of a normal distribution. For example if a standardized data set (such as the rate of returns on the stock market) has an appreciable skewness or a kurtosis (or both) that differ appreciably from 0 and 3, then the normal probability distribution assumption may be harder to justify.

The central limit theorem (CLT) is one of the most important theorems in probability theory. It states that for n iid random variables with a sample mean $\bar{X}_n$, and if the population variance $\sigma^2 > 0$ is finite, then the sampling distribution of $\bar{X}_n$ tends to a normal probability distribution as $n \to \infty$. Explicitly, if $X_1, X_2, \ldots, X_n$ is a sequence of iid random variables, with $E(X_1) = \mu$ and $\mathrm{var}(X_1) = \sigma^2$, then:

$$\operatorname*{Lim}_{n\to\infty} \Pr\left\{\frac{(\bar{X}_n - \mu)\sqrt{n}}{\sigma} \leq z\right\} = \Phi(z)$$

where $\Phi(z)$ is the CDF (cumulative distribution function) of a standard normal probability distribution (i.e. of mean zero and variance 1). It is used extensively, as in many cases its assumptions are reasonable. A logarithmic transformation of the exponential of a normally distributed random variable leads to the lognormal (to be seen below and profusely used in finance) while the inverse Gaussian distribution is a "counting" distribution of normally distributed events, given by:

$$f(x) = \frac{\theta}{\sqrt{2\pi x^3}} \exp\left\{-\frac{\theta}{2x}\left(\frac{x-\mu}{\mu}\right)^2\right\}, x > 0, \theta > 0, \mu > 0$$

Some continuous distributions are summarized in the table 3.4 below.

Table 3.4 Continuous Distribution

Normal	Lognormal	Gamma	Beta
$\frac{\exp\left[-\frac{1}{2}\frac{(x-\mu)^2}{\sigma^2}\right]}{\sqrt{2\pi\sigma}}$	$\frac{1}{x\sqrt{2\pi\sigma^2}} \exp\left[-\frac{\ln^2(x/x_0)}{2\sigma^2}\right]$	$f(x) = \frac{x^{\alpha-1}e^{-x/\beta}}{\beta^\alpha \Gamma(\alpha)}$	$\frac{x^{p-1}(1-x)^{q-1}}{B(p,q)}$
$\mu, \sigma > 0$		$x > 0, \alpha > 0, \beta > 0$	$0 \leq p, q \leq 1$
μ	$x_0 e^{\sigma^2/2}$	$x\beta$	$p/(p+q)$
σ^2	$x_0^2 e^{\sigma^2}\left(e^{\sigma^2} - 1\right)$	$x\beta^2$	$\frac{pq}{p(p+q)^2(p+q+1)}$
Skewness 0	$E(x^3) = x_0^3 e^{9\sigma^2/2}$		$E(x^3) = \frac{\Gamma(\alpha+\beta)\Gamma(\alpha+3)}{\Gamma(\alpha+\beta+3)\Gamma(\alpha)}$
Moments	$E(x^n) = x_0^n e^{n^2\sigma^2/2}$		$E(x^n) = \frac{\Gamma(\alpha+\beta)\Gamma(\alpha+n)}{\Gamma(\alpha+\beta+n)\Gamma(\alpha)}$
MGF : $M_S(u) = e^{u^2/2}$		$M(u) = (1 - u\beta)^{-\alpha}$	$M(u) = 1 + \Sigma_{k=1}^{\infty} \prod_{i=0}^{k-1} \frac{(\alpha+i)}{\alpha+\beta+i} \frac{u^k}{k!}$

3.6.1 *The Lognormal Probability Distribution*

The Lognormal probability distribution arises by a functional transformation of normally distributed random variables. Let $z = \ln(x)$ have a normal probability distribution mean and variance μ, σ^2 then x has a lognormal probability distribution given by:

$$f_{LN}(x|\mu,\sigma^2) = \begin{cases} \frac{1}{x\sigma\sqrt{2\pi}} \exp\left\{-\frac{1}{2\sigma^2}(\ln x - \mu)^2\right\} & 0 < \infty \\ 0 & x \le 0 \end{cases}$$

Its expected mean and variance are: $E(x) = e^{\mu+\sigma^2/2}$ and $\mathrm{var}(x) = e^{2\mu+\sigma^2/2}\left(e^{\sigma^2} - 1\right)$. In addition, it can be shown by elementary calculations that its third and fourth central moments (i.e. about the mean) are:

$$E(x - Ex)^3 = e^{3\mu+3\sigma^2/2}\left(e^{3\sigma^2} - 3e^{\sigma^2} + 2\right)$$

Hence its index of skewness is:

$$\xi = \frac{E(x - Ex)^3}{(\mathrm{var}(x))^{3/2}} = \frac{\left(e^{3\sigma^2} - 3e^{\sigma^2} + 2\right)}{(e^{\sigma^2} - 1)^{3/2}} \quad \text{or } \xi = (\lambda^3 - 3\lambda + 2)(\lambda - 1)^{-3/2}, \lambda = e^{\sigma^2}$$

The index of skewness is therefore independent of the distribution mean. Deriving it with respect to λ the exponential variance, we note that it is increasing.

3.6.2 *The Exponential Distribution*

The exponential distribution relates to both the geometric probability distribution and the Poisson distribution. It is given by:

$$f(x|\mu) = \begin{cases} (1/\mu)e^{-x/\mu} & x \ge 0 \\ 0 & x < 0 \end{cases}$$

Its cumulative distribution is $F(x|\mu) = 1 - e^{-x/\mu}$ if $x \ge 0$ while its mean and variance are respectively μ and μ^2. The exponential distribution occurs in numerous situations. For example, in a Poisson counting process, the time between events has in fact an exponential probability distribution. The exponential distribution is also memory-less, with all events statistically independent and therefore its usefulness in memory base models is of limited value. Other distributions (such as the Weibull distribution to be defined later on) have as particular cases the exponential distribution.

A mixture of the exponential utility by a normally distributed random variable is recurring often in risk related problems.

3.6.3 *The Gamma Probability Distribution*

The Gamma probability distribution is a two parameters probability distribution defined as follows:

$$f(x|\alpha,\beta) = \begin{cases} \frac{x^{\alpha-1}e^{-x/\beta}}{\beta^{\alpha}\Gamma(\alpha)} & \text{if } x \geq 0 \\ 0 & \text{if x} < 0 \end{cases}, \quad E(x) = \alpha\beta, \mathrm{var}(x) = \alpha\beta^2$$

where the Gamma function $\Gamma(\alpha)$ is given by:

$$\Gamma(\alpha) = \int_0^{\infty} x^{\alpha-1}e^{-x}dx, \quad \alpha > 0$$

The Gamma function satisfies the relationship $\Gamma(\alpha) = (\alpha - 1)\Gamma(\alpha - 1)$. This distribution is used frequently in reliability analysis and in many other situations. Its moment generating function is given by: $M(u) = \frac{1}{(1-u\beta)^{\alpha}}$ and $\ln M(u) = -\alpha \ln(1 - u\beta)$ and therefore:

$$\frac{\partial \ln M(u)}{\partial u} = \frac{\partial M(u)/\partial u}{M(u)}\bigg|_{u=0}. = \frac{\partial M(u)/\partial u}{1} = \frac{\alpha\beta}{1 - u\beta}\bigg|_{u=0}. = \alpha\beta$$

and therefore (as we will be able to prove subsequently), when $\alpha = n$ is an integer, it results from summing (i.e. the convolution) n random variables all of which have an exponential probability distribution with parameter β.

3.6.4 *The Beta Probability Distribution*

The Beta probability distribution belongs to a large family of bounded distributions. For simplicity, let $x \in [0, 1]$ be this interval. The Beta probability distribution is then defined as follows:

$$f(x) = \frac{1}{B(p,q)}x^{p-1}(1-x)^{q-1}, \; B(p,q) = \frac{\Gamma(p)\Gamma(q)}{\Gamma(p+q)} \text{ and } 0 \leq x \leq 1$$

where p and q are two parameters defining the distribution mean and variance, and $\Gamma(.)$ is a gamma function,

$$E(x) = \frac{p}{p+q}, \mathrm{var}(x) = \frac{pq}{p(p+q)^2(p+q+1)}$$

For example, x, may be a probability of a risk occurring, a proportion of patients that have recovered following a specific test procedure, and their like. Its parameters define a large number of potential results. In addition, it has a number of interesting analytical properties that render it of particular interest. For example, instead of the distribution of x, consider its odds, namely,

$$y = \frac{x}{1-x}$$

Then if x has a Beta probability distribution as stated above, the probability of y is also Beta type distribution called the Beta Prime distribution. Such a distribution belongs to a family of distribution called Pearson type VI and is given by:

$$h(y) = \frac{1}{B(p,q)} y^{p-1}(1+y)^{p+q}$$

In other words, if one were to define engineer a system whose characteristic if statistically modeled by its odds, then using the Beta Prime distribution would imply as well that its underlying distribution is the Beta Probability Distribution.

3.6.5 *Binomial Default with Learning*

Say that at time t=0, the probability of a default is p. Let the prior probability distribution of p be a Beta probability distribution with parameters β, γ or $p \sim B(\beta,\gamma)$ with a mean $p_0 = E(p) = \frac{\beta}{\beta+\gamma}$. An instant of time later say that the event occurs, in which case, we revise the probability that such an event occurs. Applying Bayesian learning, the new probability mean estimate will be:

$$E(p_{1,1}) = \frac{\beta+1}{\beta+\gamma+2}$$

Similarly, if the event does not occur, then:

$$E(p_{0,1}) = \frac{\beta}{\beta+\gamma+1}$$

Generally, if after n periods, m events have occurred and therefore n-m times did not occur, the posterior probability of has a binomial probability distribution with parameters:

$$p_{m,n} \sim B(\beta_{m,n}, \gamma_{m,n}), \beta_{m,n} = \beta + m,\ \gamma_{m,n} = \gamma + n$$

with $$E(p_{m,n}) = \frac{\beta+m}{\beta+\gamma+m+n}$$

3.6.6 *The Logistic Distribution*

Let p be the probability of an event. For example, the probability of a default, the probability that a loan is not paid, etc. Define by (1-p)/p the odds of such an event.

Assume that the ln(odds) of such experiments can be regressed on a number of explanatory variables, namely:

$$\ln\left(\frac{p}{1-p}\right) = \sum_{i=1}^{n} w_i x_i + \varepsilon_n$$

These regressions imply that p is a random variable with a Logistic Probability Distribution.

$$\frac{p}{1-p} = e^{\sum_{i=1}^{n} w_i x_i + \varepsilon_n} \text{ and therefore } p = \frac{e^{\sum_{i=1}^{n} w_i x_i + \varepsilon_n}}{1 + e^{\sum_{i=1}^{n} w_i x_i + \varepsilon_n}} \text{ and } 1 - p = \frac{1}{1 + e^{\sum_{i=1}^{n} w_i x_i + \varepsilon_n}}$$

3.6.7 *The Linear Exponential Family of Distribution*

The linear exponential family of distributions generalizes a number of probability distributions. It is written as follows:

$$f(x) = \frac{p(x)e^{r(\lambda)x}}{q(\lambda)}$$

For example, if $r(\lambda) = -1/\lambda, q(\lambda) = \lambda^{\alpha}$ and $p(x) = x^{-(1-\alpha)}/\Gamma(\alpha)$, we obtain then a Gamma Probability distribution.

3.7 Extreme Distributions and Tail Risks

Extreme probability distributions are particularly important as they emphasize the distribution of potentially significant (and thereby risky events when these are losses or risk consequences). Let a system distribution be given by $F(x)$. Consider next a sample of n tests, $\{S_1, S_2, \ldots, S_n\}$, assumed to be identically and independently distributed (a strong assumption in practical risk problems), expressing the largest of all losses (say in a number of Hurricanes in a season), let its largest observed value be: $Y_n = S_{(n)} = \text{Max}\{S_1, S_2, \ldots, S_n\}$ and let $S_{(j)}$ be the j_{th} order statistic, with $S_{(1)} < S_{(2)} < S_{(3)} < \ldots.. < S_{(n)}$ over n events. Let $F_j(S),\ j = 1, 2, \ldots n$ denote the density function of $S_{(j)}$. The distribution function of the maximum $Y_n = S_{(n)}$ is given by:

$$F_n(S) = P\{S_{(n)} < S\} = P\{all\ \ S_i < S\} = [F(S)]^n$$

Similarly for the minimum statistic: $Y_1 = S_{(1)} = \min\{S_1, S_2, ..., S_n\}$, we have,

$$F_1(S) = P\{S_{(1)} < S\} = 1 - \Pr\{\ S_{(1)} > S\} = 1 - \Pr\{\text{ all } S_i > S\}$$
$$= 1 - [1 - F(S)]^n$$

Generally, for the jth statistic,

$$F_{(j)}(S) = \sum_{j=i}^{n} \binom{n}{i} [F(S)]^i [1 - F(S)]^{n-i}$$

Therefore, by deriving with respect to S, we have the distribution:

$$f_j(S) = \frac{n!}{(j-1)!(n-j)!} f(S)[F(S)]^{j-1}[1 - F(S)]^{n-j}$$

In particular for the maximum and the minimum statistics, we have the following marginal probability distributions: $f_n(S) = nf(S)[F(S)]^{n-1}$ and $f_1(S) = nf(S)[1 - F(S)]^{n-1}$

Recording of the High and the Low provide two data samples that can be used to estimate the distribution. Since these statistics are taken from the same sample, we have a joint bi-variate distribution given by (see Tapiero 2010a).

$$f_{kj}(S_1, S_2) = \frac{n!}{(j-1)!(n-k)!(n-j)!} f(S_1)f(S_2)[F(S_1)]^{j-1}[F(S_2) - F(S_1)]^{k-j-1}$$
$$[1 - F(S_2)]^{n-k} - \infty < S_1 < S_2 < \infty$$

and therefore, for the joint distribution of the High and the Low, we have k = n and j = 1:

$$f_{12}(S_1, S_2) = nf(S_1)f(S_2)[F(S_2) - F(S_1)]^{n-2}\ [1 - F(S_2)], -\infty < S_1 < S_2 < \infty$$

The conditional probability of the high given a low for example is given by: $f_{2:1}(S_2|S_1) = f(S_1, S_2)/f_1(S_1)$ while the first two moments of the range are:

$$E(S_2 - S_1)^m = \int_{S_2 > S_1}^{\infty} \int_{-\infty}^{S_2} (S_2 - S_1)^m f_{n2}(S_1, S_2) dS_1 dS_2$$
$$= n \int_{S_2 > S_1}^{\infty} \int_{-\infty}^{S_2} (S_2 - S_1)^m f(S_1) f(S_2)[F(S_2) - F(S_1)]^{n-2}$$
$$[1 - F(S_2)] dS_1 dS_2; \quad m = 1, 2$$

These results depend on the statistical independence of the sampled data. For some distributions, explicit and analytical results can be obtained. Otherwise, it is necessary to turn to numerical and simulation techniques. For large samples, limit distributions for the High and Low order statistics can be defined.

Assume large samples with High (and Low) statistics with probability distributions $F_{Y_n}(R) = [F_R(R)]^n$. At the limit, when n is large, this distribution converges to a family of extreme distributions. These distributions are given by three families:

Gumbel: $F_{1,Y}(y) = \exp(-e^{-y}),\ y \in IR$: The Gumbel Distribution

$$\text{Weibull: } F_{2,Y}(y) = \begin{cases} \exp(-(-y)^{-\alpha}) & \text{for } y < 0 \quad (\alpha < 0) \\ 0 & \text{for } y \geq 0 \end{cases} \quad \text{: The Weibull Distribution}$$

$$\text{Frechet: } F_{3,Y}(y) = \begin{cases} 0 & \text{for } y \leq 0 \\ \exp(-y^{-\alpha}) & \text{for } y > 0 \ (\alpha > 0) \end{cases} \quad \text{The Frechet Distribution}$$

These three distributions can be represented as a single distribution with the following parameters:

$$F_\xi(y) = \exp\left\{-\left(1 + \frac{\xi y}{\sigma}\right)\right\}^{-1/\xi} \quad \text{with } \sigma > 0, \{\xi > 0, \xi < 0, \xi \to 0\}$$

3.7.1 *Approximation by a Generalized Pareto Distribution*

The family of extreme distributions considered above can be approximated by a Generalized Pareto Distribution (GPD) which has "fat tails" (as we shall see subsequently). Such an approximation is particularly useful to express the tail risk of extreme distributions by the GPD tail. Namely, we have then:

$$1 - F_{3\alpha} = 1 - x^{-\alpha}, x \geq 1,\ 1 - F_{2\alpha} = 1 - (-x)^{-\alpha}, -1 \leq x \leq 0,$$
$$1 - F_{1\alpha} = 1 - e^{-\alpha}, x \geq 0$$

and generally,

$$1 - F_\xi = 1 - \left(1 + \frac{\xi x}{\sigma}\right)^{-1/\xi}, x \geq 0$$

3.7.2 *The Weibull Distribution*

The Weibull distribution is a two parameters extreme distribution, given above as follows:

$$f(x|a,\beta) = a\beta x^{a-1}e^{-\beta x^a},\ x \geq 0,\ \beta,\ a > 0;\ F(x) = 1 - e^{-\beta x^a}$$

with a mean, variance and a hazard rate:

$$E(x) = \beta^a \Gamma\left(\frac{1}{a}+1\right),\ \text{var}(x) = \beta^{2a}\left[\Gamma\left(\frac{2}{a}+1\right) - \Gamma^2\left(\frac{1}{a}+1\right)\right],$$
$$h(x) = a\beta x^{a-1}, \beta > 0$$

The hazard rate in particular has both increasing and decreasing risk properties, since:

$$\frac{\partial h(x)}{\partial x} = \begin{cases} > 0 & \text{If } a > 1 \\ < 0 & \text{If } 0 < 1 \end{cases}$$

Note that when $a = 0$, this distribution reduces to the exponential probability distribution. These properties are used to model equipment failure as a function of age, patients' illnesses as a function of age, etc.

Explicitly, consider a system reliability (life time greater than a parameter a, for sure) modeled by:

$$f(t) = c\mu((t-a)\mu)^c \exp\{-((t-a)\mu)^c\}, t \text{ and } F(t) = 1 - e^{-((t-a)\mu)^c}$$

The reliability is thus, $R(t) = 1 - F(t) = e^{-((t-a)\mu)^c}$ while the hazard rate is: $h(t) = c\mu((t-a)\mu)^{c-1}$. Of course, if the parameter is random with a probability distribution $g(\mu)$ the hazard rate is given by: $E_\mu h(t|\mu)$

3.7.3 *The Burr Distribution*

The Burr probability distribution can result from a mixture of the Weibull distribution and therefore requires more parameters to specify or estimate than the Weibull. However, the Burr distribution includes many and standard distributions as special cases. Let the Weibull probability distribution be:

$$f(\tau|\beta) = \alpha\beta\tau^{\alpha-1}e^{-\beta\tau^\alpha}, \tau \geq 0, \alpha, \beta > 0$$

and say that β has a Gamma probability distribution, $g(\beta) = \delta^\nu \beta^{\nu-1}e^{-\delta\beta}, \beta > 0,\ \delta,\ \nu > 0$. Then, the resulting mixture is the Burr distribution:

$$f(\tau) = \int_0^\infty f(\tau/\beta)g(\beta)d\beta = \frac{\alpha\nu\delta^\nu\tau^{\alpha-1}}{(\tau^\alpha+\delta)^{\delta+1}}, \tau \geq 0, \alpha, \delta, \nu > 0$$

For such a distribution the cumulative distribution function (and therefore the reliability of events modeled by such a distribution) is analytical and given by:

$$\Re_a(T) = \text{Prob}(\tau \geq T) = 1 - \left(1 + \frac{\tau^{\alpha}}{\delta}\right)^{-\nu}$$

Example: The Weibull and the Burr Distributions Approximations The two parameters (c, ζ) Weibull probability Distribution and its cumulative are given by:

$$f(\tau) = \frac{c}{\zeta}\left(\frac{\tau}{\zeta}\right)^{c-1} e^{-(\tau/\zeta)^c}, \ \tau \geq 0, \ \zeta, \ c > 0; \ F(\tau) = 1 - e^{-(\tau/\zeta)^c},$$

while the mean and the variance are:

$$E(\tau) = \zeta\Gamma\left(\frac{1}{c} + 1\right), \ \text{var}(\tau) = \zeta^2\left[\Gamma\left(\frac{2}{c} + 1\right) - \Gamma^2\left(\frac{1}{c} + 1\right)\right]$$

Now assume that an implied call option price is defined parametrically by the Weibull distribution. Namely,

$$\hat{C}_{it} = e^{-R_f(T-t)} \int_{K_i}^{\infty} (S_T - K_i)\hat{f}_t(S_T)dS_T$$

$$= e^{-R_f(T-t)} c \int_{K_i}^{\infty} \left(\frac{S_T}{\zeta}\right)^c e^{-(S_T/\zeta)^c} dS_T - e^{-R_f(T-t)} K_i e^{-(K_i/\zeta)^c}$$

If we set:

$$\left(\frac{S_T}{\zeta}\right)^c = u, \ c\left(\frac{1}{\zeta}\right)\left(\frac{S_T}{\zeta}\right)^{c-1} dS_T = du \text{ then,}$$

$$\hat{C}_{it} = e^{-R_f(T-t)} c \int_{\left(\frac{K_i}{\zeta}\right)^c}^{\infty} ue^{-u} dS_T - e^{-R_f(T-t)} K_i e^{-(K_i/\zeta)^c}$$

For the PUT option, we have similarly:

$$\hat{P}_{it} = e^{-R_f(T-t)} \int_0^{K_i} (K_i - S_T)\hat{f}_t(S_T)dS_T = e^{-R_f(T-t)} K_i\left[1 - e^{-(K_i/\zeta)^c}\right]$$

$$- e^{-R_f(T-t)} \zeta \int_0^{\left(\frac{K_i}{\zeta}\right)^c} u^{\frac{1}{c}} e^{-u} du,$$

Note that:

$$\int_0^{u_i} u^{\frac{1}{c}} e^{-u} du = \gamma\left(1+\frac{1}{c}, u_i\right) \text{ and } \int_0^{\left(\frac{K_i}{\zeta}\right)^c} u^{\frac{1}{c}} e^{-u} du = \gamma\left(1+\frac{1}{c}, \left(\frac{K_i}{\zeta}\right)^c\right)$$

and therefore:

$$\hat{P}_{it} = e^{-R_f(T-t)} K_i\left[1 - e^{-(K_i/\zeta)^c}\right] - e^{-R_f(T-t)} \zeta\gamma\left(1+\frac{1}{c}, \left(\frac{K_i}{\zeta}\right)^c\right)$$

The underlying price is then

$$\hat{S}_t = e^{-R_f(T-t)} E_{RND,W}(S_T) \text{ and } \hat{S}_t = e^{-R_f(T-t)} \zeta\Gamma\left(\frac{1}{c}+1\right)$$

Note that with the Weibull distribution on hand, the hazard rate is:

$$h(T) = \frac{c}{\zeta}\left(\frac{T}{\zeta}\right)^{c-1} \text{ and therefore, } \frac{c}{\zeta}\left(\frac{T}{\zeta}\right)^{c-1} = -\frac{\frac{\partial^2 C_t}{\partial K^2}(S_t, t)}{\frac{\partial C_t}{\partial K}(S_t, t)}$$

Several authors like to use the Burr distribution because it includes as special cases numerous and well known distributions. In his case, we have:

$$F(S_T, T) = 1 - \frac{1}{(1+{S_T}^c)^q}, S_T \geq 0, c, q > 1; \quad f(S_T, T) = \frac{qc{S_T}^{c-1}}{(1+{S_T}^c)^{q+1}}$$

or using the following notation:

$$f(S_T) = \frac{\alpha\upsilon S_T^{\alpha-1}}{(S_T^\alpha + \delta)^{\delta+1}}, \quad E({S_T}^m) = \frac{1}{\alpha}\delta^{(m+\alpha)/\alpha} B\left(1+\frac{m}{\alpha}, \delta - \frac{m}{\alpha}\right)$$

which can be used to fit an available data set to the distribution and optimize to obtain parameters estimates. Similarly for a Burr III distribution we have:

$$F_{BR}(S_T, T) = \left[1 - \frac{1}{(1+(S_T/\beta)^c)}\right]^\alpha, S_T \geq 0, c > 0, \alpha > 0, \beta > 0$$

while the probability distribution is:

$$f(S_T, T) = \frac{c\alpha {S_T}^{c\alpha-1}({S_T}^c + \beta^c) - c\alpha {S_T}^{c(\alpha-1)}}{({S_T}^c + \beta^c)^{1+\alpha}}$$

Thus by simply minimizing the sum of squared differences between a model premia conditional upon the parameters of the distribution and the observed option premia, an estimate of the approximate risk neutral distribution can be obtained (for references see Rodriguez 1977 and Tadikamalla 1980). The problem with this technique however, is that it is mostly appropriate for an estimation of "a specific risk neutral distribution" for a specific instant of time rather than to estimate the evolution of the risk neutral distribution over a stochastic process.

3.8 Simulation

When models are overly complex and resist analytical or a numerical treatment, an appropriate alternative consists in simulating the model. Simulation is an experimental approach based on the algorithmic generation of random numbers that are models to simulate specific probability distributions. Such "experimentation" allows one to answer "what if" questions pertaining to model parameters and assumptions the model may assume.

Simulation is the oldest analysis tool known to man—that is, the representation of the real world by numbers and other symbols that can be readily manipulated. To gain a better grasp of the real world, games such as chess to simulate battles, backgammon to simulate racing, and other games to simulate hunting and diplomacy were already invented in antiquity. Today, a modern game like monopoly simulates complex engineering designs, flights to the stars, the birth and death of stars, financial markets, etc. The modern use of simulation can be traced to Von Neumann and to Ulam in the late 1940s when applying it to the solution of nuclear shielding problems that were either too expensive for physical experimentation, or too complicated for treatment by known techniques. The application of simulation to risk related problems is immense reducing its randomness to specific events using random numbers generators (Monte Carlo Techniques) and manipulated to replicate the underlying and particular characteristics that the model is defined by. Figure 3.5 summarizes such an application to assess the default loss distribution of a portfolio. Assumptions regarding the distribution of losses, their simultaneous

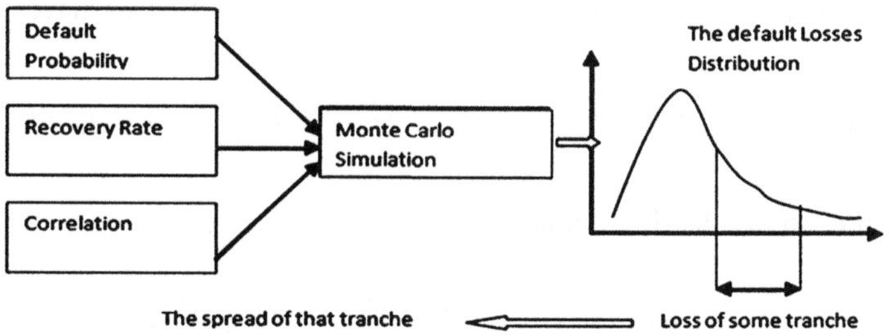

Fig. 3.5 The principles of Monte Carlo simulation

occurrence, their recovery rates, their sharing etc. Using Monte Carlo simulation, the portfolio aggregate loss probability distribution can be generated and used for risk analysis and management to evaluate losses and other calculations.

Example: The Game of Craps The game of craps consists of the following. Two fair dice are rolled:

1. If their sum is either 7 or 11, then you win;
2. If their sum is 2, 3 or 12, then you lose
3. For any other result, say x, the pair of dice are rolled again and again until either x or 7 results. You then win with an x, but lose with a 7.

Using Monte Carlo simulation we can repeat a certain number of games, say N, and count the number of wins, say WIN. The probability of winning is then given by the statistical "estimate" (No. ofWIN)/N. The term "estimate" is used since repeating each game is an experiment.

To simulate the game, we have to generate on the computer the role of a pair of dices—the game randomness generator. To simulate the roll of one die, define the probability of each of its potential outcome 1, 2, 3, 4, 5 and 6, all of which have equal probability. To generate a single roll, random numbers between "0" and "1" (or any 0 an 100, 0 or 1,000 etc.) and a table affecting any such number (obtained by a random number generator) to a specific event. In our case, the probability of any outcome is $1/6 = 0.166$ or 0.167, ensuring that the sum over all 6 outcomes equals 1. An allocation rule for this particular case is given by the following Table 3.5 based random number between 0 and 999:

Table 3.5 assigns to any random number 000-166 to the first outcome 1, 167-332 to the second outcome 2, and so on. For example, the number 062 assigns a roll of the die (outcome) 1, 035 assigns again the outcome 1, 221 assigns the outcome 2, and so on. As a result, of this assignment process, we end up with a sequence of simulated outcomes 1,1,2,4,4,... that are used in conjunction with all other (non-random) elements the model is constructed from. The generation of such numbers provides us with what we might expect to get if we were to actually roll the die. Evidently, a different sequence would invariably result if we were to take another batch of random numbers. But this in no way invalidates the fact that we have a representative set of outcomes which we can use in the simulation of our game.

Table 3.5 The allocation of probabilities to the role of a dice

Roll	Probability	Cum. Prob.	R.N Interval
1	0.167	0.167	000–166
2	0.167	0.334	167–333
3	0.167	0.500	334–499
4	0.167	0.667	500–666
5	0.167	0.834	667–833
6	0.167	1.000	834–999

Table 3.6 The simultaneous roll of a pair of dice

(a) The sample outcomes

Value	Roll outcomes	No.	Probabilities	Cumulatives
2	(1,1)	1	0.028	0.028
3	(2,1),(1,2)	2	0.055	0.083
4	(3,1),(2,2),(1,3)	3	0.084	0.167
5	(4,1),(3,2),(2,3),(1,4)	4	0.110	0.277
6	(5,1),(4,2),(3,3),(2,4),(1,5)	5	0.139	0.416
7	(6,1),(5,2),(4,3),(3,4),(2,5),(1,6)	6	0.168	0.584
8	(6,2),(5,3),(4,4),(3,5),(2,6)	5	0.139	0.723
9	(6,3),(5,4),(4,5),(3,6)	4	0.110	0.833
10	(6,4),(5,5),(4,6)	3	0.084	0.917
11	(6,5),(5,6)	2	0.055	0.972
12	(6,6)	1	0.028	1.000

(b) Assigning random numbers

Value	Probability	Cumulative Prob.	R.N. Assignment
2	0.028	0.028	000–027
3	0.055	0.083	028–082
4	0.084	0.167	083–166
5	0.110	0.277	167–276
6	0.139	0.416	277–415
7	0.168	0.584	416–583
8	0.139	0.723	584–722
9	0.110	0.833	723–832
10	0.084	0.917	833–916
11	0.055	0.972	917–971
12	0.028	1.000	972–999

In Table 3.6 above we summarize an alternative allocation of probabilities to the simultaneous throw of a pair of die. For example, to obtain a three, we can obtain a 1 and 2 or a 2 and 1 in a simultaneous throw. As whose probability is 0.055, etc. Such an allocation would require for the simultaneous throw of two die only one (pseudo) random number instead of two (if each dice would be considered separately. This reduces of course the amount of computational time to simulate the simple craps game. While such considerations are not particularly important when simulating simple risk models, it becomes important when models are both large and extremely complex.

Monte Carlo simulation underlies simulation languages and numerous software used for risk analysis.

References and Additional Reading

Additional probability reference sources include both fundamental books, such as Hacking Ian (2006) book on the Emergence of Probability (as well as a series of books on the history and the philosophies of probability), Feller's 2 volumes (1971),

Johnson and Kotz (1969, 1970a) and (1970b) as well as numerous reference books on probability distributions and their properties. In addition, reference books such Abramowitz and Stegun (1965), Gradstein and Ryzhik, (1965) provide an extensive tabulation of some distribution and mathematical functions and integrals which are used to define complex risk probability models, Peizer and Pratt (Part I)as well as Pratt (part II) for normal approximations for binomial, F, Beta and other common, related tail probabilities, Rodriguez (1977) for a guide to the Burr Type XII distributions.

Applications books such as Barlow and Proschan (1965), Bolland and Proschan (1994), and Shaked and Shantikumar (1994) on Reliability, Barrois (1834) on Actuarial statistics, Gerber (19790, on Insurance and Actuarial problems, Kullback (1959) (on Information Statistics and Entropy Functions) are a sample of books in addition to the multitude of references in all professions that are based on elements of probability and their applications.

For example, the following references consulted in various fields, based on the use of probability moments included books and papers by Corrado and Su (1996), Daykin et al. (1994), Denuit (2001), Everitt and Hand (1981), Tadikamalla (1980), Wald and Wolfowitz (1940), Wiggins (1992), Willasen (1981), Yang and Zhang (2000), Brandt and Diebold (2006), on a no arbitrage approach to range based estimation of return covariances and correlations.

Chapter 4
Multivariate Probability Distributions: Applications and Risk Models

Abstract Multiple, simultaneous and dependent sources of risks are common to risk modeling. To model their manifestations, we use multivariate probability distributions and models to express their dependence and their interactions. The purpose of this chapter is to summarize a number of approaches to multivariate probability modeling and measurement of dependence. These include statistical and functional models, Bayesian techniques, families of multivariate probability distributions and copula. Both short term and long-run memory and fractal models are relegated to Chap. 5.

4.1 Introduction

Multivariate probability distributions are used to model multiple sources of risks and their co-dependence. A number of approaches include:

- *Statistical functional dependence*. It is used to model the relationships between a given set of random variables linked as testable models. Such models include multivariate probability distributions as well as stochastic processes when these are inter-temporal models.
- *Conditional* (*latent and common risk*) *models* used to express the dependence of one variable or a set of variable on others which we can express through application of Bayesian calculus or through randomization of a probability distribution parameters.
- *Functional and algebraic modeling* (*Copulas*) of multivariate probability distributions using their marginal distributions.
- *Short-term memory models*. These are "bifurcating probability models" (also called persistent models) changing future probability events as a function of a preceding event realization. For example, if a bank defaults, would such an event change the probability that another bank will default? Does a specific event affect the probability of such an event recurring? (see Chap. 5)

C.S. Tapiero, *Engineering Risk and Finance*, International Series in Operations Research & Management Science 188, DOI 10.1007/978-1-4614-6234-7_4,

- *Long-run memory and fractal models*. These models account for a nonlinear time variance evolution. These models are based on empirical observations that the linear time growth hypothesis of variance growth in normal distributions is too limiting (see Chap. 5).

Our purpose in this chapter is to elaborate on the rationale of some multivariate modeling approaches and outline some essential approaches to multivariate risk modeling.

4.2 Measures of Co-variation and Dependence

The concept of dependence permeates "uncertainty and thereby risk models" in both latent and explicit manners. Dependence underlies the occurrence of risk events through causal or explicit statistical co-variations. Statistical dependence is mostly modeled as a co-variation between two random variables and parametrized by a linear correlation. It may be defined by observing that paired comparisons tend (more or less) to increase or decrease "together"; It expresses a reaction to common (or latent) risk factors as well as be conditional on the outcomes of past events (short or long memory). In this section we shall elaborate elements of dependence and emphasize their importance to risk measurement and risk exposure. Risk management in this sense has to account explicitly for the effects of dependence.

4.2.1 Statistical and Causal Dependence: An Oil Example

The price of oil affects the price one pays for gasoline at the pump as well as its many derived uses. When its market price increases, the price of "oil derivatives" (such as plane tickets, transportation costs etc.) may increase as well. For example, let $\tilde{U}(T)$ be the random price of oil and let $\tilde{V}(T|U(T))$ be the conditional price of a Gallon at the pump conditional to a specific value $U(T)$. The probability of this price is then given by the conditional probability distribution $f_{V|U}(\tilde{V}(T)|U(T))$. If $\tilde{U}(T)$ is a random variable with probability distribution $f_U(\tilde{U}(T))$, the joint (and bi-variate) probability distribution is:

$$f_{V,U}(\tilde{V}(T),\tilde{U}(T)) = f_{V|U}(\tilde{V}(T)|\tilde{U}(T))f_U(\tilde{U}(T))$$

If these prices are independent, then: $f_{V,U}(\tilde{V}(T),\tilde{U}(T)) = f_V(\tilde{V}(T))f_U(\tilde{U}(T))$ and therefore,

$$\frac{f_{V,U}(\tilde{V}(T),\tilde{U}(T))}{f_V(\tilde{V}(T))f_U(\tilde{U}(T))} \neq 1 \quad \text{if dependent}$$
$$\frac{f_{V,U}(\tilde{V}(T),\tilde{U}(T))}{f_V(\tilde{V}(T))f_U(\tilde{U}(T))} = 1 \quad \text{if independent}$$

In practice multivariate relationships are difficult to model and estimate. Petroleum products are the "lifeblood" of the economy. Further, there are few competing supply sources that the US can rely on as their risks (Nuclear), environmental impact (Coal) and costs (Sun) render their uses relatively marginal. A shortage of oil can therefore have dire economic and predictable direct and indirect consequences on its many users. If the US could increase appreciably the amount of oil or its substitutes (such as Gas, more nuclear power), or would develop and implement technologies that can reduce oil consumption, the price of oil will decline and may also have important effects on the economy as a whole since so many elements of all economic activities depends on the price of oil. For example, say that only one product depends directly on the price of oil and let there be a large number of other products that are affected by the price–oil correlation. Such a situation leads then to a price "contagion" due to a common dependence on the price of oil.

Causal dependence is due to either the effects of one causal factor onto another (whether due to a rare event or not such as the Tsunami in Japan in 2011) or through co-variations (for example, their co-variation effects on global supply chains, stock markets etc.) can lead to catastrophic risk events. A financial market index with assets that are "price co-dependent" may lead as well to a very sharp drop in the index. For example, if the DOW market index is based on 300 companies but with 3 Hi tech firms (say Apple, Google and Microsoft) being by far the largest part of the index price, then it is sufficient for a new technology or a new set of regulations to cause to a sudden drop in the internet economic sector and to an appreciable drop in the DOW index. For example, consider the random variables, $\tilde{x}_1(t)$ and $\tilde{x}_2(t)$ with means and variances $\left(\hat{x}_i(t), \sigma_i^2(t)\right)$, $i = 1, 2$ and each with its (uni-variate) probability distributions $f_i(\tilde{x}_i(t))$, $i = 1, 2$. Say that they are dependent. How can we express their dependence? Namely, how can we characterize the joint probability distribution that defines the variables associations based on some of their moments or their marginal distributions? Such a question raises challenging modeling and estimation problems. A statistical co-variation is not causal, as it expresses a statistical co-movement (and therefore it is quantitatively symmetric) while a causal-co-movement expresses an "asymmetric" co-variation, with one movement depending on the other but not its inverse. Some of the approaches used include defining parameters of co-dependence as well as seeking families of joint distributions are varied, based on their joint outcomes and the identification of causal factors. Below we consider some explicit examples.

4.2.2 *Statistical Measures of Co-dependence*

A number of statistical measures of dependence include:

1. Pearson's correlation coefficient (or linear correlation)
2. Kendall's τ
3. Spearman's rho
4. Medial correlation coefficient.

The Linear correlation: is the familiar correlation of two random variables defined below and expresses the statistical observation that two random variable may move in probability in the same directions (a positive correlation) or in counter-directions (a negative correlation):

$$\rho\sigma_X\sigma_Y = E((x - E(x))(y - E(y)))$$

For example, consider the price of two currencies. Does a currency decline or increase because of the rating of its Treasury bills (Sovereign bonds) or due to the price of its (indexed) financial market? Does the health of a patient improve because of exercises? Is diet without exercising or both diet and exercises, needed to maintain one's health and how does diet, exercise and a health index each relate to each other? Does a car breakdown because of insufficient maintenance? Does a car fail because of driving conditions? These questions and many others imply that risks arising from multiple factors may occur individually or simultaneously due to their dependence. The number of possibilities to consider (where there are many risk factors) is generally extremely large. For example, for n risk factors, there may be $\binom{n}{2} = \frac{n!}{(n-2)!2!}$ co-variations and even more tertiary and higher order variations. For two (n = 2) factors we have a single parameter which is the variable co-variation. It is defined as follows: $\tilde{x}_1$ and $\tilde{x}_2$ are correlated if expectation of their product is not null, or:

$$\begin{cases} E(\tilde{x}_1\tilde{x}_2) = 0 & \text{Uncorrelated} \\ E(\tilde{x}_1\tilde{x}_2) \neq 0 & \text{Correlated} \end{cases}$$

If each of these variables has a known mean and a known variance, then by definition of the covariance,

$$E(\tilde{x}_1\tilde{x}_2) = \hat{x}_1\hat{x}_2 + \rho\sigma_1\sigma_2$$

where $\rho \in [-1, +1]$ defines an index of correlation. If $\rho = (0 >, < 0)$ a correlation index indicates respectively independence, growing and decreasing variance due to co-dependence. In this case, not that since $E((x - E(x))(y - E(y))) = \rho\sigma_X\sigma_Y$ or $E(xy) = E(x)E(y) + \text{cov}(x, y)$. As a result, if y is constant then $\text{cov}(x, y) = 0$ and generally, the correlation of linear combinations of random variables yield:

$$\text{cov}\left(\sum_{i=1}^{m} a_i x_i, \sum_{j=1}^{n} b_j y_j\right) = \sum_{i=1}^{m}\sum_{j=1}^{n} a_i b_j \text{cov}(x_i, y_j)$$

Correlation is thus an indicator of dependence, although two variables can be dependent and yet have no correlation. This is the case as co-variation can be measured in different ways as we shall see subsequently.

Example: A Market Index and Co-variation A market index consists of a selected set of n stocks each of which has a relative weight. The stock index is then the weighted sum:

$$\tilde{S}_{Index}(t) = \sum_{i=1}^{n} w_i \tilde{S}_i(t)$$

where w_i are weights associated to the price of a commodity, a firm, a region economic activity i whose price at time t is a random variable $\tilde{S}_i(t)$. The index rate of return is then:

$$\tilde{R}_{Index}(t) = \frac{\Delta\tilde{S}_{Index}(t)}{\tilde{S}_{Index}(t)} = \sum_{i=1}^{n} w_i \frac{\Delta\tilde{S}_i(t)}{\tilde{S}_i(t)} \quad \text{and} \quad \tilde{R}_{Index}(t) = \sum_{i=1}^{n} w_i \tilde{R}_i(t)$$

And therefore, the mean rate of return and the returns variance of the weighted aggregate index are:

$$E\left(\tilde{R}_{Index}(t)\right) = \sum_{i=1}^{n} w_i E\left(\tilde{R}_i(t)\right) \quad \text{var}\left(\tilde{R}_{Index}(t)\right) = \text{var}\left(\sum_{i=1}^{n} w_i \tilde{R}_i(t)\right)$$

If individual risk factors are correlated, say with correlation ρ_{ij}, defining their statistical dependence. The index variance is:

$$\text{var}\left(\tilde{R}_{Index}(t)\right) = \sum_{i=1}^{n} w_i^2\, var\left(\tilde{R}_i(t)\right) + 2\sum_{i=1}^{n}\sum_{j\neq 1}^{n} w_i w_j \rho_{ij} \sqrt{\text{var}\,\tilde{R}_i(t)}\sqrt{\text{var}\,\tilde{R}_j(t)}$$

If all returns are positively correlated, then holding on to an index would seem to be more risky than constructing one's own index, better suited to one's needs. However, if some returns are negatively correlated, the variance will be smaller and therefore the index will be considered "less risky". These simple and obvious observations underlie the approach to risk diversification that seek to reduce the variance by constructing an index from returns that are on the whole profitable (in expectation) and have a smaller variance than other investment alternatives. In practice, however, an index may be subject to common (macroeconomic) risk factors, say ξ with probability distribution $f_\xi(\xi)$. In other words,

$$\tilde{R}_{index}(t|\xi) = \sum_{i=1}^{n} w_i \tilde{R}_i(t|\xi) \quad \text{and} \quad \tilde{R}_{index}(t) = \sum_{i=1}^{n} w_i \int_{x \in \Psi} \tilde{R}_i(t|x) f_\xi(x) dx$$

In this case, all stocks are co-dependent, both due to their co-variation and due to their common risk factors.

Example: Competing Default Risks "Competing risks" are defined by multiple sources of risks, each of which may lead to default or to an adverse consequence. For example, a heart attack due to a number of reasons (risk factors). For example, say that a system defaults due to two causes and let $T_i \geq 0,\ i = 1, 2$ be the time indicators at which a default occurs and let their joint cumulative probability distribution be $F(T_1, T_2)$. The question of interest is how can default be defined in terms of both time indicators? (in other words, which risk factor is likely to manifest itself first). For univariate risks, the hazard risk is uniquely defined by the probability distribution "of default" and it survival function $1 - F(t) = S(t)$:

$$h(t) = \frac{f(t)}{1 - F(t)} = \frac{f(t)}{S(t)}$$

In a multivariate context, there is no unique relationship between the "default rate" and the time indicators. For the bi-variate case, we have by definition:

$$\begin{aligned} h(t_1, t_2) &= \underset{\Delta t_1 \to 0, \Delta t_2 \to 0}{Lim} \left\{ \frac{1}{\Delta t_1 \Delta t_2} P(t_1 \leq T_1 \leq t_1 + \Delta t_1,\ t_2 \leq T_2 \leq t_2 + \Delta t_2 | T_1 > t_1, T_2 > t_2) \right\} \\ &= \frac{f(t_1, t_2)}{1 - F(t_1, t_2)} = \frac{f(t_1, t_2)}{S(t_1, t_2)} \end{aligned}$$

To obtain a survival function of time t only some additional information is needed. Explicitly, while for univariate risk models we have $h(t) = -\frac{\partial}{\partial t} \ln(S(t))$, in the bivariate case, we have for each time indicator:

$$h_i(t_1, t_2) = -\frac{\partial}{\partial t_i} \ln(S(t_1, t_2))$$

where $h_i(t_1, t_2)$ denotes the default hazard rate of the ith risk indicator. A general expression for the survival function (of an asset, not defaulting by time t for both these times) is therefore,

$$S(t_1, t_2) = S_1(t_1)S_2(t_2) \exp\left(H(t_1, t_2)\right) \text{ where}$$

$$H(t_1, t_2) = \ln\left(\frac{S(t_1, t_2)}{S_1(t_1)S_2(t_2)}\right) = \ln\left(\frac{1 - F(t_1, t_2)}{(1 - F_1(t_1))(1 - F_2(t_2))}\right)$$

This hazard rate is not uniquely defined however. A-priori, we can postulate a parametric joint distribution and express it with specified marginals and then use the default (price) spreads to estimate the implied parameters in such a distribution.

Example: Linear Regression Statistical techniques such as "Linear Regression" (as well as most empirically based estimation techniques) are based on regressing explanatory variables on a dependent one. Explicitly let, $\{y, x_i,\ i = 1, \ldots, n\}$ be a set of variables and let $\{y_k,\ k = 1, 2, \ldots, T\}$ and $\{x_{ik},\ i = 1, \ldots, n;\ k = 1, \ldots, T\}$ be data sets of the independent and the explanatory variables. Assume a linear regression relationship such as:

$$y = a_0 + a_1 x_1 + \ldots + a_n x_n + \varepsilon_n$$

where ε_n is a residual that depends on a number of variables, their variance and co-variance. The meaningfulness of regression estimates depends on the assumptions we can (legitimately) make regarding the explanatory variables $(x_1, \ldots, x_n)$, their dependence and the error variable ε_n. For example, if the explanatory variable are dependent (i.e. correlated), the greater the variance of the dependent variable and therefore the riskier the pronouncements of its predictions. Similarly, for a given set of estimates $(\hat{a}_1, \ldots, \hat{a}_n)$ the regression residual $\hat{\varepsilon}_n$ is also meaningful, whose variance and co-variation with explanatory variables are to be considered as well (see Durbin–Watson Statistics in regression analysis). There is both an extensive literature and empirical experience for such estimation that addresses the many intricate problems of co-variations and risk estimators. The purpose of these estimators is to predict and at the same time qualify statistically these estimates.

Kendal's Tau:

When a distribution is elliptic, linear correlation between random variables cannot be used (Szego 2002) and therefore a more general form of dependence may be needed. Such an approach is given by the definition of concordance (a positive dependence) which is defined in terms of the following. Let two distinct observations (x', y') and (x'', y'') of a vector (X, Y) of continuous random variables. These observations are said to be concordant (or positively dependent) if:

$$(x' - x'')(y' - y'') > 0, \text{ i.e. if } x' > x'' \text{ then } y' > y''$$

And discordant (negatively dependent) if:

$$(x' - x'')(y' - y'') < 0, \text{ i.e. if } x' > x'' \text{ then } y' < y''$$

This definition underlies the Kendall τ as a measure of concordance. It is calculated in the following manner. We consider n pairs of observations $(x^i, y^i), i = 1, \ldots, n$ from a vector (X, Y) of continuous random variables. The number of paired comparisons in such a sample is $\binom{n}{2}$ each of which can be either concordant or discordant. Let c be the number of concordant observations and d be the number of discordant observations $\left(c + d = \binom{n}{2}\right)$. Kendall's τ is then defined by:

$$\tau = \frac{c - d}{c + d} = (c - d)/\binom{n}{2}$$

Similar to the correlation coefficient, Kendall's τ provides a scalar measure of the degree of dependence, or more precisely the degree of monotonic dependence. Consider again for example two currencies, say the EURO and the Yuan price in US Dollars. Over a period of time (of stable Foreign Exchange Rates), a $\binom{n}{2}$ paired comparison of all prices will indicate how many have moved together in the same direction (equals c) with respect to the dollar and how many have not (equals d). On that basis, the Kendall's τ can be calculated. For example, over t = 100 experimental periods, the price of oil and airlines tickets has moved together 62 times. Thus, c = 62, d = 38 and therefore, $\tau = (62 - 38)/100 = 0.24$.

Spearman Rho $\rho s(X, Y)$:
Say that (X_1, Y_1), (X_2, Y_2) and (X_3, Y_3) are three independent continuous random vectors with common joint distribution function $H(U, V)$ (whose margins are (F, G)). In this case, Spearman's rho is defined by:

$$\rho s(X, Y) = \frac{Cov(U, V)}{\sqrt{Var(U)}\sqrt{Var(V)}}$$

where observations are ranked. Both Kendall's tau and Spearman's rho provide best alternatives to the linear correlation coefficient as a measure of dependence for non-elliptical distributions. The extension to higher dimensions are obvious, we simply write pairwise correlations in an $n \times n$ matrix.

Medial Correlation Coefficient:
A Medial correlation coefficient for a pair X, Y of continuous random variables is defined by:

$$\beta(X,Y) = \mathrm{P}\{(X - X_M)(Y - Y_M) > 0\} - \mathrm{P}\{(X - X_M)(Y - Y_M) < 0\}$$

where X_M and Y_M are medians of X and Y. This coefficient is similar to Kendall's tau. It measures the dependence at the "center" of the distribution and is bounded by -1 and 1.

4.3 Multivariate Discrete Distributions

The Multinomial Distribution: The multinomial distribution is a generalization of the binomial distribution, although only in a limited sense. Explicitly, say that we have n types of Bernoulli outcomes $\tilde{x}_i$ and let $p(\tilde{x}_i = 1) = p_i$ be the independent probabilities of each outcome type with $\sum_{i=1}^{n} p_i = 1$. In this case, the probability in N trials of obtaining $\tilde{r}_i, i = 1, \ldots n$ outcomes defines the multinomial distribution given by:

$$p(r_1, r_2, \ldots, r_n) = \frac{N!}{r_1! r_2! \ldots, r_n!} p_1^{r_1} p_2^{r_2} \ldots p_n^{r_n}, \quad \sum_{j=1}^{n} r_j = N$$

$$E(r_i) = Np_i, \mathrm{var}(r_i) = Np_i(1 - p_i), cov(r_i, r_j) = -Np_i p_j$$

This distribution unlike its "sister", the multivariate binomial distribution assumes that the underlying (parameters) Bernoulli probabilities p_i, p_j are fixed and thus independent. When the underlying probabilities p_i, p_j are potentially dependent, we have then a multivariate Bernoulli distribution.

The Multivariate Bernoulli Distribution: For convenience, consider first a bi-variate Bernoulli process which we subsequently generalize to a multivariate binomial distribution. The uni-variate process assumes as stated earlier two values, say 1 or 0 which we denote by $\tilde{x}_i \in \{1, 0\}$ with $p(\tilde{x}_i = 1) = p_i$. Consider next two dependent $(\tilde{x}_i, \tilde{y}_j)$, $i = 0, 1; j = 0, 1$ events with four states, $\langle \{1, 1\}, \{1, 0\}, \{0, 1\}, \{0, 0\} \rangle$ each with a joint probability $\{p_{ij} = p(\tilde{x}, \tilde{y})\}$. A general expression for dependent bi-variate Bernoulli random variables $(\tilde{x}, \tilde{y})$ is thus:

$$p(x,y) = \{p_{11}\}^{xy} \{p_{10}\}^{x(1-y)} \{p_{01}\}^{(1-x)y} \{p_{00}\}^{(1-x)(1-y)},$$
$$(x,y) \in [1,0] \quad \text{and} \quad p_{11} + p_{10} + p_{01} + p_{00} = 1$$

And generally for a K vector of multivariate Bernoulli probability distribution (note the change in notation):

$$p(x_1, x_2, \ldots, x_K) = \{p_{11\ldots1}\}^{\prod_{j=1}^{K} x_j} \{p_{011..1}\}^{(1-x_1)\prod_{j=1}^{K} x_j} \ldots\{p_{00\ldots0}\}^{\prod_{j=1}^{K}(1-x_j)}$$

The probability generating function of say a bivariate Bernoulli is thus:

$$P^*(z_1, z_2) = \sum_{x_1=1,0} \sum_{x_2=1,0} p(x_1, x_2) z_1^{x_1} z_2^{x_2}$$

or:

$$P^*(z_1, z_2) = \sum_{x_1=1,0} \sum_{x_2=1,0} \{p_{11}\}^{x_1 x_2} \{p_{10}\}^{x_1(1-x_2)} \{p_{01}\}^{(1-x_1)x_2} \{p_{00}\}^{(1-x_1)(1-x_2)} z_1^{x_1} z_2^{x_2}$$

And therefore,

$$P^*(z_1, z_2) = p_{00} + p_{10} z_1 + p_{01} z_2 + p_{11} z_1 z_2$$

The moments of the distributions are found by the PGF (Probability Generating Function) above:

$$\frac{\partial P^*(z_1, z_2)}{\partial z_1} = \sum_{x_1=1,0} \sum_{x_2=1,0} x_1 p(x_1, x_2) z_1^{x_1-1} z_2^{x_2},$$

$$\frac{\partial^2 P^*(z_1, z_2)}{\partial {z_1}^2} = \sum_{x_1=1,0} \sum_{x_2=1,0} x_1(x_1 - 1) p(x_1, x_2) z_1^{x_1-2} z_2^{x_2}$$

$$\frac{\partial P^*(z_1, z_2)}{\partial z_2} = \sum_{x_1=1,0} \sum_{x_2=1,0} x_2 p(x_1, x_2) z_1^{x_1} z_2^{x_2},$$

$$\frac{\partial^2 P^*(z_1, z_2)}{\partial {z_2}^2} = \sum_{x_1=1,0} \sum_{x_2=1,0} x_2(x_2 - 1) p(x_1, x_2) z_1^{x_1} z_2^{x_2-2}$$

While,

$$\frac{\partial^2 P^*(z_1, z_2)}{\partial z_1 \partial z_2} = \sum_{x_1=1,0} \sum_{x_2=1,0} x_1 x_2 p(x_1, x_2) z_1^{x_1-1} z_2^{x_2-1}$$

At $(z_1, z_2) = (1, 1)$, we have for example:

$$\frac{\partial P^*(1,1)}{\partial z_1} = E(x_1), \frac{\partial^2 P^*(1,1)}{\partial {z_1}^2} = E(x_1(x_1 - 1))$$

$$\frac{\partial P^*(1,1)}{\partial z_2} = E(x_2), \frac{\partial^2 P^*(1,1)}{\partial {z_2}^2} = E(x_2(x_2 - 1)), \quad \frac{\partial^2 P^*(1,1)}{\partial z_1 \partial z_2} = E(x_1 x_2)$$

which leads to the following moments:

$$E(x_1) = p_{10} + p_{11}, \text{var}\ (x_1) = (p_{10} + p_{11})(1 - (p_{10} + p_{11}))$$
$$E(x_2) = p_{01} + p_{11}, \text{var}\ (x_2) = (p_{01} + p_{11})(1 - (p_{01} + p_{11}))$$
$$E(x_1x_2) = p_{11}$$

Since the sum of probabilities for all events is equal to one, we can write: $p_{11} = 1 - p_{00} - p_{10} - p_{01}$. Further, setting $1 - p_{00} - p_{01} = p_1$ and $1 - p_{00} - p_{10} = p_2$ we obtain:

$$E(x_1) = p_1 = 1 - p_{00} - p_{01}, \text{var}\ (x_1) = p_1(1 - p_1)$$
$$E(x_2) = p_2 = 1 - p_{00} - p_{10}, \text{var}\ (x_2) = p_2(1 - p_2)$$

As well as:

$$E(x_1x_2) = p_{11} = E(x_1)E(x_2) + \rho\sqrt{\text{var}(x_1)\, var(x_2)}$$

where ρ is the correlation of the bi-variate Bernoulli distribution. Thus,

$E(x_1x_2) = p_{11} = p_1 - p_{10} = p_2 - p_{01} = p_1p_2 + \rho\sqrt{p_1p_2(1 - p_1)(1 - p_2)}$ and $\text{cov}(x_1, x_2) = \rho - p_1p_2$

For example, say that $p_1 = 0.2$, $p_2 = 0.1$ and $\rho = 0.3$ then we have the following system equations:

$$p_{11} = 0.02 + 0.3\sqrt{0.02 * 0.8 * 0.9} = 0.02 + 0.036 = 0.038$$
$$p_{10} = 0.2 - 0.038 = 0.162, \ \ p_{01} = 0.1 - 0.038 = 0.062,$$
$$p_{00} = 0.7 + 0.038 = 0.738$$

And therefore, the PGF is:

$$P^*(z_1, z_2) = 0.738 + 0.162z_1 + 0.062z_2 + 0.038z_1z_2$$

While for an independent process we have $E(x_1x_2) = p_{11} = p_1p_2$ and,

$E(x_1x_2) = p_1 - p_{10} = p_2 - p_{01} = p_1p_2$ and therefore $p_{10} = p_1(1 - p_2)$ and $p_{01} = p_2(1 - p_1)$

Example: Let x_1 be a Bernoulli (bank inspection by a regulators) event and let x_2 be the bank to be non-compliant to regulation. Assume that both events are statistically dependent defined by a bivariate Bernoulli distribution as indicated in the table below.

	Regulation compliant	Not compliant
No reg. inspect.	p_{00}	p_{01}
Regul. inspect.	p_{10}	p_{11}

For a bank, risk is defined by the event of being both inspected and non-compliant which occurs with probability p_{11}. The probability of being non-compliant is in this case $p_{01} + p_{11}$ while for the inspector-regulator, the probability of an inspection is $p_{10} + p_{11}$. The regulator risk is again the inspection cost incurred plus the cost of not detecting a non-compliant bank. We can generalize this problem to a sampling inspection of say n inspections (or inspecting n articles in a regulation book). The probability distribution of detecting, say r outcomes, has then a bi-variate binomial distribution whose probability generating function is:

$$G^*(z_1, z_2) = (p_{11} + p_{01}z_1 + p_{10}z_2 + p_{00}z_1z_2)^n,$$

With $E(x_1) = n(p_{01} + p_{00})$ and $E(x_2) = n(p_{00} + p_{10})$ and $E(x_1x_2) = n(n-1)$ $[p_{00} + (p_{01} + p_{00})(p_{10} + p_{00})]$. But, since $E(x_1x_2) = E(x_1)\,E(x_2) + \rho\sigma_1\sigma_2$ we have: $\rho = \frac{E(x_1x_2) - E(x_1)E(x_2)}{\sigma_1\sigma_2}$ where $\sigma_1\sigma_2$ can be calculated by using the second derivatives of the probability generating function.

Example: Consider a stock price and say that we define two events, one at time T and the other at time T + 1. The event is defined as the stock price has increased. On the basis of a sample of observations (say 100) we conclude that p_{00} is the proportion of the sample observation with which the stock went up both at time T and at time T + 1. While p_{01}, is the sample proportion that the stock went up at T and then declined at T + 1, and so on. The probability distribution of two such events under the specific conditions stated here may then be modeled as a bi-variate binomial distribution.

	T + 1 Price increase	T + 1 Price decrease
T Price increase	p_{00}	p_{01}
T price decrease	p_{10}	p_{11}

An extension to three periods, say a memory of two periods T − 1 and T and the next period T + 1 will provide then the following $2^3 = 8$ probabilities: p_{000}, $p_{001}, p_{010}, p_{011}, p_{100}, p_{101}, p_{110}, p_{111}$ whose probability generating function is:

$$P^*(z_1, z_2, z_3) = \{p_{000} + p_{001}z_3 + p_{010}z_2 + p_{011}z_2z_3 + p_{100}z_1 + p_{101}z_1z_3 + p_{110}z_1z_2 + p_{111}z_1z_2z_3\}$$

It is left as an exercise to calculate the marginal means, their variances and their correlations.

Example: Consider the average $(x_1 + x_2)/2$. The expected average probability and its variance are:

$$E\left(\frac{x_1+x_2}{2}\right) = (p_1+p_2)/2$$

$$\begin{aligned}\text{var}\left(\frac{x_1+x_2}{2}\right) &= \frac{1}{4}(\text{var}\,(x_1) + \text{var}\,(x_2) + 2\,\text{cov}(x_1,x_2)) \\ &= \frac{1}{4}(p_1(1-p_1) + p_2(1-p_2) + 2(\rho - p_1p_2))\end{aligned}$$

The implications of these results to risks are weighty. When events are dependent, their variance may increase or decrease. Assume for example a system consisting of two identical components engineered so that a critical failure occurs if both fail. In this case, the probability that 0, 1 or 2 fail is:

$$\binom{2}{i}(x)^i(1-x)^{2-i}, \; i = 0, 1, 2$$

Where the mean and the variance are (as will be proved generally, below):

$$E(i) = 2E(x), \quad \text{var}\,(i) = 2E(x)[1-E(x)] + 2(2-1)\text{var}\,(x)$$

If we set $E(x) = p, \quad \text{var}\,(x) = \sigma^2$, then $E(i) = 2p, \text{var}(i) = 2p[1-p] + 2(2-1)\sigma^2$. The probability that the system fails is thus $(\tilde{x})^2$ and therefore in expectation it is $E(\tilde{x}^2)$. Now assume for comparison that the two components are statistically dependent, what will be the consequences for the probability of both items failing?

Example: Let $(\tilde{x}_1, \tilde{x}_2, \ldots., \tilde{x}_n)$ be, n bi-variate random variables, defined as follows:

$$f(\tilde{x}_i = 1 | p_{1,i}) = p_{1,i}$$
$$f(\tilde{x}_i = 1, \; \tilde{x}_j = 1 | p_{2,ij}) = p_{2,ij}, \; i, \; j = 1, \ldots n; \; i \neq j$$
$$\ldots\ldots\ldots$$
$$f(\tilde{x}_i = 1, \; \tilde{x}_j = 1, \ldots.., \tilde{x}_n = 1 | p_{k,i,j\ldots n}) = p_{k,i,j\ldots n}$$

where $\{p_k\}$ is a parameter set (similar to that defined above). Since this is a binary variable, we have:

$$\begin{aligned}E(\tilde{x}_i | p_{1i}) &= p_{1i}, \quad E(\tilde{x}_i\tilde{x}_j | p_{2,ij}) = p_{2,ij} \\ \text{cov}(\tilde{x}_i, \tilde{x}_j | p_{1i}, p_{1j}p_{2,ij}) &= p_{2,ij} - p_{1i}p_{1j}; \quad i \neq j \text{ and} \\ \rho_{ij}(\tilde{x}_i, \tilde{x}_j) &= \frac{p_{2,ij} - p_{1i}p_{1j}}{\sqrt{p_{1i}(1-p_{1i})p_{1j}(1-p_{1j})}}\end{aligned}$$

For a portfolio of three and more loans (or a reliability system with multiple components), we may calculate the appropriate parameters recursively. The number of parameters required to represent such dependence, between two, three, four and more random variables will require an increasing number of parameters which renders such an approach unrealistic. For this reason, other approaches can be devised to represent and capture the essential characteristics of dependence and at the same time maintain the model tractable.

Example: Bernoulli Default Dependent Probability Models Multivariate Bernoulli (and thus binary) models meeting the necessary conditions for dependence stated above, can be constructed in various ways. Define the event $y_1 = 1, 0$ of a default at some future time by a probability $y_1 = 1, 0 \quad p(y_1 = 1) = \alpha_1$. Consider instead two dependent events and set the following probabilities $f(y_1, y_2|\alpha_1, \alpha_2, \rho)$ where $(\alpha_1, \alpha_2, \rho)$ are a set of parameters. Four possibilities arise defined by: $(y_1, y_2) = (1, 1;\ 1, 0;\ 0, 1;\ 0, 0)$. The parameter ρ relates to a potential dependence of (y_1, y_2). Say that:

$$f(y_1, y_2|\alpha_1, \alpha_2, \rho) = \frac{(\alpha_1)^{y_1}(1-\alpha_1)^{1-y_1}(\alpha_2)^{y_2}(1-\alpha_2)^{1-y_2}\rho^{y_1 y_2}}{1 + \alpha_1\alpha_2(\rho - 1)}$$

The joint probability thus assumes as indicated earlier four values:

$$f(y_1 = 1, y_2 = 1|\alpha_1, \alpha_2, \rho) = \frac{(\alpha_1)(\alpha_2)\rho}{1 + \alpha_1\alpha_2(\rho - 1)},$$
$$f(y_1 = 1, y_2 = 0|\alpha_1, \alpha_2, \rho) = \frac{(\alpha_1)(1-\alpha_2)}{1 + \alpha_1\alpha_2(\rho - 1)}$$
$$f(y_1 = 0, y_2 = 1|\alpha_1, \alpha_2, \rho) = \frac{(1-\alpha_1)(\alpha_2)}{1 + \alpha_1\alpha_2(\rho - 1)},$$
$$f(y_1 = 0, y_2 = 0|\alpha_1, \alpha_2, \rho) = \frac{(1-\alpha_1)(1-\alpha_2)}{1 + \alpha_1\alpha_2(\rho - 1)}$$

with a marginal probability for the first variable given by:

$$f(y_1 = 1|\alpha_1, \alpha_2, \rho) = \frac{(\alpha_1)(\alpha_2)\rho}{1 + \alpha_1\alpha_2(\rho - 1)} + \frac{(\alpha_1)(1-\alpha_2)}{1 + \alpha_1\alpha_2(\rho - 1)} = \frac{\alpha_1(1 + \alpha_2(\rho - 1))}{1 + \alpha_1\alpha_2(\rho - 1)} = \alpha_1$$

and of course $f(y_1 = 0|\alpha_1, \alpha_2, \rho) = 1 - \alpha_1$. Calculations provide the probability $f(y_2|\alpha_1, \alpha_2, \rho)$ as well as their conditional $f(y_1|y_2, \alpha_1, \alpha_2, \rho)$ and $f(y_2|y_1, \alpha_1, \alpha_2, \rho)$. If $\rho = 1$, then the probability distribution of the joint events indicates a statistical independence since:

$$f(y_1, y_2|\alpha_1, \alpha_2, \rho) = (\alpha_1)^{y_1}(1-\alpha_1)^{1-y_1}(\alpha_2)^{y_2}(1-\alpha_2)^{1-y_2}$$

Thus the parameter ρ, an indicator of dependence is,

$$f(1,1|\alpha_1,\alpha_2,\rho) = \frac{\alpha_1\alpha_2\rho}{1+\alpha_1\alpha_2(\rho-1)}$$

Note that if $0 < \rho = \frac{f(1,1)(1-\alpha_1\alpha_2)}{(1-f(1,1,))}$, it will indicate a positive dependence since the probability of the one increases the probability of the other. The conditional probability distributions of one event on the other can be shown to be also of the "binomial type" since:

$$f(y_2|y_1,\alpha_1,\alpha_2,\rho) = \frac{(\alpha_2)^{y_2}(1-\alpha_2)^{1-y_2}\rho^{y_1y_2}}{1+\alpha_2(\rho^{y_1}-1)} = \pi_2^{y_2}(1-\pi_2)^{1-y_2}$$

which points out to Bernoulli events with probabilities:

$$\pi_1 = f(y_2=1|y_1=1,\alpha_2,\rho) = \frac{\alpha_2\rho}{1+\alpha_2(\rho-1)} \quad \text{and}$$

$$\pi_2 = f(y_2=1|y_1=0,\alpha_2,\rho) = \alpha_2$$

The expected default (or non-default) of the second, given the first are thus:

$$f(y_2=1|1,\alpha_1,\alpha_2,\rho) = \frac{(\alpha_2\rho)^{y_2}(1-\alpha_2)^{1-y_2}}{1+\alpha_2(\rho-1)} = \rho\frac{\alpha_2}{1+\alpha_2(\rho-1)};$$

$$f(y_2=0|1,\alpha_1,\alpha_2,\rho) = \frac{1-\alpha_2}{1+\alpha_2(\rho-1)}$$

expressing clearly the dependence between these two variables.

Example: The probability of two default events at the same time is given by 0.05 while the probability of a first default is 0.08 and the second has a probability of 0.10. As a result, we have:

$$0.05 = f(1,1|\alpha_1,\alpha_2,\rho) = \frac{\alpha_1\alpha_2\rho}{1+\alpha_1\alpha_2(\rho-1)} \quad \text{and}$$

$$\rho = \frac{f(1,1|\alpha_1,\alpha_2,\rho) - f(1,1|\alpha_1,\alpha_2,\rho)\alpha_1\alpha_2}{\alpha_1\alpha_2(1-f(1,1|\alpha_1,\alpha_2,\rho))} = \frac{0.05-(0.05)0.08}{0.08(1-0.05)} = \frac{0.046}{0.08(0.95)}$$

which provides an estimate for the dependence parameter and therefore all the relevant probabilities.

4.3.1 Estimating the Bi-variate Bernoulli Parameters

Now assume that a data set consists of the following record of defaults:

$$(0,0),(0,0),(0,0),(1,1),(1,0),(1,0),(0,1),(0,0),(1,0)$$

How can we use this data set to calculate the underlying parameters of the joint distribution function? To do so, we can simply apply the Ln Likelihood function:

$$Ln\prod_{j=1}^{n} f(y_{1j},y_{2j}|\alpha_1,\alpha_2,\rho) = Ln\prod_{j=1}^{n}\left\{\frac{(\alpha_1)^{y_{1j}}(1-\alpha_1)^{1-y_{1j}}(\alpha_2)^{y_{2j}}(1-\alpha_2)^{1-y_{2j}}\rho^{y_{1j}y_{2j}}}{1+\alpha_1\alpha_2(\rho-1)}\right\}$$
$$=\sum_{j=1}^{n}\begin{pmatrix} y_{1j}\ln(\alpha_1)+(1-y_{1j})\ln(1-\alpha_1)+ \\ y_{2j}\ln(\alpha_2)+(1-y_{2j})\ln(1-\alpha_2)+ \\ y_{1j}y_{2j}\ln\rho \end{pmatrix} - n\ln(1+\alpha_1\alpha_2(\rho-1))$$

and maximizing with respect to the joint distribution parameters. Or,

$$\underset{\alpha_1,\alpha_2,\rho}{Max}\sum_{j=1}^{n}\begin{pmatrix} y_{1j}\ln(\alpha_1)+(1-y_{1j})\ln(1-\alpha_1)+ \\ y_{2j}\ln(\alpha_2)+(1-y_{2j})\ln(1-\alpha_2)+ \\ y_{1j}y_{2j}\ln\rho \end{pmatrix} - n\ln(1+\alpha_1\alpha_2(\rho-1))$$

Example: Assume a portfolio consisting of 50 loans each with an expected probability of default given by:

$\alpha_i = \{0.02, 0.05, 0.07, 0.02, 0.01, ...\}$. Default loans are separated into two classes, high risk (with individual default probabilities of 0.04 and higher and low risk, with default probabilities less than 0.04). These classes indicate that the mean default probabilities of loans in the high and the low risk categories are 0.06 and 0.03 respectively while their variance equals in both cases the mean in their risk class. Assume at first that these variances are null while the default correlation is 0.4. In this case, we note that the first class (high) risk is defined by the random variable y_1 whose parameter is $\alpha_1 = 0.06$ while the second class (low risk) is a random variable y_2 whose parameter is $\alpha_2 = 0.03$ while the correlation is:

$$E(y_1) = \alpha_1 = 0.06;\;\; E(y_2) = \alpha_1 = 0.03$$
$$E(y_1y_2) = \alpha_{12};\;\; \rho_{12}(y_i, y_j) = 0.4 = \frac{\alpha_{12}-\alpha_1\alpha_2}{\alpha_1-\alpha_1\alpha_2} = \frac{\alpha_{12}-0.018}{0.02-0.018}$$

with $\alpha_{12} = 0.018 + 0.002 * 0.4 = 0.018 + 0.0008 = 0.0188$ defining the last parameter.

Mixtures of the Bivariate Bernoulli Distribution
Consider again the PGF of the bivariate binomial:

$$P^*(z_1,z_2) = (p_{00} + p_{10}z_1 + p_{01}z_2 + p_{11}z_1z_2)^n = [q(z_1,z_2)]^n$$

Say that parameter n is also a parameter whose probability generating function is $G^*(z_3)$. Namely,

$$G^*(z_3) = \sum_{n=0}^{\infty} {z_3}^n g(n)$$

As result, the mixture binomial distribution has a PGF which is given by

$$\sum_{n=0}^{\infty} P^*(z_1,z_2|n)g(n) = \sum_{n=0}^{\infty} (q(z_1,z_2))^n g(n) = G^*(q(z_1,z_2))$$

This expression can be used to calculate the moments of the resulting distribution.

However, if the Bernoulli distribution depends on a common risk factor ξ whose probability distribution is $f(\xi)$ where each of its probabilities depends on this factor, the resulting PGF is then:

$$\wp^*(z_1,z_2) = \int_{\Re} P^*(z_1,z_2|\xi)f(\xi)d\xi$$
$$= \left(\begin{array}{r} \int_{\Re} (1 - p_{11}(\xi) - p_{01}(\xi) - p_1(\xi))f(\xi)d\xi + z_1 \int_{\Re} p_{10}(\xi)f(\xi)d\xi \\ +z_2 \int_{\Re} p_{01}(\xi)f(\xi)d\xi + z_1z_2 \int_{\Re} p_{11}(\xi)f(\xi)d\xi \end{array} \right)$$

Note that each of these probabilities can be defined in terms of their marginal means and their correlation.

Finite mixtures of multivariate Bernoulli distributions are applicable in diverse fields to model a population of binary, multivariate measurements in terms of a few latent classes (see Carreira-Perpinan and Reinals 2000; Everitt and Hand 1981).

Problem: In general reliability systems with n components require at least k functioning components for non-default, the probability of operating (and therefore no default) has a far greater variance due to the dependence of the components—a dependence arising from operating in a common environment, subject to common external disturbances etc. Similar situations arise with loans and mortgage portfolios that bear credit risks as well as diseases that depend on common factor (say diabetes). Common risk elements that induce a dependence are various macroeconomic factors, common disasters (such as Hurricanes) etc. In such cases, the risk of investors holding such portfolios can be extremely large (as this was the case with credit derivatives and Mortgage Backed Securities and market prices of real estate). To calculate the risk arising from such dependence, a multi-variate binomial (or multi-factor risk dependent) distributions would be needed.

4.3.2 *The Bivariate Binomial Distribution*

The sum of dependent Bernoulli dependent random variables yields the bi-variate binomial distribution whose PGF is (by the convolution theorem):

$$P^*(z_1, z_2) = (p_{00} + p_{10}z_1 + p_{01}z_2 + p_{11}z_1z_2)^n$$

With its moments appropriately calculated. In particular, its marginal means are as stated before:

$$\left\{\frac{\partial P^*(z_1, z_2)}{\partial z_1}|z_1 = 1, z_2 = 1\right\} = n(p_{10} + p_{11}) = E(x_1),$$
$$\left\{\frac{\partial^2 P^*(z_1, z_2)}{\partial {z_1}^2}|z_1 = 1, z_2 = 1\right\} = n(n-1)(p_{10} + p_{11})^2 = E(x_1(x_1 - 1))$$
$$\left\{\frac{\partial P^*(z_1, z_2)}{\partial z_2}|z_1 = 1, z_2 = 1\right\} = n(p_{01} + p_{11}) = E(x_2),$$
$$\left\{\frac{\partial^2 P^*(z_1, z_2)}{\partial {z_2}^2}|z_1 = 1,\ z_2 = 1\right\} = n(n-1)(p_{01} + p_{11})^2 = E(x_2(x_2 - 1))$$

As well as:

$$\left\{\frac{\partial^2 P^*(z_1, z_2)}{\partial z_1 \partial z_2}|z_1 = 1, z_2 = 1\right\} = n(n-1)(p_{10} + p_{11})(p_{01} + p_{11}) = E(x_1x_2)$$

The multivariate binomial distribution in this particular case can be calculated explicitly using both its PGF or using combinatorial arguments. In this case, it can be verified that:

$$B(x_1 = i, x_2 = j)$$
$$= \sum_{k=0} \frac{n!}{i!(j-i)!(k-i)!(n-j-k+i)!}(p_{11})^i (p_{10})^{k-i} (p_{01})^{k-j} (p_{00})^{n-k-j+i}$$

Convergence to other discrete multivariate distributions (such as the multivariate Poisson distribution) can be found in Bairamov and Gultekin (2010).

"Derivatives" of dependent Bernoulli models (as it were the case with the univariate Bernoulli model), are numerous. Among them are a Multinomial probability distribution which is based on independent Bernoulli distributions as stated earlier, the Multi-variate Binomial Probability Distribution whose moments were calculated above and various approximations leading to multi-variate Poisson and of course the Multivariate Normal. Since many models based on dependent multivariate distributions are unwieldy, Copula are used (see below) while their co-variation structure and parameters are estimated based on simulation and the analytical treatments of Copulas.

4.3.3 *The Multivariate Poisson Probability Distribution*

A multivariate Poisson distribution can be formulated as follows:

$$p(r_1, r_2, \ldots, r_n) = \frac{e^{-\sum_{i=1}^{n}\lambda_i} \lambda_1^{r_1}\lambda_2^{r_2}\ldots\lambda_n^{r_n}}{r_1!r_2!\ldots,r_n!}, \quad \text{with } E(x) = \text{var}(x) = \lambda$$
$$(r_1, r_2, \ldots, r_n) = 0, 1, 2, \ldots.; \quad \lambda_i > 0, \quad i = 1, \ldots, n$$

Note that (see Chap. 3) the probability generating function of the univariate Poisson is:

$$P_X(z) = \sum_{x=0}^{\infty} z^x \frac{\lambda^x e^\lambda}{x!} = e^{\lambda(z-1)}$$

Similarly as we proceeded in the construction of a bivariate binomial distribution using the bivariate Bernoulli probability generating function, we may construct bivariate Poisson distributions with their bivariate PGF. For example, set: $P_X(z_1, z_2) = e^{\lambda_1(z_1-1)+\lambda_2(z_2-1)+\rho(z_1-1)(z_2-1)}$ and therefore at $z_1 = z_2 = 1$ we have:

$$\frac{\partial P_X(z_1, z_2)}{\partial z_1} = \lambda_1 = E(x_1), \quad \frac{\partial P_X(z_1, z_2)}{\partial z_2} = \lambda_2 = E(x_1), \quad \frac{\partial^2 P_X(z_1, z_2)}{\partial z_1 \partial z_2} = \lambda_2\lambda_1 + \rho$$
$$= E(x_1x_2)$$

which provides a definition for its correlation.

4.4 The Multivariate Normal Probability Distribution

The bivariate normal probability distribution is the joint distribution of two continuous random variables (*X*,*Y*) having a joint probability density function:

$$f(x,y) = \frac{1}{2\pi\sigma_x\sigma_y\sqrt{1-\rho^2}} e^{-\frac{1}{2(1-\rho^2)}\left[\left(\frac{x-\mu}{\sigma_x}\right)^2 - 2\rho\frac{x-\mu}{\sigma_x}\cdot\frac{y-\nu}{\sigma_y} + \left(\frac{y-\nu}{\sigma_y}\right)^2\right]}, -\infty, y < +\infty$$

where the marginal distributions are normal with mean and variance given by $N(\mu,\sigma_x^2)$ and $N(\nu,\sigma_y^2)$ while the correlation is ρ. In general, the vector normal probability distribution is given by:
$f(\mathbf{x}) = \frac{1}{(2\pi)^{-\frac{n}{2}}}|\mathbf{A}|^{\frac{1}{2}}e^{-\frac{1}{2}(\mathbf{x}^T-\boldsymbol{\mu}^T)\mathbf{A}(\mathbf{x}^T-\boldsymbol{\mu}^T)}$ with a characteristic function: $\Phi(t) = E\left(e^{it^T\mathbf{x}}\right) = e^{-\frac{1}{2}\mathbf{t}^T\mathbf{A}^{-1}\mathbf{t}+i\mathbf{t}^T\boldsymbol{\mu}}$

For the bivariate case, we have:

$$\mathbf{A} = \begin{bmatrix} \sigma_1^2 & \rho\sigma_1\sigma_2 \\ \rho\sigma_1\sigma_2 & \sigma_2^2 \end{bmatrix} \text{ and } \mathbf{A}^{-1} = \begin{bmatrix} \frac{1}{\sigma_1^2(1-\rho^2)} & \frac{\rho}{\sigma_1\sigma_2(1-\rho^2)} \\ \frac{\rho}{\sigma_1\sigma_2(1-\rho^2)} & \frac{1}{\sigma_2^2(1-\rho^2)} \end{bmatrix}$$

with a characteristic function:

$$\Phi(t_1,t_2) = e^{-\left(\sigma_1^2 t_1^2 + 2\sigma_1\sigma_2 t_1 t_2\right)}$$

The multivariate normal probability distribution underlies a great many risk models and is therefore extremely important.

4.5 Other Multivariate Probability Distributions (*Statistics and Probability Letters*, 62, 203, 47–412)

The Bivariate Beta (*with positive covariance*):
Consider a bi-variate beta probability distribution, i.e. whose marginal distributions have Beta probability distribution. Let (x, y) be a vector of Beta distributed random variables, with marginals:

$$B(x:a,c) \quad \text{and} \quad B(y:a,c),\ 0 \le x,y \le 1$$

Define the following four parameters distribution:

$$f(x,y:w) = \frac{x^{a-1}y^{b-1}w^{a+b+c-1}e^{-wq}}{(1-x)^{a+1}(1-y)^{b+1}\Gamma(a)\Gamma(b)\Gamma(c)}, \qquad q = \frac{1-xy}{(1-x)(1-y)}$$

An integration over $w > 0$ yields a bivariate probability distribution defined by:

$$f(x,y) = \frac{x^{a-1}y^{b-1}(1-x)^{b+c-1}(1-y)^{a+c-1}}{B(a,b,c)(1-xy)^{a+b+c}}$$

The first two moments of this distribution are then:

$$E(x) = \frac{a}{a+c}, \quad E(y) = \frac{b}{b+c};$$
$$\mathrm{var}(x) = \frac{ac}{(a+c)^2(a+c+1)}, \quad \mathrm{var}(y) = \frac{bc}{(b+c)^2(b+c+1)}$$

While the cross moments are:

$$E(x^k y^m) = {}_3F_2(a+k, b+m, s, s+k, s+m : 1)$$
$$\text{where} \quad s = (a+b+c), \quad d = \frac{B(s,k)B(s,m)}{B(a,k)B(b,m)\Gamma(s)}$$

Thus, if a bi-variate Bernoulli probability distribution has a probability generating function:

$$P^*(z_1, z_2 | p_{01}, p_{10}) = (p_{00} + p_{10}z_1 + p_{01}z_2 + (1 - p_{00} - p_{10} - p_{01})z_1z_2)$$

With p_{00} assumed known but (p_{10}, p_{01}) have a bivariate Beta probability distribution, then the Probability generating function of the mixed process is:

$$E\{P^*(z_1, z_2|p_{01}, p_{10})\} = p_{00} + (1-p_{00})z_1z_2$$
$$+ \int_0^1\int_0^1 (p_{10}z_1(1-z_2) + p_{01}z_2(1-z_1))f(p_{01}, p_{10})dp_{01}dp_{10}$$

and therefore

$$P^*(z_1, z_2) = p_{00} + (1-p_{00})z_1z_2 + z_1(1-z_2)E(p_{10}) + z_2(1-z_1)E(p_{01})$$

Thus, if $E(p_{01}) = \frac{a}{a+c}$, $E(p_{10}) = \frac{b}{b+c}$, we obtain:

$$P^*(z_1, z_2) = p_{00} + (1 - p_{00})z_1 z_2 + z_1(1 - z_2)\frac{a}{a+c} + z_2(1 - z_1)\frac{b}{b+c}$$

While the mixture bi-variate binomial model has the following probability generating function:

$$P^*(z_1, z_2|n) = \left[p_{00} + (1 - p_{00})z_1 z_2 + z_1(1 - z_2)\frac{a}{a+c} + z_2(1 - z_1)\frac{b}{b+c}\right]^n$$

The Bivariate Lomax:
The bivariate Lomax distribution is a fat tails distribution given by the following with each of its marginal defined by power distributions. It is stated explicitly by:

$$1 - F(x_1, x_2) = (1 + \lambda_1 x_1 + \lambda_2 x_2 + \lambda_3 x_1 x_2)^{-q}, \; x_1 > 0, \quad x_2 > 0, \quad q > 0,$$
$$0 \le \lambda_3 \le \lambda_1 \lambda_2 (1 + q)$$

The Bivariate Power Distribution:
The Power probability distribution is a fat tail distribution and therefore of particular importance in risk modeling. Its extension to a bivariate format is then:

$$1 - F(x_1, x_2) = (1 - \lambda_1 x_1 - \lambda_2 x_2 - \lambda_3 x_1 x_2)^q, 0 \le x_1 \le (\lambda_1)^{-1},$$
$$0 \le x_2 \le \frac{1 - \lambda_1 x_1}{\lambda_1 + \lambda_3 x_1} 0, \quad q > 0, \quad 0 \le \lambda_3 \le \lambda_1 \lambda_2 (1 + q)$$

These bi-variate distributions are a sample a large set of potential models. A number of publications have developed far more their definitions and their properties. Some references include Johnson and Kotz 1969, 1970a, b; Joe 1997; Joe and Hu 1996.

4.6 Dependence and Copulas

Copulas are algebraic structures, $C(.,)$ that model the joint CDF (Cumulative Distribution Function) using its marginal CDFs. These models are used extensively in practice. They are used in reliability models, in insurance and in financial credit risk and derivatives portfolios to assess the risk properties of these portfolios. The term "copula" comes from the Latin, and refers to connecting or joining together. It is closely related to the more familiar English words "couple" or "copulate." "Copulas" are algebraic structures that define the way in which multiple and dependent random variables relate to each other through their marginal cumulative distributions. There exists a large range of potential copulas, which renders the problem of selecting one particular copula to be fitted to a data set, a challenging task.

Simplistically, a Copula is defined by the marginal distributions of two risk factors S and M each with Cumulative Density Functions (CDF): $F_S(x)$ and $F_M(y)$ while their joint CDF is $F_{SM}(x,y)$ with:

$$\{F_S(x), F_M(y)\} \underset{C}{\Rightarrow} \{F_{SM}(x,y) \equiv C(F_S(x), F_M(y))\}$$

written as follows,

$$\{F_{SM}(x,y)\} \triangleq \mathsf{C}(\mathsf{u},\mathsf{v}) \text{ with } \mathsf{u} \equiv \mathsf{F_S}(\mathsf{x}), \quad \mathsf{v} \equiv \mathsf{F_M}(\mathsf{y})$$

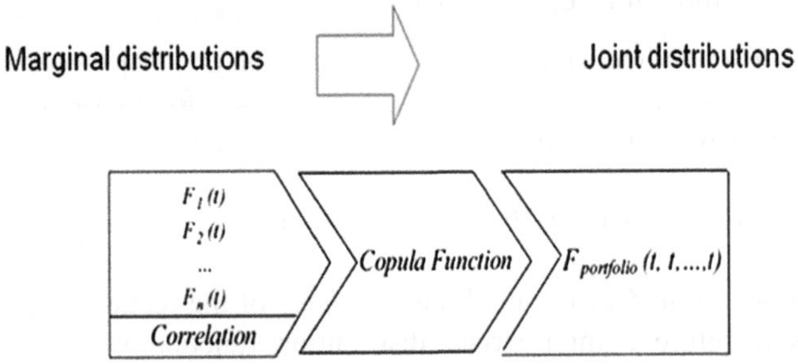

A two-dimensional copula $C(u,v)$ with u and v denoting its marginal CDF has of course the same properties of a CDF or $0 \le C(u,v) \le 1$; $0 \le u \le 1$, $0 \le v \le 1$, $C(0,u) = C(u,0) = 0$ and $C(1,u) = C(u,1) = u$.

Due to their practical importance, Copulas have been the subject of considerable research and application. There are numerous copulas, designed to meet a broad variety of modeling and statistical needs. For example the following three copulas exhibit specific statistical properties:

$$\begin{cases} 1. \quad \text{Mixture Copula} & C(u,v) = (1-\rho)uv + \rho \min(u,v),\ \rho > 0 \\ 2. \quad \text{Co-monotonic Copula} & C_U(u,v) = \min(u,v) \\ 3. \quad \text{Counter-monotonic Copula} & C_L(u,v) = (u+v-1)^+ \end{cases}$$

A mixture copula refers to a mixture of two probability distributions while co-monotonic copulas point to probability distributions that move in tandem (or vice versa for counter-monotonic copulas). Other copulas can be defined ex post to better fit observed empirical records, or any joint distributions. Their applications are indeed very large, and numerous publications document their characteristics (e.g., Nelsen 1999; Cherubini et al. 2004; Embrechts et al. 2002, 2003, 2009; McNeil et al. 2005; Patton 2007).

The Fréchet–Hoeffding Bounds Inequality: Consider the functions M and W defined on $[0,1]^2$ as follows: $M(u,v) = \min\{u,v\}$ and $W(u,v) = \max\{u+v-1, 0\}$. Then for any two-copula C and any u, v in [0,1], $W(u,v) \le C(u,v) \le M(u,v)$

Functions M and W are Fréchet–Hoeffding lower and upper bounds. Furthermore, the upper bound satisfies the definition of a copula, but the lower bound does not for dimensions $n \geq 3$. If the copula of two random variables X and Y is $M(u, v)$, then they are said to be co-monotonic.

Copula Invariance Theorem: Let $X_1, \ldots, X_n$ be n *continuous* random variables with copula C. Then if $\alpha_1(X_1), \ldots, \alpha_n(X_n)$ are strictly increasing functions of $X_1, \ldots, X_n$ respectively, then random variables $Y_1 = \alpha_1(X_1), \ldots, Y_n = \alpha_n(X_n)$ have exactly the same copula C.

This result shows that the full dependence between n random variables is completely captured by the copula, independently of the shape of the marginal distributions.

A formal definition of a copula is given below:

Definition 1 A (two-dimensional) *copula* is a function C: $[0,1]^2 \rightarrow [0,1]$ such that
(C1) $C(0, x) = C(x, 0) = 0$ and $C(1, x) = C(x, 1) = x$ for all $x \in \mathrm{I}$;
(C2) C is *2-increasing*: for $a, b, c, d \in \mathrm{I}$ with $a \leq b$ and $c \leq d$,

$$V_C([a, b] \times [c, d]) = C(b, d) - C(a, d) - C(b, c) + C(a, c) \geq 0$$

The function V_C in (C2) is called the *C-volume* of the rectangle $[a, b] \times [c, d]$. A copula is therefore a multivariate distribution function whose margins are uniformly distributed on [0,1]. Existence of a copula is stated by Sklar's theorem which is summarized by the following:

Theorem (Sklar's Theorem): Let H be a two-dimensional distribution function with marginal distribution functions F and G. Then there exists a two-dimensional copula function C such that for all x, y in R,

$$H(x, y) = C(F(x), G(y)).$$

Conversely, for any distribution functions F and G and any copula C, the function H defined above is a two-dimensional distribution function with marginal distributions F and G. Furthermore, if F and G are continuous, C is unique.

From Sklar's Theorem we see that univariate marginal cumulative distributions can be separated, and their dependence structure be represented by a copula. Given a joint distribution function H with continuous marginal distributions F and G, as in Sklar's Theorem, a corresponding copula $C(u, v) = H(F^{-1}(u), G^{-1}(v))$, can be constructed where F^{-1} is the *cadlag* (*i.e. a right continuous distribution function*) *inverse* of F, given by $F^{-1}(u) = \sup\{x | F(x) \leq u\}$, for any u, v in [0,1]. For example, the product Copula $C(u, v) = uv$, called the *product copula* is continuous in random variables X and Y in C if and only if X and Y are independent.

Example: The Bivariate Bernoulli Copula The purpose of this example is to consider the simple Bernoulli model and through a copula obtain a broader set of co-variant models. The Bernoulli model, as seen earlier assumes two values (0,1), each with probabilities $1-p$ and p. Consider two marginal Bernoulli distributions. The joint (dependent, independent) distribution assumes four potential values (00, 01, 10, 11) and probabilities $(p_{00}, p_{01}, p_{10}, p_{11})$ with marginal probabilities $(1-p_1, p_1)$and $(1-p_2, p_2)$. The cumulative marginal distribution assumes as well two values:

$$F_1(i=0,1) = (1-p_1, 1) = (u(0), u(1))$$

$$F_2(j=0,1) = (1-p_2, 1) = (v(0), v(1))$$

while the Copula is as stated above defined by $C(F_1(i), F_2(j)) \equiv C(u(i), v(j))$. Let a bi-variate Bernoulli model be co-monotonic and select the copula:

$$C(F_1(i), F_2(j)) \equiv C(u\,(i), v\,(j)) = \min(u(i), v\;(j))$$

Then the following values are obtained:

$$\begin{aligned} C(F_1(0), F_2(0)) &\equiv C(1-p_1, 1-p_2) = \min(1-p_1, 1-p_2); \\ C(F_1(0), F_2(1)) &\equiv C(1-p_1, 1) = 1-p_1 \\ C(F_1(1), F_2(0)) &\equiv C(1, 1-p_2) = 1-p_2;\;\; C(F_1(1), F_2(1)) \equiv C(1,1) = 1 \end{aligned}$$

Say that the marginal Bernoulli distribution has probabilities $p_1 = 0.3, p_2 = 0.6$, thus:

$$\begin{aligned} &C(F_1(0), F_2(0)) \equiv C(1-p_1, 1-p_2) = C(0,7,0.4) = 0.4; \\ &C(F_1(1), F_2(0)) \equiv C(1, 1-p_2) = C(1, 0.4) = 0.4 \\ &C(F_1(0), F_2(1)) \equiv C(1-p_1, 1) = C(0.7, 1) = 0.7; \\ &C(F_1(1), F_2(1)) \equiv C(1,1) = 1 \end{aligned}$$

Consider next the mixed Copula
$C(F_1(i), F_2(j)) \equiv (1-\rho)F_1(i)F_2(j) + \rho\,\min(F_1(i), F_2(j))$ with:

$$\begin{aligned} C(F_1(0), F_2(0)) &\equiv (1-\rho)(1-p_1)(1-p_2) + \rho\,\min(1-p_1, 1-p_2) \\ &= (1-\rho)(0.7\times 0.4) + 0.4\rho \\ C(F_1(0), F_2(1)) &\equiv (1-\rho)(1-p_1) + \rho(1-p_1) = (1-p_1) = 0.7 \\ C(F_1(1), F_2(0)) &\equiv (1-\rho)(1-p_2) + \rho(1-p_2) = (1-p_2) = 0.4 \\ C(F_1(1), F_2(1)) &\equiv (1-\rho) + \rho = 1 \end{aligned}$$

The cumulative distribution is thus:

$$F(0,0) \equiv 0.28 + 0.12\rho;\;\; F(1,0) \equiv 0.4;\;\; F(0,1) \equiv 0.7;\;\; F(1,1) \equiv 1$$

While the individual probabilities are:

$$\begin{aligned}
&f(0,0) \equiv 0.28 + 0.12\rho;\; f(1,0) \equiv F(1,0) - F(0,0) = 0.12(1-\rho)\\
&f(0,1) \equiv F(0,1) - F(0,0) = 0.42 - 0.12\rho\\
&f(1,1) \equiv F(1,1) - f(0,0) - f(1,0) - f(0,1)\\
&\qquad = 1 - 0.28 - 0.12\rho - 0.12(1-\rho) - 0.42 + 0.12\rho = 0.18 + 0.12\rho
\end{aligned}$$

Example: The Gumbel Copula, the Highs and the Lows Extreme (Weibull, Gumbel, Frechet) probability distributions are used to model the high and the low of identically distributed random samples. They are also used to model extreme risk distributions. Say that the joint cumulative probability distribution of the highs and the lows are modeled now by type A Gumbel distributions $[u(x), v(y)]$ with joint distribution:

$$F(x,y) = u(x)v(y)\exp\left(-\theta\left[\frac{1}{\ln u(x)} + \frac{1}{\ln v(y)}\right]^{-1}\right)$$

where $u(x)$ and $v(y)$ are marginal Gumbel cumulative distributions. Note that for such a distribution,

$$u(x) = [x]^n \; v(y) = 1 - [1-y]^n$$

and since the high and the low are drawn from the same probability distribution, we have $y = x = F(S)$. Thus, the bivariate copula, where $u = u(x),\; v = v(y)$, is:

$$C(u, v|x) = uv\,\exp\left(-\theta\left[\frac{1}{\ln\, u} + \frac{1}{\ln\, v}\right]^{-1}\right)$$

And therefore,

$$C(x,y|S) = [x]^n(1 - [1-y]^n)\exp\left(-\theta\left[\frac{1}{n\,\ln[x]} + \frac{1}{\ln(1-[1-y]^n)}\right]^{-1}\right),\quad x = y$$
$$= F(S)$$

which provides a joint cumulative distribution that can be tested against empirical data. Alternatively, a type B Gumbel given by (note that another parameter m has been inserted) is given by:

$$F(x, y) = u(x)v(y)\exp\left(\left[\left(\ln\frac{1}{u(x)}\right)^m + \left(\ln\frac{1}{v(y)}\right)^m\right]^{\frac{1}{m}}\right) \quad \text{where } m > 1,$$

$$m = \frac{1}{\sqrt{1-\rho}}$$

with ρ being a coefficient of correlation. We can of course calculate the explicit joint probability distribution, using the fact that $f(x, y) = \frac{\partial^2 F(x, y)}{\partial x \partial y}$ and simulate such a joint distribution using Monte Carlo techniques and calculate its moments (mean, variance, covariance, skewness, kurtosis, etc.).

4.6.1 Copulas and Dependence Measures

Measure of dependence can be calculated using Copulas. For the familiar correlation of two random variables defined by: $\rho\sigma_X\sigma_Y = E((x - E(x))(y - E(y)))$ we have:

$$\rho(X, Y) = \frac{1}{\sqrt{Var(X)}\sqrt{Var(Y)}} \int_0^1\int_0^1 [C(u, v) - uv]dF_1{}^{-1}(x)dF_2{}^{-1}(y)$$

The Kendal's Tau is as stated above the probability of concordance minus the probability of discordance. This "coefficient" is also bounded by -1 and 1 and is given by:

$$\tau(X, Y) = 4\int_0^1\int_0^1 C(u, v)dC(u, v) - 1$$

Spearman Rho: In this case, it is defined as

$$\rho s(X, Y) = 12\int_0^1\int_0^1 [C(u, v) - uv]dudv = \frac{Cov(U, V)}{\sqrt{Var(U)}\sqrt{Var(V)}} = r(F(X), G(Y))$$

Finally, the *Medial Correlation Coefficient* for a pair X, Y of continuous random variables measures the dependence at the "center" of the distribution and is bounded by -1 and 1.

$$\beta(X, Y) = 4C\left(\frac{1}{2}, \frac{1}{2}\right) - 1$$

4.6.2 Copulas and Conditional Dependence

Random variables are dependent both horizontally (among themselves) and conditionally, presenting complex relationships. In some cases, their modeling can be simpler to formulate based on the conditional distributions. An alternative model to simple inter-temporal relationships (exponential smoothing, autoregressive, mean reversion, etc.) may be the use of conditional copulas. For example, given two factors, it might be easier to use a joint copula in terms of the (statistical) conditional distributions of the factors' distributions. For example, postulate a joint cumulative distribution for two factors (S and C) by a bivariate copula:

$$C(F_S(S_1), F_C(c_1)) = \int_{-\infty}^{S_1} \int_{-\infty}^{c_1} f(x, y)dxdy$$

It might be more intuitive to express C conditional on another factor distribution S rather than specifying directly their joint distribution. To do so, we proceed as follows: Let the first derivative of a bivariate copula with respect to its first (risk factor) argument u be $C_1(u, v) = \frac{\partial C(u,v)}{\partial u}$ or

$$C_1(u, v) = \frac{\partial C(u, v)}{\partial u} = C_{2|1}(v|u) = C_{2|1}(F_C(c_1)|F_S(S_1))$$

This conditional copula defines the conditional cumulative probability distribution of $(c_1|S_1)$ or

$$F_{c_1|S_1}(c_1) = C(F_C(c_1)|F_S(S_1))$$

The conditional probability distribution of C and that of S is thus:

$$\frac{\partial F_{c_1|S_1}(c_1)}{\partial c_1} = f(c_1|S_1)$$

In practice, additional knowledge, both qualitative and quantitative, is needed to define a copula. To this end, families of copulas have been studied, providing a library of relationships one may select from. Next, an example developed by Embrechts et al. (2002, 2003, 2007) is used to derive copulas from their condition (model) copulas.

Example (Conditional Copulas (Embrecht)): Say that a copula is given by $C(u, v) = [u^{-\alpha} + v^{-\alpha} - 1]^{-\frac{1}{\alpha}}, \quad \alpha > 0$ and calculate the derivative with respect to v. This results in a conditional copula, $C_{2|1}(v|u) = \frac{\partial C}{\partial u} = u^{-\alpha-1}[u^{-\alpha} + v^{-\alpha} - 1]^{-\left(\frac{1}{\alpha}+1\right)}$.

Now set the conditional copula $q = C_{2|1}(v|u)$ and solve for v. We obtain then the conditional density function given u,

$$v = \left\{-1 + \left(u^{-\alpha} q^{-\frac{\alpha}{1+\alpha}}\right) - u^{-\alpha}\right\}^{-\frac{1}{\alpha}} = \left\{-1 + u^{-\alpha}\left(q^{-\frac{\alpha}{1+\alpha}} - 1\right)\right\}^{-\frac{1}{\alpha}}$$

In other words, in terms of security prices x_1 and x_2 we have, by definition of $v = F_2(x_2), \quad u = F_1(x_1)$,

$$F_2(x_2) = \left\{-1 + (F_1(x_1))^{-\alpha}\left(q^{-\frac{\alpha}{1+\alpha}} - 1\right)\right\}^{-\frac{1}{\alpha}}$$

An explicit conditional copula is thus:

$$q = F_{2|1}(x_2|x_1) = \{-1 + (F_1(x_1))^{\alpha}[(F_2(x_2))^{-\alpha} - 1]\}^{-\frac{1+\alpha}{\alpha}}$$

while the conditional probability distribution is given by:

$$\begin{aligned} f_{2|1}(x_2|x_1) &= \frac{\partial F_{2|1}(x_2|x_1)}{\partial x_2} \\ &= (1+\alpha)\frac{f_2(x_2)}{F_2(x_2)}\left(\frac{F_1(x_1)}{F_2(x_2)}\right)^{\alpha}\{-1 + (F_1(x_1))^{\alpha}[(F_2(x_2))^{-\alpha} - 1]\}^{-\left(\frac{1+2\alpha}{\alpha}\right)} \end{aligned}$$

In other words, given x_1, the probability distribution function of x_2 is given by the equation above. Generally, if we define the k-dimensional Copula, with:

$$\begin{aligned} C_1(u_1) = u_1,\ C_2(u_1,u_2),\ldots\ldots, C_n(u_1,u_2,u_3,\ldots,u_n) \text{ with } C(u_1,u_2,u_3,\ldots,u_n) \\ = C_n(u_1,u_2,u_3,\ldots,u_n), \end{aligned}$$

the conditional copula is then:

$$C_k(u_k|u_2,u_3,\ldots,u_{k-1}) = P\{U_k \le u_k|U_2 = u_2,\ U_3 = u_3,\ldots,U_{k-1} = u_{k-1}\} \text{ and}$$

$$C_k(u_k|u_2,u_3,u_{k-1}) = \frac{\dfrac{\partial^{k-1} C_k(u_1,u_2,\ldots,u_k)}{\partial u_1,\ldots\ldots\partial u_{k-1}}}{\dfrac{\partial^{k-1} C_{k-1}(u_1,u_2,\ldots,u_{k-1})}{\partial u_1,\ldots\ldots\partial u_{k-1}}}$$

Thus, for k = 2, we have as seen earlier:

$$C_2(u_2|u_1) = \frac{\dfrac{\partial C_2(u_1,u_2)}{\partial u_1}}{\dfrac{\partial C_1(u_1)}{\partial u_1}} = \frac{\dfrac{\partial C_2(u_1,u_2)}{\partial u_1}}{\dfrac{\partial}{\partial u_1}u_1} = \frac{\partial C_2(u_1,u_2)}{\partial u_1}$$

Such models are, of course, only modeling tools that allow us to construct broader and more comprehensive models of dependence to reflect both our appreciation of factors' co-variation and their empirical evidence. For extensions of such an approach, we refer the reader to Embrechts et al. (2003) and Nelsen (1999).

Additional Sources and References

Risk and dependence models are necessarily multivariate and are therefore more difficult to analyze and define their underlying properties. Multivariate probability distributions may be found in books such as Joe 1997 on Multivariate Models and Dependence Concepts (see also Joe and Hu 1996), Aas et al. 2009; Bairamov and Gultekin 2010; Rodriguez 1977 and in Statistical Handbooks.

Malvergne and Sornette (2002) have demonstrated the transformational relationship of distributions to the normal probability distribution.

Bisawa and Wang (2002) have considered in particular the bivariate binomial distribution. Chib and Greenberg 1998, analyze multivariate probit models, Denuit and Lamber (2005) provide constraints on concordance measures in bivariate discrete data, while John Peter (1962) provides a tolerance region for multivariate distributions applicable to multivariate risk models. Fang et al. (1987) book on Symmetric Multivariate and Related Distributions provide an exhaustive list of such distributions and their properties. In 1988 Marshall and Olkin characterize families of multivariate distributions while Oakes 1994 defines families of multivariate survival distributions.

Specific applications to financial risk models have been pointed out by Stein 1973 and to Stochastic Programming problems by Tintner and Sengupta 1975.

Nelsen 1999 and an extensive research conducted at the Federal Polytechnic University in Zurich by Embrechts et al. has contributed immensely to the mathematical development of Copulas (Sklar 1973) as a modeling approach to multivariate distributions (Embrechts et al. 2001; Embrechts 2000a, b; Embrechts et al. 1997, 2001) Additional references include Cherubini et al. (2004), Frey and McNeil (2001), Everitt and Hand (1981), Anderson and Sidenius (2004) extending the normal copula, Li (1999, 2000a, b) who has been used profusely on default correlation to price credit derivatives and Lindskog (2000) on modeling copulas, Aas et al. (2009) construct a Pair copula of multiple dependence with application to insurance, Nikoloupoulos and Karlis (2008) study Multivariate logit copula model with application to dental data, that can also be applied to financial default models, Patton (2006) models asymmetric exchange rate dependence.

Chapter 5
Temporal Risk Processes

Abstract This chapter provides an introduction to inter-temporal probability processes used commonly in modeling risk processes. The chapter begins with the questions "what is time," "what is memory?" how are their definitions used to construct quantitative and temporal models. Elementary models such as probability models with the Markov property, random (binomial) walks, Poisson processes and continuous state and time stochastic processes are both presented intuitively and applied to many risk problems. These stochastic processes are then extended to more complex situations including long run memory models (fractal models), short memory models as well as to models departing from the basic Markov property and random walks models. While some of these models require a more advanced quantitative background than assumed for Chaps. 3 and 4, their applications are used to highlight both their importance and their implications to financial and risk models.

5.1 Time, Memory and Causal Dependence

An appropriate question we may begin with is: What is time? What is memory? And how do the two conjure to define dependence, change and define temporal risk processes? Greek philosophers associated time with change. Aristotle's (in *Methaphysica*) for example, defined time as objective and numerable associated to change and movement. In the Middle Ages, Saint Augustine (*Confessions*, Book X), suggested a more subjective approach to time and concluded that we are always in the present: The present of the present; The present of the future and The present of the past. Thus, at any instant of time our actions depend on the "here and now" of "our memory" of the past and of our expectation for the "hereafter." Some of our actions account for this explicitly, always implicitly. In other words, we may say that each one of our acts is framed in a perspective of the now, the memory of our past and our future expectations. Technically, we may use our record of the past (the memory) to assess our present (in which case a present state is the filter of our past),

C.S. Tapiero, *Engineering Risk and Finance*, International Series in Operations Research & Management Science 188, DOI 10.1007/978-1-4614-6234-7_5,

and construct forecasts of the future. These future forecasts need not be based on our past, but rather be based on some preferences (for example, as implied in the Arrow-Debreu theory for future assets pricing). In either situation, we have different approaches to conceive the here and now, from its past and/or its future. For the two, past and future to meet at the here and now, a process of change, a temporal risk model, expressing the time dependence of past and future and tested in light of observable events.

There are various concepts of time, however, each reflecting a belief and wisdom of its day. Martin Luther claimed that there is no meaning to time except in a spiritual sense, in terms of life in the hereafter. Time in this sense is "timeless." Kant brought to us the notion that time and space are not concepts but they are "forms" of intuition, and so on. Bertrand Russel, living in a great era of physical and scientific discoveries uses time, or rather the irreversibility of time, to determine time as a causal mechanism for explaining change. He states:

> *. . .it is possible to obtain ostensive definitions of the words for temporal relations: "preceding", "succeeding", "before", "after", "earlier", "later". When these words have come to be understood, we can understand such sentences a "A precedes B" even when A and B are not part of one specious present, provided we know what is meant by "A" and what is meant by "B".*

Thus, if we "causally order" a set of events, we have in essence described their inter-temporal dependent effects, their past and future meeting at each event as it occurs. Bertrand Russel, pursuing further the notion of time, in his "Time and Experience" distinguished two sources of our knowledge of time.

> *One is the perception of succession within one specious present, the other is memory. Remembering can be perceived, and is found to have a quality of greater or less remoteness, in virtue of which all my present memories can be placed in a time order. But this time is subjective, and must be distinguished from historical time. Historical time has to the present th relation of "preceding", which I know from the experience of change within one specious present. In historical time, all my present memories are now, but in so far as they are veridical they point to occurrences in the historical past (Bertrand Russel, Time and Experience, p. 216).*

Such an approach reflects too the underlying ideas expounded by Saint Augustine in his Confessions. To develop such ideas in a scientific sense, psychologists such as the School of Janet, 1920, Fraisse (1957) and many others have devised an experimental approach seeking to better understand time and memory and in particular in modeling the interactions of time-memory-change.

Piaget's ("La Notion du Temps chez l'Enfant"—The Notion of Time in Infants), indicated that time is tied to the notion of memory and parallels to our perception and awareness of elapsed time. Without memory there is no perception of time and therefore a coupling of time and memory is essential to construct the mechanisms that explains how change unfolds (or in other words, to construct the process of change). For example, if a process is temporally unrecorded, then time, as a dimension along which our analysis is made and extrapolated, is simply not defined. Conversely, without temporal sequentiality or time dependence, memory does not exist. In other words time, memory and expectations are intrinsically related.

While time is that variable recording the evolution of a process, memory and expectations are defining the incremental process of how (and why) a process evolves over time. For example, if a financial market is in equilibrium, it will not move from this equilibrium and therefore it will have no memory, nor change, since there will be no cause for change—no memory and no expectations. This means that a market in equilibrium has no temporal causality. When a market is in disequilibrium, it may be due to expectations or to a memory—either that of the memory of the equilibrium it seeks to return to, or the expectation of some other equilibrium (note that there may be multiple equilibriums or a dynamic equilibrium, reaffirmed as new information and a new environment occur).

For example, when financial markets are subjected to shocks, innovations and other interferences leading these markets to be out of equilibrium, then a persistent change will be induced seeking to regain a state of equilibrium at which financial markets can be again quoted uniquely and exchanges realized. In this sense, memory, which frames the past, and expectations framing the future, induce change and therefore the passage from one state to another.

These concerns are associated importantly to how we perceive and deal with uncertainty and thus their importance to risk modeling. For example, for finance while current prices are unpredictable, a "rational expectation of future prices" is constructed to define a price which is then inverted to define the future as a function of present prices one may observe. In this particular case, the past is no guide to the future and therefore predicting and trading with technical charts is misguided. Of course, some traders, anchored in their experiences would prefer to use "charting" techniques, reflecting a future price as an extrapolated-memory of the past with an unpredictable future. Lacking a comprehensive theory that can reconcile the past and the future and outlines the process of change, theories based on memory and expectations will continue to compete.

Below we shall consider a number of specific and quantitative inter-temporal risk models. We shall consider a number of models including Markov models, long run and short run memory models.

5.2 Time and Change: Modeling (Markov) Random Walk

In a memory-less process, there is no change, for every event is independent. For example, the inter-event time in a Poisson Process whose probability distribution is exponential, there is no memory as the time between events is defined by a memory-less probability distribution. In a memory based process, change is imbedded in a memory (a filter of past information)—whether objective or subjective and expectation for a process moving to another equilibrium state. For example, an insured who recently had an accident is likely to be in the forthcoming future more careful. Similarly, an aging part may have the memory of both its past stress and the work it has been subjected to—both of which may alter its reliability. Memory and forgetting can have far more intricate effects however. Advertising by firms are planned to counter the forgetting effects of consumers and remind them of past associations and positive experiences as well as construct new expectations to stimulate their wants. Numerous probability models are constructed to replicate such processes and to predict the future risks and uncertainties they may point to or generate. Such models make specific assumptions about their temporal and statistical dependence. They include Markov Chains, random walk as well as Jump models, fractal models and short memory models.

Random walks are defined by the aggregation of independent random events modeled by:

$$x_{t+1} = x_t + \Delta x_t$$

where $\Delta x_t = \tilde{\varepsilon}_t$ is a series of independently and identically distributed random variables. Markov Chains consist similarly of independent (transition) events, defined over a number of finite (or infinite) states. When events are memory dependent, higher order processes than random walks and Markov ones are needed to model their behavior. For example, if the probability of an event is a function of both its current state and its previous movement, then such a process has a higher order memory. In practice these characteristics are used to profit. In particular some financial traders use a stock price momentum to buy more of a stock (while contrarians, do exactly the opposite). In this case, momentum implies that a price growth feeds a price growth. Such processes, generalize random walks, are also called persistent random walks as we shall see subsequently at the end of this chapter. Such models can lead to contagious effects observed in various areas (such as health, finance, spread of rumors, etc.). Such memory has an important effect on change and thereby on volatility and risk. When events are positively time correlated, volatility increases and alters the expectation for change or its uncertainty. We consider first random walks.

5.2.1 Modeling Random Walks

In its simplest form, the discrete state random walk counts a series of Bernoulli experiments. Other models include processes generated by Gamma probability models (called Gamma Processes) as well as Pareto Levy probabilty models based on an underlying independent Pareto Levy distributions (introduced in Chap. 3). A background and a review of such models and their study can be found in an extensive literature (see additional resources and references at the end of this chapter). Below, we construct elementary stochastic models including the random-Bernoulli walk and its extensions to Markov Chains and to Normal (Brownian) probability limit models (modeled as a stochastic differential equations). In addition, we consider random-Poisson walks, and Queueing models. Finally, we conclude with Fractals, Range Processes, the Hurst Index and short memory processes.

5.2.2 *Stochastic and Independent Processes*

A stochastic process is a pair $\{x, t \geq 0\}$ or also written $x(t)$ with $f(x, t)$ denoting the probability distribution that the process time path assumes a real value x at time t. Such processes are independent increment processes. Explicitly, $\Delta x(t)$, $\Delta x(t) = x(t+1) - x(t)$ is an independent increment if $\Delta x(t)$ and $\Delta x(s)$ are statistically independent. It is also a stationary process and has a linear variance growth, namely, $E(x(t)) = tE(x(t)) + (1-t)E(x(0))$, $\text{var}(x(t)) = t\text{var}(x(t)) + (1-t)\text{var}(x(0))$. The proof is as follows: Let $f(t) = E(x(t)) - E(x(0))$. Stationary increments imply that:

$$\begin{aligned} f(t+s) &= E(x(t+s)) - E(x(0)) = E(x(t+s) - x(t)) + E(x(t) - x(0)) \\ &= E(x(s) - x(0)) + E(x(t) - x(0)) = f(t) + f(s) \end{aligned}$$

The only solution of this equation is the time linear function $f(t) = tf(1)$. It means that assuming an underlying random walk risk model, its variance is necessarily time linear. This contrasts to models such as long run memory models as well as short memory ones, where evolution of the variance may be time nonlinear.

Expectations in stochastic processes are called "adapted" when we have a filtration $\Im_t$ which defines and is defined with respect to the information (in this case, the past history of the stochastic process). If $\Im_t$ is known at that time, the process is no longer random at that particular time, while future realizations are random. This property is common to a number of random increment processes such as the Poisson and Wiener (Normal) processes as we shall see subsequently. These processes have no memory, i.e. past variations do not affect present or future ones.

5.2.3 *The Bernoulli-Random Walk: A Technical Definition*

Let $\Delta\xi(t) = \pm\Delta x$, be the outcomes of a Bernoulli (bi-state) random events, each with probability p and $1-p$ respectively. Thus,

$$\Delta\xi(t) = \begin{cases} +\Delta x & w.p.\ p \\ -\Delta x & w.p.\ 1\text{-}p \end{cases}$$

Since each jump is essentially a Bernoulli experiment with probability of success, p, the probability of y successes in n independent trials is binomial and given by:

$$P_y^n = \binom{n}{y} p^y (1-p)^{n-y}, \quad y = 0, 1, 2, 3, \ldots n$$

If y is the number of positive jumps, then $n - y$ is the number of negative jumps (where $\Delta\xi = -\Delta x$). The distance d covered in n jumps is therefore equal to: $d = [y - (n - y)]\Delta x$. The mean distance and its variance are given by $E(d)$ and $\mathrm{var}(d)$, with $q = 1 - p$. Or, $E(d) = n(p - q)\Delta x$; $\mathrm{var}(d) = 4npq(\Delta x)^2$. This leads to the unrestricted random walk process. Inserting restrictions on the number of states or on specific states will lead to different random walks.

When the time intervals between jumps are very small and their numbers increases the random walk converges to a normal probability distribution. Namely, let Δt and Δx be very small increments. In a time interval [0, Δt], the number of events is n given by $n = [t/\Delta t]$. When Δx is a small time increment, then (with $t/\Delta t$ integer):

$$E(d) = \frac{t(p-q)\Delta x}{\Delta x}, \quad \mathrm{var}(d) = \frac{4tpq(\Delta x)^2}{\Delta x}$$

Assuming existence of limits, we set the mean and variance limits to be $\Delta x/\Delta t$ and $(\Delta x)^2/\Delta t$, $\Delta x \to 0$ or, $\lim\limits_{\Delta t \to 0} \frac{\Delta x}{\Delta t}(p - q) = 2C$, $\lim\limits_{\Delta x \to 0} \frac{(\Delta x)^2}{\Delta t} = 2D$. Introduce the variable transformation $p = 1/2 + (C/2D)\Delta x$, the continuous time expected mean displacement $m(t)$ and variance $\sigma^2(t)$ are: $m(t) = 2Ct$; $\sigma^2(t) = 2Dt$.

Finally, by *the law of large numbers*, we obtain a displacement whose distribution is the normal probability distribution with mean and variance $m(t)$ and $\sigma^2(t)$ respectively, or

$$f(x,t) = \frac{1}{\sqrt{2\pi}\sigma(t)} \exp\left\{-\frac{1}{2}\frac{[x - m(t)]^2}{\sigma^2(t)}\right\} \quad \text{where} \quad (f(x,t) \to 0 \text{ as } x \to \pm\infty).$$

To represent the process as a stochastic process we write: $\Delta x = \Delta m(t) + \sigma \Delta W(t)$ with $\Delta m(t) = 2C\Delta t, \sigma = \sqrt{2D}$ where $\Delta W(t)$ are normally and identically distributed random variables with mean 0 and variance Δt. In continuous time, such an equation is often written by:

$$dx(t) = 2Cdt + \sqrt{2D}dW(t)$$

Such an equation is called an Ito-stochastic differential equations. Its evolution is entirely independent of its past history. A position at an instant of time depends only on its previous position. Such assumptions, while mathematically convenient do not always represent the evolution of observed processes. The stringency of the assumptions required to construct such processes, embed therefore the limitation of risk models. Such processes can be generalized however to account for a greater variety of risk models.

5.2.4 The Trinomial Random Walk

The trinomial random walk is defined by the following model:

$$x_{t+1} = x_t + \varepsilon_t; \quad y_t = x_t \ \varepsilon_t = \begin{cases} +1 & w.p. \ \ p \\ 0 & w.p. \ \ r \\ -1 & w.p. \ \ q \end{cases}$$

where ε_t is a trinomial probability distribution given (p, r, q) whose cumulative distribution is $(p, p + r, 1)$. It is simple to demonstrate that the process mean and variance are:

$$E(X_t) = (p - q)t; \quad \mathrm{Var}(X_t) = [r(1 - r) + 4pq]t$$

and in particular, the standardized random variables tends in probability to a zero mean, unit variance normal random variable.

$$\frac{X_t - t(p - q)}{\sqrt{[r(1 - r) + 4pq]t}} \sim N(0, 1)$$

Assuming that individual transitions are independent, the probability of m positive jumps, n negative jumps in T periods, has a multinomial probability distribution given by:

$$P(m, n, T - m - n) = \frac{p^m T!}{m!} \frac{q^n}{n!} \frac{r^{T-m-n}}{(T - m - n)!}$$

with the first two moments given by:

$$E(m) = Tp, \ E(n) = Tq, \ E(T - m - n) = T(1 - p - q)$$

as well as

$$\mathrm{Var(m) = Tp(1 - p), Var(n) = Tq(1 - q), cov(m, n) = -Tpq.}$$

5.2.5 Random Walk as a Difference Equation

Consider a gambler whose wealth is n, has the chance in any one try to win \$1 with probability p and a probability q of losing. Given that there can only be two states at any one time, the wealth probability is defined in terms of two states $\mathrm{n} - 1$ and $\mathrm{n} + 1$. The probability of a wealth is then:

$$P_n = pP_{n-1} + qP_{n+1}, \ p + q = 1$$

Whose solution is a function of two parameters $P_n = a + b(p/q)^n$ where a and b are two arbitrary constants that depend on the boundary conditions of the game.

If we assume that the game terminates when there is no more money to gamble with, then the boundary conditions are: $P_0 = 1, P_N = 0$ and the solution for the parameters (a, b) leads to:

$$P_n = \frac{(p/q)^N - (p/q)^n}{(p/q)^N - 1} \text{ for } p \neq q \text{ and } P_n = \frac{N-n}{N} \text{ for } p = q$$

Of course, by changing the boundary conditions, different probabilities will be obtained. In this sense, a process probability distribution, is defined by the transition probabilities and by the restrictions (boundaries) imposed on it. Each transition depends on its present state and not on its outcome (it has therefore no memory of previous events). If we seek the wealth probability at a given trial time t, then, by recursion, the following probability results:

$$P_n(t+1) = p_{n-1}P_{n-1}(t) + p_{n+1}P_{n+1}(t) = \sum_{i=n-1,n+1} p_i P_i(t)$$

where $\{p_{n-1}, p_{n+1}\}$ are now state dependent probabilities, expressing the dependence of the transition probability on its state. A general expression for such models based on say, N states, can be written for convenience as follows:

$$P_n(t+1) = \sum_{j=1}^{N} p_{jn} P_j(t)$$

where p_{jn} is the probability of being at time t in state j and transiting to state n in the time interval $[t, t+1]$. These are Markov models which are often used to construct risk models.

5.2.6 *The Random-Poisson Continuous Time Walk*

We consider next the same process but in continuous time and obtain a Poisson process. It is based on two assumptions: (1) Events are independent and (2) Events occur one at a time. If the event rate (i.e. the number of average events occurring in a given time interval) is also constant, the Poisson probability distribution results. Consider two subsequent instants of time $[t, t+dt]$ and say that at time $t+dt$ there are n events occurrences. Given all events can only occur one at a time, this state can arise from two potential previous states at time t, given by n and $n-1$. The probability of an event occurring in the time interval is λdt in $[t, t+dt]$ as a result, since these probabilities are independent, the probability of n occurrences at time t + dt, $P_n(t+dt)$ is explicitly given by:

$$P_n(t+dt) = P_n(t)[1-\lambda dt] + P_{n-1}(t)[\lambda dt], \quad n = 1, 2, 3, \ldots.$$
$$\text{and } P_0(t+dt) = P_0(t)[1-\lambda dt]$$

The first set of equations represents the transition probabilities to an "n" state while the second equation represents the transition into the zero state. These equations can be written as a systems of linear differential equations which we write as follows:

$$dP_n(t)/dt = -P_n(t)\lambda + P_{n-1}(t)\lambda, \quad n = 1, 2, 3, \ldots. \text{ and } dP_0(t)/dt = -P_0(t)\lambda$$

It is a simple exercise to show that the solution of these equations leads to a Poisson probability distribution:

$$P_n(t) = e^{-\lambda t}[\lambda t]^n/n! \text{ with } F_m(t) = \sum_{n=0}^{m} P_n(t) = \sum_{n=0}^{m} e^{-\lambda t}\frac{(\lambda t)^n}{n!} = 1 - \int_0^{\lambda t} \frac{x^m e^{-x}}{m!} dx$$

whose mean and variance are equal and given by $E(n) = \text{var}(n) = \lambda t$. Further, $\int_0^{\lambda t} \frac{x^m e^{-x}}{m!} dx$, is the Gamma integral.

The Poisson process has the independence and memory-less properties as well as an inter-event exponential probability distribution (compare to the random walk whose interevent time has a geometric distribution). For example, if a Poisson process is used to count the number of insurance claims, then these claims must be statistically independent. The memory-less property further requires that if an event has already occurred at time b, then the probability that the next event occurs is independent of that first event. Explicitly, let T be the next time an event occurs. We want to prove that the probability: $P(T > a + b|T > b)$, that an event occurs at a $+b$ is independent of b, the first time occurrence and demonstrate that this inter-event time is only a function of a (which has an exponential probability distribution). By elementary calculus of conditional probabilities,

$$P(T > a + b/T > b) = \frac{P(T > b \text{ and } T > b + a)}{P(T > b)} = \frac{P(T > b + a)}{1 - F(b)}$$
$$= \frac{1 - F(b + a)}{1 - F(b)}$$

If the distribution is exponential, this is given by:

$$P(T > a + b/T > b) = \frac{1 - F(b + a)}{1 - F(b)} = \frac{1 - \left[1 - e^{-\lambda(a+b)}\right]}{1 - \left[1 - e^{-\lambda b}\right]} = e^{-\lambda a} = 1 - F(a)$$

As a result, we can conclude that: "At any time", after an event has occurred, the residual time for the next event to occur, has an exponential probability distribution. The mean residual time is equal to $1/\lambda$. This residual time is independent of the event which has occurred at time b. This implies then that the Poisson process has no memory as we stated above.

5.2.7 *The Continuous Time Continuous State Approximation*

Consider next the same equation with extremely small time interval and small movements . In small intervals of time, we have:

$$P_n(t+\Delta t) = P_{n-\Delta n}(t)p + P_{n+\Delta n}(t)q, \quad q = 1-p$$

Δn is a small increment in the growth or loss of wealth in the time interval Δt. Replacing these equations by Taylor series approximations, we have:

$$\begin{aligned} P(t,n) + \frac{\partial P(t,n)}{\partial t}dt &\approx pP(t,n) - p\frac{\partial P(t,n)}{\partial n}dn + \frac{p}{2}\frac{\partial^2 P(t,n)}{\partial n^2}(dn)^2 \\ &+ (1-p)P(t,n) + (1-p)\frac{\partial P(t,n)}{\partial n}dn + \frac{(1-p)}{2}\frac{\partial^2 P(t,n)}{\partial n^2}(dn)^2 \end{aligned}$$

And therefore,

$$\begin{aligned} P(t+\Delta t,n) &= P(t,n) + \frac{\partial P(t,n)}{\partial t}dt \\ P(t,n-\Delta n) &= P(t,n) - \frac{\partial P(t,n)}{\partial n}dn + \frac{1}{2}\frac{\partial^2 P(t,n)}{\partial n^2}(dn)^2 \\ P(t,n+\Delta n) &= P(t,n) + \frac{\partial P(t,n)}{\partial n}dn + \frac{1}{2}\frac{\partial^2 P(t,n)}{\partial n^2}(dn)^2 \end{aligned}$$

which is reduced to:

$$P(t,n) + \frac{\partial P(t,n)}{\partial t}dt = P(t,n) - (p-q)\frac{\partial P(t,n)}{\partial n}dn + \frac{1}{2}\frac{\partial^2 P(t,n)}{\partial n^2}(dn)^2$$

And therefore, to:

$$\frac{\partial P(t,n)}{\partial t} = -(p-q)\frac{\partial P(t,n)}{\partial n}\frac{dn}{dt} + \frac{1}{2}\frac{\partial^2 P(t,n)}{\partial n^2}\frac{(dn)^2}{dt}$$

This equation is a partial differential equation where $\frac{dn}{dt}$ denotes a rate of change in n while $\frac{(dn)^2}{dt}$ denotes a quadratic variation of n. One is used to denote a mean rate of change while the other a variance rate of change. Such a formulation is representative of a functional approach to the random walk in continuous time. Such formulations assume many forms. For example, let $\mu(t,n) = \frac{dn}{dt}$, $\sigma^2(t,n) = \frac{(dn)^2}{dt}$ which indicates that the actual rate of change and its variance are a function of n, and leading to nonlinear equations we can use to represent more complex models (for example, growth and contagion models) in terms of a nonlinear partial differential equation.

5.2.8 *The Poisson-Jump Process and its Approximation as a Brownian Model*

We noted earlier that the Poisson process is an independent process and therefore it consists of *stationary independent increments* which are *identically distributed.* That is, for an increment Δx_t, we have $E(\Delta x_t) = \lambda\ \Delta t$ where λ is the mean rate of a Poisson distribution:

$$P(\Delta x_t = n) = \exp(-\lambda \Delta t)(\lambda \Delta t)^n/n!,\ \ \mathrm{n} = 0,1,2,3,\ldots.$$

An alternative representation of the Poisson process consists in setting:

$$\Delta x_t = \lambda\ \Delta t + \sigma\ \tilde{\mu}\ (\Delta t)$$

For the Poisson process, we have the special case $\lambda = \sigma^2$ and $\tilde{\mu}\ (\Delta t)$ is a Poisson noise standardized in the following manner. We let $\tilde{\mu}(\Delta t) = \mu\ (\Delta t) - \Delta t$ where $\mu\ (\Delta t)$ is a random variable with parameter Δt, thus:

$$\mathrm{Prob}[\tilde{\mu}(\Delta t) = 0] = 1 - \Delta t + o(\Delta t); \mathrm{Prob}[\tilde{\mu}(\Delta t) = 1]_t$$
$$= \Delta t + o(\Delta t);\ \ \mathrm{Prob}[\tilde{\mu}(\Delta t)\ \geq\ 2] = o(\Delta t)$$

or $\mathrm{E}\{\tilde{\mu}(\Delta t)\} = 0$ and $\mathrm{var}(\tilde{\mu}(\Delta t)) = \Delta t$. When Δt becomes very small, we define formally a stochastic differential equation for the Poisson process;

$$dx_t = \lambda dt + \sigma\tilde{\mu}\ (dt);\ \ \sigma^2 = \lambda$$

Evidently, $\tilde{\mu}dt$ represents now a discrete state process, taking on values of 0 and 1 only. A generalization of this approach will lead to a class of stochastic models we shall call, jump processes. It is also evident that if we approximate the Poisson process by the Wiener process, we shall have equations of the form:

$$dx_t = \lambda dt + \sigma dW$$

where dW is a Wiener process. The representation of a discrete random process (such as Poisson), by a continuous one (diffusion approximation) is made however, for convenience. Generalizations are straight forward leading to jump processes (see Tapiero 1998b).

5.2.9 *The Multiplicative Bernoulli-Random Walk Model*

The multiplicative random walk model is defined as follows:

$$S_{t+1} = \begin{cases} (1+h)S_t & \text{with probability } p \\ (1-\ell)S_t & \text{with probability } 1-p \end{cases}$$

where S_t denotes for example, the price of a stock at time t and S_0 is the initial price. In this case, the price at time t, is given by the binomial probability distribution, or,

$$S_t = (1+h)^i(1-\ell)^{t-i}S_0 \quad \text{with probability} \binom{t}{i}p^i(1-p)^{t-i}$$

where i is the number of times the price has increased with probability p and $t-i$ is the number of times it has moved with probability $(1-p)$. That is:

$$0 \leq p = \frac{[(1+r)-(1-\ell)]}{(1+h)-(1-\ell)} = \frac{r+\ell}{h+\ell} \leq 1.$$

The expected values are then calculated by the following (note that these are not market prices, but just values):

$$\widehat{S}_t = \sum_{i=0}^{t}\binom{t}{i}((1+h)p)^i[(1-\ell)(1-p)]^{t-i}S_0,$$

$$E(S_t^k) = \sum_{i=0}^{t}\binom{t}{i}\left(p(1+h)^k\right)^i\left((1-p)(1-\ell)^k\right)^{t-i}S_0^k$$

For the variance, $\text{var}(S_t) = E(S_t^2) - \widehat{S}_t^2$. By letting $a = Hp$ and $b = L(1-p)$, we obtain of course the binomial sum with $a+b \neq 1$. $\hat{S}_t = \sum_{i=0}^{t}\binom{t}{i}a^ib^{t-i}S_0$.

5.2.10 *The BD Model in Continuous Time with Distributed Times Between Jumps*

It is possible to generalize the Birth-Death random walk in a number of ways. Say that the probability of a positive jump is p and the probability of a negative jump is q, $(p+q=1)$. However, the time between jumps has an exponential distribution with mean $1/\mu$. In this case, the probability of being in state x at time t, $P(x,t)$, is given by "the exponential memory process":

$$P(x,t)=\int_0^t \mu e^{-\mu z}[pP(x-1,t-z)+qP(x+1,t-z)]dz,\ p+q=1$$

which can be written as follows (where $\tau=t-z$):

$$P(x,t)=-\int_0^t \mu e^{-\mu(t-\tau)}[pP(x-1,\tau)+qP(x+1,\tau)]d\tau$$

Taking a derivative with respect to time t, we obtain:

$$\begin{aligned}\frac{dP(x,t)}{dt}&=-\frac{d}{dt}\int_0^t \mu e^{-\mu(t-\tau)}[pP(x-1,\tau)+qP(x+1,\tau)]d\tau\\&=\mu e^{-\mu(t-t)}[pP(x-1,t)+qP(x+1,t)]\\&\quad+\mu\int_0^t \mu e^{-\mu(t-\tau)}[pP(x-1,\tau)+qP(x+1,\tau)]d\tau\end{aligned}$$

and therefore:

$$\frac{dP(x,t)}{dt}=-\mu P(x,t)+\mu pP(x-1,t)+\mu qP(x+1,t)$$

where the initial conditions are:

$$P(0,0)=1,\ \ P(x,0)=0,\ \forall x\neq 0$$

Note that this corresponds to a birth-death process which is given by the following probabilities:

$$\begin{cases} x_{t+\Delta t} = x_t + \varepsilon_t \quad \text{where} \\ \varepsilon_t = \begin{cases} +1 \ \text{w.p.} & \mu p \Delta t \\ 0 \ \text{w.p} & 1\text{-}\mu\Delta t \\ -1 \ \text{w.p.} & \mu q \Delta t \end{cases} \end{cases}$$

and therefore, (using our previous result for the birth death random walk):

$$E(x_t) = t\mu(p-q); \quad \text{var}(x_t) = t\left[\mu(1-\mu) + 4\mu^2 pq\right]$$

which leads to the approximate stochastic differential equation:

$$dx = \mu(p-q)dt + \sqrt{\mu(1-\mu) + 4\mu^2 pq}\, dw$$

The process defined above can be thought of as an exponential pausing time. For example, if the time between exchange rates changes is exponential, then the model above might be used to represent profitably the evolution of rates (taking into account the time spent when no change is allowed). The pausing time may take other forms however. In many applications this might be relevant. Market forces, central banks intervention in currency markets to stabilize exchange rates and short term interest rates as well as transaction costs of various sorts, may individually and collectively dampen exchange and interest rates variations. This means that the probability of remaining on the same state in a given time interval is appreciable, (expressing memory effects as well), or that the time between successive transitions is also appreciable. In this sense, the process may be "sticky," hanging on to its current state as we have seen for the exponential memory-less pausing time. For this reason, the BD model captures an important dimension on process variation.

Say that the time between transitions, τ, has an Erlang probability distribution, then the probability of reaching state n at time t is given by:

$$P(n,t) = \int_0^t \Phi(t-\tau)[qP(n+1,\tau) + pP(n-1,\tau)]d\tau, \; p+q=1 \quad \text{or}$$

$$P(n,t/m) = \frac{1}{\lambda^m \Gamma(m)} \int_0^t (t-\tau)^{m-1} e^{-(t-\tau)/\lambda}[qP(n+1,\tau) + pP(n-1,\tau)]d\tau$$

which is equivalent to:

$$\frac{dP(n,t/m)}{dt} = -\frac{1}{\lambda}\Delta P(n,t/m), \;\; \Delta P(n,t/m) = [P(n,t/m) - P(n,t/m-1)]$$

while another derivative yields,

$$\lambda^2 \frac{d^2P(n,t)}{dt^2} = \Delta P(n,t/m) - \Delta P(n,t/m-1)$$

with

$$P(n,t/1) = \frac{1}{\lambda}\int_0^t e^{-(t-\tau)/\lambda}[qP(n+1,\tau) + pP(n-1,\tau)]d\tau$$

which is the memory-less model. Although the study of this equation is somewhat more difficult, it can be studied using Transforms techniques. We see that although in the mean, the BD process evolves proportionately to $(p-q)$, it's variance is smaller than the equivalent process without any memory. In this sense, if a process exhibits a smaller variance than expected (and it is based on the usual random walk), then it is likely to be a BD random walk.

5.3 Inter-Event Times and Run Time Stochastic Models

Assume a random walk and consider the probability of a success (for example a positive event) p, $q = 1 - p$. Such processes, define the run time till a predetermined or a number of predetermined events occur. In risk models for example, such processes are used to assess the amount of time for cumulative losses and gains reach a given level. For example, for a reserve of 200,000,000 set aside by a bank based on the VaR regulation, how long would it take for the cumulative loss and gains to reach the reserve set aside. For simplicity consider a process of simple Bernoulli events, and let n be the number of trials till a default event occurs. Its random variable y_n has then a geometric probability distribution:

$$y_n = \begin{cases} 0 & n = 0 \\ pq^{n-1} & n = 1,2,3,4,5,\ldots. \end{cases}$$

The probability distribution of obtaining r successes is therefore given by the convolution of the random variable $\{y, n \geq 0\}$. First we calculate the distribution probability generating function (PGF):

$$y^*(z) = \sum_{n=1}^{\infty} z^n pq^{n-1} = \frac{pz}{1-qz}, \quad |z| < 1$$

The rth convolution PGF is given by the PGF rth product, or:

$$Y_r^*(z) = [y^*(z)]^r = \left[\frac{pz}{1-qz}\right]^r$$

A series expansion of the denominator yields,

$$Y_r^*(z) = p^r z^r \begin{bmatrix} 1 + rqz + \frac{r(r+1)}{2!} q^2 z^2 + \frac{r(r+1)(r+2)}{3!} q^3 z^3 + \ldots\ldots \\ \ldots\ldots. + \frac{r(r+1)\ldots.(n-1)}{(n-r)!} q^{n-r} z^{n-r} + \ldots\ldots.. \end{bmatrix}$$

which we rewrite simply by:

$$Y_r^*(z) = p^r z^r \left[1 + rqz + \frac{r(r+1)}{2!} q^2 z^2 + \ldots\ldots.. + \binom{n\text{-}1}{n\text{-}r} q^{n-r} z^{n-r} + \ldots\ldots.. \right]$$

Since the probability of the *rth* event is the *rth* coefficient of the expanded series, the probability distribution of obtaining r successes in n trials is given by the negative binomial distribution (NBD).

$$Y_r = \begin{cases} 0 & n = 0 \\ \binom{n\text{-}1}{n\text{-}r} q^{n-r} z^{n-r} & n = r, r+1, r+2, r+3, \ldots\ldots\ldots \end{cases}$$

In Chap. 11 we shall consider a number of control problems related to regulation and the controls of regulated using such a negative binomial distribution. Some of these consist in defining a control strategy of a regulated when regulation consists of a large number of items to be controlled. For example, financial regulation may involve thousands of specific items to be controlled which render full controls (i.e. all regulation items) overly expensive. The geometric probability distribution and its generalization, the negative binomial distribution may in some cases be used to construct such strategies. Explicitly, say that the prior assumption of the regulator is that there is a probability p that the regulated is non-compliant. Thus, the probability of controlling successive regulation items and finding one non-compliant has a geometric probability distribution whose parameter n is the random number of controls till one non-compliance is detected. In this case, a policy may be defined in terms of a parameter such that, if m regulation items have been found to be compliant, and none are non-compliant, then the regulated may be deemed to be compliant and given a "clean bill of regulation-health." If $n < m$, then of course, the regulated is found to be non-compliant and further controls may be deemed necessary or not and the regulated penalized.

5.4 Randomized Random Walks and Related Processes

5.4.1 The Randomized Random Walk Distribution

As a special case, assume that the time between jumps in a random walk is an exponential distribution with a common density e^{-t}. In other words, the time between jumps is regulated by a jump Poisson process while the jump size remains

+1 with probability p and -1 with probability q, independent of each other and of the Poisson process. We have seen earlier that for random walks, the probability of being in position r after n jumps, is:

$$\binom{n}{\frac{1}{2}(n+r)} p^{\frac{1}{2}(n+r)} q^{\frac{1}{2}(n-r)} = \binom{r+2\nu}{r+\nu} p^{r+\nu} q^{\nu}; \quad 2\nu = n-r$$

Further, Using the Poisson distribution, it is also well known that at time t, the probability that there were exactly n jumps (which is equal $2\nu + r$) is given by $e^{-t}t^n/n!$. Thus, a mixture of the binomial distribution by the Poisson distribution leads to the desired probability distribution of the random walk, or

$$e^{-t} \sum_{\nu=0}^{\infty} \frac{t^{r+2\nu}}{(r+2\nu)!} \binom{r+2\nu}{r+\nu} p^{r+\nu} q^{\nu} = \sqrt{(p/q)^r} e^{-t} I_r(2t\sqrt{pq})$$

where I_r is a Bessel function, representing the probability of being in state r at time t. This function is explicitly given by the polynomial function:

$$I_r(x) = e^{-t} \sum_{\nu=0}^{\infty} \frac{1}{\nu!} \frac{1}{\Gamma(\nu+r+1)} \left(\frac{x}{2}\right)^{r+2\nu}$$

Note that when the probability of p is random, it may be estimated by a logistic probability model:

$$\ln\left(\frac{\tilde{p}}{1-\tilde{p}}\right) = \sum_{i=1}^{k} a_i x_i + \tilde{\varepsilon}$$

where $\tilde{\varepsilon}$ is an error, a standard normally distributed random variable. In this case, the probability distribution of being in state r is given by:

$$E_p\left\{\sqrt{(\tilde{p}/(1-\tilde{p})^r} e^{-t} I_r\left(2t\sqrt{(\tilde{p}(1-\tilde{p})}\right)\right\}$$

5.4.2 *Binomial-Lognormal Process*

By the same token, we can determine a binomial-lognormal price process when the time between jumps is exponential. As stated in Chap. 3, let the multiplicative random walk,

$$x_{n+1} = \begin{cases} Hx_n & \text{with probability } p \\ Lx_n & \text{with probability } 1-p \end{cases}$$

where x_n denotes the price at the nth jump with $x_0 = 0$, the initial price. In this case, the price (position of the random walk) at time t, is given by the binomial probability distribution, or

$$x_n = H^i L^{n-i} x_0 \text{ with probability } \binom{n}{i} p^i (1-p)^{n-i}$$

with i the number of "positive jumps" with probability p and $n-i$ the number of "negative jumps" with probability $(1-p)$. Now, if the probability distribution between jumps has an exponential probability distribution $(1/\lambda)e^{-t/\lambda}$, then, the probability that there are i jumps in n trials has a Poisson probability distribution with mean λn, or

$P(i;\lambda) = (\lambda n)^i e^{-\lambda n}/i!$ and therefore x_n is :

$$x_n = x_0 e^{-\lambda n} L^n \Phi(n), \text{ with probability : } e^{-\lambda n}(1-p)^n \Gamma(n,\beta),$$

$$\Gamma(n,\beta) = \sum_{i=0}^{n} \binom{n}{i} \frac{\beta^i}{i!}, \quad \beta = \lambda n \left(\frac{p}{1-p}\right)$$

with $\Phi(n) = \sum_{i=0}^{n} \alpha^i / i!$, $\alpha = \lambda n (H/L)$.

The expected value is then calculated explicitly by the following:

$$\hat{x}_n = e^{-\lambda n}(L(1-p))^n x_0 \Gamma(n,\delta), \quad \delta = \lambda n H p / L(1-p)$$

Again, note that in case the probability p is given by the logistic distribution, we have:

$$E_p e^{-\lambda n}(1-p)^n \sum_{i=0}^{n} \binom{n}{i} \frac{\left(\lambda n\left(\frac{p}{1-p}\right)\right)^i}{i!} = e^{-\lambda n} \sum_{i=0}^{n} \binom{n}{i} \frac{(\lambda n)^i}{i!} E_p\left(p^i (1-p)^{n-i}\right)$$

5.5 Markov Chains

The origin of Markov chains is due to Markov, a Russian mathematician who first published in the Imperial Academy of Sciences in St. Petersburg in 1907, a paper studying the statistical behavior of the letters in Onegin, a well known poem of

Pushkin. Assume that a process is denoted by its states, $i = 1, 2, \ldots n$. For example, $i = 0, 1, 2, \ldots M$ might denote the state of a treasury confined to a set of acceptable levels 0, 1, 2,. . .,M while $i = 0, 1$ might denote two states, a claim (a risk event) has occurred or not etc. When we construct a Markov chain, it is possible to switch from one state to another with a given probability, expressing the random character of the process. For a two states process, we have four possible transitions which are given below.

00: Initial state "0," subsequent state "0"
01: Initial state "0," subsequent state "1"
10: Initial state "1," subsequent state "0"
11: Initial state "1," subsequent state "1"

For example, the first number, say "0," may stand for a no insurance claim in a specific year, while the second number may stand for the claim state in a subsequent year. Switching from one state to another is then defined by the conditional transition probability p_{ij}. In matrix form we have: $\mathbf{P}$, also called the Markov transition matrix.

$$\mathbf{P} = \begin{bmatrix} p_{11} & p_{12} & p_{13} \\ p_{21} & p_{22} & p_{23} \\ p_{31} & p_{32} & p_{33} \end{bmatrix}, \quad p_{ij} \geq 0,\ i = 1, \ldots, n;\ j = 1, \ldots, n \text{ and } \sum_{j=1}^{n} p_{ij} = 1$$

Generally, the matrix $\mathbf{P^T}$ denotes the transition probabilities from one state to another in T steps. As a result, $\mathbf{P^T} = \mathbf{PP^{T-1}} = \mathbf{P^{T-1}P}$ which is also the *Chapman-Kolmogorov* equation which we can write as follows:

$$p_{ij}(T) = \sum_{k=0}^{n} p_{ik}(t) p_{kj}(T - t) \text{ or } \mathbf{P^T} = \mathbf{P^t P^{T-t}}$$

with $p_{ij}(t)$, the transition probability from i to j in t steps.

$$p_{ij}(T) = \sum_{k=0}^{n} p_{ik}(1) p_{kj}(T - 1) \quad \text{Or :} \quad \mathbf{P^T} = \mathbf{P^{T-1}P}$$

which is the second part of the above equation. In two steps, we have then:

$$p_{ij}(2) = \sum_{k=0}^{n} p_{ik}(1) p_{kj}(1) = \sum_{k=0}^{n} p_{ik} p_{kj}$$

Further, if the number of steps is large, the transition probabilities are called ergodic transition probabilities and are given by the equilibrium probabilities (assuming they exist):

$$\pi_{ij} = \operatorname*{Lim}_{t\to\infty} P_{ij}(t) \text{ or } \pi = \operatorname*{Lim}_{t\to\infty} \mathbf{P^t}$$

An application of the Chapman-Kolmogorov matrix multiplication formula, yields

$$\mathbf{P}\pi = \pi\mathbf{P} = \pi, \sum_{\mathrm{i}=1}^{\mathrm{n}} \pi_i = 1; \quad \pi_{\mathrm{i}} \geq 0$$

The implications of these equations however are important in practice since they reveal the states and the probabilities a process will tend to. Thus, if we compare two Markov chains, we can study their long run effects by their ergodic probabilities.

Example Consider an insured who is assumed at any 1 year to be in one of the two states: (0, 1). The first state "0" states that no claim is made within the year while "1" states that the insured had an accident and has therefore put in a claim. When in a given year the insured has no claim, the probability that a claim is made in the following year is α while the probability that no claim is made when the insured has previously claimed is β. The claims Markov chain is then:

$$\mathbf{P} = \begin{bmatrix} 1-\alpha & \alpha \\ \beta & 1-\beta \end{bmatrix}$$

The nth period Markov chain, corresponding to the switching probabilities in n periods is thus given by calculating the nth power of the switching probabilities matrix. In other words, it is a simple exercise to show (by induction) that:

$$\mathbf{P}^n = \frac{1}{\alpha+\beta}\begin{bmatrix} \beta & \alpha \\ \beta & \alpha \end{bmatrix} + \frac{(1-\alpha-\beta)^n}{\alpha+\beta}\begin{bmatrix} \alpha & -\alpha \\ -\beta & \beta \end{bmatrix}$$

Note that when the number of periods becomes very large, we obtain:

$$\overline{\mathbf{P}} = \frac{1}{\alpha+\beta}\begin{bmatrix} \beta & \alpha \\ \beta & \alpha \end{bmatrix}$$

Of course, this matrix can be calculated directly by inverting **P**, or by solving the matrix equalities $\pi = P\pi, \ \pi_1 + \pi_2 = 1, \ \pi_i \geq 0$.

5.6 Applications

5.6.1 The Sums of Poisson Distributed Events Is Also Poisson

If several Poisson processes have means given by $\lambda_i t, \ i = 1, 2, \ldots n$ their sum is also a Poisson process whose mean is the sum of the individual process means. The proof of this property is straightforward by application of the convolution theorem. In the same spirit, "fractionating" a Poisson process whose mean is λ by a Bernoulli distribution with parameter $p, \ 0 < p < 1$ yields also a Poisson process with parameter λp. For example, say that the number of claims of an insurance firm has a Poisson distribution with mean λt. Let p be the probability that the claim is not justified. In such a case, show that the number of claims that are justified has a Poisson distribution whose mean is $(1 - p)\lambda t$.

5.6.2 Collective Risk and the Compound Poisson Process

In actuarial mathematics, a risk is a random variable on some probability space consisting of the claim size, and the number of claims. Thus, a risk is an independent sequence $N, \ y_1, y_2, \ldots, y_N$ where N is the number of claims (given by a Poisson distribution for example) and the claims y_i which are assumed statistically independent. In this case, the risk defined by the total claims is given by: $X = \sum_{i=1}^{N} y_i$. For example, if individual claims are limited to a maximal amount of a, then the risk is:

$$X_a = \sum_{i=1}^{N} \min(y_i, a) \quad \text{while } Y = X - X_a = \sum_{i=1}^{N} \max(y_i - a, 0)$$

It may be insured by a reinsurer (who assumes this risk for a given premium which is negotiated between the first insurer and the reinsurer). Credit unions are also, in some ways, similar to collective risk (mutual) insurance firms. They provide credit to consumers, particularly those with low incomes who are unable to pay the substantially higher fees and interest rates charged by banks. The mathematical analysis of such claim processes can be treated by the compound Poisson process defined by $\{X(t), t \geq 0\}$ which is represented by:

$$X(t) = \sum_{i=1}^{N(t)} y_i$$

where $N(t)$ is a Poisson process. The study of such processes is easily performed by using the probability generating function. Note that the PGF for the Poisson distribution is given by:

$$N^*(z) = \sum_{i=0}^{\infty} \frac{(\lambda t)^i e^{-\lambda t}}{i!} z^i = e^{-\lambda t} \sum_{i=0}^{\infty} \frac{(\lambda z t)^i}{i!} = e^{-\lambda t} e^{\lambda z t} = e^{-\lambda t(1-z)}$$

Thus, the probability generating function of the compound Poisson process is given by:

$$X^*(z) = \sum_{i=0}^{\infty} \prod_{j=1}^{i} y_j^*(z) \frac{(\lambda t)^i e^{-\lambda t}}{i!} = \sum_{i=0}^{\infty} [y^*(z)]^i \frac{(\lambda t)^i e^{-\lambda t}}{i!}$$
$$= \sum_{i=0}^{\infty} \frac{(\lambda t y^*(z))^i e^{-\lambda t}}{i!} = e^{-\lambda t(1-y^*(z))}$$

Using the formulas for mean and variance in terms of the probability generating function of random sums (as detailed in the mathematical appendix), we obtain:

$$E(X(t)) = E(N(t))E(y) = \lambda t E(y)$$

$$\mathrm{Var(X(t))} = \mathrm{Var(N(t))E^2(y)} + \mathrm{Var(y)E(N(t))} = \lambda \mathrm{t}\left[\mathrm{Var(y)} + \mathrm{E^2(y)}\right] = \lambda \mathrm{tE(y^2)}$$

The proof is straightforward. Let the logarithm of the probability generating function be given by: $\Phi(z) = \log(X^*(z)) = -\lambda t(1 - y^*(z))$. Thus, the mean is found by evaluating at $z = 1$, $\partial\Phi(z)/\partial z = \lambda t \partial y^*(z)/\partial z$ or $E(X(t)) = \lambda t E(y)$ as stated above. Similarly, the variance is found by:

$$\mathrm{Var(X(t))} - \mathrm{E(X(t))} = \left.\frac{\partial^2 \Phi(\mathrm{z})}{\partial \mathrm{z}^2}\right|_{\mathrm{z}=1}$$

where $\partial^2\Phi(z)/\partial z^2 = \lambda t \partial^2 y^*(z)/\partial z^2$ which yields the previous result for the variance. Thus, if we were to consider again an insurance claim process where claim events occur at the Poisson rate λ with a probability distribution for claim magnitudes given by $F(.)$ with mean and variance given by: $E(y)$, $\mathrm{Var(y)}$, then the mean and the expectation of claims and their variance over a period of time $(0, t)$ are given by $E(X(t))$, $\mathrm{Var(X(t))}$, as calculated above. Note that if the claims process has a self participation clause of a and insurance covers all claims up to and including the amount b, the mean and the variance of the claim process are instead:

$$E(X(t)) = E(N(t))E(y) = \lambda t \left[\int_a^b z dF(z) + b(1 - F(b))\right]$$

and

$$\begin{aligned}\mathrm{Var}(\mathrm{X}(\mathrm{t})) &= \mathrm{Var}(\mathrm{N}(\mathrm{t}))\mathrm{E}^2(\mathrm{y}) + \mathrm{Var}(\mathrm{y})\mathrm{E}(\mathrm{N}(\mathrm{t})) \\ &= \lambda \mathrm{t}\left[\int_{\mathrm{a}}^{\mathrm{b}} \mathrm{z}^2 \mathrm{dF}(\mathrm{z}) + \mathrm{b}^2(1 - \mathrm{F}(\mathrm{b}))\right]\end{aligned}$$

5.6.3 *Time VaR*

In order to introduce the concept of Time VaR, consider a portfolio whose value at time t is $W_t(a)$ where a is a vector of parameters denoting the investment profile of the portfolio in assets with various returns and risks. For simplicity, denote by $t = 0$ the current time with a current value $W_0(a)$. Consider next a time interval $t \in [0, T]$ over which the Value at Risk is set to a proportion $0 \leq \xi \leq 1$ of the portfolio current value and let P_{VaR} be the VaR risk exposure. In other words, assuming that the portfolio process will evolve according to an empirically or theoretically defined probability distribution in the time interval $t \in [0, T]$, then the probability of a loss greater than the VaR is specified by the risk exposure P_{VaR}, or $\Pr(W_T(a) - W_0(a) \leq -VaR_T) = P_{VaR}$ with $VaR_T = \xi W_0(a)$ and equivalently:

$$\Pr\left[W_T(a) \leq (1 - \xi)W_0(a)\right] = P_{VaR}$$

Note that in this approach a portfolio loss greater than the VaR could occur before the final time and yet at the final time the VaR condition could be met. For this reason, a Time VaR approach will be used which maintains continuously (or discretely) over time the risk exposure of the portfolio consistent with the VaR in the specified time interval. To do so, let, τ be the first instant (the first passage time) that a portfolio reaches a $VaR_t = \xi_t W_0(a)$ at time t, or:

$$\tau = Inf\{t \geq 0,\ W_t(a) \leq (1 - \xi_t)W_0(a)\}$$

whose cumulative probability distribution is $F_t(\tau)$. Then, the probability that a portfolio violates the loss risk exposure Value at Risk prior or at time to T is specified by the risk constraint:

$$F_T(T) \leq P_{VaR},\ VaR_T = \xi_T W_0(a) \text{ and } \tau = Inf\{t \geq 0, W_t(a) \leq (1 - \xi_T)W_0(a)\}$$

A portfolio that meets such specification (ξ_T, T, P_{VaR}) can of course be designed ex-ante by the selection of an investment profile of parameters (a). Similarly, given an investment portfolio profile (a), the risk exposure (P_{VaR}) and time T, the VaR that meets such a risk exposure is $\xi_T W_0(a)$ found by a solution of ξ_T above.

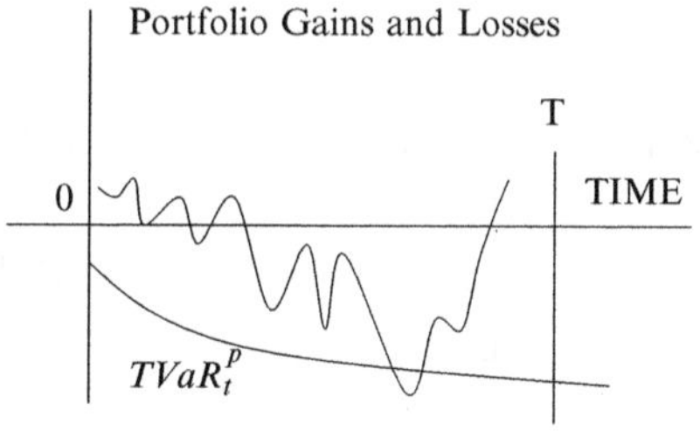

Fig. 5.1 Time VaR

In a temporal setting, when the portfolio gains and losses are recorded continuously (or at discrete times), a loss smaller than the $VaR_T = \xi_T W_0(a)$ might be encountered prior to time T that violates the portfolio risk exposure. To circumvent such a possibility, we suggest the construction of a time VaR (TVaR), found by a solution for ξ_t with $VaR_t = TVaR_t = \xi_t W_0(a)$ in

$$F_t(t|\xi_t) = P_{VaR};\ 0 \le t \le T \text{ and } VaR_t \le VaR_T$$

In this context, a Time VaR provides a continuous control of the portfolio risk exposure rather than a control between two instants of time. This is represented in Fig. 5.1 for example, where the TVaR line is a control limit used to track the portfolio at risk exposure $p = P_{VaR}$.

To demonstrate these results, consider first a portfolio whose linear growth rate over a given period of time is known and given by α and subject to a Brownian (normal) motion noise with variance σ^2. Thus, if we set a fixed interval of time T, the probability distribution of the portfolio appreciation (or loss if negative) is given by the Normal probability distribution with mean αT and variance $\sigma^2 T$. In this case, for the traditional VaR approach we have $W_T(a) = W_0(a) + \tilde{x} \le (1 - \xi_T)W_0(a)$ or $\tilde{x} \le -\xi_T W_0(a)$ where $\tilde{x}$ has a normal distribution $N(\alpha T, \sigma^2 T)$ and $F_N(-VaR_T|\alpha T, \sigma^2 T) = P_{VaR};\ VaR_T = \xi_T W_0(a)$ and $F_N(-VaR_T|\alpha T, \sigma^2 T)$ is the cumulative normal probability distribution with mean and variance $(\alpha T, \sigma^2 T)$. Inversely, let the portfolio appreciation rate (or loss rate) be fixed and given by X and let the time for the portfolio to appreciate (depreciate) by such an amount be random. The probability distribution of such a time is then given by an inverse Gaussian distribution explicitly given by:

$$g(\tau) = \frac{1}{\sqrt{2\pi\sigma^2\tau}} e^{-(X-\alpha\tau)^2/2\sigma^2\tau}$$

The cumulative distribution $G(t|X, \alpha, \sigma^2)$ can be calculated as well (Johnson and Kotz 1970a, p. 137). Of course, if we set the VaR for T periods to $VaR_T = \xi_T W_0(a)$, then the risk exposure in a Time VaR approach, yields the condition:

$$G(T|-VaR_T, \alpha, \sigma^2) = P_{VaR},\ VaR_T = \xi_T W_0(a)$$

where $G(.|.,.)$ is the cumulative of the inverse Gaussian distribution. Further, the VaR consistent with this risk specification in the time interval $t \in [0,T]$ is given by a solution for ξ_t^* below.

$$G\left(t\middle|-\xi_t^* W_0(a), \alpha, \sigma^2\right) = P_{VaR},\ TVaR_t = \xi_t W_0(a)$$

In the management of a portfolio risk exposure, $TVaR_t = \xi_t^* W_0(a)$ provides in the time interval $t \in [0,T]$ a mechanism for an online control of the portfolio consistent with the VaR risk exposure specification for the time interval $t \in [0,T]$. It can be applied as well to a number of situations such the control of individual traders in a Market Making Firm, the control of investors-speculators in a Fund of Fund etc. Of course, an extension to other portfolio processes (whether in discrete or continuous time) are also possible. Below, we consider in particular a trinomial portfolio process for which complete results are given for both Time VaR and Volatility at Risk exposure.

5.6.4 A Portfolio Trinomial Process

Let a trinomial portfolio value process which can decrease by a unit over a period of time with known probability s or decrease by a unit with probability q. Let $r = 1 - s - q$ be the probability that the portfolio does not change in value and let L_t be a trinomial profits-loss model. Thus, the parameters (s, r, q) determine the trinomial portfolio whose process is given by:

$$W_{t+1} = W_t - L_t; W_0 = W_0(a);\ L_t = \begin{cases} +1 & w.p. \quad s \\ 0 & w.p. \quad r \\ -1 & w.p. \quad q \end{cases}$$

If $r = 0$, we have as a special case the asymmetric random walk. While for $s = q$ we have a symmetric birth-death process that corresponds to an investment in an asset that has the same probabilities of increasing and decreasing. Let VaR_T^p be the value at risk corresponding to a risk exposure $p = P_{VaR}$ over the time interval $[0,T]$ and let $\tau(VaR_T^p)$ be the first time that the wealth process declines to the value at risk $Var_T^p = \xi_T W_0(a)$. That is:

$$\tau(K_T) = Inf\left(t > 0, \sum_{i=1}^{t} L_i \geq K_T\right),\ K_T = VaR_T^p = \xi_T W_0(a) \geq 1$$

Let $F_{\tau(K_T)}(t) = P(\tau(K_T) \leq t)$ be the cumulative distribution of the first passage time to a $VaR_T^p = \xi_T W_0(a) \geq 1$ loss. As a result, the time VaR risk constraint is

$F_{\tau(K_T)}(T) \leq P_{VaR}$. An explicit probability distribution of $\tau(K_T)$ is given by the following proposition which is proved below.

Proposition

$$P(\tau(K_T) \leq t) = K_T\left(\sqrt{\frac{s}{q}}\right)^{K_T} \sum_{K_T \leq i \leq j \leq t} \frac{1}{i}\binom{i}{\frac{K_T+i}{2}}(\sqrt{sq})^i \binom{t}{j}(1-r)^{j-i} r^{t-j}$$

and therefore, for a time interval [0, T], $K_T = VaR_T^p = \xi_T W_0(a)$ is given by a solution of the following equation:

$$F_{\tau(K_T)}(T) = K_T\left(\sqrt{\frac{s}{q}}\right)^{K_T} \sum_{K_T \leq i \leq j \leq T} \frac{1}{i}\binom{i}{\frac{K_T+i}{2}}(\sqrt{sq})^i \binom{T}{j}(1-r)^{j-i} r^{T-j} = P_{VaR}$$

Further, a computational expression for $\alpha_n = P(\tau(k) \leq n),\ n \geq k$ is given recursively by:

$$\alpha_k = s^k;\quad \alpha_{n+1} = \alpha_n + k\left(\sqrt{\frac{s}{q}}\right)^k \rho_n;\quad \rho_n$$
$$= \sum_{k \leq i \leq n+1} \frac{1}{i}\binom{i}{\frac{k+i}{2}}(\sqrt{sq})^i \binom{n}{i-1} r^{n-i+1}$$

Proof: (see below).

Note that the risk parameters $(VaR_T^p = K_T, p = P_{VaR})$ define completely the Time VaR risk constraint at time T for a given portfolio profile (expressed by the trinomial probabilities). If we maintain this risk specification, then a solution for the Time VaR as a function of time $t < T$ is needed. Proceeding as in the previous section, we will seek a solution for K_t in:

$$F_{\tau(K_t)}(t) \leq P_{VaR},\quad VaR_t^p = K_t$$

Further, if the $VaR_t^p = K_t$ is fixed, we can express risk control parameter in terms of a $TVaR_K^p$ where $F_{\tau(K)}(TVaR_K^p) \leq P_{VaR}$. The specification of any two determines the third. Thus, we may specify the portfolio risk in terms of VaR_t^p, p and solve for $TVaR_K^p$ or vice versa, specify $TVaR_K^p$, p and calculate the corresponding Value at Risk. Finally, the design of an investment portfolio, i.e. selecting the portfolio parameters (s, q, r) optimally can be found as well by specifying and maximizing the expected utility of the portfolio holdings by the selection of the trinomial parameters and subject to a TVaR risk constraint (for related problems, see for example, Gourieroux et al. 2000; Basak and Shapiro 2001).

Proof of the Proposition The proof of our proposition consists in the following. We use first a result of Feller (1971, p. 352, 1977, 2nd edition) for the pure random walk and subsequently, generalize this result to a trinomial process. Finally, we also provide an iterative scheme for calculating the probability in the proposition. Using Feller's (1971), equation (4.14), p. 32, the probability distribution of stopping time at a, given by:

$$P(\tau(a) = n) = \frac{a}{n}\binom{n}{\frac{a+n}{2}} s^{\frac{n+a}{2}} q^{\frac{n-a}{2}}$$

At present we define two random walks $(\hat{y}_n)_{n\geq 0}$ and $(\hat{x}_n)_{n\geq 0}$ in $\Re_+$ and independent such that:

$$P(\hat{y}_{n+1} - \hat{y}_n = 1) = 1 - r,\ \ P(\hat{y}_{n+1} - \hat{y}_n = 0) = r$$

$$P(\hat{x}_{n+1} - \hat{x}_n = 1) = \frac{s}{1-r},\ \ P(\hat{x}_{n+1} - \hat{x}_n = 0) = \frac{q}{1-r}$$

Then, $x_n = \hat{x}(\hat{y}_n))$ is a random walk with parameters (s, q, r). Set $\hat{y}^{-1}$, the inverse of $\hat{y}$:

$$\hat{y}^{-1}(k) = \inf\{n;\ \ \hat{y}(n) \geq k\}$$

Then: $\tau(a) = \hat{y}^{-1}(\hat{\tau}(a))$ with

$$\tau(a) = \inf\{k \geq 0, x_k \geq a\};$$
$$\hat{\tau}(a) = \inf\{k \geq 0, \hat{x}_k \geq a\}$$

The proof is as follows. First note that:

$$\text{(a) } k < \hat{y}^{-1}(\hat{\tau}(a)) \Rightarrow \hat{y}^{-1}(k) < \hat{\tau}(a) \Rightarrow \hat{x}(\hat{y}(a)) < a$$
$$\text{(b) } k = \hat{y}^{-1}(\hat{\tau}(a)) \Rightarrow \hat{y}^{-1}(k) = \hat{\tau}(a) \Rightarrow \hat{x}(\hat{y}(a)) = a$$

Furthermore, we have: $\hat{y}^{-1}(k) > n \Leftrightarrow \hat{y}(n) < k$ which is equivalent to: $\hat{y}^{-1}(k) \leq n \Rightarrow \hat{y}^{-1}(n) \geq k$. We calculate next the probability $P(\tau(a) \leq n)$. First $P(\tau(a) \leq n) = P(\hat{y}^{-1}(\hat{\tau}(a)) \leq n) = P(\hat{\tau}(a) \leq \hat{y}(n))$. However, $\hat{\tau}(a)$ and $\hat{y}(n)$ are two random variables, independent and $\hat{y}(n)$ is a binomial distribution $B(n, 1-r)$ wherefrom, applying the mixture:

$$P(\tau(a) \leq n) = \sum_{a\leq i\leq j\leq n} P(\hat{\tau}(a) = i)P(\hat{y}(n) = j)$$
$$= \sum_{a\leq i\leq j\leq n} \frac{a}{i}\binom{i}{\frac{a+i}{2}}\left(\frac{s}{1-r}\right)^{\frac{i+a}{2}}\left(\frac{q}{1-r}\right)^{\frac{i-a}{2}}\binom{n}{j}(1-r)^j(r)^{n-j}$$

which is reduced to the proposition, or:

$$P(\tau(a) \le n) = a\left(\sqrt{\frac{s}{q}}\right)^a \sum_{a \le i \le j \le n} \frac{1}{i}\binom{i}{\frac{a+i}{2}}(\sqrt{sq})^i \binom{n}{j}(1-r)^{j-i}(r)^{n-j}$$

This equation can be reduced to a recursive form where: $\alpha_n = P(\tau(a) \le n)$, $n \ge a$. Therefore,

$$\begin{aligned}
&\alpha_{n+1} = P(\tau(a) \le n+1) = P(\hat{\tau}(a) \le \hat{y}(n+1))\\
&P(\hat{\tau}(a) \le \hat{y}(n+1), \hat{y}(n+1) - \hat{y}(n) = 0)\\
&P(\hat{\tau}(a) \le \hat{y}(n), \hat{y}(n+1) - \hat{y}(n) = 0) + P(\hat{\tau}(a) \le \hat{y}(n) + 1, \hat{y}(n+1) - \hat{y}(n) = 1)
\end{aligned}$$

or a more convenient form: $\alpha_{n+1} = r\alpha_n + (1-r)P(\hat{\tau}(a) \le \hat{y}(n)+1)$. However,

$$P(\hat{\tau}(a) \le \hat{y}(n)+1) = P(\hat{\tau}(a) \le \hat{y}(n)) + P(\hat{\tau}(a) = \hat{y}(n)+1),$$

$$\begin{aligned}
\alpha_{n+1} &= r\alpha_n + (1-r)P(\hat{\tau}(a) \le \hat{y}(n)+1) = P(\hat{\tau}(a) \le \hat{y}(n)+1)\\
&= \sum_{a \le i \le n+1} P(\hat{\tau}(a) = i)P(\hat{y}(n) = i-1)\\
&= \sum_{a \le i \le n+1} \frac{a}{i}\binom{i}{\frac{a+i}{2}}\left(\frac{s}{1-r}\right)^{\frac{i+a}{2}}\left(\frac{q}{1-r}\right)^{\frac{i-a}{2}}\binom{n}{i-1}(1-r)^{i-1}r^{n-i+1}\\
&= \frac{a}{1-r}\left(\sqrt{\frac{s}{q}}\right)^a \sum_{a \le i \le n+1} \frac{1}{i}\binom{i}{\frac{a+i}{2}}(\sqrt{sq})^i\binom{n}{i-1}r^{n-i+1}
\end{aligned}$$

Note that $\alpha_n = 0$, for all n < a, since we require at least n steps for the process to cross the level a $\ge$ 0 and therefore, initially, $\alpha_a = P(\tau(a) = a) = s^a$ and for n > a, $\alpha_a = s^a$ and

$$\alpha_{n+1} = \alpha_n + a\left(\sqrt{\frac{s}{q}}\right)^a \rho_n;\ \rho_n = \sum_{a \le i \le n+1} \frac{1}{i}\binom{i}{\frac{a+i}{2}}(\sqrt{sq})^i\binom{n}{i-1}r^{n-i+1} \quad \text{Q.E.D}$$

5.7 Risk Uncertainty, Rare Events and Extreme Risk Processes

The assumption that we can formalize our lack of knowledge by a probability distribution may be presumptuous. Kalman (1994) for example, states that for the majority of observed phenomena of randomness in nature (always excluding games of chance) there is little or no experimental evidence in favor of a conventional probability. Kolmogorov, as well as a number of information theorists (Jaynes 1957a, b) suggested that randomness is defined by both non-uniqueness and non-regularity. For example, a die of six faces has non-uniqueness. Further, expansion of

$\sqrt{2}$ or of π provide an infinite string of numbers that appear irregularly, and can therefore be thought of as non-regular (and therefore 'random'). These properties are often used in Monte Carlo simulation as indicated in Chap. 3. By the same token, The Nobel Laureate, Born, in his 1954 inaugural address stated that randomness occurs when 'determinacy lapses into indeterminacy' without any logical, mathematical, empirical or physical argumentation, preceding thereby an important research efforts on chaos. Statements such as: 'we might have trouble forecasting the temperature of coffee 1 min in advance, but we should have little difficulty in forecasting it an hour ahead' by Edward Lorenz (1966), a weather forecaster and one of the co-founders of chaos theory, reinforced the many dilemmas we must deal with in modeling inter-temporal risk models. A similar statement can be repeated for many observations" such as "We might predict in the long run a stock market S&P index but we cannot predict tomorrows' stock prices"; "We might predict whether a patient (with a reasonable probability) will have a certain disease (based on current medical diagnosis technology) but we cannot predict whether he will be struck by this disease tomorrow" (although advances in DNA research provide the hope of a better understanding of the triggers of some diseases).

In forecasting, the accuracy of predictions does not always turn out to be any better than those of weather (chaotic) forecast. Studies comparing small and large econometric models showed that there was no difference between the two and therefore, the specificity of models and their complexity does not lead necessarily to better forecasts. Accrued evidence, theoretical and empirical points out further that the assumptions made by risk models are in practice violated and thus their application ought to be both more critical and more sensitive to a broader set of results (and therefore be more open to uncertainty). Co-dependence between many risk factors—whether causal or not, whether short and long term, undermines for example existence of simple risk models. Rather, there are dynamic and structural processes, changing both the risks and the environment in which we function that are difficult to ascertain. Can we assume that "noise" and statistical errors have normal probability distributions? These are some of the issues we confront in practice. Bounded rationality in physics assumes for convenience that a stable (equilibrium) have a number of characteristics including: Existence of an equilibrium (or fixed point or a stationary state), or equivalently an insensitivity to initial conditions in the long run; Periodicity and Structural stability which allows stable systems to be amenable to analysis. When "random" events do not share all these properties, predicting their future manifestation may turn out to be unwieldy or at best implied in indeterminacy.

Commensurate statistical analysis and mathematical modeling of such events, is of course bounded by what we are able to conceive and model. A number of tools have been forthcoming in the last decades that have improved appreciable risk assessment. Techniques such as ARCH and GARCH (see Engle 1987, 1995; Bollerslev 1986, 1990; Bollerslev et al. 1992, 1994) have been developed to model and to estimate simultaneously both expected values and their variance. These are fundamental techniques in financial econometrics used to estimate a realized volatility (as opposed to implied volatility, embedded in the current

Fig. 5.2 Black swans in the realm of the possible

information of future prices). These models point out that the time variation of an empirical volatility provides also estimates of tails risk models. *In these models, conditional variance of the data is permitted to change through time as new data is added to the information set.* In other words, tails are induced by time variations and volatility randomness and therefore reveal an aspect of "uncertainty." Additionally, a number of risk models have been based on adopting skew and extreme values probability models (as noted in Chap. 3) and providing greater flexibility in modeling dependence between multiple risks factors (as it is the case with Copulas, indicated in Chap. 4). These approaches seek to depart from assumptions and models based on the normal probability distribution. These models emphasize tails and Pareto-Stable Levy processes, fractal (long run "memory" with a nonlinear growth of variance) processes (see Mandelbrot et al. 1968, 1979, 2006) as well short term (contagion and persistent) memory processes. These models provide a fertile ground for modeling risk events accounting for "greater irregularity" in the evolution of temporal processes and the bifurcation of models as actual observations unfold.

Nassim Taleb in his 2007 book Black Swans (see also Taleb 2008), suggested to map randomness and decision making under risk and uncertainty into a quadrant with two classes of randomness and decisions. Decisions are referred to as "simple" or "binary" lead to decisions such as "very true or very false," "matters or does not matter." By the same token, statistical tests in the control of quality may state, "A product is fit for use or the product is defective." In health care, "The patient is healthy" or "The patient is not healthy." Statements of the type "true" or "false" can then be stated with some confidence as a Bernoulli probability within a probability interval. A second type of decisions is more complex, emphasizing both its likelihood of occurrence and its consequences. "Randomness" is defined as well in two layers: "Predictable events" or with finite and well behaved variance or distributions (i.e. risk models) and "Unpredictable events." For example, random walks converging to Normal/Gaussian-Poisson processes are first layers events while "Fractals" and infinite variance (Pareto-stable and chaotic) models or Lévy stable distributions (or Levy processes that have leptokurtic distributions and potentially infinite variance as will be defined below) belong to the second layer. "Uncertainty" is then embedded in a nonlinear functional time growth of variance. Figure 5.2 highlights Black Swans to emphasize that they do exist, even though there may not be many. A greater awareness of Black Swans, may eventually lead to mitigating their risk consequences.

Domain/application	Simple decisions	Complex decisions
Thin tailed "Gaussian-Poisson" Distributions	I. Classical Statistics	II. Complex Statistics
Heavy tailed, Fractal Distributions	III. Complex Statistics	IV. Extreme fragility

The statistical representation and the modeling of rare, complex, extreme and uncertain events with consequential extreme risks are challenging however. Approaches we refer to may be categorized and based on the following risk models:

- Fat tail distributions using skewness as a departure from normality, kurtosins as a preliminary departure from thin tails (normal) distributions.
- Infinite (or extremely large) variance distributions, accounting.
- Pareto Levy and Stable Distributions with mathematiccal characteristics partly tractable and applicable, accounting for the reproducibility of their pronbability distribution as well as a far greater variety of potential skewed probability distributions (both analytical and empirical).
- Fractals processes to account for the non-linearity of the evolution of variance in inter-temporal stochastic processes. These models are often coined "long-run" memory stochastic processes.
- Persistent, Contagion or bifurcating random walk processes (or Short Memory Models). These models account for the specific events that occur and their effects on the subsequent evolution of the underlying stochastic process.

These approaches provide an extremely large set of possibilities (albeit difficult to treat mathematically) to construct risk models. Below, we consider some essential technical elements to present their usefulness in risk modeling. Again, numerous references on current and past studies of such models are indicated at the end of this chapter.

5.7.1 Hurst Index, Fractals and the Range Process

The range process is defined by the difference between the largest and the smallest in a consecutive series of events. Unlike the sample range defined in Chap. 3, based on a sample of identically and independently distributed observations, such a range is defined over a time series. Such a process is particularly useful as it provides an additional statistic to measure the variability of a "time series". It also complement volatility estimates. For example, when a series "becomes contagious," the evolution of the range will have a tendency to grow faster than expected otherwise when events are drawn from a "stable" series. Further, if volatility increases over time, the range will also grow providing thereby a signal to volatility growth and vice versa. Finally, when an extraordinary event occurs, say an outlier, the range too will increase suddently, to remain thereafter more stable than would be expected otherwise. For these reasons the range process is useful. Combining the range and the volatility into a single statistic provides an estimate of an index for "chaos or contagion" of a time series. This R/S statistics is applied to detect volatility shifts as well as document long run memory (in the Mandelbrot sense), an index of "chaos" in financial time series, etc. Bloomberg financial stations for example, provides such a statistic for financial time series, named as well the Hurst exponent (Hurst 1951) or the R/S index. Mandelbrot and Van Ness (1968), Mandelbrot and Wallis (1968,1969), Mandelbrot and Taqqu (1979), and a number of co-workers have essentially used such an index to quantify the statistical bias arising from self similarity power laws in time series (fractal models). In other words, it expresses the degree of power non-linearity in the variance growth of series as stated earlier.

Calculations for the range and the R/S statistic are made as follows. Let a series of observations (samples) be fixed to length N, the sample range given over a series $y_{t,\,N} = y_{t+1}, y_{t+2}, \ldots, y_{t+N}$ by:

$$R_{t,N} = \mathrm{Max}\{\mathrm{y}_{\mathrm{t,N}}\} - \mathrm{Min}\{\mathrm{y}_{\mathrm{t,N}}\}$$

while the sample standard deviation is calculated by:

$$S_{t,\,N} = \sqrt{\frac{\sum_{i=1}^{N} (y_{t+i} - \widehat{y}_{\mathrm{tlN}})^2}{N-1}}$$

where $\widehat{y}_{t,N}$ is the sample average. A regression of the logs of Hurst equation (R/S) $= (Const * N)^H$ provides an estimate of the Hurst exponent, or:

$$\ln\left(\frac{R_N}{S_N}\right) = a + bH; \;\; b = \log(\alpha N)$$

where N is the sample length, H is the Hurst exponent, R the sample's range, S the sample's standard deviation and finally α is a constant. For random (Wiener) processes in particular, the Hurst index turns out equal to 0.5. While for any values larger than 0.5 obtained in a regression, it may indicate a fractal model or "long run dependence." Use of the Hurst index should be careful and critical however.

The origins of Hurst exponent are due to Hurst in 1951 who began working on the Nile River Dam project and studied the random behaviour of the Nile water level and the influx of water from rainfall over the thousand years data that have been recorded (since Pharaos' times). His observation was made that if the series were random, the sample range would increase with the square root of time. Such a result is confirmed by many time series and theoretically as well for normal processes. Hurst noted explicitly that most natural phenomena follow a biased random walk and are thus characterized by the parameter H expressing series' dependence (coined by Mandelbrot the "Joseph Effect" who interpreted Pharao's dream, see also Mandelbrot 1972; Mandelbrot and Taqqu 1979; Mandelbrot and Van Ness 1968; Beran 1994). Explicitly, say that the correlation C between disjoint increments of a temporal series is given by $C = 2^{2H-1} - 1$. Thus, if $H = 0.5$, the disjoint intervals are uncorrelated. For $H > 0.5$, the series are correlated, exhibiting a memory effect since past observations increase future ones as stated above. For $H < 0.5$, these series are called "anti-persistent." Such analyses require large samples N however which might not be always available. For this reason, such analyses are used when series are long such as sunspots, water levels of rivers, intra-day trading stock, market ticker data etc.

An attempt to represent these series, expressing a "persistent behaviour" (or alternatively a nonlinear variance growth) was reached by Mandelbrot and Wallis (1969) who introduced a fractional Brownian motion, denoted by $B_H(t)$ (see also Green and Fielitz 1977, 1980 for applications in finance). It is basically a continuous random function, defined as a moving average of past increments of ordinary Brownian motion, denoted as $B(t)$. In this function the variance does not grow linearly over time but grows at a rate, a function of the Hurst index. If $H < 0.5$, the long run dependence is negative (or in the words of Hurst, it is an anti-persistent time series) while for $0.5 < H < 1$, the dependence is positive. Fractional Brownian motion is based on the self-similarity of the power law for such processes which means that the increment for a time interval s are in distribution proportional to the standard deviation elevated at the power H, or s^H, or:

$$B_H(t+s) - B_H(t) \underset{i.d.}{\rightarrow} s^H[B(t+1) - B(t)]$$

where i.d. means in distribution. Furthermore, the increments variance is:

$$E[B_H(t+s) - B_H(t)]^2 = s^{2H}E[B(t+1) - B(t)]^2$$

This means that the variance for any time interval s is equal to s^{2H} times the variance for the unit interval. In this sense, fractional Brownian motion can be fractioned into Brownian motions. Of course, it is now obvious that for H = 0.5, the variance will grow at a linear time rate (as it is the case for random walks and the Brownian motion) and it is non-linear otherwise. In this sense, assuming a relationship between the Hurst exponent and the notion of long run dependence in time series, an estimate of the one indicates the other. For inter-temporal processes (whether risk or otherwise), such observations are extremely important. First and foremost, long run dependence violates the basic assumptions made regarding "normal" processes (for example, rates of returns). For these reasons, they can be used to test the validity of such models. Second, the Hurst index can be used as a "herd or contagion effect" index applied to stocks, contagious diseases, environmental states, or other time series, meaning that series volatility that have a tendency to grow, will grow faster over time if the index is greater than 0.5 and vice versa if the index is smaller than 0.5. For these reasons, the R/S index has also been associated to revealing the evolution of time series that may become increasingly unpredictable.

5.7.2 R/S and Outliers Risks

The R/S statistic over a i.i.d. samples has already been recognized in the detection of outliers. Irwin in 1925 pointed out that when the standard deviation is known, the following statistic (see Barnett and Lewis 1994) for testing upper outliers in a sample of size n where $x_{(i)}$ is the ith ordered statistic (see also Chap. 3):

$$\left[x_{(n)} - x_{(n-1)}\right]/\sigma \text{ and} \left[x_{(n-1)} - x_{(n-2)}\right]/\sigma$$

Subsequently, a number of related approaches have been devised. Of particular interest is the range/spread (standard deviation) statistic defined by the R/S statistic (based on i.i.d. statistical samples however):

$$\left[x_{(n)} - x_{(1)}\right]/s$$

Neyman and Scott (1971) introduced further, the concept of outlier prone and outlier resistant families of distributions which is based on the ratio of the two extremes of a samples and the range of that sample.

The study of R, R/S remains challenging.

Feller as early as 1951 remarked that it is difficult to compute the range distribution in a symmetric random walk. Imhof (1985) as well as Vallois (1995, 1996) and Vallois and Tapiero (1995a, b, 1996, 1997a, b, 1998, 2001), Tapiero and Vallois (1996, 2000) have renewed interest in this process and have characterized the basic statistical properties of the range process which can be used fruitfully for the study of volatility in stochastic processes.

In summary, the range process is important as a statistic for inter-temporal risk models for a number of reasons. First, it provides a statistic for a process second order variation. A process which exhibits extremely large variation can be detected by estimating the process range. Of course, a process volatility provides such information but its estimates are slower in measuring bursts in volatility growth. Second, by noting that the standard deviation process is more robust than the range process, a combination of the two processes can provide some useful statistic based on the speed at which they evolve whenever the process parameters change.

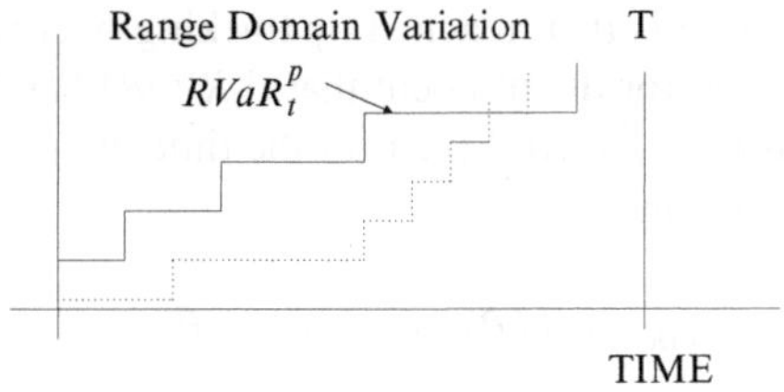

Fig. 5.3 Portfolio range variation and inverse range process

5.7.3 RVaR, TRVaR and Volatility at Risk

Application of the range to estimate volatility in i.i.d. samples are used in control charts as well as in finance to estimate the real volatility (Parkinson 1980). Explicitly, using Feller's result (1971) Parkinson provided a process volatility estimator using range statistics and given by:

$$\hat{\sigma}^2(p) = \frac{1}{(4\ln 2)Tn}\sum_{i=1}^{n} R_i$$

where $R_i = \max_{i\in I_i} X(t) - \min_{i\in I_i} X(t)$ is the ith sample range of the time series $X(t)$, n is the number of intervals I_i, $i = 1, 2,\ldots,n$ over which the range is estimated, T is the length of the interval while $X(t)$ is assumed to be a normal process with volatility σ. An adjustment to this estimate, based on the transformation of Parkinson's samples is suggested by Kunitomo (1995), providing thereby an estimate which is equivalent to the estimates based on samples variances. Such analyses do not consider however underlying stochastic processes but rather i.i.d. samples. Their statistical meaningfulness was not proved to be efficient however. Consider instead an underlying stochastic process (a random walk with drift, a trinomial stochastic model and their like) with known statistical characteristics and consider instead the inverse range process. To do so we consider a *TRVaR* (a Time Range VaR) as a measurement of volatility risk exposure. Explicitly, let *RVaR* (Range Volatility at Risk) be the amplitude "allowed" range variations for an underlying process, and let *TRVaR* be the passage time to such an amplitude (assuming a predictable process) and let the risk exposure of the volatility at risk be denoted by (P_{RVaR}) (Fig. 5.3). Finally, let $\theta(RVaR)$, be a random variable (the inverse range process), denoting the theoretical (and therefore predicted) first time that the *RVaR* is reached for a predictable reference process. Then, the volatility at risk exposure over the time interval $\theta \in [0, T]$ can be specified by the quantile volatility risk.

$$\Pr(\theta(RVaR) \leq TRVaR_T) \leq P_{RVaR}$$

A stochastic specification for $\theta(RVaR)$ for a known underlying process provides the "control" amplitude $RVaR_T$ consistent with the volatility at risk exposure.

Volatility at Risk Exposure is then reached by tracking over time the evolution of the process range. Assume for the moment that $\theta(RVaR)$ has a known cumulative probability distribution $F_{\theta(.)}(.|\alpha_T, \sigma_T)$ then for the time interval [0, *T*], the *RVaR* is given by solving for *RVaR* in:

$$F_{\theta(RVaR_T)}(TRVaR_T|\alpha, \sigma) = P_{RVaR}$$

with $\theta(.)$'s distribution assumed known or calculated. Further, for a set of time consistent set of control parameters over range amplitudes, we have:

$$F_{\theta(RVaR_t)}(TRVaR_t|\alpha, \sigma) = P_{RVaR};\ \ t \leq T$$

Alternatively, we may let the amplitude vary by $(RVaR - m), m = 1, 2, 3, \ldots,$ $RVaR - m$ and determine the $(TRVaR_m)$ corresponding to the given risk exposure, or:

$$F_{\theta(RVaR_T - m)}(TRVaR_m|\alpha, \sigma) = P_{RVaR}; m = 1, 2, 3, \ldots RVaR - 1$$

Although the analytical calculation of these distributions is in general quite difficult, Monte-Carlo simulation of the underlying process can be used to construct volatility at risk control parameters. In the next section we shall consider a special (applicable) case we deal with analytically. In order to simplify its presentation, essential results are summarized with their proof referred to papers of Vallois and Tapiero (1995a, b, 1996, 1997a, b, 1998, 2001) (Fig. 5.3).

Example Let the range process over a time interval $(0, t)$ be given by $\{\Re_t, t \geq 0\}$

$$\Re_t = \mathrm{Max}[\mathrm{W}_0, \mathrm{W}_1, \mathrm{W}_2, \ldots, \mathrm{W}_t] - \mathrm{Min}[\mathrm{W}_0, \mathrm{W}_1, \mathrm{W}_2, \ldots, \mathrm{W}_t]$$

Since $\{\Re_t, t \geq 0\}$ is a growing process, we can study its inverse process $\theta(k)$, the passage time process for the range to grow at or above the specified amplitude *k*;

$$\theta(k) = Inf(t \geq 0; \Re_t \geq k)$$

Clearly, $\{\Re_n < k\}$ is equal to $\{\theta(k) > n\}$ and therefore the probability process law of $\Re_t$ can be studied instead through its passage time. For the trinomial model (see Chap. 4), let $\lambda = q/s$ and define by $(ARRL_k, VRRL_k)$ the average time for the process to reach an amplitude *k* and its variance. These moments are given by the following proposition:

Proposition *Consider the trinomial random walk, then:*

$$ARRL_k = E(\theta(k)) = \frac{1}{s(\lambda - 1)}\left(\frac{(k+1)^2}{\lambda^{k+1} - 1} - \frac{k^2}{\lambda^k - 1} - \frac{1}{\lambda - 1} + k\right) \text{if } \lambda \neq 1$$

$$ARRL_k = \mathrm{E}(\theta(k)) = \frac{k(k+1)}{4s} \quad \text{if } \lambda = 1 \text{ and } \lambda = q/s.$$

The variance of the range run length of a trinomial portfolio process is given in terms of the following functions:

$$D(k) = \frac{k^2}{\lambda^k - 1}, \quad G(\mathrm{k}) = \frac{k^3}{(\lambda^k - 1)^2}, \quad D_1(k) = k + D(k+1) - D(k) + \frac{1}{1-\lambda}$$

$$D_2(k) = 4[(k+1)D(k+1) - kD(k) + G(k+1) - G(k)] - D_1(k)^2 + k(k+1) - \frac{4\lambda}{(1-\lambda)^2}$$

and

$$VRRL_k = \mathrm{var}[\theta(k)] = \frac{1}{1-r}\frac{1+\lambda}{1-\lambda}D_1(k) + \frac{1}{(1-r)^2}\left(\frac{1+\lambda}{1-\lambda}\right)^2 D_2(k)$$

and in particular,

$$VRRL_k = \underset{\lambda \to 1}{\mathrm{Lim}}\, \mathrm{Var}\,(\theta(k)) = \frac{k(k+1)}{12(1-r)^2}(6r + (k-1)(k+2))$$

Proof See Vallois and Tapiero (1996c).

To verify these results note that for a symmetric random walk that (Vallois and Tapiero 1995a, b)

$$\mathrm{var}(\theta(k)) = \frac{k(k-1)(k+1)(k+2)}{12}$$

The explicit probability distribution of the inverse range process is difficult to determine however. Since the variance is of order squared the average time, for the amplitude to reach a given level k, it may be reasonable to assume that is has "fat tails" and thus use conveniently the Weibull distribution or some of its mixtures. Explicitly, assume the following two parameters Weibull distribution for the range inverse process (see also Chap. 3):

$$f_{\theta(k)}(\tau) \approx \frac{c}{\zeta}\left(\frac{\tau}{\zeta}\right)^{c-1} e^{-(\tau/\zeta)^c}, \ \tau \geq 0, \ \zeta, \ c > 0; \ F_{\theta(k)}(\tau) = 1 - e^{-(\tau/\zeta)^c},$$

$$ARRL_k = E(\tau) = \zeta\Gamma\left(\frac{1}{c}+1\right), \ VRRL_k = \mathrm{var}(\tau) = \zeta^2\left[\Gamma\left(\frac{2}{c}+1\right) - \Gamma^2\left(\frac{1}{c}+1\right),\right]$$

which can be solved for the parameters (ς, c) in terms of the computed $(ARRL_k, VRRL_k)$. First note that

$$\frac{VRRL_k}{(ARRL_k)^2} = \frac{\Gamma\left(\frac{2}{c}+1\right)}{\Gamma^2\left(\frac{1}{c}+1\right)} - 1$$

which can be solved for c and by substitution for (ς). Thus, for a fixed amplitude k:

$$F_{\theta(k)}(TRVaR_k) = 1 - e^{-(TRVaR_k/\zeta)^c} \le P_{TRVaR} \rightarrow TRVaR_k \le \zeta\left[\ln\left(\frac{1}{1-P_{TRVaR}}\right)\right]^{1/c}$$

which can be written as a product of the ARRL, or:

$$TRVaR_k \le ARRL_k \frac{\left[\ln\left(\frac{1}{1-P_{RVaR}}\right)\right]^{1/c}}{\Gamma(1+1/c)}$$

In particular, for a symmetric trinomial portfolio process, we have:

$$ARRL_k = \frac{k(k+1)}{2(1-r)} = \zeta\Gamma\left(\frac{1}{c}+1\right),$$
$$VRRL_k = \frac{k(k+1)}{12(1-r)^2}[6r+(k-1)(k+2)] = \zeta^2\left[\Gamma\left(\frac{2}{c}+1\right) - \Gamma^2\left(\frac{1}{c}+1\right)\right]$$

which we use to calculate the process parameters,

$$\zeta = \frac{k(k+1)}{2(1-r)\Gamma(1+1/c)}; \quad 1 + \frac{[6r+(k-1)(k+2)]}{3k(k+1)} = \left[\frac{\Gamma\left(\frac{2}{c}+1\right)}{\Gamma^2\left(\frac{1}{c}+1\right)}\right]$$

For large amplitudes, we also note that:

$$\frac{4}{3} \approx \left[\frac{\Gamma(1+2/c)}{\Gamma^2(1+1/c)}\right] \rightarrow \Gamma^2(1+1/c) = 0.75\Gamma(1+2/c)$$

which is independent of k and r. In this case, $1/c = 0.555$ and therefore,

$$TRVaR \le \frac{1}{0.931}\left[\frac{k(k+1)}{2(1-r)}\right]\left[\ln\left(\frac{1}{1-P_{RVaR}}\right)\right]^{0.555}$$

This means that for risk parameters $P_{RVaR} = 0.1$, the process test value $TRVaR$ equals $0.3225(ARRL_k)$. In other words,

$$TRVaR_{RVaR-i} = 0.3225(ARRL_{RVaR-i}) = 0.3225\left(\frac{(RVaR-i)(RVaR-i+1)}{2(1-r)}\right)$$

Table 5.1 Time VaR TRVaR and range amplitude

r	k = 3	k = 4	k = 5
0.1	0.5208	0.538	0.542
0.2	0.536	0.545	0.55
0.3	0.553	0.555	0.557
0.4	0.568	0.564	0.562
0.5	0.584	0.574	0.568

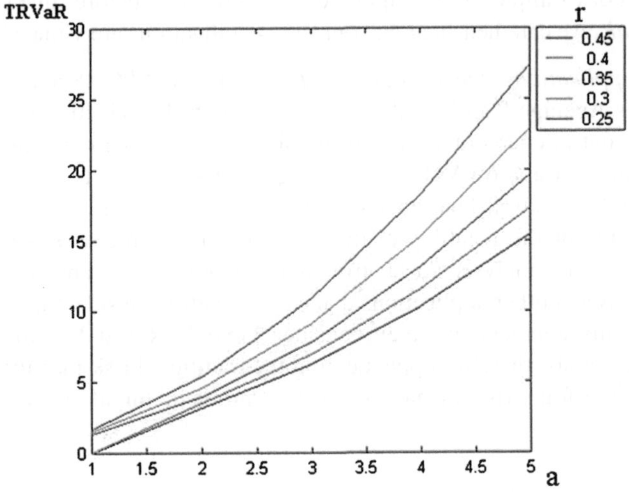

Fig. 5.4 The TRVaR-amplitude graph for a risk specification of 0.05

TRVaR computations for $P_{RVaR} = 0.2, 0.3, 0.4$ yields as well coefficients given by 0.4915, 0.634 and 0.7745. Of course, for small amplitude values (say 5, 6, 7, etc.) the parameter 1/c will be a function of k and r. It will be relatively insensitive to these values and will tend to be equal between 0.52 and 0.56. A set of calculation for the parameter c for small amplitudes are included (Table 5.1 and Fig. 5.4).

For example, for a test based on amplitudes of $k = 5$ and $r = 0.4$, we have a critical test value *TRVaR* which is given by:

$$TRVaR = \left[\frac{30}{(1.2)(0.92622)}\right]\left(\ln\left[\frac{1}{1 - P_{RVaR}}\right]\right)^{0.562} = 26.991\left(\ln\left[\frac{1}{1 - P_{RVaR}}\right]\right)^{0.562}$$

Of course, precise tables can be constructed by using simulation. Using a MATLAB program, we calculated as well the following graphs that highlight the risk parameters sensitivities. In this graph, volatility at risk expressed for amplitudes from $k = 1$ to 5 are calculated in terms of the probability of process stability. Explicitly, the graph provides a solution to the following equation:

$$F_{\theta(RVaR-m)}(TRVaR_m|\mu,\sigma) = P_{RVaR};\ m = 1,2,3,\ldots RVaR - 1;\ \mathrm{k} = RVaR - m;$$

with a risk specification of 0.05. When $r = 0$, for example, we have a pure random walk while for $r = 0.2$, we have a trinomial model with a probability of increasing or decreasing wealth which equals to 0.4. Throughout this example, we used a test probability of 0.05 and thus obtained the *TRVaR* corresponding to the process parameters. For example, if an amplitude of 3 is reached before time 6, violating the test *TRVaR*, we can then state that our risk specifications are violated (Fig. 5.4).

Additional applications can be developed. Traditional SPC (Statistical Process Control) uses Average Run Lengths statistics to control production processes and to construct control charts of various sorts. Similarly financial regulation based on Capital Adequacy Ratio (or VaR) framing require that controls be instituted in real time and therefore be based on run lengths tests for both the time to exhaust a given capital set aside or the volatility-contagion of financial processes. Although, we have applied our analysis to a trinomial model, there are a number of generalizations and other applications that can be readily considered. For example, in portfolios management, we may consider Time VaR and Volatility (Range) sensitive constraints, thereby supplementing the traditional risk management tools available to portfolio risk managers. Such a problem might be formulated as follows:

$$\underset{a\in\Im}{\mathrm{Max}}\ W_T(a)\ \text{Subject to}:$$
$$\Pr(W_T(a) \le (1-\xi)W_0(a)) \le P_{VaR}\ \text{and}\ F_{\theta(RVaR_T)}(TRVaR_T|\alpha_a,\sigma_a) \le P_{RVaR}$$

where the underlying portfolio value growth process is a function of the parameters (α_a, σ_a) which are a function of the portfolio investment profile a. The price of loss and volatility risk exposure for an individual investor seeking to maximize the final portfolio value would then be specified by the optimization problem's Lagrange multipliers.

5.7.4 *The Generalized Pareto Distribution (GPD)*

The motivation for these models is to better account for fat tail processes. These distributions have a regular variation at infinity, or:

$$\underset{t\to\infty}{\mathrm{Lim}} \frac{1-F(tx)}{1-F(t)} = x^{-\xi}, \xi > 0$$

For example, a particularly useful case that satisfies this property is the Pareto distribution given by:

$$f(t) = \alpha s^{\alpha} t^{-\alpha-1},\ \ F(t) = 1 - s^{\alpha} t^{-\alpha} \text{ and } \operatorname*{Lim}_{t\to\infty} \frac{1 - F(tx)}{1 - F(t)} = \operatorname*{Lim}_{t\to\infty} \frac{s^{\alpha}(xt)^{-\alpha}}{s^{\alpha} t^{-\alpha}} = x^{-\alpha}, \alpha > 0.$$

Pareto Stable models have been popular in the Finance modeling literature ever since Fama's thesis supervised by Mandelbrot. A standard form of the Pareto distribution is given by:

$$f(x) = \alpha\theta^{\alpha}(x + \theta)^{-(1+\alpha)},\ x \geq 0.$$

The survival function $S(x)$ is then given by $1 - F(x)$ and can be shown to be $S(x) = \theta^{\alpha}(x + \theta)^{-\alpha}$. As a result, the hazard rate is:

$$h(x) = \frac{\alpha}{x + \theta}$$

Generally, Pareto Stable Distributions (PSD) are a class of probability distributions defined stable as follows: If say two random variables x_1 and x_2 are distributed as $a_1 u + b_1$ and $a_2 u + b_2$, their sum $x = x_1 + x_2$ is also of the form $au + b$. These probability distributions are characterized by general (natural logarithm) ln characteristic functions $\varphi(z) = E(e^{ixz}), i = \sqrt{-1}$ with three parameters m, c, α given by (see Feller 1971 for example):

$$\ln \varphi(z) = imz - c|z|^{\alpha}\left\{1 - \frac{iz}{|z|}\operatorname{tang}\left(\frac{\alpha\pi}{2}\right)\right\}, i = \sqrt{-1}$$

These distributions have first a field of attraction, thus the distribution of their sums differ only by location parameters. Second, they conform to some time series that exhibit the fat tail property. An explicit (albeit computationally intensive formulation) expression for Pareto Stable distributions is given by Feller (1977) by:

$$\text{For } x > 0,\ \ 0 < \alpha < 1:\ \ f(x:\alpha,\gamma) = \frac{1}{\pi x}\sum_{k=1}^{\infty}\frac{\Gamma(\alpha k + 1)}{k!}(-x)^{\alpha}\sin\left(\frac{k\pi}{2}(\gamma - \alpha)\right)$$

$$\text{For } x > 0,\ \ 1 < \alpha < 2:\ f(x:\alpha,\gamma) = \frac{1}{\pi}\sum_{k=1}^{\infty}\frac{\Gamma(\alpha^{-1}k + 1)}{k!}(-x)^{k}\sin\left(\frac{k\pi}{2\alpha}(\gamma - \alpha)\right)$$

$$\text{For } x < 0: f(-x:\alpha,\gamma) = f(x:\alpha,-\gamma) \text{ with } |\gamma| \leq \begin{cases} \alpha & \text{if } 0 < \alpha < 1 \\ 2 - \alpha & \text{if } 1 < \alpha < 2 \end{cases}$$

Some special and well known cases include:

(a) The normal probability distribution
(b) The generalized Pareto distribution indicated above
(c) The Cauchy (infinite variance) probability distribution
(d) Levy processes

We consider below, Normal and Pareto Levy Stable Distributions.

5.7.5 *The Normal Distribution and Pareto Levy Stable Distributions*

Assume that $\{R_i\}, i = 1, 2 \ldots, n$ are the i.i.d. random variables with known mean μ and known finite variance σ^2. The standardized random variable $\tilde{R}_n$ of a sample of i. i.d. random variables $\{R_i\}, i = 1, 2 \ldots, n$ is defined by:

$$\tilde{R}_n = \frac{1}{\sigma\sqrt{n}}\left\{\sum_{j=1}^{n} R_j - n\mu\right\}$$

is known to converge, by the central limit theorem to the standard normal distribution. It is both stable and reproducible (since the sum of a normal distribution has also a normal distribution). The proof of such an assertion is easily demonstrated by using the Characteristic Function given by: $\Psi_1(\lambda) = E\left(e^{i\lambda x}\right)$. Since these samples random variables are i.i.d., we have the sample characteristic function:

$$\Psi_n(\lambda) = (\Psi_1(\lambda))^n \text{ as well as } \Psi_{nk}(\lambda) = (\Psi_n(\lambda))^k = (\Psi_1(\lambda))^{nk}.$$

In all such cases, the characteristic function maintains its same functional form—giving it the name of a stable distribution. Thus, functionally the distribution is said to be infinitely divisible.

A slight generalization leads to the definition of Pareto-Levy processes. Let $\{R_i\}$, $i = 1, 2 \ldots, n$ be a random event at time i with characteristic function $\xi_1(\lambda)$. And let $R_n = \sum_{j=1}^{n} R_j$, have a characteristic function $\xi_n(\lambda)$. Unlike the normal probability case, we would like to find a characteristic function that meets the following conditions:

$$\xi_n(\lambda) = [\xi_1(\lambda)]^n \text{ and } [\xi_1(\alpha(n)\lambda)] = [\xi_1(\lambda)]^n$$

In other words, both the sum and the product of the random variable multiplied by some parameters (a function of the summing parameter) are distribution invariant. In this case, we can proceed as follows (see Seville Nanjana, MIT Lecture Notes, April, 2006). First note that:

$$[\xi_1(\alpha(1)\lambda)] = [\xi_1(\lambda)]^1, \ \ \xi_1(\alpha(1)\lambda) = \xi_1(\lambda) \ \text{ where } \alpha(1) = 1.$$

Next let the ln characteristic function be $\Lambda(\lambda) = \ln[\xi_1(\lambda)]$. Thus, $\ln[\xi_1(\alpha(n)\lambda)] = n\ln[\xi_1(\lambda)]$ implies $\Lambda(\alpha(n)\lambda) = n\Lambda(\lambda)$. Deriving with respect to n we have:

$$\Lambda(u) = n\Lambda(\lambda), \ \ u = \alpha(n)\lambda \ \text{ and}$$

$$\frac{d\alpha(n)}{dn}\frac{d\Lambda(u)}{du}\lambda\Lambda(u) = \Lambda(\lambda), \ \ u = \alpha(n)\lambda \ \text{ or at n} = 1:$$

$$\lambda\frac{d\alpha(1)}{dn}\frac{d\Lambda(\alpha(1)\lambda)}{du} = \Lambda(\lambda) \ \text{ and } \ \frac{d\Lambda(u)}{du} = \frac{d\Lambda(\lambda)}{d\lambda} = \frac{\Lambda(\lambda)}{\lambda\frac{d\alpha(1)}{dn}}$$

A solution to this differential equation is then of the type:

$$\Lambda(\lambda) = \begin{cases} a|\lambda|^\alpha & \lambda > 0 \\ b|\lambda|^\alpha & \lambda < 0 \end{cases} \ \text{ And thus by symmetry } \ \xi(\lambda) = \begin{cases} e^{a|\lambda|^\alpha} & \lambda > 0 \\ e^{b|\lambda|^\alpha} & \lambda < 0 \end{cases}$$

Since, $\xi(-\lambda) = \xi(\lambda)^*$, the following characteristic function results,

$$\xi(\lambda) = \exp\{(c_1 + ic_2\text{sgn}(\lambda))|\lambda|^\alpha\} \ \text{ with } c_1, c_2 \in \Re$$

The characteristic function above is a special case of the Levy-Kintchine formula for finite mean distributions for Pareto Stable distributions with the following characteristic function:

$$\xi(\lambda) = \exp\left\{-a|\lambda|^\alpha\left(1 - i\beta\tan\left(\frac{\alpha\pi}{2}\right)\text{sgn}(\lambda)\right)\right\}$$

where β relates to the distribution's skewness. If the distribution is symmetric, it is then a Levy distribution which is given as stated above by:

$$\beta = 0 \text{ and } \xi_L(\lambda) = \exp\{-a|\lambda|^\alpha\}$$

The motivation for Levy processes arises because of our own concern that a two moments distribution does not capture the richness that data indicates. In particular, data skewness, contradicting the assumption of normally distributed data. Levy processes thus provide an opportunity to characterize such stable processes.

5.8 Short Term Memory, Persistence, Anti-persistence and Contagion

Defaults may be time and information dependent. When a default occurs (an information for subsequent events) it may alter and increase the probability of subsequent defaults—a contagion then occurs. For example, an insured whose health has been impaired 1 year may be more prone to claim subsequently. Such persistence recurs in many instances. Are fears regarding financial markets persistent? (*The Financial Times*, February 7, 2008, p. 28). Most actuarial counting approaches assume that events are independent. In other words, prior events do not alter the basic probabilities laws and randomness that determine the occurrence of subsequent events. To circumvent this lacuna, the traditional and statistical credibility theory approach in insurance evaluates the objectivity and the subjectivity of a risk source and devises a statistical "learning" mechanism that allows the updating of the underlying claim probability. Using Bayesian statistics for example, credibility theory divides risk events into a number classes each with a propensity for the event to occur (default) which are updated using subjective prior estimates of risk classes and an accrued experience—the observed process history. The goal of credibility theory is then to set up an experience rating system. Unlike credibility theory, this "persistent" approach presumes that there may be an inherent persistence in an underlying process that will dictate the probability laws (and therefore the randomness) with which subsequent events occur. The probability of a subsequent event will then be determined by the past memory (in our case, the single past event being defaulting or not) rather than be determined by a statistical estimator based on the accrued evidence of past defaults. Explicitly, while credibility theory seeks to integrate "experience" in estimating the propensity of an event to occur, persistence defines redefines the probability of a subsequent event. Credibility however, uses the fact that an event occurred (or a lack of it) to revise its probability (say by applying Bayes theorem) with a credibly (larger) probability. In this sense these approaches differ fundamentally.

To construct an elemental persistence process, assume for example that default in a period 2 depends on whether (say a loan) has defaulted previously or not (for technical notes related to this problems see also Vallois and Tapiero 2007, 2009, as well as Patlak 1953; Weiss and Rubin 1983; Balinth 1986; Claes and Van den Broeck 1987; Weiss 1994, 2002; Pottier 1996). Denote the event of default in any one period be defined by (0, 1). The event "0" states that no default occurred while a "1" states indicates that a default has occurred. As a result, we construct the following potential events and their probabilities:

$$\begin{array}{c c} & \begin{array}{cc} \text{ND} & \text{D} \end{array} \\ \begin{array}{c} \text{ND} \\ \text{D} \end{array} & \left[\begin{array}{cc} 1-\alpha & \alpha \\ \beta & 1-\beta \end{array} \right] \end{array}, \quad 0<\alpha,\ \beta<1$$

In other words, when a loan defaults in a given year, then in the following period, the probability of default is $\beta > 0$. When there is no default at one period, then the probability of defaulting in the next period is α. This defines a simple two-states Markov chain, given by:

$$\mathbf{P} = \begin{bmatrix} 1-\alpha & \alpha \\ \beta & 1-\beta \end{bmatrix}, \quad 0 < \alpha,\ \beta < 1$$

If we denote by y_t the random event can assume at time t then the probabilities of these random events are:

$$\begin{cases} P(y_t = 1|y_{t-1} = 0) = \alpha,\ P(y_t = 0|y_{t-1} = 0) = 1 - \alpha \\ pP(y_t = 1|y_{t-1} = 1) = 1 - \beta, P(y_t = 0|y_{t-1} = 1) = \beta \end{cases}$$

or

$$[P(y_t = 0) \quad P(y_t = 1)] = \begin{matrix}[P(y_{t-1} = 0) & P(y_{t-1} = 1)]\end{matrix} \begin{bmatrix} 1-\alpha & \alpha \\ \beta & 1-\beta \end{bmatrix}$$

Over a period of time t, the total number of default loan events is therefore given by:

$$x_t = \sum_{j=0}^{t} y_j$$

The number of defaults are conditional on the initial event (the current memory of default or no default), denoted by: $y_0 = 0$ or $y_0 = 1$.

In this memory-based persistent default process, we can calculate the probability process characteristics.

5.8.1 *Mathematical Calculations*

The probability generating function $G(\lambda, t) = E\{\lambda^{x_t}\}$ of the number of loans that have defaulted in a time interval (0, t) where $\rho = 1 - \alpha - \beta,\ \rho \in]-1, 1[$ was shown to be equal to (see Vallois and Tapiero 2007):

$$G(\lambda, t) = \frac{G(\lambda, 1)}{\sqrt{\delta}}\Psi_t - \lambda\rho\frac{G(\lambda, 0)}{\sqrt{\delta}}\Psi_{t-1},\ t \geq 1$$

with

$$G(\lambda,0) = P(x_0 = 0) + \lambda P(x_0 = 1)$$
$$G(\lambda,1) = (1 - \alpha + \lambda\alpha)P(x_0 = 0) + \lambda(1 + (\lambda - 1)(\alpha + \rho))P(x_0 = 1)$$

where

$$a = 1 - \alpha + \lambda\alpha,\ \ \delta = (a + \lambda\rho)^2 - 4\lambda\rho,\ \ \Psi_t = \mu_0^t - \mu_1^t$$
$$\mu_0 = \frac{1}{2}\left(a + \lambda\rho + \sqrt{\delta}\right),\ \ \mu_1 = \frac{1}{2}\left(a + \lambda\rho - \sqrt{\delta}\right)$$

where $P(x_0 = 0)$ is the probability that initially no loan defaulted while $P(x_0 = 1)$ denotes the probability that there was a default.

This generating function provides the moments of the number of defaul in a given time interval with an underlying persistent default process as well as the probabilities of the number of defaults in the given time interval. Further, it clearly points out to the effects of the persistence index $\rho = 0$ on the default process. When $\rho = 0$ then $\beta = 1 - \alpha$ and whatever the previous outcome (whether a default has occurred or not), the subsequent probability of default is α while that of no default is $1 - \alpha$. When the persistence index is positive, $\rho > 0$ then $\beta = 1 - \alpha - \rho$. That is, if a default occurs at some time, then the probability that a default occurs subsequently has a smaller probability. And vice versa, when the persistence index is negative $\rho < 0$ and the underlying default process would point out to a "contagious" default process (for example, with a portfolio's loans defaulting at an increasing rate). Inversely, for $\rho > 0$ it will indicate for example that the portfolio of loans has a built-in "incentive effect," reducing a default probability in a given period following a default made in the previous one.

Our result allows an explicit calculation of the moments of the persistent counting process, defined recursively as shown below. Note that:

$$\frac{\partial G}{\partial \lambda}(\lambda,t) = E\left(x_t\lambda^{x_t-1}\right),\ \ \frac{\partial^2 G}{\partial \lambda^2}(\lambda,t) = E\left(x_t(x_t-1)\lambda^{x_t-2}\right),$$
$$\frac{\partial^3 G}{\partial \lambda^3}(\lambda,t) = E\left(x_t(x_t-1)(x_t-2)\lambda^{x_t-3}\right)$$

and so on for higher order terms. Using these terms and setting $\lambda = 1$, we obtain the necessary equations which allow the calculation of the mean, the variance, the kurtosis and other moments of the persisting default process distribution. It is also useful to derive a number of special and well known cases to confirm the validity of our results. First note that when $\lambda = 1$ then $G(1,t) = 1$ as expected. Further, when there is no persistence (i.e $\rho = 0$), we have $G(\lambda,t) = \frac{G(\lambda,1)}{a}\Psi_t$, $\mu_0 = a$, $\mu_1 = 0$ and $\Psi_t = a^t$ and therefore: $G(\lambda,t) = \{P(x_0 = 0) + \lambda P(x_0 = 1)\}a^t$.

In particular, if initially, $P(x_0 = 0) = 1$, then $G(\lambda, t) = a^t$ which corresponds as expected to the Probability Generating Function of a binomial distribution for counting the number of default over a given period of time. However, if $P(x_0 = 1) = 1$, then $G(\lambda, t) = \lambda a^t$. A convenient recursive expression for the generating function is found by setting $\Psi_t = {\mu_0}^t - {\mu_1}^t$ where μ_0 and μ_1 solve: $\mu^2 - \mu(\lambda(\rho + \alpha) + 1 - \alpha) + \lambda\rho = 0$ and verify the second order equation: $\Psi_{t+2} - (a + \lambda\rho)\Psi_{t+1} + \lambda\rho\Psi_t = 0$. As a result, the probability generating function $G(\lambda, t)$ satisfies as well the second order recursive equation given by:

$$G(\lambda, t+2) = (1 - \alpha + \lambda(\alpha + \rho))G(\lambda, t+1) - \lambda\rho G(\lambda, t)$$

Deriving this equation with respect to λ with $\lambda = 1$ we obtain a recursive expression for the moments of the default process. Concentrating our attention on the first moments only, derivatives yield the following recursive equation:

$$\begin{aligned}
G(\lambda, t+2) &= (1 - \alpha + \lambda(\alpha + \rho))G(\lambda, t+1) - \lambda\rho G(\lambda, t) \\
\frac{\partial^k G(\lambda, t+2)}{\partial\lambda^k} &= (1 - \alpha + \lambda(\alpha + \rho))\frac{\partial^k G(\lambda, t+1)}{\partial\lambda^k} - \lambda\rho\frac{\partial^k G(\lambda, t)}{\partial\lambda^k} \\
&\quad + k(\alpha + \rho)\frac{\partial^{k-1} G(\lambda, t+1)}{\partial\lambda^{k-1}} - k\rho\frac{\partial^{k-1} G(\lambda, t)}{\partial\lambda^{k-1}}, \quad k = 1, 2, 3, \ldots.
\end{aligned}$$

With initial conditions:

$$\begin{aligned}
&G(\lambda, 0) = P(x_0 = 0) + \lambda P(x_0 = 1) \\
&G(\lambda, 1) = (1 - \alpha + \lambda\alpha)P(x_0 = 0) + \lambda(1 + (\lambda - 1)(\alpha + \rho))P(x_0 = 1) \\
&\frac{\partial G(\lambda, 0)}{\partial\lambda} = P(x_0 = 1), \quad \frac{\partial G(\lambda, 1)}{\partial\lambda} = \alpha P(x_0 = 0) + (1 + (2\lambda - 1)(\alpha + \rho))P(x_0 = 1) \\
&\frac{\partial^2 G(\lambda, 0)}{\partial\lambda^2} = 0, \quad \frac{\partial^2 G(\lambda, 1)}{\partial\lambda^2} = 2(\alpha + \rho)P(x_0 = 1) \\
&\frac{\partial^j G(\lambda, 0)}{\partial\lambda^j} = 0, \quad \frac{\partial^j G(\lambda, 1)}{\partial\lambda^j} = 0, \quad \text{for } j \geq 3
\end{aligned}$$

At $\lambda = 1$, we can write these expressions in the following manner which simplifies their numerical solution:

$$\begin{aligned}
G(1, t+2) &= (1 + \rho)G(1, t+1) - \rho G(1, t) \\
\frac{\partial^k G(1, t+2)}{\partial\lambda^k} &= (1 + \rho)\frac{\partial^k G(1, t+1)}{\partial\lambda^k} - \rho\frac{\partial^k G(1, t)}{\partial\lambda^k} \\
&\quad + k(\alpha + \rho)\frac{\partial^{k-1} G(1, t+1)}{\partial\lambda^{k-1}} - k\rho\frac{\partial^{k-1} G(1, t)}{\partial\lambda^{k-1}},
\end{aligned}$$

While the initial conditions stated above in equation and leading to:

$$
\begin{aligned}
&G(1,0) = 1, \ \ G(1,1) = 1 \\
&\frac{\partial G(1,0)}{\partial \lambda} = P(x_0 = 1), \ \ \frac{\partial G(1,1)}{\partial \lambda} = \alpha + (1+\rho)P(x_0 = 1) \\
&\frac{\partial^2 G(1,0)}{\partial \lambda^2} = 0, \ \ \frac{\partial^2 G(1,1)}{\partial \lambda^2} = 2(\alpha+\rho)P(x_0 = 1) \\
&\frac{\partial^j G(1,0)}{\partial \lambda^j} = 0, \ \ \frac{\partial^j G(1,1)}{\partial \lambda^j} = 0, \ \ \text{for } j \geq 3.
\end{aligned}
$$

Similarly, we can calculate the probabilities of the persistent default process by setting $\lambda = 0$ in the derivatives of the generating functions. In this case, the probabilities are given by:

$$
p_i(t) = \frac{1}{i!}\frac{\partial^i G(\lambda,t)}{\partial \lambda^i}, i = 0,1,2,3,\ldots,t \text{ with :}
$$

$$
\begin{aligned}
&G(0,t+2) = (1-\alpha)G(0,t+1) \\
&\frac{\partial^k G(0,t+2)}{\partial \lambda^k} = (1-\alpha)\frac{\partial^k G(0,t+1)}{\partial \lambda^k} + k(\alpha+\rho)\frac{\partial^{k-1} G(0,t+1)}{\partial \lambda^{k-1}} - k\rho\frac{\partial^{k-1} G(0,t)}{\partial \lambda^{k-1}},
\end{aligned}
$$

With the initial conditions:

$$
\begin{aligned}
&G(0,0) = P(x_0 = 0), \ \ G(0,1) = (1-\alpha)P(x_0 = 0) \\
&\frac{\partial G(0,0)}{\partial \lambda} = P(x_0 = 1), \ \ \frac{\partial G(0,1)}{\partial \lambda} = \alpha P(x_0 = 0) + (1-\alpha-\rho)P(x_0 = 1) \\
&\frac{\partial^2 G(0,0)}{\partial \lambda^2} = 0, \ \ \frac{\partial^2 G(0,1)}{\partial \lambda^2} = 2(\alpha+\rho)P(x_0 = 1) \\
&\frac{\partial^j G(\lambda,0)}{\partial \lambda^j} = 0, \ \ \frac{\partial^j G(\lambda,1)}{\partial \lambda^j} = 0, \ \ \text{for } j \geq 3.
\end{aligned}
$$

These equations define a numerical approach to calculating both the moments and the probabilities of a persistent process. A more direct approach will be outlined subsequently however.

Explicit results for the first two moments are provided below with proofs found directly from these equations. Let $x_0 = 0$, then the expected number of default loans and its second moment are:

$$
E(x(t)) = (1+\rho)E(x(t-1)) - \rho E(x(t-2)) + \alpha, \ \ t \geq 2
$$

$$
E\left(x(t)^2\right) = (1+\rho)E\left(x(t-1)^2\right) - \rho E\left(x(t-2)^2\right) + \frac{\alpha}{1-\rho}\left[2\alpha t + \rho + 1 - 2\left(\frac{\alpha + (1-\alpha)\rho^t - \rho^{t+1}}{1-\rho}\right)\right], \ \ t \geq 2
$$

In these equations, note that we have as expected (Vallois and Tapiero 2007):

$$E(x(t)) = \frac{\alpha}{1-\rho}\left[t+1-\frac{1-\rho^{t+1}}{1-\rho}\right]$$

This clearly indicates the nonlinear time effects (and the potential contagions) of persistence in such default processes. A verification can be reached by setting:

$$\rho = 0 \text{ and } E\left(x(t)^2\right) = E\left(x(t-1)^2\right) + \alpha[1 + 2\alpha(t-1)].$$

Summing for $1 \le t \le n$, we obtain $E\left(x(n)^2\right) = n\alpha(1+\alpha(n-1))$. Since $x(n)$ is binomial, we have $E(x(n)) = n\alpha$, $\mathrm{Var(x(n))} = n\alpha(1-\alpha)$ and thereby: $E\left(x(n)^2\right) = n\alpha(1-\alpha) + n^2\alpha^2 = n\alpha(1+(n-1)\alpha)$ as expected.

Explicit expressions for the persistent counting probabilities can be determined as well using the recursive probability generating functions. In this case, we calculate $(p_k(t); 0 \le k \le t)$ by recurrence. Initially these are specified by:

$$p_0(0) = 1,\ p_0(1) = 1-\alpha, p_1(1) = \alpha$$

Further, $(p_k(t+2); 0 \le k \le t+2)$ is defined as a function of $(p_k(t+1); 0 \le k \le t+1)$ and $(p_k(t); 0 \le k \le t)$ by using the recursive equation

$$p_k(t+2) = (1-\alpha)p_k(t+1) + (\alpha+\rho)p_{k-1}(t+1) - \rho p_{k-1}(t),$$

for all $0 \le k \le t+2$, and by convention, we set $p_k(t) = 0$ if $k < 0$.

Subsequent calculations will indicate the underlying process probabilities. In particular, we have for the first three probabilities:

$$p_0(t) = (1-\alpha)^t$$
$$p_1(t) = \alpha(1-\alpha)^{t-2}((1-\alpha-\rho)t+\rho),\ t \ge 1$$
$$p_2(t) = \alpha\left\{\begin{array}{l}(\alpha+\rho)(1-\alpha)^{t-2}\\ +\frac{1}{2}(1-\alpha)^{t-4}(1-\alpha-\rho)(t-2)(\alpha(1-\alpha-\rho)t+\alpha+2\rho-\alpha^2+\alpha\rho)\end{array}\right\},\ t \ge 2$$

Of course, when there is no persistence, this is reduced as expected to:

$$p_0(t) = (1-\alpha)^t$$
$$p_1(t) = \alpha t(1-\alpha)^{t-1},\ t \ge 1$$
$$p_2(t) = \frac{1}{2}\alpha^2(1-\alpha)^{t-2}t(t-1), t \ge 2$$

5.8.2 Persistence and the Probability of Losses in a Contagion

Next assume that every default has a loss probability distribution $\tilde{Z}_j$. The total loss over a time interval (0, t) is thus a Compound Loss process given by $\xi_t = \sum_{j=1}^{x_t} \tilde{Z}_j$ with mean $\mu = E(\tilde{Z}_i)$ and known variance $\mathrm{var}(\tilde{Z}_i)$. For independent loss processes, the mean loss up to time t and its variance are:

$$E\left(\sum_{i=1}^{x_t} \tilde{Z}_i\right) = E(x_t)E(\tilde{Z}_i) \quad \text{and} \, \mathrm{var}\left(\sum_{i=1}^{x_t} \tilde{Z}_i\right) = \left[E(\tilde{Z}_i)\right]^2 \mathrm{Var(x_t)} + \mathrm{Var(Z_i)E(x_t)}.$$

However, this is not the case if a process is probability dependent (persistent). A more general expression is needed. Explicitly, consider the following random default:

$$S_t = \sum_{i=0}^{t} \tilde{Z}_i^0 1_{\{y_i=0\}} + \sum_{i=0}^{t} \tilde{Z}_i^1 1_{\{y_i=1\}} \quad \text{where} \quad x_t = \sum_{i=0}^{t} y_i$$

where $\tilde{Z}_i^0$ is a "normal expense" occurring in any regular period (defined by the fact that no specific default has occurred) while $\tilde{Z}_i^1$ is a "large loss due to a default" (of course, if $\tilde{Z}_i^0 = \tilde{Z}_i^1$ then $S_t = \xi_t$ as stated above). We assume that $\{\tilde{Z}_i^0, \tilde{Z}_i^1\}$ are random variables independent of each other and independent of the Markov (persistent) claims. In this case, the compound claim mean and variance and the claim probability generating function are given by the following:

$$E(S_t) = (t+1)E(\tilde{Z}_1^0) + \left(E(\tilde{Z}_1^1) - E(\tilde{Z}_1^0)\right)E(x_t)$$

$$\begin{aligned} E(S_t^2) &= (t+1)\left[\mathrm{var}(\tilde{Z}_1^0) + (t+1)\left(E(\tilde{Z}_1^0)\right)^2\right] + E(x_t^2)\left[E(\tilde{Z}_1^1) - E(\tilde{Z}_1^0)\right]^2 \\ &\quad + E(x_t)\left[\mathrm{var}(\tilde{Z}_1^1) - var(\tilde{Z}_1^0) + 2(t+1)E(\tilde{Z}_1^0)\left(E(\tilde{Z}_1^1) - E(\tilde{Z}_1^0)\right)\right] \end{aligned}$$

While their Laplace Transform is:

$$E(e^{-\lambda S_t}) = \left[E\left(e^{-\lambda \tilde{Z}_i^0}\right)\right]^{t+1} G(z,t), \quad z = \frac{E\left(e^{-\lambda \tilde{Z}_1^1}\right)}{E\left(e^{-\lambda \tilde{Z}_1^0}\right)}, \quad \lambda \geq 0$$

where $G(z,t)$ is a probability generating function:

$$G(z,t) = E(z^{x_t}) = \sum_{i=0}^{t+1} z^i P(x_t = i).$$

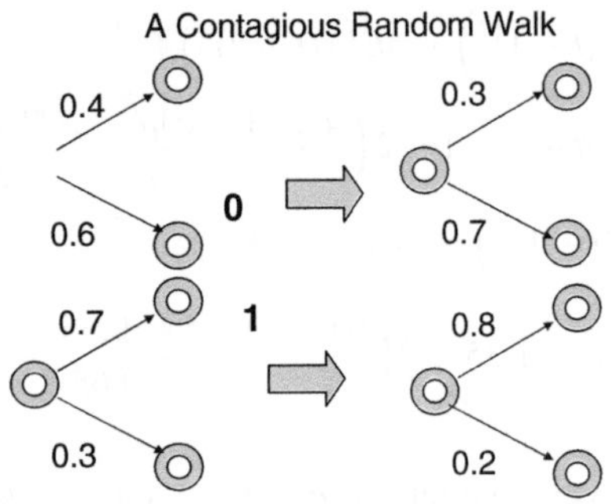

Fig. 5.5 Short memory processes

Note that when $\tilde{Z}_1^0$ is null, then, we find that:

$$E(S_t) = E(\tilde{Z}_1^1)E(x_t) \text{ and } \mathrm{Var}(S_t) = E(x_t)\mathrm{Var}(\tilde{Z}_1^1) + \mathrm{Var}(x_t)\left(E(\tilde{Z}_1^1)\right)^2$$

These results can then be used to assess cumulative losses in contagions and other dependent processes. Note however that the variance of this process has increased over time due to the process persistence. Further, based on these moments, risk exposure can be determined more precisely which would use these two moments as a first approximation (although higher order moments can be calculated as well, with the generating function of the persistent Compound Process).

Example: Some Memory Models In Fig. 5.5, note that the following situation occurs. If an event (=1) has occurred with probability $0.4 < 0.6$, then the probability that the next event is equal 1 is decreasing. However if te probability of obaiting a 0 is 0.6, then the next event probability is increasin to 0.3 (Fig. 5.5).

Example Assume that in the underlying persistent process defined by probabilities $(\alpha, 1-\beta)$ we have, $\alpha = 0.05$ and $1-\beta = 0.40$. In other words, when a default occurs the probability of another default is 0.40 while when no default occurs, the probability of a default is only 0.05. As a result, $\rho = 1-\alpha-\beta = 0.40 - 0.05 = 0.35$ and therefore the expected number of defaults over a time interval (0, t) is:

$$E(x(t)) = \frac{\alpha}{1-\rho}\left[t+1-\frac{1-\rho^{t+1}}{1-\rho}\right] = \frac{0.05}{1-0.35}\left[t+1-\frac{1-0.35^{t+1}}{1-0.35}\right]$$
$$= \frac{0.01}{0.13}\left[t+1-\frac{1-0.35^{t+1}}{0.65}\right]$$

By the same token the second moment is:

$$\begin{aligned} E\left(x(t)^2\right) &= (1+\rho)E\left(x(t-1)^2\right) - \rho E\left(x(t-2)^2\right) + \\ &\quad \frac{\alpha}{1-\rho}\left[2\alpha t + \rho + 1 - 2\left(\frac{\alpha + (1-\alpha)\rho^t - \rho^{t+1}}{1-\rho}\right)\right] \\ &= (1.35)E\left(x(t-1)^2\right) - \rho E\left(x(t-2)^2\right) + \\ &\quad \frac{0.05}{0.65}\left[0.1t + 1.35 - 2\left(\frac{0.05 + 0.95 * (0.35)^t - (0.35)^{t+1}}{0.65}\right)\right], \quad t \geq 2 \end{aligned}$$

Next consider the reverse situation, $\alpha = 0.10$ and $1 - \beta = 0.05$. In this case, $\rho = 0.05 - 0.10 = 0.05$. What would be the mean number of defaults and their variance? How do such results differ? (show that the first *points* out to persistence and the other to anti-persistence).

References and Additional Reading

Time has been and remains both a philosophical concept and the means to organize our thoughts in a tractable and temporal perspective. Aristotle, Saint Augustine, Einstein, Janet (see Fraisse 1957) and so many others have sought to grasp the meaning of time and reveal its future. In my paper 1978a, time is used in its modeling context and used to reconcile the passing of time, the memory of past events, and their interaction with its change and the unpredictable future. The simplest of all such models are of course stochastic processes, whether memoryless or well structured Markov memory models (see also Tapiero 1994c on stochastic modeling). The complexity of these factors leads mostly to intractable and chaotic evolutions (Lorenz 1966; May 1974, 1976) and to the evolution of mathematical techniques (chaos theory for example) that seek to reduce this intractability to a recognizable quantitative framework. In economics and finance, authors such as Grandmont and Malgrange (1986) on nonlinear dynamics, Medio (1992) on chaotic dynamics and an impressive number of research papers published under the banners of econophysics and financial physics (see the many publications in *Physica A*) are providing multiple avenues in stochastic and nonlinear modeling and making an appreciable contribution to risk engineering and to finance.

Stochastic models are used in many field. This is the case in physics, Einstein (1906, 1956); in engineering, Stratonovich (1963); in actuarial science and stochastic models in general such as for example, Feller (1951–1971), Cox and Miller (1965), Levy (1937), Lundberg (1909, 1940), Prabhu (1965, 1980), Seal (1969, 1978), Tetens (1786), Klugman et al. (2000) on Loss Models, Lax, Wei and Xu, on random processes in physics and finance, Shaked and Shantikumar (1994) on reliability and many others. In addition to an extensive literature on stochastic processes in insurance (for example, Embrechts 2000a, b; Embrechts et al. 1997, 2001, 2002, 2003, 2009). Optimization of such models are reached by applications of stochastic dynamic programming and control (e.g. Bensoussan 1982 on control, Bensoussan and Tapiero 1982 on impulse control, Bismut 1975, 1978 on the Growth and intertemporal allocation of risks based on a duality in optimal stochastic control (see also Bismut 1976, on Theorie Probabiliste du Controle des Diffusions).

Applications in Operations and risk engineering abound as well. Queuing Theory is a branch of discrete events stochastic processes (Saaty 1961; Gross and Harris 1985; Basawa and Prabhu 1981; Basawa and Rao 1980), on the statistical analysis of queuing processes and application of quality control techniques to queues operations, Hsu and Tapiero (1987a, b, 1988a, b, 1992, 1994), Hsu et al. (1993), Liu et al. (1990) on general queues, Tapiero (1982a, b) applications in advertising stochastic models and in marketing, including repeat purchase models and information advertising content (Tapiero 2000a), queue in environmental control problems (Tapiero 2005a), Barnett and Toby (1994), on *Outliers in Statistical Data.*

Stochastic models consist fundamentally of elementary probability events. In Chap. 3 we noted for example that the Bernoulli process underlies a broad number of probability models. Such a model, expanded to its multivariate form, and sequenced in some logical manner can lead to complex processes that may be analyzed either numerically or by simulation (Carreira-Perpinan and Reinals 2000 on mixtures of multivariate Bernoulli probability models, Neural computation de la Selva 1988 on correlated random walks, Galambos 1978 on extreme statistics, Leadbetter et al. 1983 on extremes in related sequences and Rubenstein 1981 on simulation, Eberlein (2001) on Levy processes). In physics texts such models abound and increasingly are applied to financial problems (Lax et al. 2006 on Random Processes in Physics and their Applications in Finance), Cartea and Howison (2003) on Limits of Levy-Stable Processes and their Applications to Option Pricing, Bouchaud and Potters (2003), on applications of stochastic models of physics to financial risks and derivatives pricing. Their applications are varied, we referred to Fujiwara (2004) on Zipf law in firms bankruptcy, Garlaschelli et al. (2005) on the scale free nature of market investment networks, Saito et al. (2007) interfirm relationships, etc.

Just as the binomial distribution and the negative binomial (Pascal) probability distribution were used to calculate the number of "success" events occurring in a given time interval and the random number of events to be repeated to obtain a given number of "success" events, in a temporal perspective, we also use run processes. For example, Mood (1940) provided some basic results on the distribution of runs, Chow et al. (1971a, b), Wolfowitz (1943), Wald (1947, 1961) on sequential analysis and decision function, Robbins and Siegmund (1971) on Optimal stopping time, Siegmund (1985), Fou ans Koutras (1994), Capocelli and Ricciardi (1972) on inverse first passage times in stochastic differential equations (with subsequent papers extending their results).

In economics ARCH and GARCH models (Engle 1987, 1995; Bollerslev et al. 1992, 1994, on ARCH and GARCH models, Nelson and Foster 1994 for example) have led to our constructing multivariate stochastic processes (essentially mean and the volatility stochastic variations—one dependent on the other).

Parkinson and subsequently Yang and Zhang used hi and lows (the range process) in time series to assess this volatility. Stochastic models in finance are used extensively to assess the price of options. Other stochastic models based on Gamma processes were introduced by Baxter (2007) with their application in finance by Carr and Madan as well as a number of recent papers that suggest that we construct stochastic models of models correlations (see De la Selva and Lindenberg 1988 on correlated random walks for a discrete events approach to such models).

Range processes statistics complement variance statistics. Feller (1951) provided initial theoretical results on the range process (*The asymptotic distribution of the range of sums of random variables*) while Hurst (1951) study of the water level of the Nile based on records taken over thousands of years has led Mandelbrot to introduce fractal models and to the mathematical study of the Hurst index (a Range to Standard deviation statistic). Mandelbrot (1997, 2004), Mandelbrot and Taqqu (1979), Granger (1980), Beran (1994) on the statistics of long run memory model, Mandelbrot and Wallis (1968), Otway (1995) as well as Tapiero and Vallois with many subsequent applications beginning with Mandelbrot and his students. These applications include: Booth et al. (1982), Cheung (1983), Fung and Lo (1995), Green and Fielitz (1977, 1980), on applications of RS analysis in FX markets, Mandelbrot and Van Ness (1968), Rogers and Satchell (1991) on Hi and Los and variance estimates etc.

In economics, this index was used as "an index of chaos," and to study various facets of economic time series. For example, Diebold and Rudenbusch (1989) use the R/S analysis in monetary economics, in financial contagion, Martınez-Jaramillo et al. (2010) on contagion and systemic risks, Haubrich and Lo (1989) on long run dependence and business cycles, Helms et al. (1984) on memory in commodity markets, Hsieh (1991) on chaos in financial markets, Lo (1991, 1997) on fat tails in stock market, Anderson (1957) on contagions in collective risk theory etc. Booth et al. (1982, 1972, 1997, 2004), Cheung (1983), Diebold and Rudenbusch (1989), Lo (1991), Mandelbrot (1963, 1972, 1997, 2004) are some applications to FX markets, commodities and economic problems.

In continuous stochastic models, Imhof (1985) provided the first results for the range in a Brownian motion while Vallois (1995, 1996), Vallois and Tapiero (1995a, b, 1996, 1997, 2001), Tapiero and Vallois (1997a, b) provide explicit results on range processes (see also Siebenaler 1997; Siebenaler et al. 1997; Tapiero and Vallois 1996, 2000).

Stochastic models of Contagion and Persistence are also varied, some include a seminal book on random walks and persistence by Weiss and Rubin (1983), Domokos and Thoth (1984), Persistent random walks in a one dimensional and random environment, Balinth (1986), Mzasoliver et al. (1992), on the continuum limit of a two dimensional persistent random walk, Amaral et al. (1997) on persistence model in Physics, Claes and Van den Broeck (1987) the random walk with persistence, Stanley et al. (1996), Pottier (1996) persistence in biased random walks and in a random environment, Weiss (1994), seminal book on *Aspects and Applications of the Random Walk*, Weiss (2002), Some applications of persistent random walks and the telegrapher's equation, Weiss and Rubin (1983), De la Selva et al. (1988), Correlated random walks (Pottier 1996; Patlak 1953).

My own research with Pierre Vallois (Vallois and Tapiero 1997a, b, 2009) and the recent seminal paper by Herman and Vallois (2010a), who provide both a summary and explicit convergence results for persistent stochastic processes leading to both limit Brownian motions and telegraphic processes (second order partial differential equations in both state and time). These models provide an important foundation to contagious stochastic processes. Additional and related studies include Tapiero, C.S. and P. Vallois on The Inter-Event Range Process and Testing for Chaos in Time Series, Vallois P. and C.S. Tapiero (1997a), on Range Reliability in Random Walks.

Chapter 6
Risk Measurement

Abstract The purpose of this chapter is to introduce traditional approaches to the measurement of risk applied to financial and risk engineering problems. These approaches are gradually challenged by a growing sophistication, a greater access to data and axiomatic foundations and metrics to value and price risk. Financial approaches are contributing to risk practice by using money as a common metric to value risk and performance. While this chapter focuses on risk measurement subsequent Chaps. 7–10 will focus on risk valuation, risk economics and the economics of uncertainty.

6.1 Introduction

This chapter defines approaches to risk measurement. It seeks to reduce measurements (whether data based or not) into structured "countable states" to assess and to mitigate defined risks. To do so, preferences and accountability dictate the manner in which data is measured, reduced to risk models and used for risk valuation and pricing. A variety of risks and risk objectives originate in multiple fields, each with its own perception and definition of risk, generating a large number of measurements applied to meet their own experience and needs. For business consulting firms, risk measurement is the process by which companies determine the risks they deem important for their Business plans and operations. They construct and use models to determine what level of risk is associated to the many internal layers that make up a business (within an organizations) and externally (their clients, their suppliers, environmental factors etc.).

Data combined with hypothetical models, either justified by statistical techniques or by an "internal conventional wisdom" may be reduced to risk metrics. Such reduction, while it provides comfort in numbers, need not characterized fully risks. "Comfort in numbers" may instead induce risks when they are qualified as certain and used to reach decisions based on false premises. How certain are we of our measured facts? They may lead one to underscore uncertain and appreciable

C.S. Tapiero, *Engineering Risk and Finance*, International Series in Operations Research & Management Science 188, DOI 10.1007/978-1-4614-6234-7_6,

risks! When "certainty" sweeps an "awareness of uncertainty", it may also lead to the intolerance of ambiguity and to certainty risks. For example, some social systems built on the certainty of their right and their might have evolved into predatory social systems based on the intolerance of all others. In finance, the certainty of financial models to price assets led investment firms to assume disastrous economic bets. The "certainty" of one's right, negating all other rights can therefore be self-destructive. In this sense, measurement, risk models and their use are neither a search of certainty (although risk averse investors seek such certainty), nor are they the end of risks that lay ahead. Presuming a state of certainty thus induces risks. A "Certainty Model" however, may be needed to induce one to make a decision or act. Herbert Simon, refers to the importance of a "Locus of Uncertainty Absorption" as a source of power, on the one hand absorbing information and reinterpreting it to others by reducing it to a "defined certainty" (my words). For example, would your absolute beliefs in a future increase in the price of gold lead you to buy more of it? Would awareness of tail risks (with substantial consequences) let you bet on a rare outcome? The fundamental theory of finance (see Chaps. 7 and 8) has essentially "done away with risk" by setting a number of rational assumptions and financial models based on the conclusion that current and observed market prices embeds all the information about the asset and its risk. The current price is then the unique true price, while future market prices are embedded into this current certainty. In such situations, financial markets are ultimately intelligent, reducing future risk consequences to a certain present price! Such markets however are based on risk models and are therefore partial models of the uncertainty. In this sense, while risk measurement is essential to construct risk models, these models are hypothesis expressing the best we can do, given the data-measurements we have defined and collected to reject or not the hypotheses that underlie such models and assess their risks.

Measurements are used for multiple purposes. They can help detect future states, measure "deviations from expectations", estimate financial consequences as well as induce, support and mitigate the risks of a decision process. Measurement of potential risks may also provide a warning about future risks and comfort in what we know, or at least delineates (by implication) risk events we may be able to deal with. Measurements may be tailored to a person or to a financial institution, to a Government policy, to "a safety agency" etc., each of with its own particular attitude and valuation of risk. For example, on the one hand, risk regulation (whether financial, environmental, safety, health, liability risks, drug development etc.) enforces compliance and on the other it is also an incentive to the regulated to comply with regulation.

Figure 6.1 above summarizes some elements motivating risk measurement and risk models and their purposes.

Risk measurements based on objective (and statistical sampling of) data, on the consensus of experts or educated guesses by experienced parties or visionaries, provide an infrastructure of knowledge and knowhow to focus risk concerns. For example, application of the VaR metric to regulating the banking and the insurance system leads to focusing on reserves as a principal tool to control risks.

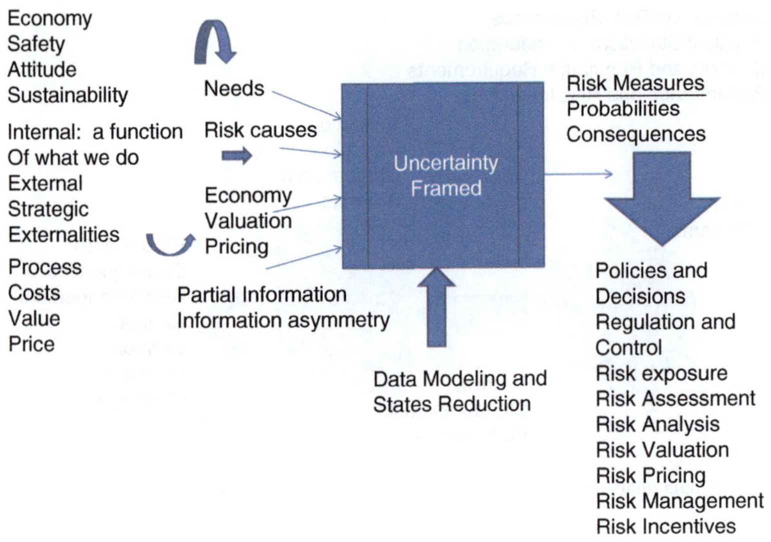

Fig. 6.1 Selected dimensions in risk measurement

This diverts a bank from attending to other risks and thus turns out to be ineffective in managing other risks. In fact while measurements are benign, their use embeds risks.

Risk measurements, while essential have thus unintended consequences. Some of the following questions address potential problems:

- Is a prediction conditioned by risk measurement self-fulfilling?
- Do measured risks migrate? i.e. measured in one place, move to another stealth place!
- Do measurements measure what they claim they measure?
- Do measurements bound our risk awareness?
- Are the measurements of risk derivatives accounted for?
- Are risk externalities defined and accounted for?
- Are strategic risks identified and accounted for?

Examples abound. Recording crime in an area will have crime travel to another; increasing regulation controls in one sector, will shift the regulated to other sectors; augmenting the security at airports, will lead to open new channels to terror attacks; over regulating businesses may lead to their migration and to a loss of jobs, etc.

Can one be sure that a car safety depends on one factor or multiple and interacting factors. Using one measure of safety may ignore another, potentially a better predictor of safety? Are derivatives risks accounted for? Risks, thus, are like Octopuses, a head and many arms, deriving from their head, dependent and contagious, with the one increasing causally the risk of the other. Losing one's money or one's health may lead to other losses or amplifying one's own losses.

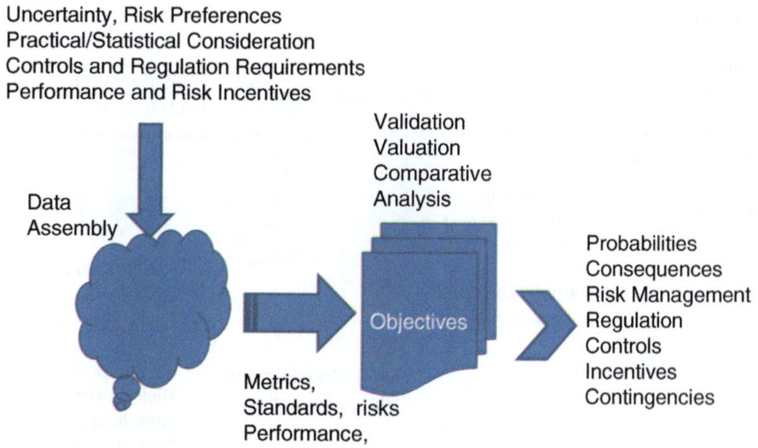

Fig. 6.2 The data analysis process

During the MBS crisis in 2008, some people lost their jobs and subsequently their home to Banks as they became unable to pay their monthly mortgages. In other words, risks travel both horizontally to unrelated risks and vertically in amplifying their risks'. In this sense, risk measurements are mostly incomplete, neglecting some aspects of risks and representing only some statistical manifestations of such measurements.

Risk measurements by regulators—financial and environmental, or class room tests measurements provide incentives to improve one's score (or at least study harder for the test period). However while measurements are specific, their applications are multi-faceted, reflecting both our many needs to manage risks and the importance of their consequences. Measurements, whether objective or subjective, ordinal and quantitative are thus used for the following reasons:

- To define needs that expresses explicitly or implicitly risk consequences.
- To reconcile risk measurements with well stated objectives.
- To signal an intent and thereby motivate risk mitigation.

In the latter case, measurements may cater to "one objective" (such as compliance to regulation, financial losses, comparative performance, making money and not losing it, etc.); to multiple objectives (such as seeking both a high performance and no loss); to "strategic objectives" (such as competitive scores, meeting the objectives of a number of parties, each of which has its own information, preferences and agendas) as well as to "societal (Externalities) objectives" (mostly public objectives and ethical behaviors and attitudes, etc.) (Figs. 6.2 and 6.3).

Technically, risk measurements are used to:

- Detect risk causes quantified by the probabilities of adverse events (their distributions, statistical dependence, causal dependence, etc.)
- Measure consequences (and therefore value), including:

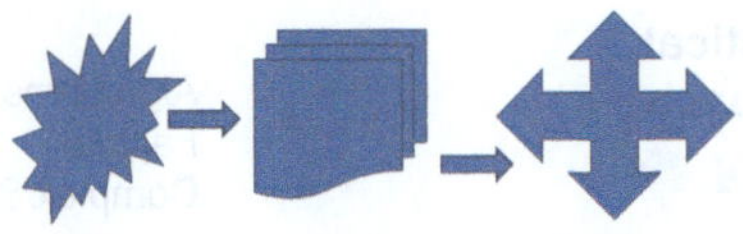

Fig. 6.3 Data reduction and scaling

  - Measures of deviations from a standard (an expectation, approved practices, a benchmark, regulated emissions and their like)
  - Measure consequences (deviations from expectations, comparative measurements of performances to best practices, non-financial political risks, health-probabilities, branding, reputation, etc.)
  - Measurements of Value (costs, individual valuation and market valuations-prices)

- Provide incentives, controls, inducing competition by "comparative" measures (e.g. Olympic trials competition).

Risk measurements are then translated by metrics that are combined and used with a risk-rationality. In this sense,

* Risk measurements * Risk Metrics * Risk rationality

are often integrated to engineer systems expressing our preferences, our information, our concerns and our needs for risk management.

6.2 Big Data and Risk Measurement

The growth of internet communications and social media has fueled a plethora of data based methods to make "some sense" of the large quantities of data produced by an "infinite" number of sources. An increasing number of software and start-up firms are proposing "black boxes" to interpret 'consumers' moods and intents using internet comments on stocks, on financial assets and economic variables. Algorithms and automatic learning machines are then created to seek and interpret images, to detect a "flow of sentiments" that claim to be related (and thereby, be early predictors) of stock markets performance, predict consumers' choices and their implied preferences. A rising tide of data driven algorithms based on "pseudo-rationalities" is thus emerging engulfing large firms and business interests. This approach competes with and complements at the same time the "statistical approach" to data measurement and interpretation. However, the use of big-data may lead to self-fulfilling prophecies, with wanted results "found" in data masses. For example, say that a firm hires a big data consulting firm to express the public attitude towards that firm and its' CEO? It would be like paying a psychiatrist to hear what ones wants to hear. In a big data set the

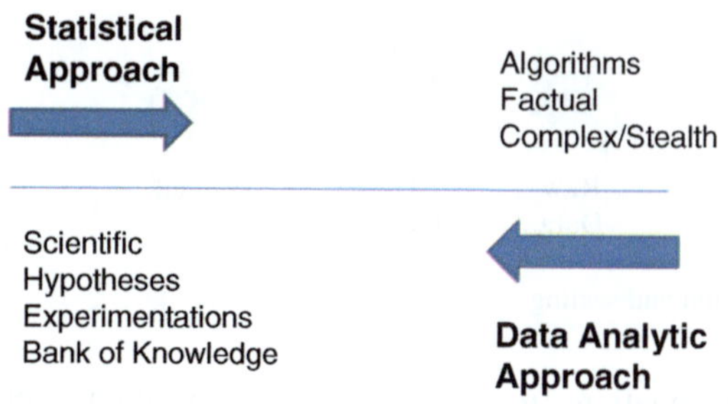

Fig. 6.4 Statistical and data analytical approaches

chances of finding what ones want are equally high. Furthermore, searching for meaning in large data sets, without theories, may be like seeking the North Pole without a compass. For these and other reasons, Big Data based on the accumulation of private information is a growing source of risk that has led already to important security problems. For example, Big Data has also made it possible to steal national and military secrets, to divulge private financial and health information etc.

The statistical approach is based on fundamental and structured hypotheses emanating from scientific statements or theories to be refuted or not, based on statistical (data) evidence. As a result, the "statistical/scientific" approach reveals "uncertainty" from a given and tested knowledge base. While, data driven "algorithms" are based on a "robotized artificial intelligence" that has the capacity to treat large data sets to reveal co-variations, "associations", "patterns" etc. based on their search (in some cases improved by built-in learning algorithms) rather than based on theory (Fig. 6.4). Thus while the "statistical/scientific" approach is an evolutionary and evolving process, based on a cycle to "hypothesize, measure, test and confirm-or-not", the data driven approach is a short cut, seeking to "know" even though it may not know why. It is an interpretation of data at any one time with a predictive artificial intelligence that decision makers might not understand as data revelations can be as complex and as stealth as artificial intelligence may be to its users. It is therefore, a statement of a current fact—a presumed certainty, rather than a recognition that all knowledge is partial—embedded in a greater uncertainty.

The British astrophysicist Arthur S. Eddington once wrote "No experiment should be believed until it has been confirmed by theory!" (*NY Times*, March 27, 2012, page D1).

In this sense, data driven measurements can be misleading and reveal about anything one may choose to discover. For example, digitalization of the Bible with a "robotized artificial intelligence" has revealed to "digital mystiques" sequences of statements revealing new prophecies to the ends of days. To disprove its validity, a digitalization of Tolstoy "War and Peace" and a search engine revealed as well "secret intents". Other books have done so, predicting even political stalemates in the Middle East and their resolution. In this sense, data driven measurement models can make sense out of non-sense even though it is in fact nonsense. Big Data has

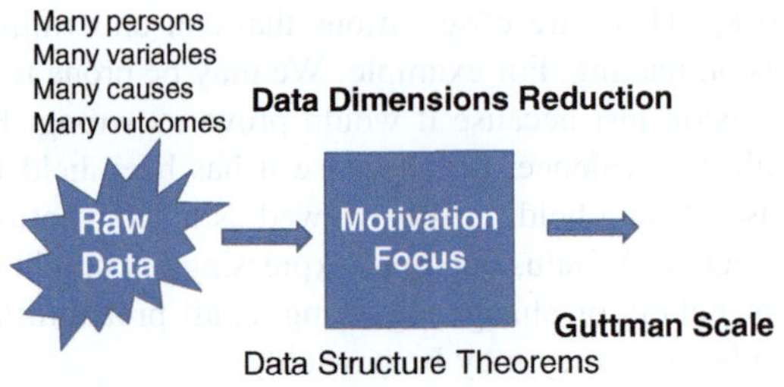

Fig. 6.5 The Guttman approach to data reduction

therefore its perils, with type II statistical errors (i.e. making a false discovery), enslaving decision makers to rationalities they do not comprehend and provides a new twist to the misuse of data measurement by seeking and finding the confirmation of one's certainty. Finally, it can turn out to be an unwieldy process victimized by the belief that a larger hay stack may help find a needle in that hay stack. Yet, big data banks, if tamed can complement the "statistical/scientific" approach to measurements by providing an opportunity to reveal new hypothesis and new directions that can set such approaches on a surer footing. For example, techniques developed by Louis Guttman has provided approaches to the measurement and the reduction of data dimensions into a scale (used profusely in the social sciences and psychology) (Fig. 6.5). A summary of his posthumous contribution summarized in Wikipedia is given below:

> Scale analysis, Facet theory and his mathematical and philosophical treatments of Factor analysis are among the important parts of his scientific legacy. Notably, Guttman first proved several fundamental theorems in matrix algebra, (that have allowed the development of algorithms for matrix dimensions reduction). Several of Guttman's contributions have been incorporated into computer packages, among them Smallest Space Analysis (SSA) and Partial Order Scalogram Analysis (POSAC).

6.3 Decision and Risk Objective Measurements

Decisions and risk objectives whether single or multiple, provide and imply policies, and vice versa, policies imply preferences (just as prices in finance imply also "common" preferences). When multiple risk objectives are confronted to one another, their reconciliation into a coherent risk management framework is generally difficult and can entail unintended consequences.

Decision objectives underlie a rationality we use to reach decisions that value both what we seek and at what risks. Such rationality can be based on a number of elementary statements (which we often call axioms) defining what a rational decision maker ought to do. It can also be behavioral, based on observations of how and why are decisions made. Numerous authors, behavioral scientists and psychologist have attracted attention to the complexity of decisions reached under uncertainty (and risk). There are observations that can undermine the axiomatic approaches to decision making. For example: We may be prone to regret, and thus avoid making a decision just because it would prove us wrong; Holding a belief plainly at odds with the evidence, just because it has been held for a long time; Anchoring—because of long held, mostly viewed as a "conventional wisdom" or some suggestion or belief; A Status quo bias, expressing our unwillingness to depart from a given set of beliefs or change; Ignoring small probabilities or extremely large outcomes just because they may be unwieldy.

These elements have also been recognized by economists and behavioral decision scientists alike. Herbert Simon indicates that decisions are defined in terms of a bounded rationality, namely all decisions are rationally incomplete, either due to practical and informational limits and the complexity of the problems we face. For example, when risks are defined in terms of countable (and thus well defined) states and their probabilities, rather than in terms of an ill defined uncertainty, any decision may necessarily result in potentially unpredictable consequences. Further, any and all models may default! For this reason, defining objectives are necessarily neglecting other objectives and thereby ignore the risks other objectives may imply. Within these constraints we proceed to define one and multiple objectives, some with their own contradictions and others at cross purposes (in which case, they will be called strategic).

Single objectives are a convenience, expressing uniquely potential, varied and contradictory wants: Reducing the probability of an oil spill but at the same time increase the cost of oil transport; Increasing the ability of a home to sustain Hurricanes with 150 miles winds but leads insurance firms to overcharge persons on demands that are not in fact necessary; Increase the credibility of diagnostic tests but also increase their costs; Preventing terrorist attacks at airports but alters the travel process rendering it unbearable; Hedge investments (or risk prevention) but reduce their profitability etc.

Multiple objectives consist of several explicit and mostly dependent objectives that require an arbitration process to reconcile their contradictions. Focusing a policy on a specific risk measurement will lead to specific outcomes implied by the risk measure while potentially increasing other risks. For example, seeking high yields and low risks; reduce the number of policemen in the street (and thereby reduce their cost) and at the same augmenting the population's insecurity; Reserves set aside to meet regulated liquidity and at the same time augment the number of approved loans to clients etc.

Multiple objectives are far more prevalent in risk measurements and management as most objectives have a "down side". For example, driving faster will bring us to destination earlier, but "speeding" has consequential risks. Decision problems of this sort are generally called "multi-criteria". Their resolutions are reached based on numerous comparative tools as well as interactive procedures in the hands of a single decision maker who evaluates also their outcomes and their multiple consequences. In some cases, preference theories are constructed to justify the use of multiple criteria into one consistent objective. For example, Markowitz

(1952), mean-variance analysis of a portfolio is based on the one hand, on a mean return objective, and on the other, on a least variance. A comparison of the infinite number of portfolios we may consider is drastically reduced to a set of "efficient portfolios" in a mean-variance sense. In health care, a patient is often confronted by several decision makers, each concerned about the patient's risks he may be most apt to judge—resulting in a treatment based on multiple viewpoints. For this reason, health care insurance firms seek a single Doctor to be responsible and to manage an insured health. In some cases, one may transform multiple objectives into a single one (for example, when all objectives are measured by money, or when multiples objectives are reduced to a sum of weighted deviations from "desirable" objectives', or when all stakeholders in a patient's health have to follow the directive of a single Doctor). When objectives are not measured in terms of identical metrics, the analysis of such objectives is more difficult. Often, financial valuation provides an efficient scale to compare such objectives, with more "real money" being always better than "less money" (with "real" accounting for risk).

Multiple Decision Makers Objectives: When decision makers have different objectives, each with the freedom to choose a decision that may affect others, decisions and risks are strategic, formulated and analyzed using games as we shall see in Chap. 11. In this sense, risk results from what others do, based on the information and the power one has over others and vice versa, the information and the power of others.

Traditional approaches reduce the objectives dimensionality by "collapsing" multiple objectives into a single one or in a scale that can evaluate approximately ones' own multivariate preferences. For businesses, one such metric is money, emphasizing thereby the importance of risk valuation and risk pricing. In some case, one would want to define a "quality of life metrics" to guide public and political decisions, but such metrics are both difficult to construct and to agree on. Below we summarize a number of objectives used by decision makers, expressing as well their risk attitude.

Expectation: Seeking the largest expected value (or the smallest opportunity loss). Risk objectives are in this case neutral. An investor would value equally the probability of a gain to those of a loss. Costs incurred as a function of risks assumed and based on defined random factors are used to measure the expected objective.

Maximax (*or the minimin*) Seeking the most among the greatest values (or vice versa, the least among the smallest). A researcher would pursue "a make or a break career experiment". One may call such a decision maker's objective risk loving.

Principle of Insufficient Reason or the Laplace Criterion for Decision Making, This principle states that unknown probabilities may be assumed to be equally likely, reflecting a state of utmost ignorance. The apparent objectivity of such an approach, ignores in fact the decision maker's risk attitude toward the future outcomes (are they large or are they small, etc.).

The Minimax Regret or Savage's Criterion : Minimize the cost of a decision's regret with regret defined as the best payoff one can obtain once the future state is (ex-post) revealed less the payoff selected by a decision ex-ante. Such an approach contrasts expected ex-ante versus expected ex-post decisions. It defines the Savage regret criterion which has inspired a number of approaches based on "disappointments models". According to Bell, disappointment is a psychological reaction to an outcome that does not meet a decision maker's expectation. The measure of disappointment is assumed to be proportional to the difference between expectation and the outcome below the expectation. Elation occurs however, when the outcome is better than its expectation.

These and other approaches seek to reconcile objectives and their risks and may be summarized by three concerns:

- The objective (including preferences)
- Information (states, risks manifestations)
- Decisions and risk models

The expected utility approach we consider in Chaps. 7 and 8 have provided a rationality to solve decision problems under risk, while the Bayesian approach to information and decision making has provided a risk engineering framework based on expectations and learning (see Chap. 2).

6.4 Risk Measurement in Various Fields

6.4.1 Medical Risk Measurement

Greater accountability in medical practice has contributed to the growth of medical data bases tailored to service a broad number of stakeholders in the medical and health care systems. These stakeholders includes patients, Doctors, Nurses, Hospitals, Health Insurance firms, Governments (both Federal and State), Medical Associations etc. all of whom have their own needs and specific means to meet these needs. All are concerned with the measurement of the quality of health care and its risks. Health risk measurements are function of both the service provider (Doctor etc.) and the patient. From a medical viewpoint, the number of factors tends to be extremely large, information partial and nonlinearities express a multi-variate dependence originating in multi-collinearity, in causalities that are not observed easily and in interactions of the patient's environment, his actions and the Doctor's-patient information asymmetry.

Medical risk measurements are thus based on seeking answers to "extremely complex" problems. The recent concern for the effects of "entitlements" (in particular Medicare in the US) on Public Budgets and the Economy in general, the concern for "health money" (and therefore risk measures associated to money) are assuming a greater importance. A convergence of risk and money in industry

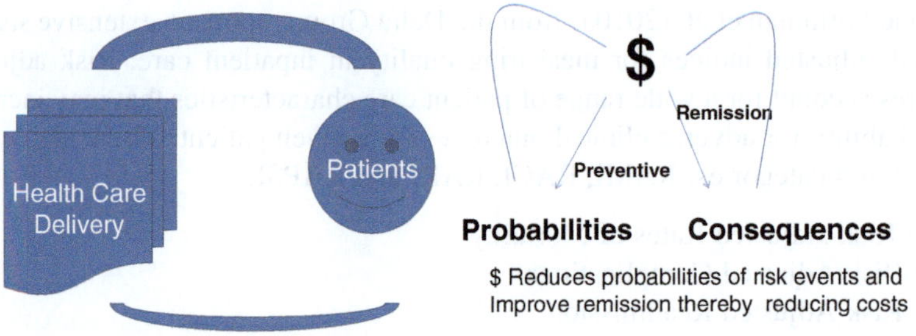

Fig. 6.6 Health care delivery: risks and consequences

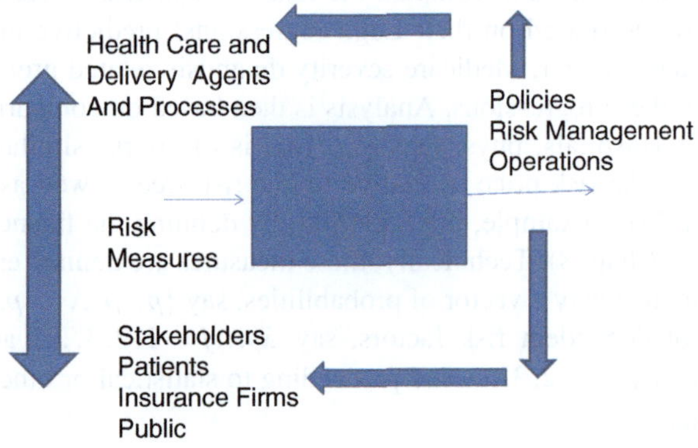

Fig. 6.7 Healthcare: many objectives, many decision makers

and in environmental problems for example, is migrating as well to health care, adopting the financial jargon for valuing and pricing risk. These groups of risk measures are used to define metrics that drive Health Care policies, and the treatment of patients, the health system, its funding and regulation. These issues at hand are however complex due to political, scientific and professional knowledge and practice that can vary from place to place, from Doctor to Doctor and from Hospital to Hospital (Fig. 6.6).

Summarily consider the following:

- Risk Factors (causes)
- Risk consequences (non-financial).
- Money (financial consequences)
- Risk Derivatives

Figure 6.7 indicates how Money invested in preventive measures reduces the probabilities of risk events while investments in quality health care, improves remission and thereby reduces costs.

Thane Forthman et al. (2010a) from the Delta Group report on extensive studies of a risk adjusted indices for measuring quality in inpatient care. Risk adjusted measures account for a wide range of patient care characteristics that may increase the probability for adverse clinical outcomes for a given patient. These indices are grouped in 4 categories: RAMI, RACI, RARI and RAPSI:

RAMI: Risk Adjusted Rates of Mortality
RACI: Risk Adjusted Complications
RARI: Risk Adjusted Readmission
RAPSI: Risk Adjusted Patient Safety Events

Using data extensive studies based on Logit models, (see Chap. 2), these probabilities are estimated using a large numbers of risk factors. These risk measures may then be used to compare the risk performance of various hospitals as well as determine (based on their Logit regressions) predictive models of risk factors such as age, gender, Medicare severity diagnosis-related group cluster etc. accounted for in these regressions. Analysis is then based on comparisons relative to a standard of Hospitals, physicians etc. This is of course similar to financial approaches where the risk price is relative to one risk free as well as relative to a "basket of assets" (for example, the S&P 500) by defining the financial Beta (see CAPM models in Chap. 8). Technically, these measures are defined empirically by estimating simultaneously a vector of probabilities, say $\{p_1, p_2, \ldots, p_n\}$ based on a large data set of dependent risk factors, say $\tilde{x}_{ij}, i, j = 1, 2, 3, \ldots.$ and their risk consequences $\tilde{y}_{ij}, i, j = 1, 2, 3, \ldots.$ and proceeding to statistical or other approaches for their analysis.

Example Human health risk measurements, unlike financial risk measurement, are mostly defined in terms of the probabilities of being exposed to hazardous substances and "health states" and their consequential health damage (which may be important for some parties such as a health care agency, a financial consequence etc.). These are expressed by a "probability" and a "health state". Of course, given the extremely large number and the complexity of these health "states", the number of possibilities (and commensurably their variety) are very large. Health precursors may consist of demographic variables, individual and genetic tendencies, behaviors, histories, physiologic changes etc. For example, is one weight a measure of health risk? Assuming for the moment that this is the case, then measures such as Body mass index (BMI); Waist circumference; Risk factors for diseases and conditions associated with obesity are sought and investigated to assess the risk of obesity measured by High blood pressure (hypertension) ; High LDL cholesterol ("bad" cholesterol) ; Low HDL cholesterol ("good" cholesterol) ; High triglycerides ; High blood glucose (sugar) ; Family history of premature heart disease ; Physical inactivity ; Cigarette smoking. In such analysis, again deviations from standard health conditions are sought to explain and prevent the risk factors obesity may lead to. This is done by comparing an individual's prognostic characteristics of health with a standard age group, thereby predicting the likelihood that a person may develop prematurely a health problem associated with a high morbidity and mortality rate.

6.4.2 RAM as Performance and Risk Measures

RAM (*Reliability, Availability and Maintainability*) defines a set of tools and approaches to managing risks in an industrial and in other environments. These tools are measurement based. Reliability R is a probability measure of a system in its up-operational state while Availability A, defines the up-operational condition of a system when needed and at any (random time). Maintainability is the probability of bringing a down component or system to its up-operational state within a certain specified amount of time and using a given set of resources. A system's capacity C, is its ability to meet its purpose at any (random) time, given its reliability. All these variables are measured using statistical data. The efficiency of RAM systems are defined in various manners and their usefulness are based on their simplicity. A simple approach consists in setting:

$$\text{Efficiency} = \frac{A}{C} = \frac{\text{Availability}}{\text{Capacity}} = \frac{\text{F(Reliability, Maintenability)}}{\text{Total Time} - \text{Planned Down Time}} = \frac{\text{F(R, M)}}{\text{T} - \text{T}_M}$$

A system rating can then be set for example by:

- $E = ADC = \textit{Availability} * \textit{Detectability} * \textit{Controllability}$
- Max (Availability) S.T. MTBSF $\geq$ T , Costs $\leq$ B
- Availability = F(Reliability, Maintenability/Reparability, Redundance)
- MTBSF = F(Reliability, Redundance)
- Costs = F(Modules-components costs, their number, maintenance costs, etc)

Example: Comparative Measurement, FMECA-Failure Mode Evaluation and Critical Analysis FMECA or Failure Mode, Effects and Criticality Analysis is an approach to evaluate the reliability and safety of a design, a product, a process or a system by considering the potential failures, the resulting effects on the product, process, system and personnel, and the criticality of these effects. It results in a procedure for determining the basic causes of failure and defining actions to minimize their effects. Its purpose is to identify problems or critical faults. Quantitatively, the application of FMECA requires that each failure mode be defined in terms of its occurrence probability (O), the probability that it is not detected (D) as well as its implication to the user (preferably in terms of costs), denoted by S. The product O*D*S may then be called the index of criticality C which is used to rank order problems to be attended to:

$$C = (O) * (D) * (S)$$

When a problem has an index which is greater than some critical value C, then the problem is dealt with and solved. Its index of criticality is then revised. Throughout FMECA, cross functional groups are used. FMECA is extremely useful because it is simple to conceptualize, even though the practical definition of probabilities (O), (D) and the calculation of costs (S) to the user, is far from simple. It is noteworthy that FMECA recognizes the importance of non-detectable factors and thus recognizes the risk implications of uncertainty.

There are various tools used together with FMECA. For example, *Fault-Tree Analysis* (or logic diagrams). Fault-Tree analysis is a useful technique applied in reliability and safety engineering. For example, it measures system safety by determining the probability than an undesired event or fault will occur. It is a graphical representation of the logic that relates certain specific events or primary failures to an ultimate undesired event. It can use Boolean (0,1) algebra as a modeling tool to represent the flows of system function failures. When systems are complex, based on network-like representations and dependencies, Boolean algebra can be used to determine the effect of specific failures (occurring with known probabilities) on systemic failure (or their probabilities). The usefulness of fault-tree analysis is extremely varied. It can be used to determine the causes of an accident, to discover failure combinations which would otherwise have not been recognized as causes of the event being analyzed and to display system configurations in design review. Failure-Tree analysis has been long practiced in Reliability Engineering (it is believed that it was initiated in the development of the Minuteman Missile) and there are many books and research papers which deal with this topic.

6.4.3 *Quality and Statistical Tracking*

Statistical tracking provides the risk assurance that performance is within specified limits. To do so, it specifies risks, it samples and analyze data. It measures and states its purpose and finally, it reaches a conclusion (e.g. reach a decision to accept or reject a measurement or conclude whether the sample evidence is inconclusive). For this purpose, a number of techniques based on statistical, Bayesian analysis as well as more advanced detection and temporal (dynamic) estimation techniques are used.

A summary of quantitative risk measures includes: Measurements to detect and define quantitatively the probability of a risk exposure and its consequences. Risk Probabilities include: Probability and Reliability; The tail risk distribution; Strategic risk and their implications (in a situation with multiple parties), Covariance risks! (for example, how is my risk independent of others or dependent on co-variations).

Events Uncertainty include: Non-countability, rare events, jump probabilities, background risk; partial information; asymmetric information; system robustness etc.

6.4.4 *Operations and Services and Risk Measurements*

Operations and Services are concerned by the direct and indirect adverse consequences of outcomes and events resulting from "what they do" and "external" events (whether theoretically predicted or not) and their effects on the firm performance, its employees, its clients and society. Operations and Services' risks are intimately related to quality—one used to measure, to define and to manage the other. An understanding of what we mean by quality and its relationship to risk is thus important. For example, the traditional industrial notion of quality emphasizes a negative approach to quality—the risk of deviating from an agreed on standard of manufacture or reference operational standards. Although such a notion is well defined and quantifiable, it need not be the true measure of quality, in particular when we refer to the quality of financial services. The evolution of "quality" and its risks have been broadened over time to include a far greater concern for the value and the price of quality, rather that the concerns for the "art" and the pursuit of "excellence" that quality requires. The rise of money as a common denominator for valuing risk is a development emanating from our market economies and the pursuit of profits that is necessarily a bounded rationality as it ignores the many elements that are either difficult or impossible to value and price.

6.5 Bayesian Decision Making: EMV and Information

Bayesian decision making is based on the premise that what we do has consequences. It encompasses extensive developments, theoretical and practical. Let there be a set of potential actions (decisions, policies, etc.) $\{a_1, a_2, \ldots, a_m\}$ and let there be a risk model consisting of n potential (countable) states, each of which has associated known and complete probabilities $\{p_1, p_2, \ldots, p_n\}, p_2 \geq 0, \sum_{i=1}^{n} p_i = 1$. To each pair of "state and probability" there correspond a consequence (whether deterministic or random) which we denote by $\{\xi_{ij} | p_i, a_j\}$. The Expected Monetary Value (EMV) of any specific action is then denoted by:

$$EMV_j = \sum_{i=1}^{n} E(\xi_{ij}) \text{ while } EMV^* = \underset{j \in \{1,2,\ldots,m\}}{\text{Max}} EMV_j$$

Define by EPPI the expected monetary value under perfect information which is defined as follows:

$$EPPI_j = EPPI^* = \underset{i}{\text{Max}}\, E(\xi_{ij})$$

This objective measure assumes then that once an event occurs, the decision maker can choose the best decision to take (for example trading on insider information). Since events are random, an expectation ex-post over observed events defines the EPPI. The EMV and the EPPI is then used to define a cost—a risk due to having selected a decision which is not the best. This is the Expected Opportunity

Loss (EOL) and in expectations EOL=EPPI–EMV. Their relationship is given by the following elementary equations:

$$EPPI = EMV + EOL \quad \text{or} \quad EPPI^{*} - EMV_j = EOL_j$$

and at the optimum when the greater EMV is selected as an action, we have:

$$\text{EPPI} = \text{EPPI}^{*} = \text{EMV}^{*} + \text{EOL}^{*}$$

Consider for example a 3X4 decision matrix with deterministic consequences:

	$\{p_1\}$	$\{p_2\}$	$\{p_3\}$	$\{p_4\}$
$\{a_1\}$	ξ_{11}	ξ_{12}	ξ_{13}	ξ_{14}
$\{a_2\}$	ξ_{21}	ξ_{22}	ξ_{23}	ξ_{24}
$\{a_3\}$	ξ_{31}	ξ_{32}	ξ_{33}	ξ_{34}

The expected values associated to each action as well as the optimal (maximal) action are then:

$$EMV_j = \sum_{i=1}^{4} p_{ji}\xi_{ji} \text{ and } EMV^{*} = \underset{j}{\text{Max}} \sum_{i=1}^{4} p_{ji}\xi_{ji}$$

Consider next the case of perfect foresight (say corresponding to insiders trading). For every action i, the decision maker knows the best consequence denoted by ξ_i^{*}, namely: $\xi_i^{*} = \underset{j}{\text{Max}}\, \xi_{ji}$ and the EPPI is the expected value of all these "best" outcomes, or:

$$EPPI = \sum_{i=1}^{4} p_i\xi_i^{*} = \sum_{i=1}^{4} p_i\left(\underset{j}{\text{Max}}\, \xi_{ji}\right)$$

Lacking a perfect foresight, for each decision $\{a_j\}, j = 1, 2, 3$, the "risk" of not taking the best action is given for all states i by $\xi_i^{*} - \xi_{ji}$. Of course, if the best decision corresponding to the occurrence of a state i is selected, then $\xi_i^{*} - \xi_{ji} = 0$ and otherwise, $\xi_i^{*} - \xi_{ji} > 0$. As a result, the expected loss or EOL associated to selecting a specific action j is:

$$EOL_j = \sum_{i=1}^{4} p_i\left(\xi_i^{*} - \xi_{ji}\right) \text{ while the Least } EOL_j \text{ is} : EOL^{*} = \underset{j=1,2,3}{\text{Min}} \sum_{i=1}^{4} p_i\left(\xi_i^{*} - \xi_{ji}\right)$$

From these equalities we easily derive then the equation stated above:

$$EPPI = EMV_j + EOL_j \quad \text{and } EPPI = EMV^{*} + EOL^{*}$$

In a Bayesian decision framework, we note therefore that the most one can do is the EPPI and therefore maximizing the EMV also implies that EOL (the cost of partial foresight) is minimized. Of course, this is somewhat limiting in practice since decision makers do not necessarily value outcomes linearly.

6.6 Multi Criteria and Ad-Hoc Objectives

Consider a vector of decisions $(d_1, \ldots, d_m)$, with each decision valued by objectives measures $(y_1(d_i), \ldots, y_n(d_i))$, $i = 1, 2 \ldots, m$. A multi objective problem $h_i(y_1(d_i), \ldots, y_n(d_i)), i = 1, 2 \ldots, m$ consists in selecting the best decision in the sense that it optimizes the set of all objectives in some manner, or:

$$\operatorname*{Max}_{i \in (1,m)} h_i(y_1(d_i), \ldots, y_n(d_i))$$

There are numerous ad-hoc approaches and techniques to solve these problems. We consider some such as:

- *AHP* (*Analytic Hierarchy Processes*) is based on a paired comparison by one or more decision makers of all objectives for a finite number of criteria. Such a technique is based on the Perron-Frobenius eigenvector in matrix algebra which consists in transforming scores collected in a matrix into eigenvalues and concentrating one's attention on the larger eigenvalues.
- *Voting rules and majority voters.* The rule leads to counterintuitive paradoxes however. For example, say that 3 decision makers select among three alternatives, yielding the rankings $\{A, B, C\}, \{B, C, A\}, \{C, A, B\}$. A majority voted for A over B, for B over C but also for C over A. In other words, the voting relationship is intransitive since $A \succ B$ and $B \succ C$ need not implies $C \succ A$ and thus violates a basic rationality axiom (called the transitivity axiom). These problems are particularly important in social choice theory (see Fishburn 1970, 1981, 1988), economics, politics, etc. For example, Arrow's paradox states that perfect democratic voting is, not just in practice, but in principle, impossible. As a result, all multi persons and multi choice decisions will always have a risk consequence—namely that the choice of an alternative may violate the preference of one or the other decision makers.
- "*Efficiency*" is defined relative to a benchmark. Such an approach is used for example when applying DEA-Data Envelopment Analysis to competing alternatives or performing entities.
- *Other approaches* to solving multi-objective optimization problems with constraints (such as linear programming) consists in reconciling (arbitrate) among the common constraints these problems have and the objectives they share (see Cogger and Yu 1985, Dyer 1990, Lootsma (1988).

Below we consider briefly both Froebenius theorem and discuss the Data Envelopment Analysis approach (used as well as a bench marking technique).

6.6.1 *Perron-Froebenius Theorem and AHP*

Consider the matrix equation notation $\mathbf{Aw} = \lambda\mathbf{w}$ where λ is an eigenvector. Perron-Frobenius theorem states that for *a squared non negative matrix there exists a non-negative entries that correspond to each of the matrix eigenvalues defining the weights we attach to each alternative*. Thus for an alternative k, we can define a score $R_k = \sum_{j=1}^{m} r_{kj}\lambda_j$ which may be used to slect the alternative with the highest score. For example, say that an alternative i is preferred to j (say, i is less risky than j) and denoted by a probability statement p_{ij} defined by a ratio of two scores assigned to two alternatives: $p_{ij} = r_i/(r_i + r_j)$ or $p_{ij} + p_{ji} = 1$ and $p_{ji}r_i - p_{ij}r_j = 0$. Say that an estimate of this probability is random given by: $p_{ij} = b_{ij} + \varepsilon_{ij}$ and therefore, $(b_{ji} + \varepsilon_{ji})r_i - (b_{ij} + \varepsilon_{ij})r_j = 0$ or $b_{ji}r_i - b_{ij}r_j + \varepsilon_{ji}r_i - \varepsilon_{ij}r_j = 0$. Thus to estimate r_i, we can minimize the error term by solving:

$$\text{Min} \sum_{i=1}^{m} \sum_{j \neq i} \left(b_{ji}r_i - b_{ij}r_j\right)^2 \text{ Subject to : } \sum_{i=1}^{m} r_i^2 = 0$$

whose solution is: $\mathbf{Dr} = \mu\mathbf{r}$ with μ an eigenvector and D has entries $d_{ii} = \sum_{i=1}^{m} b_{ik}^2$, $d_{ij} = -d_{ij}d_{ji}, i \neq j$.

Such a solution is one of a number of alternative approaches we can use to allocate "weights" to multiple objectives. Obviously, these are subjective approaches embedded in partial theories for risk preference (see references).

6.6.2 *The Data Envelopment Analysis and Benchmarking*

Benchmarking denotes a family of approaches we use to define standards with respect to which we may measure ourselves. In business and industrial management for example, common use is made of "Competitive benchmarking". It is the continuous measurement of a company's products and services, processes and practices against the standard of best competitors and other recognized leaders. These may include: Process capabilities consisting of * Customer driven measurements; * Efficiency of logistic support systems; * Demand driven sales planning; * Efficient (Lean) manufacturing; * Supplier partnering; * Business integration and coordination; * Information Technology capabilities; * Organization Capabilities etc. Each of these elements may be defined in terms specific measurements such as Teamwork, integration, aligned organization structure in case of organizational factors. Benchmarking providing measurements relative to "best practice" provides both incentives for self-improvement and a measurement of deviations from "the best in a class". Although Benchmarking is often used as a qualitative technique, it is also used quantitatively using the Data Envelopment approach (see references).

6.7 Risk Measurement Models: Axiomatic Foundations

Risk measurement models may be based on: (1) Axiomatic Foundations, characterized and measured in terms of a number of axioms presumed to define risk (2) Behavioral Foundations, based on behavioral and psychological observations of perceived risks. (3) Models of bounded rationality defining risks models that capture a broad range of probability models we can use to measure risk and (4) Utility valuation based on rationality axioms (see Chap. 7). In the following we outline some broad principles emphasizing coherent risk measures, risk-value, and deviations risk measures.

6.7.1 Coherent Risk Measures

Risk measures are based on postulates made about risk. These, may differ from field to field, leading to different measurements of risks. For an insurance and financial business, risk measurement is about risk exposure and risk premium pricing. Measurements are meant to support their essential concerns. Coherent risk measurement for example, is based on the actuarial needs of insurers defined by the following properties:

- Sub-additivity (or convexity and therefore diversification is assumed to reduce risk);
- Monotonicity;
- Homogeneity;
- Translation Invariance.

Consider for example, two portfolios A and B and let $R(A)$ and $R(B)$ be the risk measure of each. A sub-additive (convex) measure implies that aggregation reduces risk, or $R(A+B) \leq R(A) + R(B)$. This is equivalent to the convex measure that underlies the basic financial concepts of "diversification" in financial risk management or "The average of a sum (or risk) is smaller than their parts":

$$R(\lambda A + (1-\lambda)Y) \leq \lambda R(A) + (1-\lambda)R(B), \forall \lambda > 0$$

The validity of risk measurement models based on such premise depends of course on whether they truly represent insurers measurements and attitudes to risk. This is certainly the case if risks A and B are totally independent in which case, the inequality above holds at equality, or $R(A+B) = R(A) + R(B)$. Aggregation of risks into a portfolio may in some cases indicate something else—namely more rather than less risks. Imagine that A and B are dependent, does their sum create a risk measure which is greater than the sum of their part? Risks may be amplified by a co-dependence due to positive co-variation, dependence on common or latent factors as well as due to a contagion of insurance claims. Insurance firms at least in their decisions to reinsure parts or a complete portfolio indicates a contradiction which can be explained by their reluctance to increase a risk exposure.

Let the portfolio of potential insured claim be a sum of apparently independent claims. Conventional wisdom then claims that aggregation will on the average reduce "risk". If this were the case, then why would insurance firms sell (i.e. reinsure) their portfolios to other firms? Obviously, the "aggregation" of claims has raised potential of risks that are not imbedded in any particular specific insured contract. These risks may occur due to latent (common risks) or to contagion and dependence effects that manifest themselves at the aggregate level. The same principle recurs in financial regulation of firms that are extremely large and are deemed Too Big To Fail as they, on the aggregate, generate risks that can be Too Big To Bear.

Monotonicity is based on the premise that if the price of one portfolio is greater than another then its risk measure is necessarily greater. This is stated as follows: let $p(A)$ and $p(B)$ be the portfolios prices, then:

$$\{p(A) \geq p(B)\} \Leftrightarrow \{R(A) \leq R(B)\}$$

This proposition underlies the notion of no-arbitrage and the law of one price in fundamental finance (see Chap. 8). In other words, when markets are complete and assets (portfolios) are priced by a rational expectation of future returns, then the measurement of future risk is completely accounted for by the current price. This presumption might not be true when markets are incomplete which justifies the measurement of financial risks (and thereby, points out that in fact, markets are incomplete). Further, when markets are complete, theoretical finance has in fact "done away with risk and thus its measurement is irrelevant". That is there is "no need to measure risk" if it is already accounted for by a current price! Risk is measured because it is not accounted for and thus our need to mitigate its consequences! Monotonicity, while useful can also in some circumstances turn out to be misleading, i.e, risk and price may be in general monotonic, but not always.

Positive homogeneity assumes that a linear proportion, translation, of the portfolio produces an equivalent proportional change in risk or $kR(A) \geqslant R(kA),\ k > 0$. This rule leads of course to a careful assessment of what we measure. If for example, a practitioner uses the variance, then evidently such an inequality does not hold while using the standard deviation, it does holds. For some measures this premise does hold as we shall later on. This particular characteristic will be used subsequently as a measurement of value expressed in terms of returns and their standard deviation rather than their variance.

Translation invariance states that adding to a portfolio a risk free asset (say a risk free bonds, if they exist) the portfolio ought not to change its aggregate portfolio risk, or $R(A + B_f) = R(A)$. While this is an apparent statistical and financial truism it can also induce one in error. For example, traded bonds have prices that can vary from day to day as they too are subject to macroeconomic trends and default risks.

For example, consider two parties, one wealthy and the other poor, both hold the same amounts of a risky asset, but the rich holds as well a substantial investment in risk free assets. Are these risk measure equivalent? Probably not! Wealth matters (see Chap. 8 on the extended CCAPM). For example, can one compare the risk of a small loss and that of a large loss? Are these risks proportional? Probably not! (which justifies the business of lotteries with participants willing to assume small risks but rarely large risks).

These "axioms" or rather hypotheses are in fact often violated. These do not invalidate their usefulness however. It is important to appreciate that risk measures can be misleading, if not properly understood. The extensive use of Value at Risk (VaR) although generally accepted as partial measures of risks and violating basic risk measurement "axioms", are used profusely and criticized (see Szego 2002 as well as Embrechts et al. 1997, 2001; Embrechts 2000a, b, ; Rockafellar and Wets 1992, Rockafellar et al. 2002 for alternative and complementary risk measures to VaR).

6.7.2 Axiomatic Models for Deviation Risk Measurements

Deviations as risk measure is used in both statistical control techniques and in benchmarking. They are used to express "deviations" from an expectation, from a target, to express economic inequality, etc. The presumption that such deviations have risk consequences is used to justify their measurement. An axiomatic definition of such risk measurements are based on the following hypotheses regarding a deviation measure D:

- It is shift-invariant: $D(X+\alpha) = D(X)$ for any $\alpha \in \mathbf{R}$
- It is normalized such that: $D(0) = 0$ (namely a null deviation is measured as null)
- It is a positively homogeneous function: $D(\lambda X) = \lambda D(X)$ for any $X \in \mathcal{L}^2$ and $\lambda > 0$
- It is a sub-linear function: $D(X+Y) \leq D(X) + D(Y)$, for any $X, Y \in L^2$
- It is a positive function or Positivity: $D(X) > 0$ for all non-constant X, and $D(X) = 0$ for any constant X

Szego (2002) emphasizes that for "D" to be a risk measure it is also necessary that: (1) The distance between a point and itself is zero (2) The distance does not change by inverting the points and (3) Given three points, the distance between any pair cannot be larger than the sum of the distances between the other two pairs . Below we consider some examples.

6.7.3 Absolute Deviations

An expected absolute deviation is defined by $E|X - X^*|$ and measures distance. Such measures are used for example as objectives in goal programming and need not be symmetric. Explicitly, these are "objectives targets", a deviation from which have a consequence (say a linear cost). For example, let $\tilde{X}$ be a random variable and let $\tilde{X} > X^*$ express that it is better than the standard X^*. In this case, the risk consequences can be separated and summed by an asymmetric a weighted function, to express-measure "good" and "bad" deviations, or

$$w_G E|\tilde{X} - X^*|^+ - w_B E|\tilde{X} - X^*|^- \quad \text{where}$$
$$E|\tilde{X} - X^*|^+ = E\mathrm{Max}(\tilde{X} - X^*, 0) \text{ and } E|\tilde{X} - X^*|^- = EMax(X^* - \tilde{X}, 0)$$

The statistical range in a sample, the variance (standard deviations) and partial moments can also be thought of as deviation measures. Let $R(f) = x_{\max} - x_{\min}$ be the range of a statistical sample and let s_x be an estimate of the sample standard deviation, then $R(f)/s_x$ defines a range to scale measure (see chapter 5).

6.7.4 *Inequality Measures*

Pedersen and Satchell (1998) review a large number inequality measures, including the Gini coefficient: $R_G(f) = \int [1 - F(y)]dy - \int [1 - F(y)]^2 dy$ where F is a distribution of wealth. It is generalized to: $R_G(f) = \int [1 - F(y)]dy - \int [1 - F(y)]^\alpha dy,\ \alpha > 0$

6.7.5 *The Variance and the VaR*

The variance is not a coherent risk measure. This can be seen since it has neither the translation or the monotonicity properties. That is $\mathrm{var(X + a) = var(X) \neq var(X)}$ $-\mathrm{a}$ for all $a \in \mathbf{R}$, and a simple counterexample for monotonicity can be found. The standard deviation is a deviation risk measure however. Similarly, the VaR is not a risk measure if the distribution is not elliptic (but possibly symmetric).

6.7.6 *Entropy and Divergence (Distance) Metrics*

Consider two probability distributions $[p, q]$, both of which are defined in terms of countable and accountable states (and therefore define a risk model). A metric measure to discriminate between these distributions is given by "entropy functions" measuring the "deviation" of one probability distribution with respect to another. Below we consider some elementary definitions which are developed in greater detail in Chap. 10 on Uncertainty Economics. Let

$$I(p,q) = E^P\left(\ln\frac{p(x)}{q(x)}\right) \text{ and } I(q,p) = E^Q\left(\ln\frac{p(x)}{q(x)}\right)$$

where $E^P(.)$ and $E^Q(.)$ are expectation taken with respect to the probability distributions $p(x)$ and $q(x)$ respectively. If the distributions $p(x)$ and $q(x)$ define the estimates of two agents for a set of events, then $E^P(.)$ and $E^Q(.)$ is a measure of agents' perception of their "distance". Each of these measurements is called discrimination information (for example, see Kullback and Leibler 1951; Kullback 1987, 1959) and their sum is called the divergence. These measures meet the axioms of a distance metric given by:

$$J(p,q) = I(p,q) + I(q,p) = E^P\left(\ln\frac{p(x)}{q(x)}\right) + E^Q\left(\ln\frac{p(x)}{q(x)}\right)$$

For example, say that $q_i,\ i = 1,2,3,\ldots n$ is an empirical distribution and say that $p_i,\ i = 1,2,3,\ldots n$ is a theoretical distribution given by the binomial distribution: $p_i = \binom{n}{i} p^i(1-p)^{n-i}$, then:

$$J(p,q) = \sum_{i=1}^{n}\left(\binom{n}{i} p^i(1-p)^{n-i} - q_i\right)\log\frac{\binom{n}{i} p^i(1-p)^{n-i}}{q_i}$$

which defines a "distance" between the binomial distribution and the distribution q_i. For a bivariate state distribution, we have similarly:

$$I(p,q) = \sum_{j=1}^{m}\sum_{i=1}^{n} p_{ij}\log\left(\frac{p_{ij}}{q_{ij}}\right); \quad J(p,q) = \sum_{j=1}^{m}\sum_{i=1}^{n}(p_{ij} - q_{ij})\log\left(\frac{p_{ij}}{q_{ij}}\right)$$

While, for continuous distributions, we have also:

$$I(F,G) = \int\int F(x,y)\log\frac{F(x,y)}{G(x,y)}dxdy \quad \text{and}$$

$$J(F,G) = \int\int (F(x,y) - G(x,y))\log\frac{F(x,y)}{G(x,y)}dxdy$$

If the reference distribution is equi-probable (presumed to be with the « largest amount of randomness » (which need not be the distribution with the greatest risk) then for $q(x) = 1/n,\ x \in [0,n]$:

$$I(p,q) = \int p(x)\log\frac{p(x)}{1/n}dx = \log(n) + \int p(x)\log(p(x))dx = \log(n) - H(p)$$

where the function $H(p)$ defines the entropy of the distribution $p(x)$ —a "distance metric" for the distribution and the other an equi-probable distribution. This term was introduced in communication theory by Shannon and Weaver in 1949. It is due to Boltzman and Gibbs contributions in thermodynamics. The entropy is maximal when the distribution $p(x)$ is equi-probable and therefore, presumably it provides a relationships between entropy and randomness, the former measuring the amount of deviation from this randomness. In this sense, a maximum entropy distribution (given a set of assumptions which are made explicitly) is equivalent to the Principle of Insufficient Reason propounded by Laplace.

6.8 Functional and Generalized Risk Measurement Models

Functional models are designed to encompass parametrically as broad a number of potential risk measures. In 1973, Bernell K. Stone presented a class of risk (deviations) measures which include a number practical risk measures. It is specified by the following three parameters function

$$R_t(W_0, k, A; f) = \left(\int_{-\infty}^{A} |W - W_0|^k f_t(W) dW \right)^{\frac{1}{k}}$$

Particular cases are the standard deviation, $R_t(\hat{W}, 2, \infty; f)$; the Fishburn's measure and the partial moment safety first criterion given by

$$R_{\alpha,t}(f) = \int_{-\infty}^{t} (t - W)^{\alpha} f(W) dW \quad \text{and} \quad \sigma_a^{2-}(k) = \int_a^k (W - k)^2 f(W) dW$$

The quantile risk (expected tail) measure:

$$R_t(0, 1, A; f) = \left(\int_{-\infty}^{A} W f_t(W) dW \right)$$

Pedersen and Satchell (1998) provide an extension to this model to include an extremely large number of measures including measurement models in psychology, operational research and management science, on economic inequality, etc. Measurement models by Fishburn (1970, 1981) economics and on axiomatic preferences for the social sciences and economics, by Luce (1980, 1981), Luce and Weber (1986), Krantz et al. (1971), Pollatsek and Tversky (1970) in psychology, Bell (1982, 1983, 1985, 1995) emphasizing ex-post risks etc. (including the measures introduced above) turn out to be special cases.

Based on Stone's class of risks measures, Pedersen and Schatell (1998), then present an extended family of potential risk measures for some bounded function W(.) and defined in terms of 5 parameters A, b, α, θ where $\alpha > 0,\ \theta > 0$.

$$R_t(A, b, \alpha, \theta; W(.)) = \left(\int_{-\infty}^{A} \Phi(y)dy \right)^{\theta}, \Phi(y) = |y - b|^{\alpha} W(F(y))f(y)$$

Using this measure, a very large number of risk measure in several fields are outlined as special cases. We shall consider only a limited and essential number of such measurements however.

6.9 Examples and Expectations

Expectation is a simple law invariant risk measure, describing a risk-neutral behavior. It is defined as follows:

$$\mu(X) = \mathbf{E}X \text{ or } \mu(F) = \int_{-\infty}^{\infty} tdF(t)$$

Expectations can be summarized by the following equation:

$$E(X,f) = \int w(X)\Phi(f(X))dX$$

expressing both "weighted" outcomes and "weighted" of probabilities (a procedure often used by behavioral scientists for risk valuation). In this case, when the probability distribution of events is well defined, we have for example:

$$E(X) = \int Xf(X)dX : \text{ Expectation of events}$$

$$E(u(X)) = \int u(X)f(X)dX : \text{ Expectation of a utility preference function}$$

$$VaR(X) = \int_{\infty}^{-VaR} f(X)dX : \text{ The risk of a given VaR}$$

$$CVaR(X) = \int_{\infty}^{-VaR} Xf(X)dX : \text{ The VaR conditional tail risk}$$

When the distribution is not well defined, for either subjective or objective reasons, then some psychology based studies have suggested that the underlying probability distribution be redefined or standardized to a parametric form that can capture observed behaviors (see Chap. 10 on Uncertainty Economics).

6.9.1 Models Based on Ordered Distributions' Measurement

Stochastic orders provide a comparative measurement of risk distributions, establishing a stochastic order between these distributions. Such measures have been used extensively in reliability theory (Bolland and Proschan 1994; Shaked and Shantikumar 1994), insurance (for example, Embrechts et al. 2009), and also in economics as an alternative to the use of utility (Hadar and Russell 1969; Ross 1981, 1999). Essential orders are defined by the following:

- First order stochastic order dominance
- Second Order, Third Order dominance
- Hazard rate ordering
- Likelihood Ratio Ordering
- Convex and Concave ordering
- Laplace Transforms Ordering

Each of these is considered briefly. For further reference, see: Shaked and Shantikumar 1994.

Stochastic dominance: $X \geq_{st} Y$, consider two random variables X and Y denoting each a payoff. Then X is stochastically larger than Y if $P(X) \geq P(Y)$, for a a. If the cumulative probability distributions of X and Y are respectively $F_X(a) = F_X(x \leq a)$ and $F_Y(a) = F_Y(y \leq a)$, then stochastic dominance implies $\bar{F}_X(a) = \bar{F}_Y(a), \forall a$ where $\bar{F}_X(a) = 1 - F_X(a)$.

Second Order Stochastic Dominance: It is defined by: $\int_{x'}^{X} [F_X(s) - F_Y(s)]ds \geq 0, \forall x \in X$

Third Order Stochastic Dominance: It is defined by:

$$\int_{x'}^{X} [F_X(s) - F_Y(s)]ds \geq 0, \forall x \in X \text{ and } \int_{x\in}^{X} \int_{x'}^{y} [F_X(s) - F_Y(s)]dsdy \geq 0, \forall x \in X$$

Other stochastic ordering approaches include hazard rate ordering and likelihood ratio ordering. These are stronger than stochastic dominance, thus hazard rate ordering implies stochastic dominance.

Hazard Rate ordering: Define the hazard rate by: $h_X(x) = \frac{f_X(x)}{1-F_X(x)} = \frac{f_X(x)}{\bar{F}_X(x)}$. Then, a random variable X dominates (written as $\geq_h$) another Y in a hazard rate ordering sense if:

$$h_X(x) \le h_Y(y), \quad \text{or} \quad \frac{f_X(x)}{\overline{F}_X(x)} \le \frac{f_Y(y)}{\overline{F}_Y(y)}$$

Note that hazard rate ordering implies stochastic dominance but not the opposite.

Likelihood Ratio Ordering: Consider non negative random variables X and Y, we say that X is stochastically larger than Y in a sense of likelihood ratio ordering written by $X \ge_{\ell t} Y$, if $f_X(x)f_Y(y) \le f_X(y)f_Y(x), \; x \ge y$. And if the likelihood ratio is non-decreasing, or: $\frac{f_X(x)}{f_Y(x)} \le \frac{f_X(y)}{f_Y(y)}, x \ge y$ The likelihood ratio ordering is stronger than stochastic dominance and hazard rate ordering. In other words: $X \ge_{\ell t} Y \rightarrow X \ge_h Y \rightarrow X \ge_s Y \rightarrow X \ge_s^{(2)} Y \rightarrow X \ge_s^{(3)} Y$

Convex and Concave Ordering: Let X and Y be two random variables such that $E[u(X)] \le E[u(Y)]$ for all functions $u : \Re \rightarrow \Re$. Provided the expectations exist, then X is said to be smaller than Y in the convex order. Similarly, $E[u(Y)] \le E[u(X)]$ and for all concave functions if $u : \Re \rightarrow \Re$, then X is said to be smaller than Y in the concave order. The implications of convex ordering are numerous. We resume some of these by using Shaked and Shantikumar (1994) (Academic Press)

Theorem: Let X and Y be two random variables such that $E[X] = E[Y]$. Then, $X \le_{cx} Y$ if and only if: $E|X - a| \le E|Y - a|$ for all $a \in \Re$.

Laplace Transforms Ordering: Say that $u(X) = -\exp^{-sX}$ where s is positive. Then, for X and Y, two nonnegative random variables, such that, $E[e^{-sX}] \ge E[e^{-sY}]; s \ge 0$. Then X is said to be smaller than Y in the Laplace Transform order.

Chapter 7
Risk Valuation

Abstract Our concern in this chapter and in the next is to define the value of risk prospects and their price. We consider traditional and ad-hoc approaches; rational expectations and the expected utility framework. These approaches are applied to a number of cases in this chapter as well as in Chaps. 8, 9 and 10. While Chaps. 8 and 9 are concerned with risk economics, and pricing, Chap. 10 is concerned with the economics of uncertainty. This chapter provides therefore a background to the book subsequent chapters.

7.1 Value and Price

The financial approach to valuation is based on the following:

- Preference (expressing rationally a need or a want by buyers and sellers as well as their risk attitude).
- Risk models to frame uncertainty into a set of countable and accountable states.
- An inter-temporal preference for money.
- Exchange, either through competitive (free) financial markets or contractual. Exchanges may express our wants, (utilities), the wants of others, their information, willingness and an ability to pay and to share risks and rewards. In this sense, an exchange reveals and "implies" both the price at which parties are willing in a given economic environment to exchange.

The following questions highlight the importance of valuation and pricing.

- At what cost would a patient decide not to buy a certain drug, or avoid a needed care?
- How much would an individual firm or a person be willing to pay an insurer to mitigate a prospective risk? How much would an insurance firm require to insure the risk? And how can they reconcile their differences?
- How much should we pay for the safety of airports? In other words, what is the price of safety?

C.S. Tapiero, *Engineering Risk and Finance*, International Series in Operations Research & Management Science 188, DOI 10.1007/978-1-4614-6234-7_7,

- How valuable is the reliability of a system?
- How much would one be willing to pay for a warranty? Etc.

In practice, many approaches are used to define "value." Industrial management has over the years used profit functions, consisting of revenues less costs. In inventories' related models, costs are minimized with implied inventories risks embedded in shortage costs. In some cases, models are formulated with a value implied by constraints (i.e. the objective sensitivity of a value function to a constraint defined by its associated Lagrange multiplier). For a regulated firm, regulation constraints may affect its policies an incur compliance costs. These and other approaches we shall refer to provide a background to subjective valuations. This chapter is organized as follows. First we consider a plethora of traditional valuation models. Subsequently we consider foundations of pricing based on both the Walras-Arrow-Debreu economic equilibrium framework in competitive markets and thereafter a utility based approach to valuation. These are the two fundamental and complementary approaches to risk valuation and risk pricing.

Economics and finance have approached valuation of risks from two perspectives including:

- *Rational expectations*, based on implied models of exchange in competitive markets with no arbitrage.
- *Utility valuation* models, based on utility preferences and the utility of "consumption" or "wealth" now and in the future which is used to measure a utility for the demand and the (personal) price of risk.

These two approaches differ. One is based on future outcomes, negotiated and exchanged at a present time, the other based on a rationality expressing value and risk preferences. Both approaches lead to estimate prices implicitly in terms of current information as indicated in Fig. 7.1. Technically, these problems are reduced to inverse mathematical problems which we also consider in this chapter.

7.2 Rational Expectations, Martingales and the Arrow-Debreu Complete States Preferences

Financial economics, reduces risk through a trade or a series of trades, or through contractual agreements reached to share risk and define an equilibrium market price (Arrow 1963a). Games of chance, such as betting in Monte Carlo or any Casinos are popular metaphors to represent risk models of financial exchanges in stock markets, where money and financial assets (mostly financial products) are exchanged. Its historical origins can be traced to Girolamo Cardano who proposed an elementary theory of gambling in 1565 (Liber de Ludo Aleae—The Book of Games of Chance). The notion of "fair game" was then clearly stated: "The most fundamental principle of all in gambling is simply equal conditions, e.g. of opponents, of

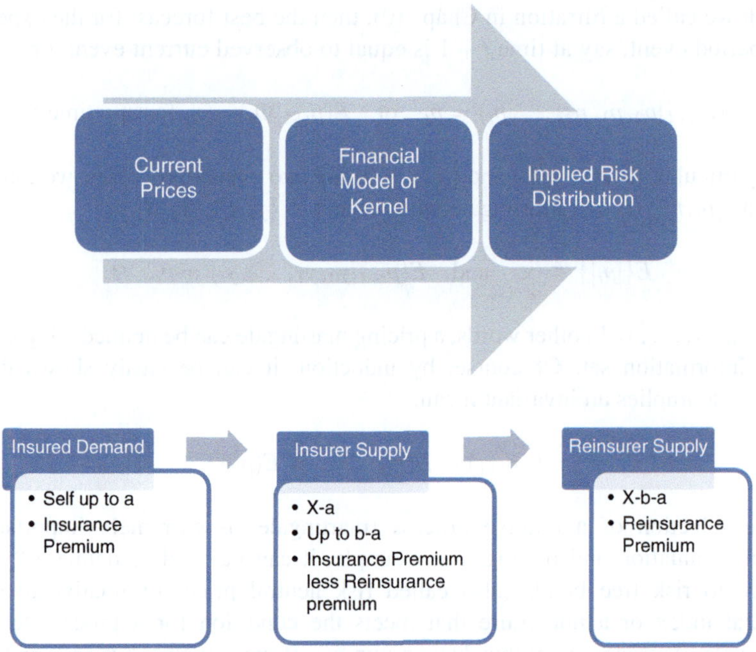

Fig. 7.1 The "Model" the price mechanism or the value of risk in finance and insurance

bystanders, of money, of situation, of the dice box, and of the die itself. The extent to which you depart from that equality, it is in your opponent's favor, you are a fool, and if in your own, you are unjust." "This is the essence of the Martingale," a mathematical property that underlies rational expectations and the concept of complete markets (although Cardano did not use the word "martingale"). The "concept of a fair game" (or Martingale), in money terms, states that such games have a null expected payoff with probability one. In other words, in the long run, such games have no gain without assuming risk. Games of Dice, Black Jack, Roulette and many other games, when they are fair, corrected for the bias each has, are thus a Martingale. Rational expectation models that seek to value chance outcomes subsume as well that under certain probability measures, chance outcomes turn out to have the properties of a Martingale. Intuitively, what does it mean?

- Tomorrow's event (say price) is today's best forecast (i.e. its expectation equals to todays observed event).
- Non-overlapping events (say prices) are uncorrelated at all lead and lags.

A technical definition of a Martingale can be resumed to the presumption that each event (such as a new price) is independent, can be summed (i.e. integrable) and its conditional expectation is time invariant (it has therefore no memory). That is, if $\Phi_t = \{p_0, p_1, \ldots, p_t\}$ are a series of recorded prices given at time t, (which we called a filtration in Chap. 10), then the best forecast for the expected next period event, say at time $t+1$ is equal to observed current event, or:

$$E(p_{t+1}|p_0,p_1,p_2,\ldots,p_t) = p_t \quad \text{or} \quad E(p_{t+1}|\Phi_t) = p_t \text{ for any time } t$$

In particular, if $\{p_t, t \geq 0\}$ and $\{y_t, t \geq 0\}$ are two events-processes, we can then say that $\{p_t, t \geq 0\}$ is a Martingale with respect to $\{y_t, t \geq 0\}$ if:

$$E\{|p_t|\} < \infty \quad \text{and} \quad E(p_{t+1}|y_0, y_1, \ldots, y_t) = p_t, \ \forall t$$

where $y_0, y_1, \ldots, y_t$. In other words, a pricing martingale can be defined with respect to an information set. Of course, by induction, it can be easily shown that a Martingale implies an invariant mean:

$$E(p_{t+1}) = E(p_t) = ... = E(p_0)$$

The selection of a relative process (martingale) is extremely important in financial valuation and pricing. For example, it can be used to define a "price" relative to risk free bonds (also called risk neutral price) or relative to some financial index or a numeraire that meets the condition for a process to be a martingale. A process may thus have a number of martingales, but only and only one such a martingale defines a market price. For this reason, finding the pricing martingale consists in fitting current and observed prices to a theoretical martingale that defines an implied pricing framework. The theoretical justification of such an approach to market prices is based on a number of assumptions that do not always hold. As stated in earlier chapters, these include:

- Countability and accountability of all state preferences
- A unique pricing martingale
- No arbitrage (namely no gain without risk)
- A law of single price, claiming that two cash flows of identical characteristics must have, necessarily, the same price (otherwise there would be an opportunity for arbitrage). Information asymmetry, insider trading etc. for example, may violate such an assumption. Technically, it also implies a unique pricing martingale
- No long or short term memory

When any or all these conditions do not hold, the "market price mechanism is said to be incomplete." Although such situations predominate risk practice, the rational expectations models are deemed robust, in the sense that incompleteness is temporary, brought back to equilibrium (completeness) by the non-sustainability of a potential for arbitrage. If arbitrage was the "rule rather than the exception," then arbitrageurs will get infinitely rich at the expense of all others (which to same extent is the case, since as we have observed in the last decades, economic inequalities have grown consistently and inexorably). Any violation of these assumptions perturbs the basic assumptions of theoretical finance, leading to incomplete markets and therefore to models that cannot value risk uniquely (and thus price). Risk economics, in a world with complex risks, strategic and global, is currently challenged by the incongruence of market pricing theories and their real manifestations in financial markets.

7.2.1 Rational Expectations Models: A Simple Quantitative Definition

Rational expectations and inter-temporal models are based on principles of fundamental finance due to the Arrow-Debreu state preference theory. Future prices are then implied in current trades of options, derivative and generally forward looking data, all of which provide a data base for estimated future risks distributions. In this approach, there is no memory of past events and predictions regarding future prices are assumed to be based on all the current information available to all investors and traders. Such information is defined in terms of a "filtration" and all prices are conditional on such a filtration, changing over time as additional information accumulates and price change (and thereby altering accordingly future prices' estimates). The Arrow-Debreu theory, underlying complete financial markets thus makes a number of important assumptions:

- There is no uncertainty but a risk framed to define the future in terms of a known set of states, $S_i(1),\ i = 1, \ldots n$ called state preferences.
- Each of these states is priced uniquely $\pi_i(1),\ i = 1, \ldots n$
- There is no arbitrage, namely no profit can be made without assuming risk and two equal risk prospects are priced equally as well.
- The current price $\pi(0)$ is defined by these future states and their prices, explicitly given by: $\pi(0) = \sum_{i=1}^{n} \pi_i(1) S_i(1)$.

In other words, all future states are currently traded and in equilibrium with each future state, having its future price. The current price is thus implied in future state preferences and their prices. Inversely, future state preferences are implied in current and observable future state prices.

For example, say that future state preferences consist of two known values $S_i(1),\ i = 1, 2$. The current price is $\pi(0) = \pi_1(1)S_1(1) + \pi_2(1)S_2(1)$ with prices $(\pi_1(1), \pi_2(1))$ unknown. Such a pricing system is incomplete since the two future prices $(\pi_1(1), \pi_2(1))$ are unknown and only one data point $\pi(0)$ is available. Therefore to "complete" this equation we need additional information that pertains to such an asset. What information can we use and what does it mean? Essentially it means that prices are set relative to an observed information we have which is defined as a function of current and observed prices that are also based on the future asset price. For example, a derivative (i.e. an option on that asset), a risk free rate (which has no risk), another and covariant asset, may be used as information sources to complete the price process etc. The richer the information the more it is possible to assess the future price.

The simplest case consists in considering an asset whose price is risk free, namely, its future states are the same. Explicitly, consider a non-default bond whose price at maturity a period hence are given by: $S_1(1) = B(1), S_2(1) = B(1)$ i.e. payouts are riskless since in all states the investor collects the same payout, and the current price equals the observed market price of the risk free bond, or $\pi(0) = B(0)$, with $B(0) = B(1)\sum_{i=1}^{N} \pi_i(1)$. This latter equation has thus introduced another equation, needed to complete the binomial pricing model. Let the risk free rate of return in such a bond be R_f, with $B(1) = B(0)(1 + R_f)$ thus,

$$B(0) = B(0)(1 + R_f)\sum_{i=1}^{N} \pi_i(1) \text{ and } \sum_{i=1}^{N} \pi_i(1) = (1 + R_f)^{-1}$$

Define the following proportion $p_i^Q(1) = \pi_i(1)/\sum_{i=1}^{N} \pi_i(1)$ where $p_i^Q(1)$ has the property of a probability distribution measure Q, $0 \le p_i^Q(1) \le 1$ and $\sum_{i=1}^{N} p_i^Q(1) = 1$ which we call a "risk neutral probability." The asset price calculated in terms of these risk neutral probabilities are then:

$$\pi(0) = \sum_{i=1}^{N}\left(p_i^Q(1)\sum_{i=1}^{N}\pi_i(1)\right)S_i(1) = \sum_{i=1}^{N}\pi_i(1)\sum_{i=1}^{N}\left(p_i^Q(1)\right)S_i(1)$$

$$= \frac{1}{1 + R_f}\sum_{i=1}^{N} p_i^Q(1)S_i(1)$$

A rearrangement of these terms yields:

$$\pi(0) = \frac{1}{1 + R_f}\sum_{i=1}^{N} p_i^Q(1)S_i(1) = \sum_{i=1}^{N} p_i^Q(1)\frac{S_i(1)}{1 + R_f} = \sum_{i=1}^{N} p_i^Q(1)\frac{S_i(1)}{B(1)}$$

Setting, $\pi(0) = \frac{S(0)}{B(0)}$, we have:

$$\frac{S(0)}{B(0)} = E^Q\left(\frac{S(1)}{B(1)}\right) \quad \text{or} \quad S(0) = \frac{1}{1 + R_f}E^Q(S(1))$$

where $E^Q(.)$ is an expectation with respect to the probability measure Q. Note that an expectation with respect to any function of the underlying future price $S(1)$ which is traded (and thereby exchanged in a financial market), say the call option $C(1) = Max(S(1) - K, 0)$ has also a price given by:

$$\frac{C(0)}{B(0)} = E^Q\left(\frac{C(1)}{B(1)}\right) = E^Q\left(\frac{Max(S(1) - K, 0)}{B(1)}\right) \quad \text{or} \quad C(0) = \frac{1}{1 + R_f}E^Q(C(1))$$

These prices, relative to the probability measure Q define a Martingale as stated earlier. Since expectation is taken with respect to all the relevant information, its expectation is conditional on that information $\Im_0$ (the filtration). In this case the pricing martingale at a time t = 0 with a filtration $\Im_0$ is denoted by:

$$\frac{S(0)}{B(0)} = E^Q\left(\frac{S(1)}{B(1)}|\Im_0\right) = E^Q\left(\frac{S(2)}{B(2)}|\Im_0\right) = \ldots\ldots = E^Q\left(\frac{S(t)}{B(t)}|\Im_0\right) = \ldots.$$

Thus, given an observed price at time t = 0 as well as its traded (and market priced) derivatives, say call and put options with m strikes, or: $\{C(0|S(0),K_j),P(0|S(0),K_j)\},\ j=1,2,\ldots m$

$$\{S(0),C(0|S(0),K_j),P(0|S(0),K_j),\ j=1,2,\ldots m\} \Leftrightarrow \{S(1),C(1),P(1)\}$$

These define 2m prices implying the future n probabilities $p_1^Q(1), i = 1,2,\ldots n$ where:

$$\frac{S(0)}{B(0)} = E^Q\left(\frac{S(1)}{B(1)}|\Im_0\right),\ \frac{C(0,K_j)}{B(0)} = E^Q\left(\frac{C(1,K_j)}{B(1)}|\Im_0\right),\ \frac{P(0,K_j)}{B(0)} = E^Q\left(\frac{P(1,K_j)}{B(1)}|\Im_0\right)$$

Applications to financial assets pricing is most common as there are competitive financial markets that provide the space and the opportunity to exchange a number of underlying assets and their derivatives. In particular, Arrow, Debreu, Lucas, Merton, Black and Scholes have provided a practical and theoretical footing for pricing derivatives. Below we consider a number of examples that do not have an open financial market (although one may conceive of markets where such exchanges can be made). The warranty price problem treated below uses the Arrow-Debreu framework. Subsequently, other problems are considered. References to such problems are outlined at the end of this chapter.

7.2.2 *The Inverse Kernel Problem and Risk Pricing*

The pricing formula set by the Arrow-Debreu rational expectations framework, in its discrete and continuous forms belong to a class of problems consisting in solving for $\Phi(y)$ given $\pi(x)$ for a kernel $K(x,y)$ in the following equation:

$$\pi(x) = \int K(x,y)\Phi(y)dy \text{ in its continuous formulation}$$

$$\pi(j) = \sum_{i=1}^{n} K(j,i)\Phi(i) \text{ in its discrete formulation}$$

Under appropriate conditions, the inverse solution is:

$$\Phi(y) = \int K^{-1}(x,y)\pi(x)dx$$

where $K^{-1}(x,y)$ is an inverse kernel. When this is the case, the integral of $K^{-1}(x,z)\ K(z,y)$ equals a Dirac Delta function $\delta(x-y)$ (equals 1 if x = y and 0 otherwise). The theoretical and mathematical problem consists in finding a unique kernel that solves such an equation. Since the kernel is essentially a data-smoothing function of the input $\Phi(y)$, there could be numerous kernels meeting such a condition. In risk pricing, such an equation corresponds to future states $\Phi(y)$ while $\pi(x)$ are prices derived from future states and priced by a kernel $K(x,y)$ associating a "weight" to each future state and for a given specific price. For simplicity, we consider a simple 2 states and 2 prices discrete problem and establish conditions for its price. The discrete kernel equation is then:

$$\begin{aligned}\pi(1) &= K(1,1)\Phi(1) + K(1,2)\Phi(2)\\ \pi(2) &= K(2,1)\Phi(1) + K(2,2)\Phi(2)\end{aligned} \quad \text{or} \quad \begin{bmatrix}\pi(1)\\ \pi(2)\end{bmatrix} = \begin{bmatrix}K(1,1) & K(1,2)\\ K(2,1) & K(2,2)\end{bmatrix}\begin{bmatrix}\Phi(1)\\ \Phi(2)\end{bmatrix}$$

Or

$$\pi = \mathbf{K}\Phi \text{ and } \Phi = \mathbf{K}^{-1}\pi \text{ and therefore } \pi = \mathbf{K}\Phi = \mathbf{K}\mathbf{K}^{-1}\pi = \pi$$

Since for a squared and invertible matrix $\mathbf{K}\mathbf{K}^{-1} = \mathbf{I}$. Such a problem in fact resumes the Arrow Debreu theory for pricing assets when markets are complete. In this framework, there are two future states prices, each priced at $\Phi(i)$, i = 1, 2 with two accountable states $K(j,i)$ corresponding to the price of an asset j if the future state i materializes. Of course if future accountable states are identical regardless of the future states (as it would be the case for a risk free bond), then for such an asset, $K(1,1) = K(1,2) = K(1)$ and its price is

$$\pi(1) = K(1)(\Phi(1) + \Phi(2)) \text{ and } \Phi(1) + \Phi(2) = \pi(1)/K(1)$$

These equations can be standardized by setting $\phi(i) = \Phi(i)/\sum_{i=1}^{2}\Phi(i)$ we also have:

$$\pi(2) = K(2,1)\Phi(1) + K(2,2)\Phi(2) \quad \text{and}$$

$$\frac{\pi(2)}{\Phi(1) + \Phi(2)} = K(2,1)\phi(1) + K(2,2)\phi(2), \quad \phi(1) + \phi(2) = 1$$

And finally,

$$\frac{\pi(2)}{\pi(1)} = \frac{K(2,1)}{K(1)}\phi(1) + \frac{K(2,2)}{K(1)}\phi(2) = \frac{1}{K(1)}E^{\phi}(K(2,i))$$

which corresponds to a simple asset price when markets are complete. In this case, if the market price of a bond is \$1, then its future price a period hence is $K(1) = \$(1 + R_f)$ and therefore the relative price of asset 2 is:

$$\pi(2) = \frac{1}{1 + R_f} E^{\phi}(K(2,i)) = \frac{1}{1 + R_f} \sum_{i=1}^{2} K(2,i)\phi_i$$

A departure from such complete equations leads necessarily to models of incomplete prices with solutions that may be deemed approximate. "Uncertainty models" include all such departures from the complete markets models, and more. A number of such situations will be considered in Chaps. 8, 9 and in particular Chap. 10. Below we consider a number of examples.

Example: Partial Market Information Say that there are only two observed prices while there are three future states. In this case, we have the price equation:

$$\pi(i) = \sum_{j=1}^{3} K(i,j)\Phi(j), i = 1, 2$$

This is a system of two equations with three unknown $\Phi(j), j = 1, 2, 3$ and therefore there may be an infinite number of potential solutions (and therefore prices). Numerous techniques can then be used. For example, say that we minimize the sum of the quadratic error approximation, written as follows:

$$\underset{\Phi(j),j=1,2,3}{Min} \sum_{i=1}^{2} (\varepsilon(i))^2 = \sum_{i=1}^{2} \left(\sum_{j=1}^{3} K(i,j)\Phi(j) - \pi(i) \right)^2$$

A solution of this quadratic optimization problem with respect to the three states yields three equations in $\Phi(j)$ which can be solved in terms of the prices $\pi(i)$. Of course, if we have an additional third asset and let its price be $\pi(3)$ as well as $K(3, 1), K(3, 2)$, $K(3, 3)$ known we have then a full information (complete state model) whose solution is unique and given by the matrix equation $\Phi = \mathbf{K}^{-1}\pi$ as calculated previously. The difference between these two estimates, calculated by $\Phi = \mathbf{K}^{-1}\pi$ and the approximate estimates $\Phi(j)$ calculated above provide then an error estimate in future state prices and thereby, an error in pricing.

Example: Moments Information Consider for convenience the standardized model defined above.

$$\pi(k) = \frac{1}{1 + R_f} \sum_{i=1}^{n} K(k,i)\phi_i, \;\; k = 1, 2, \ldots; \phi_i \geqslant 0, \sum_{i=1}^{n} \phi_i = 1$$

where the ϕ_i's have the properties of a probability distribution. To determine the ϕ_i we may again proceed in a number of ways. For example, assume a parametric probability distribution and estimate its parameters as a function of its prices $\pi(k)$. Alternatively, we can use a rationale based on the maximization of the entropy of the distribution subject to moments information. Such approaches are considered in Chap. 10.

7.3 Utility Models and Valuation

Utility valuation expresses a personal (or subjective) value embedding the individual preferences for risk and rewards. It differs from a "price" which results from an exchange in a market place where a meeting of buyers and sellers settle to exchange at a price. While utility is used to model buyer and sellers wants, it is their meeting on common terms that define their exchange price. As a result, value is derived from needs and the ability to meet these needs while the price reflects interacting and aggregate market forces that define the terms of exchange they agree on. Prices, in financial markets, are thus relative to the parties participating in these markets, the market rules and regulation, the parties' alternatives, their needs and their willingness to buy and sell. Not everything is priced however. For example, the air we breathe is a priceless asset without which we could not survive (although the presumption that it is plentiful leads to its mispricing). Lacking air is of course a risk that has no price as its value for breathing humans is infinite. Both value and price are thus intimately related to risk. If investors abhor risk, then the price of a "riskier" asset ought to be higher to compensate the buyers of such an asset. Risky prospects can have a price when markets participants can agree to an exchange of money for a risk transfer based on their wants and information. In practice therefore, people do not value money equally. An attitude towards risk and the willingness to take risks originates in a person's initial wealth, risk bearing capacity, information, emotional states and the pleasure or fear evoked by risk.

Utility functions express individuals' rationality defined by a set of axiomatic preferences. Say that $\{x,f(.)\}$ is a set of rewards (or losses) x assumed to occur with a probability distribution $f(.)$ and let the utility function assume the following premises:

Prospects can be compared
Prospects can be ranked according to their utility function
Prospects have strong independence
Prospects have transitive preferences
Prospects are indifferent if their utilities are equal

Given the probabilities of all prospects and their consequences, the axiomatic preferences stated above underlie the utility theorem stating that the expected utility provides a rational index for comparing the desirability of risk prospects, or:

$$E(u(x)) = \int u(x)f(x)dx; \ x \in X$$

The risk of a decision, is thus reflected in a "loss of utility" and vice versa (as we shall see subsequently with its price defining the risk premium). Gains are realized by a utility growth (or $\partial u/\partial x > 0$). If the utility function $u(.)$ is convex, $\partial u^2/\partial x^2 < 0$ (or concave, $\partial u^2/\partial x^2 > 0$) then by Jensen's inequality for convex functions, we have:

$$E(u(x)) \geq (\leq) u(E(x))$$

and vice versa when the utility function is concave. These mathematical properties allow a functional characterization of risk attitudes. Three cases are defined:

1. Risk aversion
2. Risk loving and
3. Risk neutrality.

Risk aversion expresses an attitude of risk avoidance and thus a preference for less risky gambles and a will to pay or forego returns for such gambles. The price to remove risk altogether is the risk premium. In contrast, a risk loving attitude is a preference for prospects that have greater "value" but more risk. Finally, risk neutrality implies that prospects are valued at their expectation, expressing and attitude oblivious to risk. For a risk averse party, the value of greater prospects with smaller probabilities decreases (due to their increased risk), which corresponds to a negative second derivative for the utility function (an assumption of concavity) and vice versa for a risk loving attitude. The mathematical properties of a utility function used to value these prospects has a certain equivalent value measured by $\bar{X}$ defined by the utility equivalence hypothesis $Eu(\tilde{x}) = u(\bar{X})$ and therefore, by inversion $\bar{X} = u^{-1}(Eu(\tilde{x}))$. The risk premium is then defined by $\pi = \hat{x} - \bar{X}$ where $\hat{x} = E(\tilde{x})$. Say that the prospect consists of its expectation plus some random element $\tilde{\varepsilon}$, $\{\hat{x} + \tilde{\varepsilon}\}$ where $E(\tilde{\varepsilon}) = 0$, its expected utility is thus: $u(\hat{x} + \tilde{\varepsilon})$. A Taylor series expansion about the mean yields

$$u(\hat{x} + \tilde{\varepsilon}) = u(\hat{x}) + \tilde{\varepsilon} u'(\hat{x}) + (\tilde{\varepsilon})^2 u''(\hat{x}) + (\tilde{\varepsilon})^3 u'''(\hat{x}) + (\tilde{\varepsilon})^4 u''''(\hat{x}) + \ldots.$$

The expected utility is thus:

$$Eu(\hat{x} + \tilde{\varepsilon}) = u(\hat{x}) + E(\tilde{\varepsilon}^2) u''(\hat{x}) + E(\tilde{\varepsilon}^3) u'''(\hat{x}) + E(\tilde{\varepsilon}^4) u''''(\hat{x}) + \ldots.$$

Since $\sigma^2 = E(\tilde{\varepsilon}^2), E(\tilde{\varepsilon}^3), E(\tilde{\varepsilon}^4)$ are the variance, skewness and kurtosis parameters of the prospects distribution, we note that value is a meaningful function all these parameters. Therefore, expected utility preferences are a function of both probability distribution and the functional form of the utility function. For example, if prospects are defined by a normal probability distribution (with variance σ^2 and a null skewness—see Chap. 2), and if we neglect all higher order moments, then $Eu(\hat{x} + \tilde{\varepsilon}) = u(\hat{x}) + \sigma^2 u''(\hat{x})$.

As noted above $Eu(\tilde{x}) = u(\bar{X})$ and $\pi = \hat{x} - \bar{X}$ as a result, $Eu(\tilde{x}) = u(\hat{x} - \pi) = u(\hat{x}) + \sigma^2 u''(\hat{x})$. Letting the risk premium be a constant and assuming that it is small relative to the prospect expected value then a Taylor series expansion yields: $u(\hat{x}-\pi) = u(\hat{x}) - \pi u'(\hat{x}) = u(\hat{x}) + \sigma^2 u''(\hat{x})$ which leads to:

$$-\pi u'(\hat{x}) = \sigma^2 u''(\hat{x}) \quad \text{or} \quad \frac{\pi}{\sigma^2} = -\frac{u''(\hat{x})}{u'(\hat{x})}$$

We call $A_P = -\frac{u''(\hat{x})}{u'(\hat{x})}$ the Arrow Pratt index of risk aversion, defining a risk premium price per a unit variance, or $\pi = \sigma^2 A_P$.

7.3.1 *Critique of Expected Utility Theory in Measuring Preferences*

Theory and practice do not always concur. There are many reasons for such a statement. Are decision makers irrational? Or are they just careless? Are they uninformed or misunderstanding, lacking the proper incentives to reach a rational decision. These may result from violating the axioms of preferences defining the "rationality" underlying expected utility theory. Empirical and psychological research has sought to test the real premises of decision makers under uncertainty and assess their potential violations. The strongest attack on expected utility theory can be found in Allais' Paradox which doubts the strong independence assumption which is needed for a consistent choice based on expected utility. It has proved to be systematically violated. "Axioms" of the EU theory renders the probability in utility expectation linear, a phenomenon which is often doubtful in practice (see also Chap. 10). The independence axiom also called the "sure-thing" principle asserts that two alternatives have a common outcome under a particular state of nature. That is, ordering alternatives should be independent of the value of their outcome. This is not always the case and counter examples abound, in particular due to Allais as stated earlier but by Ellsberg as well.

The Allais paradox requires that subjects choose first a given situation with two gambles. For example, let us confront 2 persons with two lotteries:

$$p_1 \xrightarrow{1} 1{,}000{,}000 \quad \text{and} \quad p_2 \Rightarrow \begin{cases} 0.1 & 5{,}000{,}000 \\ 0.89 & 1{,}000{,}000 \\ .01 & 0 \end{cases}$$

equivalent in expectation to:

$$p_3 \Rightarrow \begin{cases} 0.1 & 5{,}000{,}000 \\ 0.9 & 0 \end{cases} \quad \text{and} \quad p_4 \Rightarrow \begin{cases} 0.11 & 1{,}000{,}000 \\ 0.89 & 0 \end{cases}$$

In confronting 1 and 2, people chose p_1 while in confronting 1 and 3 and 4, people preferred 3 which contradicts the strong independence axiom of utility theory.

Ellsberg, however, criticizes the practice that ascribes equi-probabilities (Laplace's Principle of Insufficient Reason) when there is uncertainty and points out that regret assumes a particular importance in a multi-period context. These lead to "strange" behaviors. For example, when walking from station to station in a metro in Paris, one observe that initially one walks faster when arrival of the train cannot be seen and then when one gets closer and arrival can be seen, there is a tendency to slow down. In other words, behavior is altered when we do not see the desired prospect.

7.3.2 Examples and Problems

Binary lotteries: What is the value of a simple binary outcome $\tilde{x}$ consisting in a reward rate $R > 1$ with probability p, or a loss rate $L < 1$ with probability $1 - p$. Its expectation is $\hat{x} = pR + (1 - p)L$. Yet it appears that both individuals and firms are willing to pay more (or less) than the expected value for the prospect of a large gain, even if the probability of winning is very small. Why is it the case? However irrational it may seem at first, many people do so because they value the prospect of "winning big" even with a small probability, much more then the prospect of "losing small" with a large probability. We call such persons "risk loving" or "risk takers" to indicate their risk attitude. This uneven valuation means parties may not, necessarily, compare expected prospects equally. Parties are different, not least in their needs but also in their risk attitude, both of which reflect their preferences for risk prone prospects. This is in essence what expected utility theory attempts to do by the scaled valuation it provides. Defining the appropriate utility function is of course a challenging task. The larger the expected utility, the "better it is." For example, assuming that there is such a function, $u(.)$, two sums (R) and (L) each valued by their utility $u(R)$ and $u(L)$ equal in expectation to $pu(R) + (1 - p)u(L)$, or:

$$Eu(\tilde{x}) = pu(R) + (1 - p)u(L), \tilde{x} \in (R, L)$$

Note that in this case, the mean and the variance of this prospect are:

$$\hat{x} = pR + (1 - p)L \text{ and } \sigma^2 = p^2 R(1 - R) + (1 - p)^2 L(1 - L) - 2p(1 - p)RL$$

and as indicated above, its certain equivalent $\bar{X}$ and its risk premium π can be calculated by the following equalities: $Eu(\tilde{x}) = u(\bar{X}) = u(\hat{x} - \pi)$. Now say that another lottery $\tilde{x}'$ is defined by the parameters $\{p', R', L'\}$ with utility function $u'(\tilde{x}')$. Say that both lotteries have an equal expected utility, or $Eu(\tilde{x}) = Eu'(\tilde{x}')$.

As a result, the certain equivalent of each is $\bar{X}$ and $\bar{X}'$ with $u(\bar{X}) = u(\hat{x} - \pi), u'(\bar{X}') = u'(\hat{x}' - \pi')$ and thus:

$$\bar{X} = u^{-1}(Eu(\tilde{x})) = \hat{x} - \pi \text{ and } \bar{X}' = u'^{-1}(Eu'(\tilde{x}')) = \hat{x}' - \pi'$$

which can be used to compare the lotteries as each of these certain equivalent denotes a price equivalent for the lottery each is willing to pay. If $\hat{x} = \hat{x}'$, the risk premium each is willing to pay for the lottery is $\pi = \hat{x} - \bar{X}$ and $\pi' = \hat{x} - \bar{X}'$. If $\pi > \pi'$, the first party will be willing to pay more to remove the lottery risk, indicating that he is more risk averse than the second party.

Insurance: Do we insure our house against fire? Do we insure our belongings against theft? Should we insure our exports against currency fluctuations or against default payment by foreign buyers? Do we invest in foreign lands without seeking insurance against national takeovers? How can we prevent the risks of a demand shortfall for a product or the risks of not meeting a demand? In such situations, risk is often exchanged, or transferred from one party to the other, who in turn might transfer it to some other parties (as it is the case in insurance). In all these situations, expected utilities are compared to provide for each of the parties, a utility advantage when the exchange occurs. Namely, let a party forgo a risk free return $\bar{X}'$ for a payment $\bar{Y}$ to be received by a party holding a risky prospect whose certain equivalent is $\bar{X}$. In this case, assuming that we maintain the same utility, we have: $\bar{X} = \bar{Y} + \bar{X}'$ and therefore,

$$\hat{x} - \pi = \bar{X} = \bar{Y} + \bar{X}' = \bar{Y} + \hat{x}' - \pi' \quad \text{and} \quad \bar{Y} = (\hat{x} - \hat{x}') - (\pi - \pi') > 0$$

Assume that the utility function is an exponential concave utility function $u(\tilde{x}) = 1 - e^{-a\tilde{x}}$, $a > 0$ with $u'(\tilde{x}) = \partial u(\tilde{x})/\partial\tilde{x} = ae^{-a\tilde{x}} > 0$ and $u''(\tilde{x}) = \partial^2 u(\tilde{x})/\partial\tilde{x}^2 = -a^2 e^{-a\tilde{x}} < 0$. This implies that the greater the wealth, the greater the utility. The rate of growth of this utility is declining with increasing wealth (specified by the negative second order derivative). Its index of risk aversion is a constant or $A_P(\hat{x}) = -u''(\hat{x})/u'(\hat{x}) = a > 0$ and therefore the risk premium is $\pi = a\sigma^2 > 0$.

Problem. Show that for the logarithmic utility function $u(\tilde{x}) = \log(\beta + \gamma\tilde{x})$, $\beta > 0$, $\gamma > 0$, the index of risk aversion is decreasing with wealth since $A_P(\hat{x}) = \gamma/(\beta + \gamma\hat{x})$. For a HARA (Hyperbolic Absolute Risk Aversion) utility function it is:

$$u(\tilde{x}) = \frac{1-\gamma}{\gamma}\left[\frac{a\tilde{x}}{1-\gamma} + b\right]^{\gamma} \quad \text{and therefore } A_P(\hat{x}) = \frac{a}{b + a\hat{x}/(1-\gamma)} > 0 \quad \text{and}$$

$$\pi = \frac{a\sigma^2}{2(b + a\hat{x}/(1-\gamma))}.$$

while for the utility function $u(\tilde{x}) = \tilde{x}^{1-\gamma}/(1-\gamma)$, $A(\hat{x}) = \gamma/\hat{x}$, $\pi = \gamma\sigma^2/\hat{x}$.

Valuation of the insurance risk premium: Say that an insurance firm charges an insured loss, $\tilde{X}$, at a premium π. Should the insured take it? What is the premium that the insurer should quote? Let the insured wealth be w and let the insured prospect loss be a random variable $\tilde{X}$. Then the largest insured premium he is willing to pay is a solution of: $u(w-\pi) = Eu(w-\tilde{X})$. By the same token, for the insurer, whose utility function is $U(W)$ and whose wealth is W, the smallest premium Π he will be willing to charge is $U(W+\Pi-\tilde{X}) = EU(W)$. Thus the risk premium is some quantity π' that meets the following conditions:

$$u(w-\pi') \geq Eu(w-\tilde{X}); \quad U(W+\pi'-\tilde{X}) \geq EU(W)$$

And therefore, $\Pi \leq \pi' \leq \pi$. There may therefore be many insurance contracts for the same risks, each of which may be tailored by a discriminating insurance firm or by an insured. Premium Pricing in such cases is based on a negotiated (or not) agreement between the insurer and the insured, each privy to its own information, needs, regulatory requirements and the power each has over the other. As a result, there are many "formulas," defined mostly a-priori or based on the experience of insurers for "what premium might bring greater profit." For example, the loading factor policy is an "expected cost plus policy" that in assess the number of claims times their expected dollar value and loading this expectation by a parameter (for example, 20 %). In this case, the premium demanded equals: $P(t) = (1+L)E(N(t))E(\tilde{X})$ where $N(t)$ is the number of insured in the portfolio at a given time, $\tilde{X}$ is the aggregate claims for the portfolio given by the sum of individual claims $\tilde{X} = \sum_{j=1}^{\tilde{M}(t)} \tilde{x}_j$ where $\tilde{M}(t) \leq N(t)$ are the number of claims (a random variable) and L is the loading factor.

Other approaches recommend the use of specific utility functions, such as the logarithmic and exponential utility function. Selecting the premium policy depends therefore on the definition of the insured risk, the risk sustained by the insurance firm and the relationship between the insured and the insurer. In practice, insurance firms are businesses that ask as much money as they can and will receive. Even when they do so, they often turn to securitize their portfolios and sell it at a profit to willing financial markets. For big tickets items (or group collective insurance schemes), the premium policy is negotiated and the premium defined by negotiations between the insured and the insurer. A number of approaches used to calculate insurance premiums are outlined in Embrechts 2000a; Furman and Zatikis, and J. Goovaerts and Roger, Laeven 2006. These include for example, a weighted risk premium:

$$F_w(x) = \frac{E(\mathbf{1}\{X \leq x\}w(x))}{E(w(x))} \quad \text{with} \mathbf{1}\{S\} = \begin{cases} 1 & S \text{ is true} \\ 0 & \text{otherwise} \end{cases}$$

A number of cases include w(x) constant, w(x) = x, $w(x) = e^{\lambda x}$ (also called the Esscher Transform).

Insurance securitization: Insurance firms through aggregation of insured risks may be able to construct diversified insurance portfolios and thereby reduce the portfolio variance. Further development of insurance/finance products such as credit derivatives have contributed to a convergence of finance and insurance by transforming insurance claims portfolios into financial commodities to be sold and bought in a financial market. In such cases, insurance firms have found it profitable to either approach banks that would sell their portfolios to clients and speculators or turn directly to financial markets with product offering derived from their portfolios. In both cases, both insurance firms and banks have become intermediary in a complex financial markets where risks are difficult to assess and where buyers of such financial contracts have uncertain payoffs and claims.

Example: The Quadratic Utility Function and a Coherent Valuation Say that $\tilde{x}$, a random prospect has a quadratic utility function given by: $u(\tilde{x}) = \tilde{x} - \rho(\tilde{x} - \hat{x})^2$. It implies that expected values are preferred while variations about the expected returns are not. The parameter ρ defines an index of risk (variance) aversion. Consider the expected utility $Eu(\tilde{x}) = \hat{x} - \rho\sigma^2$, with its certain equivalent $u(\bar{X}) = E$ $u(\tilde{x})$ or $\bar{X} - \rho(\hat{x} - \bar{X})^2 = \hat{x} - \rho\sigma^2$. The risk premium measuring the price of doing away with risk is however $\pi = \hat{x} - \bar{X}$. Note that this is a quadratic equation in the risk premium π: $F(\sigma^2, \pi) = \rho[\sigma^2 - \pi^2] - \pi = 0$ whose solution is $\pi^* = -(1/2\rho)$ $\pm\sqrt{(1/4\rho)^2 + \sigma^2}$. When $\rho \to +\infty$ then $\pi^* = \sigma$ while for $\rho \to -\infty$. Its solution is: $\pi^* = -\sigma$ and therefore, the risk premium can be written as follows: $\pi^* = \lambda\sigma, \lambda \in [-1, +1]$. If λ is small, $\pi \approx 0$ and $\bar{X} = \hat{x} - \pi$ or for a quadratic utility function we have $\bar{X} = \hat{x} - \lambda\sigma$ or the "sure value" of a prospect is measured by is expectation less its risk premium measured which is proportional to the standard deviation. Such a valuation is then coherent in the sense of Artzner (1999), Artzner et al. (1997, 1999, 2000, 2001) (Chap. 6).

Example: A Portfolio Certain Equivalent and its Risk Premium Consider a portfolio consisting of financial assets, say n shares of a stock whose current price is S and a bond whose price is B. The current worth of the portfolio is: $W = nS + B$ while we estimate that its next period price is random with:

$$\tilde{W} = nS(1 + \tilde{R}) + B(1 + R_f) \text{ and } E(\tilde{W}) = nS(1 + E(\tilde{R})) + B(1 + R_f)$$

Note that $\sigma(\tilde{W}) = nS\sqrt{\operatorname{var}(\tilde{R})}$ where $\tilde{R}$ is the stock random rate of return. The expected utility is $Eu(\tilde{W}) = Eu(nS(1 + \tilde{R}) + B(1 + R_f))$. And for a quadratic utility function $u(\tilde{W}) = \tilde{W} - \rho(\tilde{W} - \hat{W})^2$, $E(\tilde{W}) - \pi = \bar{W}$, $\pi = \lambda\sigma$ and: $\bar{W} = nS_0$

$(1+E(\tilde{R}))+B_0(1+R_f)-\lambda n S_0 \sigma(\tilde{R}),\ \sigma(\tilde{R})=\sqrt{\operatorname{var}(\tilde{R})}$. Its present value, discounted at the risk free rate (since it is not a random prospect), is thus:

$$\bar{W}_0 = B_0 + \frac{nS_0\left(1+E(\tilde{R})-\lambda\sigma(\tilde{R})\right)}{1+R_f}$$

If the current price is a risk free discounted measure of the future portfolio utility, then: $\bar{W}_0 = B_0 + nS_0$. Equating these two equations, we obtain:

$$R_f = E(\tilde{R}) - \lambda\sigma(\tilde{R}) \text{ or the risk premium } \lambda\sigma(\tilde{R}) = E(\tilde{R}) - R_f$$

Example: Risk attitudes and prevention Greater aversion to risk might mean that we should invest more on maintenance to prevent breakdowns. However Dionne and Eeckouldt (1985) have pointed out that a more risk averse individual does not necessarily lead to more self-protection than a less risk averse one. Consider a decision-maker investing x in regulation compliance that reduces the probability of both being controlled, detected and penalized, to say $p(x), \partial p/\partial x < 0$. Let u(.) be a utility function with two outcomes W_L, W_{NL} defining the case of no loss in case of compliance and loss in case of no-compliance. These are defined by: $W_{NL} = W_0 - x, W_L = W_0 - l - x$. Investment to comply is found by the utility maximization, or:

$$\underset{x}{Max}\ U(x) = p(x)u(W_L) + (1-p(x))u(W_{NL}), W_{NL} = W_0 - x, W_L = W_0 - l - x$$

And therefore:

$$\frac{\partial U(x)}{\partial x} = \frac{\partial p(x)}{\partial x}(u(W_L) - u(W_{NL})) + p(x)\left(\frac{\partial u(x)}{\partial W_{NL}} - \frac{\partial u(x)}{\partial W_L}\right) - \frac{\partial u(x)}{\partial W_{NL}} = 0$$

which is a linear differential equation:

$$\frac{\partial p(x)}{\partial x}(u(W_L) - u(W_{NL})) + p(x)\left(\frac{\partial u(x)}{\partial W_{NL}} - \frac{\partial u(x)}{\partial W_L}\right) = \frac{\partial u(x)}{\partial W_{NL}}, 0 \le p(x) \le 1$$

Assuming risk neutral utility, it is reduced to:

$$\underset{x}{Max}\ U(x) = p(x)(W_0 - l - x) + (1-p(x))(W_0 - x) = -lp(x) + (W_0 - x)$$

and thereby, $-\frac{\partial p(x)}{\partial x} l = 1,\ \frac{\partial p(x)}{\partial x} < 0$. As a result $0 \le p(x) = a - \frac{x}{l} \le 1$. Now assume that the utility function is $u(W) = (W)^{1-\gamma}/(1-\gamma)$ as a result,

$$\frac{\partial p(x)}{\partial x} - p(x)\frac{(1-\gamma)({W_L}^{-\gamma} - {W_{NL}}^{-\gamma})}{{W_L}^{1-\gamma} - {W_{NL}}^{1-\gamma}} = \frac{(1-\gamma){W_{NL}}^{-\gamma}}{{W_L}^{1-\gamma} - {W_{NL}}^{1-\gamma}}$$

And therefore, the probability of a loss due to non-compliance can be calculated by the solution of the first order linear differential equation:

$$\frac{dp(x)}{dx} - p(x)\frac{(1-\gamma)((W_0 - l - x)^{-\gamma} - (W_0 - x)^{-\gamma})}{(W_0 - l - x)^{1-\gamma} - (W_0 - x)^{1-\gamma}}$$
$$= \frac{(1-\gamma)(W_0 - x)^{-\gamma}}{(W_0 - l - x)^{1-\gamma} - (W_0 - x)^{1-\gamma}}, \quad P(0) = P(W_0)$$

whose initial condition at $x = 0$ is $p(W_0)$. This equation implies that risk aversion is not sufficient to explain the investment in compliance to regulation (see Dachraoui et al. 1999 for an interpretation in terms of maintenance).

Example: Risk and Advertising Claims Advertising claims bear risks! A claim which is not met has an effect which can be detrimental. The purpose of this example is to highlight a particular (and simplistic) approach to value what we say or claim (see also Tapiero 1981, 2000a, 2005a, 2009a). Ex-ante, what we claim may turn out to be something else ex-post. For example, claiming the quality of a product that turns out not as great as its advertising claimed. A performance meeting its claim reinforces the loyalty of a client and therefore his loyalty while a performance not meeting its claim leads to disappointment. Thus, claims that underestimate performance might be "reliable," namely, mostly true, but then they might not be sufficiently attractive, while overly optimistic claims might entice a backlash (and consequently a bad reputation). In this sense, the decision to set a claim is necessarily concurrent to the decision "what to claim." Such a decision is compounded further by the fact that consumers are also a consumer of services (such as warranties, insurance, investment advice, etc.). For example, the satisfied buyer of car insurance might also buy a home and life insurance. By the same token, excessive banking fees may lead depositors to move to other banks while a satisfied consumer may use other services the bank offers. This example provides a preliminary approach to dealing with these issues by developing a default-prone risk model and valuing its claim. It is simplified to highlight the utility approach.

Say that a claim is set numerically to some value K (for example, the mileage per gallon of a car, the response time to a query, etc.). Performance however, is a function of numerous factors which can at best be characterized by a random variable $\tilde{Q}$ with a density function $f(\tilde{Q})$ (for example, lamps sold have an associated life time probability distribution). In this case, the probability $\tilde{p}$ of an "unsatisfied" consumer, experiencing a product or a service which is above or below its advertising claim, is:

$$\tilde{p} = \begin{cases} 1 & \text{If } \tilde{Q} \geq K \\ 0 & \text{If } \tilde{Q} \end{cases} \quad \text{and therefore,}$$

$$\hat{p} = \int_0^K f(x)dx = F(K) \quad \text{and} \quad \sigma_p^2 = F(K)[1 - F(K)]$$

With dissatisfaction mean and variance calculate by:

$$\hat{L} = \int_0^K (K - x)f(x)dx \text{ and } \sigma_L^2 = \int_0^K (K - x)^2 f(x)dx$$

For a sample of D consumers, let the number of dissatisfied be n then assuming statistical independence, we have a (mixture) Lexis binomial distribution with parameters $(D, \tilde{p})$ with mean and variance given by: $E(n) = D\hat{p}$, $\text{var}(n) = D\hat{p}(1 - \hat{p}) + D(D - 1)\sigma_p^2$ (see Chap. 3). If (again for simplicity, and to outline the basic financial approach to pricing advertising claims), we assume that a product or service has a marginal revenue π while product or service costs per unit are c, then, the net profit realized at say t = 1, is $\tilde{P}_1 = D\pi - c\tilde{n}$ with:

$$E(\tilde{P}_1) = D\pi - cE(\tilde{n}) \quad \text{or } E(\tilde{P}_1) = D\pi - cDF(K)$$
$$\text{var}(\tilde{P}_1) = c^2\text{var}(\tilde{n}) \quad \text{or var}(\tilde{P}_1) = c^2D^2F(K)(1 - F(K))$$

Its certain equivalent is $Eu(\tilde{P}_1) = u(CE_1)$ or $CE_1 = u^{-1}(Eu(\tilde{P}_1))$. When the propensity to meet advertising claims are dependent, due for example to a common source of (operational or external) risk, the number of dissatisfied consumers can be much greater. If we use a quadratic utility function $u(\tilde{P}_1) = \tilde{P}_1 - \rho(\tilde{P}_1 - E(\tilde{P}_1))^2$, the certain equivalent is $CE_1 = E(\tilde{P}_1) - \lambda\sqrt{\text{var}(\tilde{P}_1)}$, the risk premium is $\lambda\sqrt{\text{var}(\tilde{P}_1)}$ and the current price is $CE_0 = \frac{1}{1+R_f}CE_1$, $\lambda = [E(\tilde{P}_1) - CE_0(1 + R_f)]/\sqrt{\text{var}(\tilde{P}_1)}$. In this sense, an optimal advertising claim at time t = 0, is found by maximizing its certain equivalent which we rewrite explicitly by:

$$\frac{CE_0}{D} = \frac{1}{1 + R_f}\left(\pi - c\left(F(K) + \lambda\sqrt{F(K)(1 - F(K))}\right)\right)$$

Leading to:

$$CE_0 = \frac{1}{1 + R_f}D\left(\pi - cF(K) - 2c\frac{f(K)(F(K)(1 - F(K)))}{F(K) - f(K)}\right) > 0$$

In other words, a claim value "today" implies the claim "tomorrow" and vice versa, the consequences of an advertising claim "tomorrow" implies its value today (note that this is not a price). In this case, the sell price (the marginal revenue) per unit cost for an dissatisfied consumer is necessarily given by:

$$\frac{\pi}{c}(K)\left[1-2\frac{f(K)(1-F(K))}{F(K)-f(K)}\right]>0$$

For example, say that $f(x)=(1-\alpha)x^{-\alpha}$, $x\in[0,1]$, then $F(x)=x^{1-\alpha}$ and therefore, $F(K)=K^{1-\alpha}$ and:

$$\lambda=\left((1-\alpha)\sqrt{K^{1-\alpha}(1-K^{1-\alpha})}\right)/(K-(1-\alpha))$$

Further, $E(x)=(1-\alpha)/(2-\alpha)$. For example, if the average performance equals 0.4, then $\alpha=0.333$ and therefore, $\lambda=\left(1.32\sqrt{K^{0.66}(1-K^{0.66})}\right)/(K-0.666)$. As a result, an advertising claim of less than 0.666 is "risk loving" while an advertising claim of more than 0.666 is risk averse. In other words, even claiming the average is likely to incur greater costs. This is an expected result as the more one claims, the greater the number of dissatisfied consumers. The marginal revenue is in this case also a function of the claims and given by:

$$\frac{CE_0}{D}=\frac{1}{1+R_f}\left(\pi-c\left(K^{0.66}+\frac{1.32\left(K^{0.66}\left(1-K^{0.66}\right)\right)}{K-0.666}\right)\right).$$

7.4 Risk Prudence and Background Risk

Univariate background risks do not generalize simply to their multi-dimensional case. Such generalizations raise three different aspects to be examined:

1. Is the Arrow Pratt index of risk aversion still meaningful and retain its meaning under multiple dimensional risks?
2. Does existence of a background risk that cannot be done away with (an uninsurable risk) make the individual more or less averse to some given risk?
3. What are the multi-period risk effects of background risk? Can one shift risk from one period to the other? (and therefore trade future risks)

The first question was raised by Kihlstrom and Mirman (1974), who found that comparing risk attitudes of an individual at different levels of wealth requires that the underlying preferences meet some prerequisite for comparability. For example, say that an individual 1 prefers a sure outcome, individual 2 prefers a lottery. Is there any meaning in saying that individual 1 is more risk averse than individual 2?

The answer is no, for it would be easy to find another lottery and another sure outcome for which the difference in observable choices of the two individuals would be exactly the opposite. In fact, it is possible to have situations that a risk with its background risk would be preferable to the same risk without background risk. This is similar in spirit to the finding that sometimes more information renders one less decisive. For this reason, choices reflect differences between preference orderings rather than differences in risk attitudes. Karni (1979), further argues that multivariate risks do not typically pertain to bundle of goods, but rather to the level of income (see also the extended CCAPM in Chap. 8) and the uncertainty regarding the ultimate consumption. The limits of the Arrow-Pratt index of risk aversion was pointed out by Ross as early as 1981. Ross discovered that the Arrow-Pratt definition was not "strong" enough in another sense: In the presence of another risk which the individual could not get rid of ("background" or "uninsurable" risk, an element of uncertainty—see Chap. 10), an increase in risk aversion does not imply that the demand for any other independently distributed risky asset will decrease. Rothschild and Stiglitz' (1970), in particular showed that adding a white noise to some undesirable lottery might make the lottery desirable, which is a paradoxical result. Additional results, by Gollier (1995, 2000), Gollier and Pratt (1996), defined "risk vulnerability." This concept generalized previous definitions. It implies that the two first derivatives of the utility function are concave transforms of the original utility function. Vulnerability is both necessary and sufficient for the introduction of an unfair background risk to lower the optimal investment in any other independent risk, provided the risk to be borne is *independent* of the background risk. This concept of vulnerability has made HARA (hyperbolic absolute risk aversion) utility functions popular, to the extent that they can represent individuals who, a fortiori, are risk vulnerable individuals. *Correlated* investment risk and background risk call therefore for a more detailed characterization (see Eeckhoudt et al. 1996).

Prudence. Prudence has been defined by Kimball (1990) and Eeckoudt and Kimball (1991) in terms of the third derivative of utility. Say that $(W, \tilde{x})$ are the wealth of a person and the random payoff which results from some action respectively. If we use the expected marginal utility as a mechanism for choice, then:

$$EU'(W+\tilde{x}) > U'(W) \text{ if } U' \text{ is convex}$$

$$EU'(W+\tilde{x}) < U'(W) \text{ if } U' \text{ is concave}$$

The concept of risk premium seen above can then be applied similarly. Denote by ψ the required amount to maintain the same marginal utility levels. That is: $U'(W-\psi) = EU'(W+\tilde{x})$. Using a Taylor series expansion, we find that:

$$\psi = \frac{1}{2}\mathrm{var}(\tilde{x})\left[-\frac{U'''(W)}{U''(W)}\right]$$

The squared bracket term is called the *degree of absolute prudence*. For a risk averse decision maker, the utility second order derivative is negative ($U'' < 0$) and therefore prudence will be positive (negative) if the third derivative U''' is positive (negative), which provides an interpretation for the third derivative proposed by Kimball. Further, Kimball has also shown that if the risk premium is positive and decreasing in wealth w, then $\psi > \pi$.

7.4.1 Risk, Uncertainty and Insurance

For insurance, risk measurement is needed to support insurance firms' decisions in assessing insured contracts (i.e. probabilities and claims consequences), contracts portfolios' risks, investment risks, reserves and regulation risks, bankruptcy risks. Questions such as: What are the potential claims and on what factors do they depend? What are the potential consequences of a claim, or of a collection of claims due to a common causal factor (such as a Hurricane)? Are claims statistically predictable? Are claims insurable? Who pays for a risk event and how much? What are insured risk attitudes? Who is responsible in case a risk event materializes? What are the regulatory requirements imposed on the firm? What are the market risks that influence the firm liquidity (determined by VaR and Basel III Regulation)? What are the available means to share risk—insured (by co-participation, bonus-malus schemes)? Are they insurance or reinsurance firms (through excess-loss and stop-loss contracts)? Are they insurance portfolios, securitized and marketed directly or through intermediaries? What are the potential risks emanating from natural causes (earthquakes, weather patterns changes, etc.). These are some of the uncertain events that insurance firms may contemplate but not necessarily insure. Insurance firms essentially insure insurable risks, namely, countable events with predictable consequences based on risk models they can value. Further, they use risk sharing schemes to provide insured the incentive to be preventive (such as bonus-malus premiums) and avoid small claims that are relatively costly to treat. For insurance firms, insurance contracts are therefore framed as risk contracts while for insured, an insurance contract consists of all events not accounted for in the insurance contract (and therefore including the contract residual uncertain consequences). While the residual uncertainty is largely neglected in calculating the amount of risk an insured seeks and the premium he is willing to pay, it may in fact under estimate and under value the demand and the price insured are willing to pay for insurance.

For example, let there be an insured whose risk is defined by a potential loss $\tilde{L}$ consisting respectively of predictable and unpredictable losses $(\tilde{x}, \tilde{y})$ with $\tilde{L} \equiv \tilde{x} + \tilde{y}$. In addition we assume that initially the utility of the insured and the insurer as a function of their wealth are $u(w)$ and $U(W)$ respectively. Let the contract between the insured and the insurer pertain to predictable losses and assume that risk premium set by the insurer to be π_x. In this case, after an insurance contract is issued, the utilities of the insured and the insurer are:

$$Eu(w - \pi + \tilde{y}) \geq Eu(w + \tilde{x} + \tilde{y}) \text{ and } EU(W) \geq EU(W + \pi(1 + \tilde{R}_W) - \tilde{x})$$

where $\tilde{R}_W$ is the rate of return of the insurance firm on its investment of the reinsurance premium. Say that:

$$Eu(\tilde{w} - \pi - \tilde{y}) = Eu(\tilde{w} - \tilde{y}) - \pi E \frac{\partial u}{\partial(.)}$$
$$Eu(\tilde{w} - \tilde{x} - \tilde{y}) = Eu(\tilde{w} - \tilde{y}) - E\left(\frac{\partial u}{\partial(.)}\tilde{x}\right) + \frac{1}{2}E\left(\frac{\partial^2 u}{\partial(.)^2}(\tilde{x})^2\right)$$

And therefore,

$$\pi \leq E(\tilde{x}) + \frac{1}{2}E\left(A_P(\tilde{w} - \tilde{y})(\tilde{x})^2\right) \text{ with } -\frac{\partial^2 u/\partial(.)^2}{\partial u/\partial(.)} = A_P(\tilde{w} - \tilde{y})$$

where $A_P(w - \tilde{y})$ is the insured Arrow Pratt index of risk aversion calculated at the wealth level which includes the uncertain losses of the insured (which is not known but may be suspected). However, for an increase in $\tilde{y}$, $A_P(\tilde{w} - \tilde{y})$ increases and therefore from the equation above the willingness to pay by the insured increases. In other words, risk averse insured left with their uninsured residual uncertainty of future events may in fact be willing to pay more than the premium insurance firms extract. This situation may thus provide and "unfair" advantage to insurance firms "cherry picking" what to insure and extracting from insured a premium greater than the market price (had these contracts be securitized). Consider next the insurer:

$$EU(\tilde{W}) \leq EU(\tilde{W} + \pi(1 + \tilde{R}_W) - \tilde{x})$$
$$= EU(\tilde{W}) + \frac{\partial U(\tilde{W})}{\partial \tilde{W}}\left(\pi(1 + E\tilde{R}_W) - E\tilde{x}\right) + \frac{1}{2}\frac{\partial^2 U(\tilde{W})}{\partial \tilde{W}^2}\left(\pi(1 + E\tilde{R}_W) - E\tilde{x}\right)^2$$

or,

$$0 \leq \left(\pi\left(1 + (E\tilde{R}_W)\right) - E(\tilde{x})\right) - \frac{1}{2}A_W(\tilde{W})\left(\pi(1 + E\tilde{R}_W) - E\tilde{x}\right)^2,$$
$$A_W = -\frac{\partial^2 U(\tilde{W})/\partial \tilde{W}^2}{\partial U(\tilde{W})/\partial \tilde{W}}$$

And therefore, simply a smaller risk for the insurer

$$\frac{1}{2}A_W(\tilde{W}) \leq \frac{\pi\left(1 + (E\tilde{R}_W)\right) - E(\tilde{x})}{E\left(\pi(1 + E\tilde{R}_W) - E\tilde{x}\right)^2}$$

7.5 Expected Utility Bounds

In many instances, calculation of the expected utility is difficult. Expected utility bounds, provide a first approximation however. For risk averse investors with utility function $u(.)$ and $u''(.) \leq 0$ we can "bound" the expected utility from above. This is known as *Jensen's inequality* given by:

$$E(u(\tilde{x})) \leq u(\hat{x}) \text{ when } u''(.) \leq 0 \quad \text{and} \quad E(u(\tilde{x})) \geq u(\hat{x}) \text{ when } u''(.) \leq 0$$

And vice versa when the utility function represents a risk loving investor (i.e. $u''(.) \geq 0$). When rewards have known mean and known variance, Willasen (1981, 1990) has shown that for a risk averse decision maker, the expected utility can be bounded from below and above by:

$$u(\hat{x}) \geq Eu(\tilde{x}) \geq \hat{x}^2 u(\alpha_2/\hat{x})/\alpha_2; \;\; \alpha_2 = E(\tilde{x}^2)$$

The first inequality is Jenssen's inequality while the second provides a best lower bound on expected utility. It is possible to improve on this estimate by using the best upper and lower Tchebycheff bounds on expected utility. This inequality is particularly useful when we interpret and compare the effects of risk on the choice of financial decisions.

7.6 VaR Valuation

Value at Risk is widely practiced following its essential adoption by regulatory agencies. Risk reserves have thus become as seen earlier in Chap. 1 a measurement of risk, measuring the risk exposure of a financial institution and the amount of disposable assets it owns that can meet contingent claims on the firm's assets. It is also a simple and intuitive notion that provides a common language for risk managers. VaR measurements are defined by a body of data that allow to model a risk exposure in terms of a probability model. Figure 7.2, outlines such a risk exposure by a normal probability distribution:

A Quantile Risk exposure VaR is then defined by a risk threshold preference which we write as follows:

$$P_t[\Delta W_t(a) < -VaR_t(a, \alpha)] = \alpha$$

which denotes the conditional probability at time t of a loss smaller than the reserve VaR. VaR exposure risks are then expressed as a function of the policies one may institute to control the risk (for example, a definition of assets to be included in a portfolio) and the probability one is willing to assume by such a VaR. For example, for the normal probability model of a portfolio profits-losses distribution, we have:

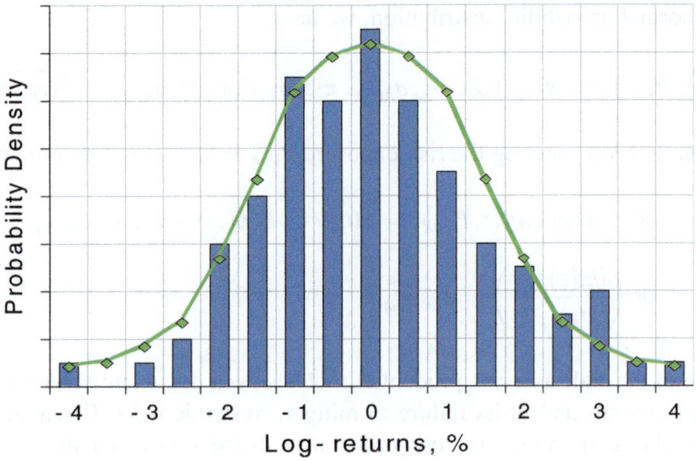

Fig. 7.2 VaR and the normal probability distribution

$$\Delta W_t \sim N(\mu_t, \Omega_t) \quad \text{then } VaR_t(a,\alpha) = -a^T\mu_t + \left(a^T\Omega_t a\right)^{1/2} Z_{1-\alpha}$$

In this case, the amount money set aside to meet contingent risks is an expression of the measured risks and our attitudes to these risks! However, note that if we write for simplicity, $\sigma = \left(a^T\Omega_t a\right)^{1/2}$ (a standard deviation) and $\lambda = Z_{1-\alpha}$ a risk coefficient, then $-VaR_t(a,\alpha) = a^T\mu_t - \lambda\sigma$ which corresponds to a quadratic utility with mean return (cost) less the risk premium with a price of risk per unit standard deviation equal to $\lambda = Z_{1-\alpha}$. In other words, VaR and Normality of the loss distribution imply a quadratic utility valuation.

The VaR can equally reflect a rational asymmetric cost objective whose optimization provides a VaR to be set aside. For example, let $\tilde{\pi}(z)$ denote the financial resources of a bank with VaR denoting the regulated reserve. We assume linear asymmetric ex-post costs.

$$Min\ \Phi_{VaR} = \alpha E[\tilde{\pi}(z) - VaR|\tilde{\pi}(z) \geq VaR] + \beta E[VaR - \tilde{\pi}(z)|\tilde{\pi}(z) \leq VaR]; \quad \beta \geq \alpha$$

or:

$$Min\,\Phi_{VaR} = \alpha \int_{VaR}^{\infty} (\tilde{\pi}(z) - VaR)dF_\pi + \beta \int_{-\infty}^{VaR} (VaR - \tilde{\pi}(z))dF_\pi;$$

An optimization with respect to VaR results in the following:

$$F_\pi(VaR, z) = P_{VaR}, P_{VaR} = \frac{\alpha}{\alpha + \beta} \quad \text{and therefore, } VaR(z, P_{VaR}) = F_\pi^{-1}(z; P_{VaR})$$

For a normal probability distribution, we have:

$$VaR(z, P_{VaR}) = -\hat{\pi}(z) + Z_{1-P_{VaR}}\sigma_{\tilde{\pi}(z)}, \quad \hat{\pi}(z) = E(\tilde{\pi}(z)), \sigma_{\tilde{\pi}(z)} = \sqrt{\text{var}(\tilde{\pi}(z))}$$

A minimum Var meeting the risk exposure P_{VaR} is then found by minimizing:

$$VaR^* = \underset{z \geq 0}{Min}\, VaR(z, P_{VaR}) = \underset{z \geqslant 0}{Min}\, V\{-E(\tilde{\pi}(z)) + Z_{1-P_{VaR}}\sigma_{\tilde{\pi}(z)}\}$$

$$\text{or } \left(\frac{\partial \hat{\pi}(z)/\partial z}{\partial \sigma_{\tilde{\pi}(z)}/\partial z}\right) = Z_{1-P_{VaR}}$$

Failure of the VaR as a risk prudential tool has been documented based on both theoretical grounds and in its failure to mitigate systemic risks. Essential critical points are the assumptions of loss probability distributions that do not always represent the risks banks have to be protected from—when using fat tail distributions, costs are excessive while when using normal probability distributions, the VaR set aside is insufficient; anti-cyclicality, requiring to set aside more money in "bad time" and less in good time, which may on the aggregate increase cyclicality, booms and busts systemic risks. In this sense VaR when used as sole risk prudential regulation tool is counter-productive. Technically, VaR may also be inefficient, too slow to detect time variations in losses resulting from markets condition. For this reason, we shall consider subsequently a Time VaR based on run time statistics for losses greater than specified by the risks we are assuming when using a VaR prudential approach.

7.7 Valuation of Operations by Lagrange Multipliers

Traditionally, operational economic and risk valuation problems are conceived as a risk-performance substitution, one valued in terms of the other. An optimization problems on cost-reliability and their substitution highlights some elements of this approach:

Min Cost	Max Reliability
Subject to : Reliability $\geq \alpha$	Subject to : Costs $\leq$ Budget

Optimal policies are determined relative to an objective (say costs) and to constraints (say reliability). The value of the latter (the constraints) relative to the former (costs) are expressed by the Lagrange multipliers associated to the problem constraints. However, these are subjective approaches, model-based, providing a self-valuation. Alternatively, some problems are assessed economically in terms of their Long Run Average Cost or returns. This assessments are justified by the Renewal Reward Theorem in Probability Theory stating that the average cycle costs and returns (provided it is a renewal cycle) are equal their long run average, namely:

$$\underset{T\to\infty}{Lim}\ \frac{1}{T}\int_{0}^{T} E\Phi\big(\tilde{X}(t)\big)dt = \frac{E\Phi\big(\tilde{X}_T\big)}{E(\tilde{T})}$$

Thus if all costs and returns (of inventory, of maintenance, of dispatching etc., see also application of long run average costs to bank's regulation and compliance in Chap. 11) can be assessed and calculated within a renewal cycle, whether deterministic or stochastic $\tilde{T}$, then their long run average equal their cycle average stated above. These are valuations allowing within a set of specific objectives to rank order alternatives. Other approaches based on financial accounting concepts include among others:

- Replacement value method: for seleccting the more advantageous cost/return alternative
- Cash flow evaluation methods inculding:
  - Average rate of return
  - Payback method
  - Internal rate of return
  - Net worth, discounted cash flow formulations etc.

These approaches are based on a derived financial valuation of operations, industrial and logistic investments etc. when a financial justification is needed.

Additional References and Additional Readings

The theory of expected utility is extensively documented. Some essential references include, Friedman and Savage (1948, 1952) (see also Chap. 6), Fishburn (1970, 1988), Pratt (1964, 1990), Pratt and Zeckhauser (1987), Caballe and Pomanski (1996) on mixed risk aversion, Karni (1979) on multivariate risk aversion, Gollier (1995), Ross (1981), Rothschild and Stiglitz (1970) on increasing risk Levy and Levy (1991), on AP measures in the multivariate case, Lintner (1965a, b), Tobin (1958), on Liquidity preferences and behavior towards risk, Markowitz (1959) on the application of Mean-variance to portfolio selection (implying in fact preferences that have a quadratic utility function). See for example Munier and Tapiero (2008) for a survey of the utility approach. Expected utility has also been criticized leading to numerous approaches based on Allais early and numerous contributions (Allais 1953, 1979a, b).

In insurance, my papers on insurance have mostly emphasized models of mutual insurance (Tapiero 1984 on mutual insurance, Jacque and Tapiero 1986, 1987,

1988a, b, 1990) with applications to insurance for computer backups [published in Communications of the ACM, 88-01: Computer Backup Pools, Disaster Recovery and Default Risk, Communications. ACM (with Kahane et al. 1988)].

Other references in insurance include for example, Bryis et al. (1989), Eeckhoudt et al. (1996), Eeckoudt and Kimball (1991), Jewell (1974) on OR in Insurance, Brockett and Xia (1995), Buhlmann (1980), Dayananda (1970) (on reinsurance), Gerber (1974, 1979), Gerber and Shiu (1994), Goovaerts et al. (2003b, c) (utility risk measures), Kaas et al. (2001) book on Modern Actuarial Risk Theory, Schonbucher (2003) on credit risks and credit derivatives thesis, McNeil et al. (2005), Gourieroux et al. (2000) (on the sensitivity analysis of VaR), Basi et al. (1998), Goovaerts et al. (1984) book on Insurance Premiums and the important book of Denuit et al. (2006), on Actuarial Theory for Dependent Risks: Measures.

The assumption of a rational and well informed decision makers have led however to a number of critical assessment of the utility approach. There are as a result, both formal and behavioral alternatives to the use of expected utility. Simon (1955, 1976) on behavioral models of rational choice, Gul (1991) on regret as well as references on behavioral decision making under uncertainty outlined in Chap. 2.

On the theoretical side, Allais (1953, 1979a, b) have provided a strong impetus to question some of the assumptions of economic models. Lacking a better theoretical alternative, the use of expected utility has underlie economic and financial models however. Muth (1961), Arrow (1951a), Lucas (1978) (see Chap. 8) have contributed however immensely to the development of complete markets theories, allowing a prediction of prices when markets are in (Walras) equilibrium. These have extended immensely the use of economic and financial models to options pricing by Black and Scholes (1973) (see also Merton 1969, 1973, 1977, 1990), Breeden and Litzenberger (1978), Cox and Ross (1978), Hull and White (2001, 2003) on CDS, (2006) on credit derivatives, Jackwerth (1999, 2000) on implied risk neutral distributions. In economic theory, Henry (1974a, b) on option of irreplaceable assets, Pyndick (1991) uncertainty and theory of renewable resources markets etc.

Applications of risk measurement and their use abound in and out of economics, finance, industrial engineering and management and elsewhere. Some of my papers include, Paroush and Tapiero (1975) on location on a line under uncertainty, Reisman et al. (1987a) on reliability pricing and quality control, and in 1987b on product failures and manufacturing reliability, and extension of our paper of 1987a in 1987b to a dynamic framework. Tapiero (2005b) on reliability design and RvaR, my 2005c paper on the use of VaR for reliability pricing, my 1987 paper on inventory control, as well as Tapiero and Kogan (2008), on risk averse order policies, my own paper 2012c on the price of safety.

Other applications include for example, Coles (1993) on the price of workers reliability, Cox and Tait (1991), on reliability, safety and risk management, Das (1995) on credit risk measurement at Morgan (1994, 1997).

Chapter 8
Risk Economics and Multi-Agent CCAPM

Abstract Utility models to price risk assets have been used following the seminal contribution of Markowitz, Sharpe, Lintner and a legion of economist and financial engineers. Models such as CAPM (Capital Assets Pricing Model) are for example linear risk models, implying a quadratic utility function which is used profusely in the financial industry. Generalization to the CCAPM (The Consumption Capital Assets Pricing Model) has further extended the theoretical and practical usefulness of such models, as well as provide an apparent relationship (in some cases) with pricing models based on Arrow-Debreu state preference theory. In this chapter this economic framework is extended, in the sense that the CCAPM may account for some endogenous factors embedded in economic aggregates (such as aggregate consumption rather than just an individual's consumer). Such an extension has not been subjected to an empirical analysis but provides a theoretical framework for assessing a number of additional factors that affect the pricing of risk. In particular, it provides an economic and financial framework to value and price assets in situations that depart from the complete market hypothesis such as debt, consumers wealth etc. Applications are considered in both this chapter and the next.

8.1 Introduction

Economic theory had long been concerned with valuation, prices and how to reconcile their many puzzles and approaches. For example, what makes certain products more or less valuable and how are their prices set? Why is water "cheap"? Why are diamonds "pricy"? Why is the price of gold unrelated to its "usefulness"? What is the value and the price of health and life? What is the price of risk and is there a price of uncertainty? How do prices depend on markets, on regulation, on rationing, on geography, etc.? Figure 8.1 summarizes a number of factors (however incomplete) that account for value and price. Value, for example, is based on individual (or subjective) preferences for both outcomes and their risks reflecting

C.S. Tapiero, *Engineering Risk and Finance*, International Series in Operations Research & Management Science 188, DOI 10.1007/978-1-4614-6234-7_8,

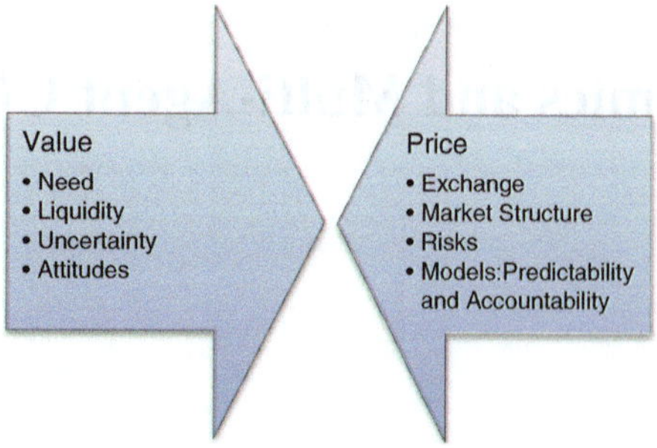

Fig. 8.1 Economic values and price

their needs, the liquidity and the uncertainty that these individual face. Price however, is implied in an exchange between one or more parties.

As a result value and price differ. Adam Smith (1776, p. 31) differentiated between these two by noting that value has two meanings, a value of consumption which underlies "needs of consumers" expressed by their utility and the value of alternatives to possession of a current good that can be purchased. These convey a value in "exchange". For example, if we own 5 pounds of tomatoes and consume them all, then our utility is in consuming these tomatoes (with no concern for future consumption). However we can exchange some of these tomatoes for either the consumption of another product which can be acquired now (and therefore has no market price risk but a consumption risk) or exchange some of the current consumption of tomatoes for the consumption of future tomatoes (and with potentially a different and uncertain price). In other words, having no pressing need for all our tomatoes but needs for potatoes provide a value obtained by an exchange of some tomatoes for some potatoes.

These types of exchanges recur in many situations. Some exchanges are made with ourselves, by delaying consumption now to invest our residual wealth for future needs and consumption or exchange one product for another (barter). Some exchanges are made with individual persons or firms in which case, values are defined based on the contractual terms that define the exchange. Exchanges can be a-personal, occurring in a market where exchanges are made at a given price—a market price, a regulated price, or exchanges by changing the consumption of a product due to derived costs, etc. For example, we may decide to reduce energy consumption (and therefore drive less, reduce the lighting of our home) and thereby reduce carbon emission. When an investment is made, we reduce the current consumption by committing our residual wealth to the uncertainties of future prices of consumption and thereby to an uncertain real consumption. For example, saving for retirement when prices of consumption at that time have changed appreciably and rendering our wealth worth much less than intended. Finally, when we go to market we buy tomatoes and potatoes at the on-going market price of the day but these prices may change from day to day.

Exchanges assume thus many forms, some are market based, and therefore impersonal, responding to interacting economic forces of needs and scarcity (or excess) that supplies and demands create, some are negotiated and contracted based on needs and an ability to pay for the exchange (and are therefore a strategic exchange). Some exchanges are made with full information by all parties concerned and some are not. Some are based on the power asymmetries that each has over the other and their ability to force an exchange at a given price or the power to retaliate. In such exchanges, uncertainty prevails due to the complexity and the timing of these exchanges. Economic theories seeking to assess the implications of such exchanges are abundant. These theories commonly distinguish between value and price, embedded in Adams Smith observations and their many refinements by Walras general equilibrium model by Arrow and Debreu's to risk markets and economic thinkers of the last century. Their fundamental elements are:

- Scarcity (supply)
- Need (or consumption) and
- Exchange (markets)

In truly competitive markets, market prices are set uniquely when an equilibrium is reached, i.e. when there is "no scarcity and no excess". In other words when demands fueled by "needs" equal their supply at the market price. When markets are not truly competitive, one or the other (scarcity or need) may determine a price, based on the relationships and the relative advantages that one of the party has over the other.

For example, when there is no scarcity but a product or commodity is administered or controlled by a party (say a monopolist) the price is based on the value of the product or commodity that the administrator establishes or based on what a particular buyer would be willing to pay. In such cases, prices might not be unique and price discrimination may be prevalent. A typical example, concerns the price of rare earths in the hand of one party (China). This may lead to prices tailored to a buyer's need and ability to pay. These situations recur in many instances. In financial services, there is no scarcity to their services, yet collusion and in some cases bank rights and rules allow them to extract non-competition profits because of their controlling position in the financial system. Such situations is amplified further by Banks (and other institutions) that have become TBTF (Too Big To Fail) or have acquired a unique control over the market mechanism. In health care, an increasingly wealthier and aging population has a greater need for "health care consumption", while the quality of health care services are by their nature scarce (due to a lack of Doctors and nurses). As a result, health care prices and costs have grown inexorably. Both the water we use and the air we breathe are very much needed. However since these two are not scarce and "uncontrolled" we never pay the price we would be willing to pay had there been a scarcity in one of the two (although, there is a growing awareness that the quality of water is becoming a scarce commodity). In pricing quality, both scarcity and need are used to define its price. Prior to the industrial revolution, quality was an "art", the product of the "artisan" whose price reflected the scarcity of his art. Subsequently, quality due to its definition based on the maintenance of industrial processes became a cost—the cost on non-quality. In the last decades, quality and management gurus have suggested that the price of quality is what the buyer is willing to pay and thus became a price that a producer can extract from a consumer. Explicitly, the price of quality is not defined by the product characteristics or production process but by what the "buyer gets from consumption" and is willing to pay for. Of course, this leads to extreme cases as well. The price of a painter such as Picasso is not in its need but only in what investors, speculators or art collectors are willing to pay for (for whatever reasons they may have).

Technically, if we can exchange, say 2 pounds of our tomatoes with another consumer willing to exchange these for 3 pounds of potatoes, then, the value of tomatoes have a value in exchange to each of the parties to the exchange. The exchange price is then 2 pounds of tomatoes for 3 pounds of potatoes. Note that the exchange may differ appreciably if the first person would seek to exchange all his 5 pounds of tomatoes for potatoes (as such an exchange is specific to the quantity and the terms of the exchange). If such an exchange occurs in a market, then the rules of the economic market within which such an exchange might occur, would reduce the tomatoes to a potato exchange to be at a fixed rate, for example, 0.8 tomatoes for 1 potato! The market price of the tomato would then be 0.8 potatoes. It is also possible that another product, say 0.6 pounds of oranges, may be exchanged for one pound of potatoes. In this case, potatoes are a "numeraire" that can be used to price the market price for exchanges.

Similarly, say that the numeraire is now cash (US dollars). Some cash may be used to consume a variety of goods now (by exchanging cash for products), while the residual cash can be saved, invested and used for later consumption. This rationale underlies insurance and investment models as well as consumption pricing models. When exchanges made for a "portfolio of products", prices are defined by what is paid for the portfolio and in returns both investment returns and their risks is obtained. If the portfolio is used for specific consumption, then an additional risk is added—the price of consumption goods. To mitigate these risks, we can proceed in many different ways. For example, we could buy goods now, store them, pay the inventory cost and be assured of consumptions a period later. Value and exchange prices are not, as stated above, limited to products. Financial services, national economic priorities etc. are confronted with the same dilemma summarized by a concern for: What is value, for whom, how much, how much to pay—for one or for all (the market price). Similarly, insurance can be used to mitigate investment risks as well as secure replacement prices in case of losses. Options are used (as a price—the option price) to mitigate potential investment losses while maintaining the possibility of investment gains. The examples below highlight some of these exchanges:

- Paying now to an insurer to reduce a future risk—the price of insurance
- Buying a branded product for the utility of consumption it provides (for example, buying a luxury car rather than a much less expensive car that would provide the same service).
- Reducing oil consumption to improve the quality of the environment
- Buying a product with a warranty for both the utility it may provide and the mitigation of future consumption risks
- Etc.

In finance, trades are often tailored to construct portfolios with specific returns and risks. An insurer is paid a premium (the price of insurance risk) by an insured to assume the cost of potential adverse future events (risks) that are contracted and defined specifically and therefore to assure their predictability. In such an exchange, the insured only mitigates the risks the insurer is willing to assume while he assumes other (uncertain or un-insurable) future events that are not specified by the insurance contract. In this sense, there are certain risks that can be exchanged (or traded) while others that cannot be exchanged. Such risks are "not diversifiable" as they cannot be bought and sold and thus they are held by their owner. When we refer to the price of risk, we do not refer to the price of "non-diversifiable", or "uninsurable", or "incomplete" risks—which are various names used to express our uncertainty that defines risks with consequences that were not accounted for. A risk model can thus be used to transform values into prices only based on "a bounded rationality" justified by consumers agreeing to an exchange price. When risk models cannot account for the complexity of transactions, for the consequences of complexity and partial and asymmetric information, power asymmetries in economic and financial markets, then risk economics falters. In these cases, uncertainty and the consequences of the unlikely prevail. The many events profusely reported in the daily news refer to these events (wrongly) as risks while they are mostly events that are not accounted for.

8.2 Economic Valuation and Pricing: Supply, Demand and Scarcity

For economists, the fundamental approach to economic valuation and pricing is "need" and "scarcity", deriving from the law of supply and demand (J.B. Says). A product which is in great demand and in short supply is scarce. Its price will therefore be greater than a product which is abundant or in excess supply. Such a premise underlies the Walrasian theory for a general economic equilibrium (Walras 1874–1877). This theory states essentially that in an exchange economy, the market price of goods tends to increase as long as there is a positive excess demand, while it tends to decrease when there is a positive excess supply. The process that moves such situations to be in equilibrium is called by Walras a "tatonnement process" (or groping process). Prices are then defined by a relative good relative to which prices are assessed (which we called a "numeraire good", a price normalized to one). For example, returns are priced relative to a risk free rate (or to a price of a basket of goods or financial assets) with the spread between the two defining a risk premium price—the price the market ascribe to the risk of returns from an asset relative to that of a risk free rate.

Walras theoretical equilibrium framework was applied to risk free products, where all products are countable and prices based on their relative need and scarcity. Arrow (1951a) and Debreu (1953) important contribution to risk economics extended Walras general competitive equilibrium to risk markets, providing thereby the means to price risk (in certain circumstances). They did so by framing future uncertainty states into denumerable states, each of which with its price responding to the laws of demand and supply. For example, defining the future in terms of two events—each with its own equilibrium price, one provides a specific profit and the other incurs a loss. An equilibrium model for pricing risk assets is then reached, based on a consumers and investors rationality to profit when they can (or no arbitroge). Such a framework has led to conventionally accepted risk pricing models and to the extraordinary growth of financial markets trading in almost any conceivable financial transaction. These risk pricing models do not indicate how such equilibrium is reached however. It is based on no-arbitrage opportunities, on predictable states, on current information and therefore shifts as new information is revealed. In such models, the market price is unique and reflects uniquely future states. In this sense current prices derived from a given set of underlying state preferences imply one another. Technically, this means that future prices are implied in current ones.

There are many situations where markets are not competitive and value is not directly associated to equilibrium prices however. In fact, firms extract more value by creating a temporary advantage (i.e. a monopoly) rather than by competing. There are also products that cost but are free (for example the cost of maintaining a free access beach). Value for such products can be ascertained in several ways. On the one hand it may be defined as a measure of well-being (or utility) when it is not available (Hicks, 1939, re-published in 1979). On the other, it may be defined in term of the many direct and indirect contracted costs to produce such products and services (whether subsidized or not). Pricing these products is incomplete since not all aspects of their exchange have been accounted for. While the Arrow-Debreu framework for pricing risk assets in complete markets is at times criticized, it has currently few viable theoretical alternatives. Explicitly, theories for pricing risk in incomplete markets are to this day, still "incomplete", unable to account uniquely for observed prices. Particular cases such as the effects of inflation, the effects of wealth inequalities, debt, borrowing and lending, rare earths etc. are considered in this chapter from the viewpoint of an extended CCAPM.

8.2.1 Valuation, Risk, and Utility Pricing: One Period Models

Our underlying assumption is that consumption has a utility but it has a price which may be a market price or a negotiated price between the consumer (the buyer) and the supplier (a firm, a financial market, etc.). Let the price of consumption be π_0 and c_0 be a consumption whose utility is $u(c_0)$. A consumer maximizing his utility defines *the price the consumer is willing to pay for consumption*, or:

$$\operatorname*{Max}_{c_0} U(c_0) = u(c_0) - \pi_0 c_0 \quad \text{and} \quad \pi_0 = \pi^*(c_0) = \frac{\partial u(c_0)}{\partial c_0}, \text{ the marginal utility}$$

when the price is fixed externally (by a supplier or by an aggregate demand for consumption the consumer has no appreciable effect on), then the price $\pi^*(c_0)$ defines the consumption at that price-defining the inverse consumption-demand function. Note that one defines the other and vice versa. When there are many consumers the price is defined by consumers' aggregate demand, the supply capacity, its technology and generally the costs of aggregate supplies. Competition may then be "competitive", since one consumer decision to consume will necessarily affect the decision to consume by another consumer (through the pricing mechanism). This situation is considered in the next section.

What one pays for consumption and what "one gets from it" may differ. For example, a luxury car may cost $100,000 but its performance may be random. In this case, the net utility of consumption is $\operatorname*{Max}_{c_0} U(c_0) = Eu(\tilde{c}_1) - \pi_1 c_0$ where $\tilde{c}_1$ is a random variable whose mean is (for simplicity) c_0 (alternatively, c_0 is a consumption claim of a seller) and where $Eu(\tilde{c}_1)$ defines the expected utility of consumption while $\pi_1 c_0$ defines what the consumer pays for such consumption. Note that for a risk averse consumer, both the expected consumption and its "risk" are embedded in the probability distribution of $\tilde{c}_1$,$f(\tilde{c}_1|c_0, \sigma_0^2)$. The price the consumer is willing to pay is then a function of both its price and the consumption risk. As a result, for a consumer whose information is $f(\tilde{c}_1|c_0, \sigma_0^2)$, the price he will be willing to pay is:

$$\pi_1 = \int \left(\frac{\partial u(\tilde{c}_1)}{\partial \tilde{c}_1} f(\tilde{c}_1|c_0) + u(\tilde{c}_1) \frac{\partial f(\tilde{c}_1|c_0)}{\partial \tilde{c}_1} \right) \frac{\partial \tilde{c}_1}{\partial c_0} d\tilde{c}_1$$

For example, let the utility be $u(\tilde{c}_1) = (\tilde{c}_1)^{1-\gamma}/(1-\gamma)$ and $\partial u(\tilde{c}_1)/\partial \tilde{c}_1 = \tilde{c}_1{}^{-\gamma}$ whose probability distribution is $f(\tilde{c}_1|c_0)$. Thus,

$$\pi_1 = \int \left(f(\tilde{c}_1|c_0) + \frac{\tilde{c}_1}{1-\gamma} \frac{\partial f(\tilde{c}_1|c_0)}{\partial \tilde{c}_1} \right) \tilde{c}_1{}^{-\gamma} \frac{\partial \tilde{c}_1}{\partial c_0} d\tilde{c}_1$$

Assume that $f(\tilde{c}_1|c_0)$ has an exponential probability distribution $f(\tilde{c}_1|c_0) = \frac{1}{c_0} e^{-\tilde{c}_1/c_0}$ whose mean is c_0, then, $\frac{\partial f(\tilde{c}_1|c_0)}{\partial c_1} = -\frac{1}{(c_0)^2} e^{-\tilde{c}_1/c_0}$. Thus, $\pi_1 = \frac{1}{c_0} \int \left(1 - \frac{\tilde{c}_1}{1-\gamma} \frac{1}{c_0}\right) \tilde{c}_1{}^{-\gamma} e^{-\tilde{c}_1/c_0} \frac{\partial \tilde{c}_1}{\partial c_0} d\tilde{c}_1$. For a given value of $\bar{c}_1$ we have $f(\bar{c}_1|c_0) - e^{-\bar{c}_1/c_0}/c_0 = 0$ with $f(\bar{c}_1|c_0)$ a constant. Thus, by implicit differentiation $d\bar{c}_1/dc_0 = (1 - \bar{c}_1/c_0)$ and:

$$\pi_1 c_0 = \int_0^\infty \left(\tilde{c}_1{}^{-\gamma} + \frac{\tilde{c}_1{}^{2-\gamma}}{(1-\gamma)c_0{}^2} - \left(\frac{2-\gamma}{1-\gamma}\right) \frac{\tilde{c}_1{}^{1-\gamma}}{c_0} \right) e^{-\tilde{c}_1/c_0} d\tilde{c}_1$$

Setting $z = \frac{\tilde{c}_1}{c_0}$ and solving the integral equation $\int_0^\infty z^{x-1} e^{-z} dz = \Gamma(x)$, we have instead of:

$$\pi_1 = \int_0^\infty \left(\frac{c_0{}^{-\gamma}}{(1-\gamma)} z^{2-\gamma} + c_0{}^{-\gamma} z^{-\gamma} - c_0{}^{-\gamma} \left(\frac{2-\gamma}{1-\gamma} \right) z^{1-\gamma} \right) e^{-z} dz$$

In other words, the price the consumer is willing to pay for an exponentially distributed consumption is:

$$\pi_1 = c_0{}^{-\gamma} \Gamma(1-\gamma) - c_0{}^{-\gamma} \left(\frac{2-\gamma}{1-\gamma} \right) \Gamma(2-\gamma) + \frac{c_0{}^{-\gamma}}{(1-\gamma)} \Gamma(3-\gamma)$$

Further, since $\Gamma(3-\gamma) = (3-\gamma)\Gamma(2-\gamma) = (3-\gamma)(2-\gamma)\Gamma(1-\gamma)$, it is reduced to:

$$\pi_1 = c_0{}^{-\gamma} \frac{\Gamma(1-\gamma)}{1-\gamma}$$

If π_1 is the market price, then inversely, current consumption equals $\left[\frac{\pi_1(1-\gamma)}{\Gamma(1-\gamma)} \right]^{-\frac{1}{\gamma}}$. This can be compared to a certain (mean) consumption whose price is $\frac{\partial u(c_0)}{\partial c_0} = \pi_0$ or $c_0^{-\gamma} = \pi_0$ or $c_0 = \pi_0{}^{-\frac{1}{\gamma}}$. In other words,

$$\frac{\pi_1}{\pi_0} = \frac{\Gamma(1-\gamma)}{1-\gamma} < 1 \quad \text{and}$$

$$\pi_0 - \pi_1 = c_0{}^{-\gamma} \left(1 - \frac{\Gamma(1-\gamma)}{1-\gamma} \right) > 0, \ 1-\gamma > \Gamma(1-\gamma)$$

The willingness to pay for an expected consumption (with an exponential probability distribution) rather than a certain consumption that equals its mean is therefore as expected less when γ an indicator of the consumer risk aversion if positive. The price spread may thus be defined as a consumption risk.

When c_0 is a consumption claim by a seller and $\tilde{c}_1(c_0)$ is the actual consumption expressing a consumption reliability of the claim (thus, the more reliable the claim—$\tilde{c}_1(c_0)$ the more a consumer may be willing to pay for). Such a model can of course be extended in a number of manners. Consumers may consume more than one product as well as be inter-temporal, consuming and saving (and investing) at a given time for future consumptions. The price defined here has not accounted for willingness to supply by a seller at a given the demand price nor have we considered demands from all other consumers that as a whole constitute a market for the product that suppliers respond to in setting the product price. These situations are considered below.

8.2.2 *Aggregate and Competing Consumption and Pricing Risks*

For simplicity, consider a single supplier and let the supplier profit be a function of aggregate demand for consumption $c_D = \sum_{j=1}^{n} c_0^j$ and thus when the supply equals the demand, $c_S = c_D$ while the supplier profit is $\pi_S c_S - \Phi(c_S)$ with $\Phi(c_S)$ the aggregate supply cost. Profit maximization yields: $\pi_S^*(c_S) = \partial\Phi(c_S)/\partial c_S$ and generally, $\pi_S^*(c_S, c_S)$ where $c_S = \min(c_S, \bar{c}_S)$ and $\bar{c}_S$ is the aggregate supply capacity. The price thus results from aggregate demand and supply economic factors. Note that c_0^j is the demand for consumption of a particular consumer whose net utility is:

$$\underset{c_0^j}{\text{Max}}\, U(c_0^j) = u_j(c_0^j) - \pi_S^*(c_D, c_S)c_0^j, \quad c_D = \sum_{j=1}^{n} c_0^j,\; j = 1, 2, \ldots, n$$

Such a problem defines a "game" of n consumers and a single (or many, competing or not) supplier, all of whom (consumers and suppliers) may be differently endowed in resources and technologies and cost structures. In this model, a price sensitive consumer or (or both) a supplier, depends on what is consumed (or supplied) and what others do. This problem in its general setting is difficult to solve and therefore we shall consider some simple problems. Subsequently, two periods models will be used to obtain on the basis of such models an extended and endogenous CCAPM pricing formula.

For simplicity, say that there are only two competing consumers and one supplier, and say that the supplier sets its price only once it receives consumers' orders. Such a situation arises for example when there is a sole producer. The formulation of such a problem is thus:

$$\underset{c_0^1}{\text{Max}}\, U_1 = u_1(c_0^1) - \pi_S^*(c_0^1 + c_0^2)c_0^1 \text{ and } \underset{c_0^2}{\text{Max}}\, U_2 = u_2(c_0^2) - \pi_S^*(c_0^1 + c_0^2)c_0^2 \text{ with}$$

$$\underset{C_0 = c_0^1 + c_0^2}{\text{Max}}\, V = \pi_S(C_0) - \Phi(C_0) \text{ Subject to : } C_0 = \overline{C}_0^1 + \overline{C}_0^2 \le K$$

where K is a supply capacity. In this case, risk arises at different levels. Consumers have no information of what each will consume and therefore the price of consumption will be random, based on both the aggregate demand and the aggregate supply capacity. Second, a demand for consumption may be met only partially due to supply constraints. Third, the producer may discriminate between consumers, charging more for consumption for the one and less for the other (such cases occur in practice and are not considered here). Uncertainties also arise due to supply channels based on various intermediaries (such as stealth supply of rare earths, financial transactions across national boundaries, etc.), raising both the complexity and the cost-price of consumption. These elements introduce strategic risks for both consumers and suppliers. Elements of these risks are considered subsequently in this and in Chaps. 9, 11, and 12 using game theory.

Example Assume for simplicity that capacity is unlimited, and consider two competing consumers, then:

$$\underset{c_0^1}{\text{Max}}\, U_1 = u_1(c_0^1) - \pi_S^*(c_0^1 + c_0^2)c_0^1 \ \text{ and } \underset{c_0^2}{\text{Max}}\, U_2 = u_2(c_0^2) - \pi_S^*(c_0^1 + c_0^2)c_0^2$$

and therefore,

$$\frac{\partial u_1(c_0^1)}{\partial c_0^1} - \pi_S^*(C_0) - \frac{\partial \pi_S^*}{\partial C_0} c_0^1 = 0 \ \text{ and } \ \frac{\partial u_2(c_0^2)}{\partial c_0^2} - \pi_S^*(C_0) - \frac{\partial \pi_S^*}{\partial C_0} c_0^2 = 0 \text{ with}$$
$$C_0 = c_0^1 + c_0^2$$

and thus the price differential:

$$\frac{\partial \pi_S^*}{\partial C_0} = \left(c_0^1 - c_0^2\right)^{-1} \left(\frac{\partial u_1(c_0^1)}{\partial c_0^1} - \frac{\partial u_2(c_0^2)}{\partial c_0^2}\right) = \frac{\pi_1^*(c_0^{*1}) - \pi_2^*(c_0^{*2})}{c_0^1 - c_0^2}$$

where $\pi_k^*(c_0^k)$ is the price the kth consumer is willing to pay assuming that prices are not a function of aggregate consumption (in which case, their consumption is c_0^{*k}). Of course, if $\pi_S^*(C_0)$ is the marginal cost of a sole supplier, that is: $\pi_S^*(C_0) = \frac{\partial \Phi(C_0)}{\partial C_0}$ then $\frac{\partial^2 \Phi(C_0)}{\partial {C_0}^2} = \left(c_0^1 - c_0^2\right)^{-1} \left(\frac{\partial u_1(c_0^1)}{\partial c_0^1} - \frac{\partial u_2(c_0^2)}{\partial c_0^2}\right)$ or:

$$\frac{\partial u_1(c_0^1)}{\partial c_0^1} = \frac{\partial \Phi(C_0)}{\partial C_0} + \frac{\partial^2 \Phi(C_0)}{\partial {C_0}^2} c_0^1$$
$$\frac{\partial u_2(c_0^2)}{\partial c_0^2} = \frac{\partial \Phi(C_0)}{\partial C_0} + \frac{\partial^2 \Phi(C_0)}{\partial {C_0}^2} c_0^2$$

If production costs are convex, $\frac{\partial^2 \Phi(C_0)}{\partial {C_0}^2} < 0$ then prices will be smaller and vice versa if productions costs are concave. A consumer consumption is therefore a function of the aggregate demand.

8.2.3 *Two Products and Derived Consumption*

Consider a "rich" consumer (who has no budget constraint) of say 2 products whose net utility is $u(c_1, c_2) - (\pi_1 c_1 + \pi_2 c_2)$ where (π_1, π_2) are the respective prices the consumer is willing to pay for each product. We have then:

$$\underset{c_1, c_2}{\text{Max}}\, U(c_1, c_2) = u(c_1, c_2) - (\pi_1 c_1 + \pi_2 c_2) \quad \text{and}$$
$$\frac{\partial u(c_1, c_2)}{\partial c_1} = \pi_1 \ \text{ and } \ \frac{\partial u(c_1, c_2)}{\partial c_2} = \pi_2$$

The simultaneous inverse demand function defines the willingness to pay for each product, $\pi_i(c_1, c_2),\ i = 1, 2$, (whether they are complementary or substitutes). The net utility yields then the willingness to pay for energy and pollution is:

$$\frac{\partial U(c_1, c_2)}{\partial c_1} = c_1 \frac{\partial^2 u(c_1, c_2)}{\partial {c_1}^2} + c_2 \frac{\partial^2 u(c_1, c_2)}{\partial c_2 \partial c_1} \quad \text{and}$$
$$\frac{\partial U(c_1, c_2)}{\partial c_2} = c_2 \frac{\partial^2 u(c_1, c_2)}{\partial {c_2}^2} + c_1 \frac{\partial^2 u(c_1, c_2)}{\partial c_2 \partial c_1}$$

By implicit differentiation,

$$\frac{dc_2}{dc_1} = -\frac{\partial U(c_1, c_2)/\partial c_1}{\partial U(c_1, c_2)/\partial c_2} = -\frac{\partial^2 u(c_1, c_2)/\partial {c_1}^2}{\partial^2 u(c_1, c_2)/\partial {c_2}^2}$$

which is the rate of substitution between these products.

Say that the first product is a risk free asset while the second is a risky one, then consumption is the number of shares bought of the risk free asset whose price is π_1 while consumption is the number of the risk asset whose price is π_2. Of course, the utility is then how much would a budget unconstrained investor, whose net utility for both asset is $u(c_1, c_2) - (\pi_1 c_1 + \pi_2 c_2)$. In this particular case, we will note that $(\pi_1 c_1 + \pi_2 c_2)$ is the price the investor is willing to pay for a portfolio (c_1, c_2). Say that $(c_1, \tilde{c}_2)$ denote the rates of return of the risk free and the risk asset, while is the utility of future returns with $u(c_1(1+R_f), c_2(1+\tilde{R}))$ where $\tilde{R}$ is the risk asset rate of return. The investor problem is then stated as follows:

$$\operatorname*{Max}_{c_1, c_2} U(c_1, c_2) = Eu\big(c_1(1+R_f), c_2(1+\tilde{R})\big) - (\pi_1 c_1 + \pi_2 c_2)$$

whose solution is:

$$\frac{\partial Eu(.)}{\partial c_1}(1+R_f) = \pi_1 \ \text{ and } \mathrm{E}\left(\frac{\partial u(.)}{\partial c_2}(1+\tilde{R})\right) = \pi_2$$

and thereby,

$$\frac{\partial Eu(.)}{\partial c_1}(1+R_f) = \ \text{ and } \frac{\pi_2}{\pi_1} = \frac{1}{1+R_f}\mathrm{E}\left(\frac{\partial u(.)/\partial c_2}{\partial u(.)/\partial c_1}(1+\tilde{R})\right)$$

Thus given the utility of a consumer, the quantities of each asset he will buy is as stated above. Inversely, given the prices a consumer is willing to pay for a risk free and a risk asset, the utility of the consumer is implied by this information. In some situations, aggregate demand for one or the other asset affects the price of that asset.

Namely, let (C_1, C_2) be their aggregate demands. Then the problem we confronted above is reduced to

$$\underset{c_1,c_2}{\operatorname{Max}} U(c_1,c_2) = Eu\big(c_1(1+R_f), c_2(1+\tilde{R})\big) - (\pi_1(C_1)c_1 + \pi_2(C_2)c_2)$$

While this formulation is not a standard approach in finance, it is used here to establish an approach to reconcile the mismatch of micro-financial decisions of an in investor (or consumer) with its macro-financial demands and supplies that contribute to defining the price that each investor will pay for one or the other assets.

For a (poorer) budget constrained consumer, we have instead the following problem:

$$\underset{c_1,c_2}{\operatorname{Max}} U(c_1,c_2) = u(c_1,c_2) \;\; \text{Subject to} : B = \pi_1 c_1 + \pi_2 c_2$$

which can be solved by the usual Lagrange optimization:

$$U(c_1,c_2) = u(c_1,c_2) + \lambda(B - \pi_1 c_1 - \pi_2 c_2) \;\; \text{and}$$
$$\frac{\partial U(c_1,c_2)}{\partial c_i} = \frac{\partial u(c_1,c_2)}{\partial c_i} - \lambda \pi_i = 0, \;\; i = 1,2$$

where $\frac{\partial U(c_1,c_2)}{\partial B} = \lambda = \frac{1}{\pi_1}\frac{\partial u(c_1,c_2)}{\partial c_1} = \frac{1}{\pi_2}\frac{\partial u(c_1,c_2)}{\partial c_2}$ denote the marginal cost of a consumption budget constraint. Since in this case, prices are a function of consumption, we have $\pi_1^*(c_1,c_2)$ and$\pi_2^*(c_1,c_2)$ as well as $\pi_1^*(c_1,c_2)\frac{\partial u(c_1,c_2)}{\partial c_2} - \pi_2^*(c_1,c_2)\frac{\partial u(c_1,c_2)}{\partial c_1}$ $= 0$ and by implicit differentiation (which differs substantially from the consumption substitution of the budget unconstrained consumer and the poor—budget constrained consumer):

$$\frac{dc_2}{dc_1} = -\frac{\dfrac{\partial \pi_1^*}{\partial c_1}\dfrac{\partial u}{\partial c_2} + \pi_1^* \dfrac{\partial^2 u}{\partial c_2 \partial c_1} - \dfrac{\partial \pi_2^*}{\partial c_1}\dfrac{\partial u}{\partial c_1} - \pi_2^* \dfrac{\partial^2 u}{\partial c_1^2}}{\dfrac{\partial \pi_1^*}{\partial c_2}\dfrac{\partial u}{\partial c_2} + \pi_1^* \dfrac{\partial^2 u}{\partial c_2^2} - \pi_2^* \dfrac{\partial^2 u}{\partial c_1 \partial c_2} - \dfrac{\partial \pi_1^*}{\partial c_2}\dfrac{\partial u}{\partial c_1}} \;\; \text{with} \;\; \pi_1^* \frac{\partial u}{\partial c_2} = \pi_2^* \frac{\partial u}{\partial c_1}$$

For example, let $u(.) = (c_1)^{1-\alpha}(c_2)^{1-\beta}/[(1-\alpha)(1-\beta)]$, then $\pi_1^* \frac{\partial u}{\partial c_2} = \pi_2^* \frac{\partial u}{\partial c_1} \rightarrow \frac{\pi_1^*}{\pi_2^*} = \frac{(1-\alpha)}{(1-\beta)}\frac{c_2}{c_1}$ and therefore,

$$\left(\frac{dc_2}{dc_1}\right)_P = -\frac{\dfrac{1}{c_1} + \dfrac{\partial\big(\ln(\pi_1^*/\pi_2^*)\big)}{\partial c_1}}{\dfrac{1}{c_2} + \dfrac{\partial\big(\ln(\pi_1^*/\pi_2^*)\big)}{\partial c_2}}$$

Since $\pi_1^*/\pi_2^* = \frac{\partial u/\partial c_1}{\partial u/\partial c_2} = \frac{c_2}{c_1}$, $\ln(\pi_1^*/\pi_2^*) = \ln c_2 - \ln c_1$, we have $\left(\frac{dc_2}{dc_1}\right)_P = -\frac{c_2}{c_1}$ which is not the same result for the "rich consumer" since:

$$\left(\frac{dc_2}{dc_1}\right)_R = -\frac{\partial^2 u(c_1, c_2)/\partial {c_1}^2}{\partial^2 u(c_1, c_2)/\partial {c_2}^2} = -\left(\frac{c_2}{c_1}\right)^2 \frac{\alpha(1-\alpha)}{\beta(1-\beta)} \text{In this case,}$$

Note that consumption substitution arises then if $\left(\frac{dc_2}{dc_1}\right)_P < \left(\frac{dc_2}{dc_1}\right)_R$ if $\frac{c_2}{c_1} < \frac{\beta(1-\beta)}{\alpha(1-\alpha)}$. What one buys and what one consumes, thus may differ for wealth constrained and unconstrained consumers.

Consider next the case with two products mean consumption (c_1^0, c_2^0) price on by their mean at (π_1, π_2) respectively and let $(\tilde{c}_1, \tilde{c}_2)$ be their real and uncertain consumptions (for example buying into a portfolio with a promised rate of return of 12 % when it may in fact be something else, buying a car with a life claim of 10 years when in fact it is something else, etc.). The utility of the two based on their claims is then:

$$\underset{c_1^0, c_2^0}{\text{Max}}\, U(c_1, c_2) = Eu(\tilde{c}_1, \tilde{c}_2 | c_1^0, c_2^0) - (\pi_1 c_1^0 + \pi_2 c_2^0)$$

We further let, $f(\tilde{c}_1, \tilde{c}_2 | c_1^0, c_2^0)$ be the joint probability distribution of the actual consumption conditional on their claims. In this case, the problem we have is how much are we willing to pay for mean claims?

$$\underset{c_1^0, c_2^0}{\text{Max}}\, U(c_1, c_2) = \iint u(\tilde{c}_1, \tilde{c}_2 | c_1^0, c_2^0) f(\tilde{c}_1, \tilde{c}_2 | c_1^0, c_2^0) d\tilde{c}_1 d\tilde{c}_2 - (\pi_1 c_1^0 + \pi_2 c_2^0)$$

A solution of this problem yields the following necessary conditions for optimality for the willingness to pay for each consumption claim:

$$\pi_1 = E\left(\frac{\partial u(\tilde{c}_1, \tilde{c}_2 | c_1^0, c_2^0)}{\partial c_1^0}\right) + \iint u(\tilde{c}_1, \tilde{c}_2 | c_1^0, c_2^0) \frac{\partial f(\tilde{c}_1, \tilde{c}_2 | c_1^0, c_2^0)}{\partial c_1^0} d\tilde{c}_1 d\tilde{c}_2$$

$$\pi_2 = E\left(\frac{\partial u(\tilde{c}_1, \tilde{c}_2 | c_1^0, c_2^0)}{\partial c_2^0}\right) + \iint u(\tilde{c}_1, \tilde{c}_2 | c_1^0, c_2^0) \frac{\partial f(\tilde{c}_1, \tilde{c}_2 | c_1^0, c_2^0)}{\partial c_2^0} d\tilde{c}_1 d\tilde{c}_2$$

which we can rewrite for convenience by:

$$\pi_1 = E\left(\frac{\partial u}{\partial c_1^0}\left(1 + \frac{u(.)}{\partial u/\partial c_1^0} \frac{\partial \ln f}{\partial c_1^0}\right)\right) \text{ and } \pi_2 = E\left(\frac{\partial u}{\partial c_2^0}\left(1 + \frac{u(.)}{\partial u/\partial c_1^0} \frac{\partial \ln f}{\partial c_2^0}\right)\right)$$

or

$$\frac{\pi_1}{\pi_2} = E\left(\frac{\partial u/\partial c_1^0}{\partial u/\partial c_2^0} \frac{\left(1 + \frac{u(.)}{\partial u/\partial c_1^0}\frac{\partial \ln f}{\partial c_1^0}\right)}{\left(1 + \frac{u(.)}{\partial u/\partial c_1^0}\frac{\partial \ln f}{\partial c_2^0}\right)}\right)$$

A solution of these two equations lead to claims and the willingness to pay for such claims given by: $c_1^0(\pi_1, \pi_2)$ and $c_2^0(\pi_1, \pi_2)$. In this case, note that if a claim is certain, then $\pi_1 = \frac{\partial u}{\partial c_1^0}$ and $\pi_2 = \frac{\partial u}{\partial c_2^0}$. When claims are not certain, then the willingness to pay will necessarily be smaller. In other words, assuming $\frac{\partial \ln f}{\partial c_i^0} < 0$ then the willingness to pay is necessarily smaller. Consider for example, a bounded consumption between 0 and 1 and let a bivariate Beta probability distribution be given by (see Chap. 4):

$$f(c_1, c_2) = \frac{{c_1}^{a-1}{c_2}^{b-1}(1-c_1)^{b+c-1}(1-c_2)^{a+c-1}}{B(a,b,c)(1-c_1c_2)^{a+b+c}}$$

where the first moments of this distribution are the claimed consumptions while their variance and covariance are an expression of the claims imprecision and their dependence:

$$E(c_1) = c_{0,1} = \frac{a}{a+c}, \quad E(y) = c_{0,2} = \frac{b}{b+c};$$
$$\text{var}(c_1) = \frac{ac}{(a+c)^2(a+c+1)}, \quad \text{var}(c_2) = \frac{bc}{(b+c)^2(b+c+1)}$$

while the cross moments is:

$$E(c_1c_2) = {}_3F_2(a+1, b+1, s, s+1, s+1 : 1) \text{ where}$$

$$s = (a+b+c), \quad d = \frac{B(s,1)B(s,1)}{B(a,1)B(b,1)\Gamma(s)} = B(a,b,c)$$

In this case,

$$\ln f(c_1, c_2) = (a-1)\ln c_1 + (b+c-1)\ln(1-c_1) + (b-1)\ln c_2 + (a+c-1)$$
$$\ln(1-c_2) - \ln B(a,b,c) - (a+b+c)\ln(1-c_1c_2)$$

$$\frac{\partial \ln f(c_1,c_2)}{\partial c_1} = -\frac{1-a}{c_1} - \frac{1-b-c}{1-c_1} + \frac{(a+b+c)c_2}{1-c_1c_2}, a<1, b<1$$
$$\frac{\partial \ln f(c_1,c_2)}{\partial c_2} = \frac{1-b}{c_2} + \frac{1-a-c}{1-c_2} + \frac{(a+b+c)c_1}{1-c_1c_2}, a<1, b<1$$

and therefore, the willingness to pay for the first product will be smaller if:

$$\frac{\partial \ln f}{\partial c_1} < 0 \text{ if } ((1-a)(1-c_1) + (1-b-c)c_1)(1-c_1c_2) > (a+b+c)c_2c_1(1-c_1)$$

To regulate consumption claims by producers extracting a high price for false claims and false advertising for example, it is necessary to render the terms $\frac{u(.)}{\partial u/\partial c_i^0}$ $\frac{\partial \ln f}{\partial c_i^0}$ as small as possible or assuring that in expectation we have: $\pi_1 = E\left(\frac{\partial u}{\partial c_1^0}\right)$ and $\pi_2 = E\left(\frac{\partial u}{\partial c_2^0}\right)$. Namely, we require that:

$$E\left(\frac{\partial u}{\partial c_1^0}\left(\frac{u(.)}{\partial u/\partial c_1^0}\frac{\partial \ln f}{\partial c_1^0}\right)\right) \approx E\left(\frac{\partial u}{\partial c_2^0}\left(\frac{u(.)}{\partial u/\partial c_1^0}\frac{\partial \ln f}{\partial c_2^0}\right)\right) \approx 0$$

In other words, consumption and consumption claims that are (necessarily) false due to the random factors that affect consumption, may be accounted for in the price of products.

8.3 The CAPM and the CCAPM

John Maynard Keynes fundamental analysis of national product, investment and consumption has provided a theoretical foundation to the valuation of consumption in terms of income. This led in 1970 to development of a CCAPM (Consumption Based Capital Asset Pricing Model), extended importantly in the 1990s to the approach coined Kernel Pricing at a Stochastic Discount Factor (SDF) (Cochrane 1989, 1991, 2000, 2005; Fama 1996; Campbell 2000, 2003; Campbell and Cochrane 1999). In this approach, consumption now is partly an exchange for a future consumption, providing both prices of consumption as well as a rationale for the valuation of risky prospects as we shall see below. It differs from the Walras-Arrow-Debreu (AD) framework as discussed previously in Chap. 7. The AD Financial framework has in principle built a risk model embedded in countable and accountable future states, all of which have a specific equilibrium price, resulting from traders wants and exchanges. Current information (the conditional filtration) may however be both insufficient (and thus incomplete) to assess these equilibrium prices. Further, the AD framework neither indicates how such equilibrium is reached and whether it may be something else than static. Similarly, financial theory does not account for heterogeneities in everything such as in the distribution of intelligence, of information etc. that lead necessarily to arbitrage. For example, the concentration of intelligence and information for the pursuit of profits, have become both regulatory and ethical issues. The strategic pursuit of funds and banks to expand profits, based on insider information, on clients preferences, their position in a global network, contribute necessarily to both inequities and inequalities between participants in financial markets. In this sense, information, intelligence and power asymmetry are not properly integrated into financial theory, except as a risk consequence to the efficiency of financial markets. Nor is the process to reach a stable equilibrium with no-arbitrage clearly assessed. In an age of globalization, and a multi-polar financial world, competing and politically motivated (by values, power and geography), these are important issues that underlie future concerns.

The CCAPM framework, by contrast, is a one sided approach providing a price a consumer is willing to pay whose preferences are defined by a utility function (and therefore risk specific) for consumption now and later. It provides a temporal substitution for a consumption between now and later. This framework provides in this case an estimate of prices implictily based on current prices, or on the basis of specific preferences. Below we shall both develop and extend the CCAPM approach to pricing value to account for factors other than just future returns. We consider the standard CCAPM model and subsequently, develop an extended CCAPM model that accountss for inflation in consumption prices as well as the wealth of the consumer, debt and other risk (competitive) models.

8.3.1 The CCAPM Model

Let there be consumer, say k, whose wealth is W_0^k and whose consumption at time $t = 0$ is c_0^k whose current price is π_0. The consumer's utility for current consumption is $u_0^k(c_0^k)$. Let $W_0^k - c_0^k$ be the residual wealth, invested at a rate of return $\tilde{R}_1^k$ resulting in a random consumption $\tilde{c}_1^k$ a time 1, where $\tilde{c}_1^k = \left(W_0^k - c_0^k\right)\left(1 + \tilde{R}_1^k\right)$ whose expected utility is $Eu_1^k(\tilde{c}_1^k)$. In this case, the consumer's current consumption, optimized over two periods is found by maximizing:

$$\operatorname*{Max}_{c_0^k} U_0^k = u_0^k(c_0^k) + \beta E\left(u_1^k(\tilde{c}_1^k)\right) \text{ Subject to} : 0 \leq \tilde{c}_1^k = \left(W_0^k - \pi_0 c_0^k\right)\left(1 + \tilde{R}_1^k\right)$$

Optimization of this risk pricing model is straightforward and yields:

$$\frac{\partial u_0^k(c_0^k)}{\partial c_0^k} = \beta E\left(\frac{\partial u_1^k(\tilde{c}_1^k)}{\partial \tilde{c}_1^k}\left(1 + \tilde{R}_1^k\right)\right)$$

or, in the CCAPM formulation:

$$1 = E\left(\tilde{M}^k\left(1 + \tilde{R}_1^k\right)\right), \quad \tilde{M}^k = \beta \frac{\partial u_1^k(\tilde{c}_1^k)/\partial \tilde{c}_1^k}{\partial u_0^k(c_0^k)/\partial c_0^k}$$

where $\tilde{M}^k$ is called the pricing kernel (see Cochrane 1989, 1991, 2000, 2005, 2007). In this model, the valuation-price of a prospective return $\left(1 + \tilde{R}_1^k\right)$ a period later is equal to $1 for a consumer k whose marginal utilities of consumption are embedded in the pricing kernel. For example, for a deterministic prospective return, given by a risk free rate R_1^f, we have $1 = E\left(\tilde{M}^k\left(1+R_1^f\right)\right)$ and therefore $\left(1+R_1^f\right)^{-1} = E\left(\tilde{M}^k\right)$ and thereby:

$$1 = \frac{1}{\left(1+R_1^f\right)} E\left(\frac{\tilde{M}^k}{E\left(\tilde{M}^k\right)}\left(1+\tilde{R}_1^k\right)\right) = \frac{1}{\left(1+R_1^f\right)} E^{\tilde{M}^k}\left(1+\tilde{R}_1^k\right)$$

where $E^{\tilde{M}^k}$ is an expectation taken with respect to a probability measure implied by the consumer's preferences. Note the similarity to the implied risk neutral pricing in the Arrow Debreu Framework.

Example: The Capital Assets Pricing Model (CAPM) The CAPM risk model consists essentially in measuring the rate of return of a risk prospect with respect to a risk free rate and a risk index. It can also be derived from the CCAPM model by letting the kernel be a linear function of the index rate of return, or $\tilde{M}_1 = a + b\tilde{R}_m$. Inserting this equation in the pricing equation we have: $1 = E\{(1+\tilde{R}_1^k)\tilde{M}_1\} = E\{(1+\tilde{R}_1^k)(a+b\tilde{R}_m)\}$ where $\tilde{R}_1^k$ is the asset rate of return. Simple manipulations point out:

$$\begin{aligned} 1 &= E\{(1+\tilde{R}_1^k)(a+b\tilde{R}_m)\} \\ &= E(1+\tilde{R}_1^k)E(a+b\tilde{R}_m) + \text{cov}\{(1+\tilde{R}_1^k),(a+b\tilde{R}_m)\}. \end{aligned}$$

Since

$$\begin{aligned} E(a+b\tilde{R}_m) = E(\tilde{M}_1) = 1/(1+R_f) \text{ with } 1 &- \text{cov}\{(1+\tilde{R}_1^k),(a+b\tilde{R}_m)\} \\ &= \frac{1}{1+R_f}E(1+\tilde{R}_1^k) \end{aligned}$$

and $1 + E(\tilde{R}_1^k) = (1+R_f)\left[1-\text{cov}\{(1+\tilde{R}_1^k),(a+b\tilde{R}_m)\}\right]$. The spread is therefore:

$$\begin{aligned} E(\tilde{R}_1^k) - R_f &= -(1+R_f)\text{cov}\{(1+\tilde{R}_1^k),(a+b\tilde{R}_m)\} \\ &= \frac{\text{cov}\{(\tilde{R}_1^k - R_f),(\tilde{R}_m - R_f)\}}{\text{Var}(\tilde{R}_m - R_f)}E(\tilde{R}_m - R_f) = \beta_k E(\tilde{R}_m - R_f) \end{aligned}$$

Setting, the spreads random variables $\tilde{\pi}_k = \tilde{R}_1^k - R_f$ and $\tilde{\pi}_m = \tilde{R}_m - R_f$, we can write:

$$\beta_k = \frac{\text{cov}\{\tilde{\pi}_k, \tilde{\pi}_m\}}{\text{Var}(\tilde{\pi}_m)} \text{ and } \hat{\pi}_k = \beta_k \hat{\pi}_m$$

Note that such a formulation of the CAPM model is based on risk premiums of both the asset and the index.

The linear pricing equation $E(\tilde{R}_1^k) = R_f + \beta_k E(\tilde{R}_m - R_f)$ given above can be tested empirical by running the linear regression (where $\tilde{R}_1^k = \tilde{k}$):

$$\tilde{k} = \alpha + \beta(\tilde{R}_m - R_f) + \tilde{\varepsilon}_k,$$

where $\tilde{k}$ is the rate of return, $\tilde{R}_m$ is the index rate of return and R_f is the risk free rate. Finally, $\tilde{\varepsilon}_k$ is assumed to be a zero mean and normally distributed random with known variance while (α, β) are two parameters estimated by linear regression. Alternatively, a first moment equality yields:

$$E(\tilde{k}) = \alpha + \beta(E(\tilde{R}_m) - R_f)$$

In particular, if the security is now a fully diversified portfolio (as a market index is expected to be), then: $E\tilde{k} = E\tilde{R}_m$ and therefore, $E\tilde{R}_m = \alpha + \beta(E\tilde{R}_m - R_f) = \alpha + (E\tilde{R}_m - R_f)$. Thus, for a diversified portfolio, we require that $\alpha = R_f$ for such a solution to be maintained. In other words, the security risk premium measured by its rate of return less the risk free rate is given by: $\tilde{k} - R_f = \beta(\tilde{R}_m - R_f) + \tilde{\varepsilon}_k$. A second moment yields: $\mathrm{var}(\tilde{k}) = \beta_k^2 \mathrm{var}(\tilde{R}_M) + \mathrm{var}(\varepsilon_k) + 2\beta_k \mathrm{cov}(\tilde{R}_M, \varepsilon_k)$. However, we also have:

$$\mathrm{var}(\varepsilon_i) = \mathrm{var}(\tilde{k} - \beta_k \tilde{R}_M) = \mathrm{var}(\tilde{k}) + \beta_k{}^2 \mathrm{var}(\tilde{R}_M) - 2\beta_k \mathrm{cov}(\tilde{k}, \tilde{R}_M)$$

Inserting $\mathrm{var}(\tilde{k})$ we have:

$$\begin{aligned}\mathrm{var}(\varepsilon_k) &= \beta_k^2 \mathrm{var}(\tilde{R}_M) + \mathrm{var}(\varepsilon_k) + 2\beta_k \mathrm{cov}(\tilde{R}_M, \varepsilon_k) \\ &\quad + \beta_k{}^2 \mathrm{var}(\tilde{R}_M) - 2\beta_k \mathrm{cov}(\tilde{k}, \tilde{R}_M)\end{aligned}$$

and therefore, $0 = \mathrm{cov}(\tilde{R}_M, \varepsilon_k) + \beta_k \mathrm{var}(\tilde{R}_M) - \mathrm{cov}(\tilde{k}, \tilde{R}_M)$ or $\mathrm{cov}(\tilde{k}, \tilde{R}_M) = \mathrm{cov}(\tilde{R}_M, \varepsilon_k) + \beta_k \mathrm{var}(\tilde{R}_M)$. However, for a portfolio which is fully diversified, $\mathrm{cov}(\tilde{R}_M, \varepsilon_k) = 0$ and therefore the security's Beta is given by:

$$\beta_k = \frac{\mathrm{cov}(\tilde{R}_M, \tilde{k})}{\sigma_M^2} = \rho_{kM} \frac{\sigma_k}{\sigma_M}$$

where ρ_{kM} is the correlation factor between the security price and the portfolio index. Note that these calculations have not assumed the error term to be normally distributed.

When the residual error has skewness, an alternative approach to calculating a Beta is through calculation of the rate of returns moments. In this case, by definition of the rate of return skew:

$$
\begin{aligned}
skew_k &= \frac{E\left(\tilde{k} - E\left(\tilde{k}\right)\right)^3}{\sigma_k^3} = \frac{E\left(\alpha + \beta\left(\tilde{R}_m - R_f\right) + \tilde{\varepsilon}_k - E\left(\alpha + \beta\left(\tilde{R}_m - R_f\right) + \tilde{\varepsilon}_k\right)\right)^3}{\sigma_k^3} \\
&= \frac{\beta^3 E\left(\tilde{R}_m - \hat{R}_m\right) + E(\tilde{\varepsilon}_k)^3}{\sigma_k^3} = \frac{\beta^3 skew_m \sigma_m^3 + skew_{\varepsilon,k}\sigma_{\varepsilon,k}^3}{\sigma_k^3}
\end{aligned}
$$

As a result, we have:

$$
\beta = \left(\frac{skew_k \sigma_k^3 - skew_{\varepsilon,k}\sigma_{\varepsilon,k}^3}{skew_m \sigma_m^3}\right)^{\frac{1}{3}}
$$

However if the residual error is normal, then its skew is null and therefore, the Beta is:

$$
\beta = \left(\frac{skew_k}{skew_m}\right)^{\frac{1}{3}} \left(\frac{\sigma_k}{\sigma_m}\right)
$$

Problem Say that x and y are normally distributed. Show that if $E(f'(y))$ exists, then (Stein 1973; Wei and Lee 1988) then $\text{cov}(x, f(y)) = E(f'(y))\text{cov}(x, y)$, $f'(y) = df/dy$. Justify as well a generalization by Losq and Chateau (1982):

$$
\text{cov}(x, f(y_1, \ldots, y_n)) = \sum_{i=1}^{n} E(\partial f / \partial y_i)\text{cov}(x, y_i)
$$

provided that all expectations exist.

8.3.2 *The Beta Model and Inflation Risk*

When there is an inflation risk, with a stochastic purchasing power $\tilde{P}$ we have:

$$
E(\tilde{R}_i \tilde{P}) = R_f E(\tilde{P}) + \beta_i E(\tilde{R}_m \tilde{P} - R_f \tilde{P})
$$

As a result,

$$
\beta_i = \frac{\text{cov}(\tilde{R}_i \tilde{P} - \tilde{R}_m \tilde{P})}{\text{var}(\tilde{R}_m \tilde{P})}
$$

To calculate explicitly these two expressions (Wei and Lee 1988), we have:

$$\begin{aligned}\operatorname{cov}(\tilde{R}_i\tilde{P}-\tilde{R}_m\tilde{P}) &= E(\tilde{R}_i)E(\tilde{R}_m)\operatorname{var}(\tilde{P})+E(\tilde{R}_i)E(\tilde{P})\operatorname{cov}(\tilde{R}_m,\tilde{P})\\ &\quad+E(\tilde{R}_m)E(\tilde{P})\operatorname{cov}(\tilde{R}_i,\tilde{P})+E^2(\tilde{P})\operatorname{cov}(\tilde{R}_i,\tilde{R}_m)\\ &\quad+\operatorname{cov}(\tilde{R}_i,\tilde{P})\operatorname{cov}(\tilde{R}_m,\tilde{P})+\operatorname{cov}(\tilde{R}_i,\tilde{R}_m)\operatorname{var}(\tilde{P})\\ \operatorname{var}(\tilde{R}_m\tilde{P}) &= E^2(\tilde{R}_m)E(\tilde{P})+2E(\tilde{P})E(\tilde{R}_m)\operatorname{cov}(\tilde{R}_m,\tilde{P})+E^2(\tilde{P})\operatorname{var}(\tilde{R}_m)\\ &\quad+\operatorname{cov}^2(\tilde{R}_m,\tilde{P})+\operatorname{var}(\tilde{P})\operatorname{var}(\tilde{R}_m)\end{aligned}$$

which can be calculated explicitly using the moments of empirical time series.

8.4 The Multi-Agent CCAPM Model: A Two Periods Model

8.4.1 *The CCAPM with Independent Prices*

The CCAPM is extended next to account for both consumption price inflation, multi-agents and their wealth endowment. We consider a consumer k, whose wealth is W_0^k and whose consumption at time t=0 is c_0^k, his utility is $u_0^k(c_0^k)$ while its current price is π_0. The maximal initial consumption a consumer can afford is $W_0^k/\pi_0 = c_0^k$. In a two period model however, a consumer residual wealth $W_0^k - \pi_0 c_0^k > 0$ (his savings) is invested. A consumer maximizing the utility of consumption with prices independent of consumers' aggregate decision is then:

$$\underset{c_0^k}{\text{Max}}\, U_0^k = u_0^k(c_0^k) + \beta E\left(u_1^k(\tilde{c}_1^k)\right) \text{ Subject to} : 0 \le \tilde{c}_1^k = \frac{\left(W_0^k - \pi_0 c_0^k\right)\left(1+\tilde{R}_1^k\right)}{\tilde{\pi}_1}$$

where $\left(1+\tilde{R}_1^k\right)$ is the return on savings and $\tilde{\pi}_1^k$ is the price of consumption in the next period. The solution of this problems leads to the following condition:

$$\frac{\partial u_0^k(c_0^k)}{\partial c_0^k} = \beta E\left(\frac{\partial u_1^k(\tilde{c}_1^k)}{\partial \tilde{c}_1^k}\frac{1+\tilde{R}_1^k}{1+\tilde{\eta}_1}\right) = 0,\ \tilde{\pi}_1 = \pi_0(1+\tilde{\eta}_1)$$

where $\tilde{\eta}_1$ is the inflation rate at time 1. The CCAPM pricing model is thus:

$$1 = E\left(\tilde{M}_1^k\left(\frac{1+\tilde{R}_1^k}{1+\tilde{\eta}_1}\right)\right) = 0,\ \tilde{M}_1^k = \beta\frac{\partial u_1^k(\tilde{c}_1^k)/\partial \tilde{c}_1^k}{\partial u_0^k(c_0^k)/\partial c_0^k}.$$

Define by R_f^{ind} the real risk free rate (i.e. the rate for a risk free and indexed to inflation bond, say a TIPS). Then, $\left(1+R_f^{ind}\right)^{-1} = E\left(\tilde{M}_1^k\right)$ and a risk neutral framework for pricing future returns of returns $1+\tilde{r}_1^k$ is:

$$1 = \frac{1}{1+R_f^{ind}} E^{M_1^k}\left(1+\tilde{r}_1^k\right),\ 1+\tilde{r}_1^k = \frac{1+\tilde{R}_1^k}{1+\tilde{\eta}_1}$$

If $f^{M_1^k}\left(\tilde{r}_1^k\right)$ is the probability distribution of the consumer rate of return, then it is implied in current prices (and their derivatives).

$$1+r_0^k = 1 \Leftrightarrow E^{M_1^k}\left(1+\tilde{r}_1^k\right)$$

Example An inflation adjusted risk free rate is given by:

$$R_f^{ind} = E^M\left(\frac{\tilde{R}_t - \tilde{\eta}_t}{1+\tilde{\eta}_t}\right)$$

where expectation is taken with respect to a probability measure defined by the pricing kernel. In this formulation, expectation of the pricing kernel is equal to the risk free, inflation adjusted discount rate. For example, say that the price of a risk free rate (non adjusted to inflation is 4 % then:

$$\left(1+R_f^{ind}\right) = E^M\left(\frac{1+0.04}{1+\tilde{\eta}_t}\right) \text{ as well as } \left(\frac{1+R_f^{ind}}{1+0.04}\right) = E^M\left(\frac{1}{1+\tilde{\eta}_t}\right)$$

and therefore, if a TIP (an inflation indexed risk free rate of return) yields 2%, or $R_f^{ind} = 0.02$, then:

$$\left(\frac{1+0.02}{1+0.04}\right) = E^M\left(\frac{1}{1+\tilde{\eta}_t}\right)$$

which indicates the expected inflation rate. And vice versa, given the rates of return to forego to insure assets to inflation, provides an implied estimate of the probability distribution of $1/(1+\tilde{\eta}_t)$ the inflation rate. If there is no inflation then we have the normal form of the CCAPM model.

$$1 = \frac{1}{(1+R_f)} E\left[\frac{\tilde{M}}{E(\tilde{M})}(1+\tilde{R}_t)\right] = \frac{1}{(1+R_f)} E^M(1+\tilde{R}_t)$$

Under this probability measure, the risk premium is null since $R_f = E^M\left(\tilde{R}_t\right)$. However under a personal probabilty measure (that defines person's forecast of future and returns risks), we have a risk premium $\prod_t$:

$$\prod_t = E^P\left(\tilde{R}_t\right) - R_f \text{ or } \prod_t = E^P\left(\tilde{R}_t\right) - E^M\left(\tilde{R}_t\right)$$

As a result, the risk premium expressing a personal valuation is equal to the expected values of both the personal and the market valuation of the rate of return. An alternative representation of this risk premium can be defined in terms of the risk aversion of the person and that of the market. This is considered later on. However when a risk premium for a commodity or an item sold can be quantified (as it is the case for the price of a warranty), then it provides an information data point regarding the potential relationships between the personal probability distribution defined by the P (personal) probability measure and the market probability measure. When additional information accrues (for example through appropriate models and derivatives), estimates of these distributions relationships are improved.

The CCAPM model can be extended to a broader set of issues and accounting for the effects of essential economic variables such as global (market) demand for consumption and its supply, debt and its price, wealth and economic inequalities, investment strategies etc. Such extensions are considered next.

8.4.2 Endogenous-Aggregate Consumption and the CCAPM

When prices are exogenous (say defined by suppliers), consumers are price takers. When prices are endogenous, they are function of both aggregate consumption and supply. In such cases, the distribution of wealth, the propensity to invest, investment strategies, prevailing market rates of returns and their risks and macroeconomic factors determine consumption and supply decisions. In this context, *prices may not be unique*, each price reflecting what the consumer will be willing to pay. For example, do the rich and the poor willing to pay the same price for the same good? By the same token, bid-ask prices do not reflect the market price but the prices that prospective investors and equity owners are willing to exchange at. In this sense, the price one is willing to pay for and the market price define the premium (of the buyer or the seller). A transaction occurs when both have a positive premium—the buyer willing to buy, the seller willing to sell. Complete market (equilibrium) prices then result from the exchange between various endowed populations of buyers and sellers, potentially with different risk preferences, needs and abilities to pay. The presumption that there is a "representative" and a homogenous set of economic agents is then, necessarily misleading. If all buyers and sellers of risk, real goods etc. were all the same, there would then be no exchange. Similarly, if pricing of future returns were the same for all investors, there would be no future market. The multi-agent (endogenous) CCAPM pricing model we consider is therefore not an alternative to CCAMP but a complement, defining additional factors that underlie both potential buyers and sellers.

Unlike the exogenous (traditional) CCAPM model, the endogenous CCAPM model provides an opportunity to assess the effects of investors' endowment (wealth), supply factors, consumers propensity to save (i.e. consume less now and save more for future consumption). In addition, it reconciles micro-factors defining the consumer and macro-economic factors that define the price of consumption.

To assess the willingness to pay of a consumer we let aggregate demand be $\tilde{C}_1 = \sum_{k=1} \tilde{c}_1^k$ while the price is a function of the aggregate demand and supply, resulting in random future prices $\tilde{\pi}_1(\tilde{C}_1, \tilde{S}_1)$. For an individual consumer, who consumes c_0^k now and $\tilde{c}_1^k$ in the future, the following problem results:

$$\underset{c_0^k}{\text{Max}}\, U_k = u_1(c_0^1) + \beta E\left(u_{k1}\left(\frac{\left(W_0^k - \pi_0 c_0^k\right)\left(1+\tilde{R}_1^k\right)}{\tilde{\pi}_1(\tilde{C}_1, \tilde{S}_1)}\right)\right);$$

$$\tilde{C}_1 \tilde{\pi}_1(\tilde{C}_1, \tilde{S}_1) = \sum_{k=1} \left(W_0^k - \pi_0 c_0^k\right)\left(1+\tilde{R}_1^k\right)$$

where $\tilde{C}_1 \tilde{\pi}_1(\tilde{C}_1, \tilde{S}_1)$ is the aggregate disposable income used to pay future consumption at time 1. As a result, the future price is a function of consumers' wealth, their investment strategy tailored to provide a random rate of return$\tilde{R}_1^k$ as well their decision to save $W_0^k - \pi_0 c_0^k$ which is necessarily a function of current prices and their decision to consume. Note that in this formulation, we assumed that the price at time t $= 0, \pi_0$ is known while rate of returns are defined by the investment strategies of consumers, that are necessarily a function of their utility, wealth etc. Finally, macroeconomic factors embedding the effects of all consumers' decisions as well as macroeconomic supply factors are combined to determine the future price. These factors are necessarily dependent. For example, a growth in consumption may induce a greater consumption of materials or commodities needed to supply aggregate consumption and therefore contribute to a supply inflation price combined with the demand inflation price (or deflation, depending on the demand-supply price process).

A number of cases arise, which we summarize by the following:

1. The general case above.
2. Supplies are rationed, such that $\tilde{S}_1 = S_1 = \bar{C}_1$ and therefore the price is $\tilde{\pi}_1(\bar{C}_1)$
3. Prices are regulated to $\tilde{\pi}_1(\tilde{C}_1, \tilde{S}_1) = \bar{\pi}_1$

8.4.3 The General Case with Independent Rates of Returns

Let a consumer optimal consumption be defined by the 2 periods utility problem defined above. The utility total differential for a specific consumer k is:

$$dU_k = \frac{\partial u_k}{\partial c_0^k} dc_0^k + \beta E\left(\frac{\partial u_{k1}\left(\tilde{c}_1^k\right)}{\partial \tilde{c}_1^k} d\tilde{c}_1^k\right)$$

And at $dU_k = 0$,

$$1 = E\left(\tilde{M}_k\left(-\frac{d\tilde{c}_1^k}{dc_0^k}\right)\right), \tilde{M}_k = \beta \frac{\partial u_{k1}\left(\tilde{c}_1^k\right)/\partial \tilde{c}_1^k}{\partial u_k/\partial c_0^k}$$

Let $\Omega\left(c_0^k, \tilde{c}_1^k\right) = \left(W_0^k - \pi_0 c_0^k\right)\left(1 + \tilde{R}_1^k\right) - \tilde{c}_1^k \tilde{\pi}_1(\tilde{C}_1, \tilde{S}_1) = 0$ and assume that aggregate consumption equals aggregate supply. Then, $\tilde{\pi}_1(\tilde{C}_1, \tilde{S}_1) = \tilde{\pi}_1(\tilde{C}_1)$ and by implicit differentiation:

$$\frac{d\tilde{c}_1^k}{dc_0^k} = -\frac{\partial\Omega_k/\partial c_0^k}{\partial\Omega_k/\partial\tilde{c}_1^k} = -\frac{\pi_0\left(1+\tilde{R}_1^k\right)}{\tilde{\pi}_1(\tilde{C}_1) + \tilde{c}_1^k \partial\tilde{\pi}_1(\tilde{C}_1)/\partial\tilde{C}_1}$$
$$= -\frac{\pi_0\left(1+\tilde{R}_1^k\right)}{\tilde{\pi}_1(\tilde{C}_1)\left(1 + \tilde{c}_1^k \frac{\partial \ln \tilde{\pi}_1(\tilde{C}_1)}{\partial \tilde{C}_1}\right)}$$

Let $\tilde{\pi}_1(\tilde{C}_1) = \pi_0(1 + \tilde{\eta}_1(\tilde{C}_1))$ with the inflation rate, a function of aggregate consumption. Replacing $\tilde{c}_1^k = \left(W_0^k - \pi_0 c_0^k\right)\left(1 + \tilde{R}_1^k\right)/\tilde{\pi}_1(\tilde{C}_1))$ provides the price of consumption. The first order necessary condition provides the pricing model:

$$1 = E\left(\tilde{M}_k\left(\frac{\left(1+\tilde{R}^k_{real}\right)}{\left(1 + \left(\frac{W_0^k}{\pi_0} - c_0^k\right)\left(1+\tilde{R}^k_{real}\right)\frac{\partial \ln \tilde{\pi}_1(\tilde{C}_1)}{\partial \tilde{C}_1}\right)}\right)\right),\ 1 + \tilde{R}^k_{real} = \frac{1+\tilde{R}_1^k}{1+\tilde{\eta}_1(\tilde{C}_1)}$$

Further, since, $\ln \tilde{\pi}_1(\tilde{C}_1) = \ln \sum_{j=1}^{n}\left(W_0^j - \pi_0 c_0^j\right)\left(1 + \tilde{R}_1^j\right) - \ln \tilde{C}_1$,

$$1 = E\left(\tilde{M}_k\left(\frac{\left(1+\tilde{R}^k_{real}\right)}{\left(1 - \left(\frac{W_0^k}{\pi_0} - c_0^k\right)\left(1+\tilde{R}^k_{real}\right)\frac{1}{\tilde{C}_1}\right)}\right)\right),\ 1 + \tilde{R}^k_{real} = \frac{1+\tilde{R}_1^k}{1+\tilde{\eta}_1(\tilde{C}_1)}$$

Since $\tilde{C}_1\tilde{\pi}_1(\tilde{C}_1) = \sum_{j=1}^{n}\left(W_0^j - \pi_0 c_0^j\right)\left(1 + \tilde{R}_1^j\right)$ and $\frac{\partial \pi_1(C_1)}{\partial C_1} = -\frac{1}{C_1}$, we have instead:

$$1 = E\left(\tilde{M}_k\left(\frac{1+\tilde{R}^k_{real}}{\left(1 - \frac{\tilde{c}_1^k}{\tilde{C}_1}\right)}\right)\right) = E\left(\tilde{M}_k\left(\frac{1+\tilde{R}^k_{real}}{\left(1 - \frac{\left(W_0^k - \pi_0 c_0^k\right)\left(1+\tilde{R}^k_{real}\right)}{\sum_{j=1}^{n}\left(W_0^j - \pi_0 c_0^j\right)\left(1+\tilde{R}^j_{real}\right)}\right)}\right)\right)$$

Let for notational convenience, $p_k = \left(W_0^k - \pi_0 c_0^k\right)/\sum_{j=1}^{n}\left(W_0^j - \pi_0 c_0^j\right)$ be the proportional disposable wealth of consumer k. Thus,

$$1 = E\left(\tilde{M}_k\left(\frac{1+\tilde{R}^k_{real}}{\left(1-\frac{p_k\left(1+\tilde{R}^k_{real}\right)}{\sum_{j=1}^{n} p_j\left(1+\tilde{R}^j_{real}\right)}\right)}\right)\right)$$

$$= E\left(\tilde{M}_k\left(\left(\left(1+\tilde{R}^k_{real}\right)+\frac{p_k\left(1+\tilde{R}^k_{real}\right)^2}{\sum_{j\neq k}^{n} p_j\left(1+\tilde{R}^j_{real}\right)}\right)\right)\right), \quad k = 1,2,....,n$$

which defines the additional real future return, $\left\{p_k\left(1+\tilde{R}^k_{real}\right)^2 / \sum_{j\neq k}^{n} p_j\left(1+\tilde{R}^j_{real}\right)\right\} > 0$. If all consumers invest in a risk free asset, inflation indexed (TIPS) bond, with a rate of return $R_f^{ind} = \tilde{R}^k_{real}$, then:

$$\frac{1-p_k}{\left(1+R_f^{ind}\right)} = E\left(\tilde{M}_k\right), \quad k = 1,2,....,n \text{ and thereby,}$$

$$1 = E\left(\frac{\tilde{M}_k}{E(\tilde{M}_k)}(..)\right) = \frac{1-p_k}{1+R_f^{ind}} E^{\tilde{M}_k}\left(\left(1+\tilde{R}^k_{real}\right)+\frac{p_k\left(1+\tilde{R}^k_{real}\right)^2}{\sum_{j\neq k}^{n} p_j\left(1+\tilde{R}^j_{real}\right)}\right),$$

$k = 1,2,....,n$

In a multi-agent CCAPM, there is no unique price for returns and thus the assumption of complete markets is violated. A price is then relative to the market size, the number of agents n, their wealth (disposable income) distribution $(p_1, p_2, \ldots, p_n)$ rather than their personal wealth. Of course their wealth will be an important factor if defining the consumers' kernel price, reflecting their utility of consumption. Alternatively, consumers risk preferences may be embedded in their investment choices. For example, say that consumers invest in a portfolio consisting of a risk free and indexed bond and in a real (inflation indexed) market index. In this case, $1+\tilde{R}^k_{real} = a_k\left(1+R_f^{ind}\right)+(1-a_k)(1+\tilde{\Re}_{real})$ and therefore, the price of a market index is defined by all investors propensity to invest in a risk free inflation indexed bond or in a inflation indexed market rate of return. Such a model may then be expressed by the following model consisting of n equations in the parameters a_k.

$$1 = \frac{1-p_k}{1+R_f^{ind}} E^{\tilde{M}_k}\left(\Phi_k\left(\tilde{\Re}_{real}\middle| R_f^{ind}, a_k, p_k, k = 1,\ldots,n\right)\right), \quad k = 1,2,....,n$$

Of course, the price a consumer will be willing to pay is defined by his implied probability distribution that values future returns. Some simplifications can be made however.

- *Geometric Distribution of Disposable Wealth*

Let $p_k = A(1-\rho)\rho^k$. Since: $A = \left(\rho(1-\rho^{n+1})\right)^{-1}$ we have: $p_k = \frac{(1-\rho)\rho^{k-1}}{1-\rho^{n+1}}$ with mean: $\mu_k = \sum_{k=1}^{n} kp_k = \frac{(1-\rho)}{1-\rho^{n+1}} \frac{\partial}{\partial\rho} \sum_{k=1}^{n} \rho^k = \frac{(1-\rho)}{1-\rho^{n+1}} \frac{\partial}{\partial\rho} \left(\frac{\rho(1-\rho^n)}{1-\rho}\right) = \frac{1-(n(1-\rho)+1)\rho^n}{(1-\rho)(1-\rho^{n+1})}$ and therefore,

$$1 = \frac{1}{1+R_f^{ind}}\left(1 - \frac{(1-\rho)\rho^{k-1}}{(1-\rho^{n+1})}\right)E^{\tilde{M}_k}\left((1+\tilde{R}_{ind}^k)\left(1+\frac{\rho^k(1+\tilde{R}_{ind}^k)}{\sum_{j\neq k}^{n}\rho^j\left(1+\tilde{R}_{ind}^j\right)}\right)\right)$$

- *Aggregate Rationing*

In this case, $\tilde{C}_1 = \bar{C}_1$ and therefore the consumer optimization problem is:

$$\underset{c_0^k}{Max}\, U_k = u_1(c_0^1) + E\left(u_{k1}\left(\beta\frac{(W_0^k - \pi_0 c_0^k)(1+\tilde{R}_1^k)}{\tilde{\pi}_1(\bar{C}_1)}\right)\right);\quad \tilde{\pi}_1(\bar{C}_1)$$

$$= \frac{\sum_{k=1} (W_0^k - \pi_0 c_0^k)(1+\tilde{R}_1^k)}{\bar{C}_1}$$

Or,

$$\frac{\partial U_k}{\partial c_0^k} = \frac{\partial u_k(c_0^k)}{\partial c_0^k} + E\left(\beta\frac{\partial u_k(\tilde{c}_1^k)}{\partial c_0^k}\frac{\partial \tilde{c}_1^k}{\partial c_0^k}\right) = 0;\; c_1^k = \bar{C}_1\frac{(W_0^k - \pi_0 c_0^k)(1+\tilde{R}_1^k)}{\sum_{j=1}\left(W_0^j - \pi_0 c_0^j\right)\left(1+\tilde{R}_1^j\right)}$$

Since,

$$c_1^k \sum_{j=1}^{n}\left(W_0^j - \pi_0 c_0^j\right)\left(1+\tilde{R}_1^j\right) - \bar{C}_1(W_0^k - \pi_0 c_0^k)(1+\tilde{R}_1^k) = 0$$

By implicit differentiation,

$$\frac{d\tilde{c}_1^k}{dc_0^k} = -\bar{C}_1 \frac{\left(1-\frac{c_1^k}{\bar{C}_1}\right)\left(1+\tilde{R}_1^k\right)}{\sum_{j=1}^{n}\left(\frac{W_0^j}{\pi_0}-c_0^j\right)\left(1+\tilde{R}_1^j\right)} = -\bar{C}_1\pi_0 \frac{\left(1-\frac{p_k}{\sum_{j=1}^{n} p_j\left(\frac{1+\tilde{R}_1^j}{1+\tilde{R}_1^k}\right)}\right)}{\sum_{j=1}^{n}\left(W_0^j-\pi_0 c_0^j\right)\left(\frac{1+\tilde{R}_1^j}{1+\tilde{R}_1^k}\right)}$$

$$1 = \bar{C}_1\pi_0 E\left(\tilde{M}_k\right)\left(\frac{\left(1-p_k/\sum_{j=1}^{n} p_j\left(\frac{1+\tilde{R}_1^j}{1+\tilde{R}_1^k}\right)\right)}{\sum_{j=1}^{n}\left(W_0^j-\pi_0 c_0^j\right)\left(\frac{1+\tilde{R}_1^j}{1+\tilde{R}_1^k}\right)}\right)$$

Again, assuming that all investors invest in a risk free asset, we have:

$$E\left(\tilde{M}_k\right) = \frac{\sum_{j=1}^{n}\left(W_0^j-\pi_0 c_0^j\right)}{\bar{C}_1\pi_0(1-p_k)} \text{ and thus } 1 = \frac{1}{(1-p_k)} E^{\tilde{M}_k}\left(\frac{\sum_{j\neq k}^{n} p_j\left(\frac{1+\tilde{R}_1^j}{1+\tilde{R}_1^k}\right)}{\left\{\sum_{j=1}^{n} p_j\left(\frac{1+\tilde{R}_1^j}{1+\tilde{R}_1^k}\right)\right\}^2}\right)$$

Thus, when supply is rationed, the price a consumer is willing to pay for future returns is defined relative to the proportional wealth consumers have and the rate of returns of their investments. In this sense, the multi-agent kernel pricing model is a sharing model defined by:

$$\frac{d\tilde{c}_1^k}{dc_0^k} = -\frac{\left(1-\frac{c_1^k}{\bar{C}_1}\right)\left(1+\tilde{R}_1^k\right)}{1+\tilde{\eta}_1(\bar{C}_1)} \text{ and } 1 = \bar{C}_1\pi_0 E\left(\tilde{M}_k\right)\left(\frac{\left(1-\frac{\tilde{c}_1^k}{\bar{C}_1}\right)\left(1+\tilde{R}_1^k\right)}{1+\tilde{\eta}_1(\bar{C}_1)}\right),\ \bar{C}_1$$

$$= \sum_{j=1}^{n} \tilde{c}_1^j$$

Since $\left(1+\tilde{R}_1^k\right) = \frac{\tilde{c}_1^k\pi_0(1+\tilde{\eta}_1(\bar{C}_1))}{\left(W_0^k-\pi_0 c_0^k\right)}$, we have the sharing rule:

$$1 = \bar{C}_1\pi_0 E\left(\tilde{M}_k\right)\left(\frac{\tilde{c}_1^k\left(1-\frac{\tilde{c}_1^k}{\bar{C}_1}\right)}{\left(\frac{W_0^k}{\pi_0}-c_0^k\right)}\right) \text{ or } \frac{1}{\bar{\tilde{C}}_0^k} E\left(\tilde{M}_k\tilde{c}_1^k\left(1-\frac{\tilde{c}_1^k}{\bar{C}_1}\right)\right)$$

$$= \frac{1}{\bar{\tilde{C}}_0^j} E\left(\tilde{M}_j\tilde{c}_1^j\left(1-\frac{\tilde{c}_1^j}{\bar{C}_1}\right)\right),\ \bar{\tilde{C}}_0^i = \frac{W_0^i}{\pi_0} - c_0^i$$

Assuming that all consumers have the same marginal utility of consumption, then:

$$0 = E\left(\tilde{M}\left\{\frac{\tilde{c}_1^k}{\bar{\tilde{C}}_0^k}\left(1-\frac{\tilde{c}_1^k}{\bar{\tilde{C}}_1}\right)-\frac{\tilde{c}_1^j}{\bar{\tilde{C}}_0^j}\left(1-\frac{\tilde{c}_1^j}{\bar{\tilde{C}}_1}\right)\right\}\right)$$

For example, let the utility of consumption of consumers be of the exponential type, with $u_k(c) = -e^{-c\lambda_k}/\lambda_k,\ u_k{}'(c) = e^{-c\lambda_k}$, then a sharing rule is defined by:

$$\frac{E\left(e^{-(\tilde{c}_1^k-c_0^k)\lambda}\tilde{c}_1^k\left(1-\frac{\tilde{c}_1^k}{\bar{\tilde{C}}_1}\right)\right)}{\bar{\tilde{C}}_0^k} = \frac{E\left(e^{-(\tilde{c}_1^j-c_0^j)\lambda}\tilde{c}_1^j\left(1-\frac{\tilde{c}_1^j}{\bar{\tilde{C}}_1}\right)\right)}{\bar{\tilde{C}}_0^j},\ j = 1,2,\ldots,k-1,k+1,\ldots,n$$

which is a function of the consumers index of risk aversions λ (assumed in this case to be equal).

If there are only two consumers, then the pricing kernel will be defined by the utilities of the consumers and of course their savings for future consumption, their investment strategy and finally current prices. A discussion of some essential results will be considered in Chap. 9.

- *Prices are Regulated*

In this case, $\Omega\left(c_0^k, \tilde{c}_1^k\right) = \left(W_0^k - \pi_0 c_0^k\right)\left(1+\tilde{R}_1^k\right) - \tilde{c}_1^k\bar{\pi}_1 = 0$ and therefore,

$$\frac{d\tilde{c}_1^k}{dc_0^k} = -\frac{\pi_0\left(1+\tilde{R}_1^k\right)}{\bar{\pi}_1} \text{ and } 1 = E\left(\tilde{M}_k\left(\frac{\pi_0\left(1+\tilde{R}_1^k\right)}{\bar{\pi}_1/\pi_0}\right)\right)$$

- *A supplier pricing*

When a supplier defines the price, equating the selling price to marginal costs, we have $\tilde{\pi}_1(\tilde{S}_1) = \partial\Phi(\tilde{S}_1)/\partial\tilde{S}_1$. In an equilibrium (supply equal demand), $\tilde{C}_1 = \tilde{S}_1$ and therefore,

$$\tilde{\pi}_1(\tilde{C}_1) = \partial\Phi(\tilde{S}_1)/\partial\tilde{S}_1 \text{ while } \partial\tilde{\pi}_1(\tilde{C}_1)/\partial\tilde{C}_1 = \partial^2\Phi(\tilde{C}_1)/\partial\tilde{C}_1{}^2$$

As a result, the multi-agent pricing model is:

$$1 = E\left(\tilde{\mathrm{M}}_k\frac{\pi_0\left(1+\tilde{R}_1^k\right)}{\frac{\partial\Phi(\tilde{C}_1)}{\partial\tilde{C}_1}\left(1+\left(W_0^k-\pi_0c_0^k\right)\left(1+\tilde{R}_1^k\right)\frac{\partial}{\partial\tilde{C}_1}\left(\frac{\partial\Phi(\tilde{C}_1)}{\partial\tilde{C}_1}\right)\right)}\right)$$

Of course in a general setting, there is both a multi-agent supply competition and a multi-agent consumption competition. Prices are then set both by demand for consumption, an ability to pay for such consumption (either because of wealth and-or because of access to inexpensive debt as we shall consider in Chap. 9) and by the capacity and the costs assumed by competing agents.

- *Example:*

Let prices be independent of aggregate consumption, $\frac{\partial \tilde{\eta}_1(\tilde{C}_1)}{\partial C_1} = 0$ and $\tilde{\eta}_1(\tilde{C}_1) \equiv \tilde{\eta}_1$ and therefore,

$$1 = \frac{1}{1+R_f^{ind}} E^{\tilde{M}_k}\left(\frac{1+\tilde{R}_1^k}{1+\tilde{\eta}_1}\right) \approx e^{-R_f^{ind}} E^{\tilde{M}_k}\left(e^{\tilde{R}_1^k - \tilde{\eta}_1}\right)$$

If both the rate of return and the inflation rate are under a probability measure $\tilde{M}_k$ normally distributed, with say means $\left(\hat{R}_1^k, \hat{\eta}_1\right)$, standard deviations $\left(\sigma_1^k, \sigma_1^\eta\right)$ and a positive correlation ρ, then under this probability measure, we have $1 = e^{-R_f^{ind}} E^{\tilde{M}_k}$ $\left(e^{\tilde{R}_1^k - \tilde{\eta}_1}\right) = e^{-R_f^{ind}} e^{\hat{R}_1^k - \hat{\eta}_1 - \frac{1}{2}\left(\left(\sigma_1^k\right)^2 + \left(\sigma_1^\eta\right)^2\right) + \rho\left(\sigma_1^k \sigma_1^\eta\right)}$ and therefore,

$$R_f^{ind} = \hat{R}_1^k - \hat{\eta}_1 - \frac{1}{2}\left(\left(\sigma_1^k\right)^2 + \left(\sigma_1^\eta\right)^2\right) + \rho\left(\sigma_1^k \sigma_1^\eta\right)$$

Note that in this case, the inflation adjusted risk free rate of return increases when the rates of return and volatilities increase.

The effects of the investment policy, the initial wealth, markets rates of returns etc. can be priced as well on the basis of this equation. Further, more complex portfolios, consisting of options will provide a richer information set which can be used to calculate the implied (price) probability distribution (and thereby obtain a broader set of information we may use to calculate the implied probability distribution). In particular, introducing the investor investment strategy, we have a price for the basket risk rate of return that is a function of the TIPS rate, non-indexed bonds as well as the going inflation rate. Explicitly, we have:

$$R_f^{ind} + \hat{\eta}_1 + \frac{1}{2}\left(\sigma_1^\eta\right)^2 = R_f + (1-a_k)(\Re - R_f) - \frac{1}{2}\left(\left(\sigma_1^k\right)^2\right) + \rho\left(\sigma_1^k \sigma_1^\eta\right)$$

And therefore, we have a quadratic equation in $1 - a_k$, the consumer portfolio investment share in the market index, or:

$$R_f^{ind} + \hat{\eta}_1 + \frac{1}{2}\left(\sigma_1^\eta\right)^2 = R_f + (1-a_k)\left(\hat{\Re} - R_f + \rho\left(\sigma_\Re \sigma_1^\eta\right)\right) - \frac{1}{2}\left((1-a_k)^2 \sigma_\Re^2\right)$$

Whose solution is: $0 \leq \left(1 - a_k^*\right) = \sqrt{a} - \sqrt{a - x}$, with:

$$a = \frac{\left(\hat{\Re} - R_f + \rho\left(\sigma_{\Re}\sigma_1^{\eta}\right)\right)^2}{4}, \quad x = -\frac{1}{2}\sigma_{\Re}^2\left(R_f^{ind} - R_f + \hat{\eta}_1 + \frac{1}{2}\left(\sigma_1^{\eta}\right)^2\right)$$

Say that $\sqrt{a-x} \approx \frac{1}{\sqrt{a}}\left(a - \frac{1}{2}x\right)$. In this case,

$$0 \leq \left(1 - a_k^*\right) \approx \frac{\sigma_{\Re}^2}{2} \frac{R_f^{ind} - R_f + \hat{\eta}_1 + \frac{1}{2}\left(\sigma_1^{\eta}\right)^2}{\hat{\Re} - R_f + \rho\left(\sigma_{\Re}\sigma_1^{\eta}\right)}$$

In other words, if $\hat{\Re} + \rho\left(\sigma_{\Re}\sigma_1^{\eta}\right) > R_f$ the propensity to invest in equity will be smaller, if the bonds corporate spread $R_f - R_f^{ind} > \hat{\eta}_1 + \frac{1}{2}\left(\sigma_1^{\eta}\right)^2$ is greater than the inflation rate plus half its variance. When the bonds spread is smaller than the inflation rate and volatility, investment in equity increases.

When prices are a function of global consumption only, then the inflation rate is also a function of global demand and prices are defined by a simultaneous solution of all consumers' consumption, their wealth, investment strategy etc. The CCAPM model is then incomplete, although it may be redefined as strategic complete as the examples and cases in Chap. 9 will highlight.

References and Additional Reading

The Walras, Arrow and Debreu, have set the foundations for pricing in an equilibrium (complete) competitive market (Arrow 1951a, b, 1963a, 1982; Arrow et al. 1996; Debreu 1959) while Keynes (1935), Cochrane (2005), Campbell and Cochrane (1999) have set up the elements of the CCAPM approach based on te utility approach to decision making under uncertainty (for example, Borch 1968; Dreze and Modigliani 1972; Epstein and Zin 1989, 1991, Sharpe on the CAPM, Adler and Dumas 1980, Losq and Chateau 1982.

Applications are numerous. For example, Esscher (1932) on the Probability Function of Collective Risk which leads to a transformation to a risk neutral probability measure (a result used by actuaries for example). Gollier (2000), Gollier and Pratt (1996), provide a study of utility and risk aversion. Duffie and Garleanu (2001), on CDO, Sen (1973) on economic inequality, Ljungqvist and Sargent (2000) on a Recurive macroeconomic Theory. Miyahara (2012) book on pricing options in incomplete markets, Culp (2006) on Structure Finance in Insurance

Pricing IPO has similarly attracted both theoretical and empirical research (Alexander 1993 on Law and avoidance in IPO, Beatty and Ritter 1986; Benveniste and Spindt 1989)

The CCAPM has provided an extensive list of publications, both on the validity of such models and based on empirical research. Some essential references include Breeden (1979), Campbell (1999, 2000, 2003), Campbell and Cochrane (1999),

Cochrane (1991, 2007), Cochrane's book on Assets pricing (2005). Other papers include Fama and French (1997), on the equity premium, Mimeo, Fama (1970a, b), on a review efficient capital markets and Fama's (1970), multiperiod consumption-investment decisions model, Almeida and Garcia (2008), on the empirical likelihood estimators for stochastic discount factors and Savov (2010). On Asset Pricing with Garbage

Numerous papers have also been published by Merton including his book in 1990 on Continuous Time Finance, 1969 on Lifetime portfolio selection under uncertainty, 1973 on Merton R.C., 1990, Continuous Time Finance book, 1973 seminal paper on the theory of rational option pricing, 1977, Optimum consumption and portfolio rules in a continuous time model and many others listed on Merton's website.

Other references on the CCAPM include, Chapman (1997), Approximating the asset pricing kernel, Dittmar (2002), Nonlinear pricing kernels, kurtosis preference and evidence from the cross section equity returns, Duffie and Zame (1989), The consumption based capital asset pricing model, Duffie's books, (1988) on Security Markets and (1992), on Dynamic Asset Pricing Theory.

Environmental applications are numerous, including Granovskii et al. (2006), on environmental and economic aspects of hydrogen production and utilization in fuel cell vehicles, Greaker (2003a), on Strategic environmental policy when the governments are threatened by relocation, as well as Greaker (2003b), on Strategic environmental policy; eco-dumping or a green strategy. Brannlund et al. (2006), on the effects on consumption and emissions, Greening et al. (2000), on a survey of energy efficiency and consumption

Chapter 9
Risk Pricing Models: Applications

Abstract This chapter provides a number of examples and problems applying the pricing models set in Chaps. 7 and 8. These models are assuming essentially that risk models are complete in the sense that states are countable and accountable while kernels are defined theoretically based on economic rationalities and practically based on parametric distributions or on ad-hoc statistical principles helpful in defining appropriate probability-state prices. A number of approaches to valuing and pricing are used to highlight their differences. Applications to debt and international assets pricing are considered as a multi-agent CCAPM model.

9.1 Debt and Risk Models

Debt problems are varied and complex. They involve individual decisions to grant credit to an individual, to a firm, to trade debt, to issue bonds (by corporations, by governments etc.). In this section, we consider a number of approaches to pricing debt risk. These include both the Arrow-Debreu framework pricing debt as an option as well as extending the CCAPM model to include debt. Debt risk models are important in market risk models that seek to securitize and price debt.

The importance of debt and its price cannot be neglected. The following data highlights for example the current debt to GNP ratios of European countries which contributes to both an increased wealth inequality as well as to unsustainable debt servicing economic issues. These ratios as well as countries acting as both borrowers and lenders increases the risk of default as well as the systemic risks due to default. Note that these countries are both creditors (lenders) and debtors, which augment their dependence and thereby the risk of default and systemic risks (if all countries are networked, each depending on the other for its financial liquidity). Below, we consider both traditional approaches to debt pricing and a CCAPM model where debt and lending are included. Such a formulation will lead to a competitive multi-agent CCAPM model.

C.S. Tapiero, *Engineering Risk and Finance*, International Series in Operations Research & Management Science 188, DOI 10.1007/978-1-4614-6234-7_9,

TOTAL DEBT AND PERCENTAGE OF GDP		
	Total Debt (in bn Euros)	Percentage of GDP
Austria	206	72
Belgium	341.0	94
France	1591	86
Germany	2061	82
Greece	329	165
Ireland	148	109
Italy	1842	121
Holland	370	65
Portugal	161	106
Spain	642	67
Sweden	144	36
Switzerland	146	52
United Kingdom	1362	80

9.1.1 *Market Risk Pricing Models for Credit Risk and Collaterals*

A Default risk arises when the debt can no longer be paid or when a loan's collateral price falls below that of the loan. Typically, structural default models specify a value process with default occurring when the process crosses some explicit threshold (for example, default may be defined when the debt to equity ratio reaches a given threshold, when a firm violates a regulation constraint, or actions are taken that turn out significant to the firms price). In this sense, default is a "stopping time" or an event defined by the evolution of a representative firm process expressing both its states and its policies (investments, speculation, etc.). In such cases, lenders recover some of the debt based on the collateral. In defaulting firms for example, bond holders access first the equity remaining at default with equity holders collecting what is left after bond holders and other obligations have been paid for.

For some firms, default can be expressed by a delayed payment, by a partial loss of its capital. Interpretation and the modeling of default are important and ought to include the many consequences of default, debt recovery and their mitigation by prior agreement covering the credit given. For example in practice we observe that bondholders may recover 20–80 % of their investment and therefore default might not mean a total loss to either equity and bond holders. Default may be expressed by a firm being down-rated, leading to costlier borrowing rates and more importantly to optional rights bond holders may have when a firm's rating is changed. In many cases a firm's corporate bond prices may fall prior and in anticipation of such down rating. For these reasons pricing default models are both complex and varied. Their importance has grown further by the securitization of credit risk baskets (such as Mortgage Backed Securities, Monoline portfolios) and hybrid optional products where default is important and to be reckoned with.

Counterparty risks are due to the malevolent actions taken by parties that are either stakeholders or competitors. For example, a borrower who does not intentionally (or not) meet his commitment to pay, say a substantial debt, may lead a firm to default. In such situations, risk is imbedded in the parties' intentionality, information, and power. Such risks are difficult to account for in a market setting however, but they are nonetheless important. For example, a bank financing an enterprise may decide for some reasons to pull its financial support leading thereby to the firms' demise or at least to its default.

Each of these default risks can be and are treated by markets' models pricing differently. For example by implied default models using the "market" pricing of similar default prone assets. In such cases, options' prices are used to obtain an implied default probability distribution reflecting the traded price (or willingness of traders and investors) that implies the probability of default and an estimate of its consequences. Alternatively, we may price the debt relative to that of the collateral. Each of these approaches are considered next.

9.1.2 The Structural-Endogenous Model and the Price of Credit Relative to its Collateral

The endogenous model is based on the presumption that default is internal, expressed by the imbalance of debt and its collateral. Merton considered such an approach for a firm of stakeholders consisting of equity (the collateral) and bond holders. In case of default, bond holders have prior claim to the residual wealth of the company. We consider first a discrete time and subsequently a continuous time model. Merton interprets the price of equity as equivalent to that of a Call option. Such an observation is based on the firm's price defined in terms of two essential variables—its value V and its bond obligation B. Equity is thus defined by (since Bondholders have priority on the firm's assets):

$$E = \begin{cases} V - B & V > B \\ 0 & V < B \end{cases} \quad \text{or} \quad E = \text{Max}(V - B, 0)$$

However, from the bond holder viewpoint, we note that the price of the firm's obligation is:

$$V_B = \begin{cases} B & V > B \\ V & V < B \end{cases} \text{ or } V_B = \text{Min}(B, V) = V - E = V - \text{Max}(V - B, 0)$$

For example, say that in a binomial model with values in the next period given by (V^+, V^-). Thus, equity and bond prices are: $(E^+, E^-) = \{(V^+ - B)^+, (V^- - B)^+\}$, $(V_B^+, V_B^-) = \{\min(V^+, B), \min(V^-, B)\}$. To price corporate firm liabilities, V_B, construct a portfolio consisting of the firm's leverage and borrowing.

Let the firm leverage be N and let b be the amount borrowed at the risk free rate. Then a replication of corporate liability is given by: $V_B = NV - b$. Thus, a period later, we have:

$$V_B = NV - b \Rightarrow \begin{cases} NV^+ - b = V_B^+ = \min(V^+, B) \\ NV^- - b = V_B^- = \min(V^-, B) \end{cases}$$

A solution for the replicating portfolio parameters are:

$$N = \frac{V_B^+ - V_B^-}{V^+ - V^-}, \quad b = \frac{V^- V_B^+ - V^+ V_B^-}{(V^+ - V^-)(1 + R_f)}$$

For no arbitrage (for cash flows with the same risks, their price ought to be the same), we have:

$$V_B = NV - b = \left(\frac{V_B^+ - V_B^-}{V^+ - V^-}\right) V - \frac{V^- V_B^+ - V^+ V_B^-}{(V^+ - V^-)(1 + R_f)} \quad \text{or}$$

$$V_B = \left(\frac{V_B^+ - V_B^-}{V^+ - V^-}\right) \left\{ V - \frac{V^- \min(V^+, B) - V^+ \min(V^-, B)}{(V_B^+ - V_B^-)(1 + R_f)} \right\}$$

For example, let the debt principal be B = 100, the risk free rate be 0.08 and the firm value is 110 which may increase or decrease to 125 and 75 respectively. To price debt we construct a portfolio NV-B replicating the corporate liabilities and subsequently calculate the corporate equity as indicated above by: $E = V - V_B$. In particular, first we have:

$$V_B = \left(\frac{100 - 75}{125 - 75}\right) \left\{ 110 - \frac{75(100) - 125(75)}{(125 - 75)(1 + 0.08)} \right\} = 72.36;$$
$$E = 110 - V_B = 110 - 72.36 = 37.64$$

In continuous time, the approach is similar. Say that a firm has N shares which are traded on the stock market at the spot price $S(t)$ with a debt owned by the bondholders whose current price is $B(t, T)$. Default occurs when the firm cannot meet the face value payment at maturity T, in which case, bondholders take over the company. Let $S(T)$ be the firm price at time T. Then:

If $V(T) < D(T) = B(T, T)/N$ the bondholders collect $NS(T)$
If $V(T) > D(T)$, the bondholder collects $D(T)$ and the equity holder collects $V(T) - D(T)$

Thus the equity holder collects $(S(T) - D(T))^+$, which implies that his hold on the firm is equivalent to that of a call option with a strike of $D(T)$. In other words, such a Call option can be price by say the Black Scholes equation. By the same token, the bond holder has an option which is equal to:

$$\mathrm{Min}(S(T),\ D(T)) = D(T) - (\ D(T) - S(T))^{+}$$

And therefore its price is:

$$V(0) = e^{-R_f T} E^{*}(D(T)) - e^{-R_f T} E^{Q}(P(T)),\ \ P(T) = (\ D(T) - S(T))^{+}$$

where $P(T)$ is the price of the put.

9.1.3 Credit Risk and Swaps: A Reduced Form or Exogenous Models

We shall begin with an example of a "Default" bond using a reduced form model. Such models make no attempt to define default as an endogenous event (arising from a low level of firm value or cash flow), but rather specify default as an exogenous event and thus do not explicitly incorporate any relationship between leverage and firm value into the model. These models were expounded on by numerous financial economists such as Jarrow and Turnbull (1995), and Jarrow et al. (1997), Schonbucher (2003) and others. The model of Jarrow et al. (1997), for example, assumes that the value of a default-free zero-coupon bond, paying one dollar at maturity at time T, is known at time t. Let B(t, T) be the price of this bond (paying one dollar at maturity time T). It is given by: $B(t,T) = e^{-R_f(t,T)(T-t)}$, $B(T,T) = 1$ where $R_f(t,T)$ is the risk free rate. Now assume a default bond (for some external reasons) that might occur in the remaining time interval (t,T). Define in particular the default event as occurring at time t with a probability distribution $f_D(\tau|\Im_t)$ where $\Im_t$ denotes the filtration of all relevant information available at time t, or.

$$\mathbf{1}_{\tau>t} = \begin{cases} 1 & \text{If default occurs in (t,T)} \\ 0 & \text{If no default occurs} \end{cases}$$

Let $B_D(t,T)$ be the price of such a default bond defined by financial markets. This price is in fact the amount of money that the bondholder collects at maturity (expressing thereby the discount applied to the bond due to a potential default). Thus, since $[1 - F_D(T|\Im_t)]$ is an implied probability denoting the probability that no default occurs, then the current price is by definition given by:

$$B_D(t,T) = E(B(t,T)\mathbf{1}_{\tau>T}) = B(t,T)[1 - F_D(T|\Im_t)] = e^{-R_f(t,T)(T-t)}(1 - F_D(T|\Im_t))$$

We can rewrite this expression in terms of a hazard rate, expressing the conditional probability of default at time t, given that there was no prior default until that time. The rationality of such an approach is that it allows us to express the money effects of a default as a discount rate amplifying the time discount of time with the time discount factor of default at some future (and random) date. This is given by:

$$h_D(t) = \frac{f_D(t)}{1 - F_D(t)} = -\frac{\partial}{\partial t} \ln[1 - F_D(t)];$$

$$-\int_t^T h_D(s)ds = \ln[1 - F_D(t)] \rightarrow 1 - F_D(t) = e^{-\int_t^T h_D(s)ds}$$

If we are at time t, the probability of no default in the time interval (t,T) is written as follows:

$$1 - F_D(T|\Im_t) = e^{-\int_t^T h_D(s)ds} \quad \text{or} \quad B_D(t,T) = e^{-\left(\int_t^T (R_f(s,T))ds + \int_t^T h_D(s)ds\right)}$$

The implied discount rate applied to the default bond is thus $\left(\int_t^T (R_f(s,T))ds + \int_t^T h_D(s)ds\right)$ with $\int_t^T h_D(s)ds$ denoting the discount due to default. The rate used is thus implied in the price of the bond, providing thereby an interpretation of the implied probability of default in terms of the price of the bond. The difference between a default and a risk free bond $B_D(t,T) - B(t,T)$ is thus the risk premium for the default bond.

Of course, the hazard rate is an implied discount rate which we can calculate parametrically by assuming a model default distribution given in term of a number of parameters, estimated using complementary financial data or a-parametrically by determining indirect sources of information relating to the default probability.

This approach to pricing default is of course fundamentally different than actuarial risk pricing. In actuarial pricing, objective statistical data and probability models are used to assess and predict mortality, default claims, etc. based on data and hypotheses based on the hypotheses made regarding the underlying process. Hedging is then sought through insurance or reinsurance or any other mechanism for risk sharing. In the financial approach to pricing default we rely instead on "financial market transactions," reflecting the bets that buyers and sellers (investors, speculators, pension funds, etc.) are making on the default asset being traded. In this sense, the default price is only a reflection of the markets participants risk attitudes and financial needs (hedging, speculating etc.) which determines the forward price of derivatives with optional rights in the future. Hedging in such a context is necessarily determined by the optional portfolios that are constructed and reflecting individual preferences for profits and risk but priced by the market.

9.1.4 *Pricing by Replication: Credit Default Spread*

Spread options can be written on all types of financial products including equities, bonds and currencies. This type of position can be purchased on large exchanges, but is primarily traded in the over-the-counter market. Some types of commodity spreads enable the trader to gain exposure to the commodity's production process. This is created by purchasing a spread option based on the difference between the inputs and outputs of the process. Common examples of this type of spread are the crack, crush and spark spreads. Credit spread options (CSO) are designed to give protection in case of spread widening. As the bond market is less liquid than the CDS market, instead of buying a put option on a bond one might want to buy a call option on a CDS. One should keep in mind that for the moment CSO's are still rather exotic instruments traded over the counter only. Moreover the contracts are very rich in detail and thus we will consider one sort only.

In credit swaps, the determination of a premium is based on the replication of equivalent risks at equal price. To do so, we consider a simple example. Let $s(x)$ be a spread, a function of default at time x, which occurs after time t. This probability is $F(x|x > t)$. This the price of a bond with spread s at time t which equals the bond price at time t to time x—the default time, up to time T, the bond maturity. In this case,

$$P(s,t) = \int_t^T B(t,x)s(x)[1 - F(x|x > t)]dx$$

Now assume that a default occurs at some future date, prior to the terminal time T where δ is a proportional recovery. Then:

$$B_D(h,t) = E^Q[(1-\delta)B(t,\tau)\mathbf{1}_{\tau\leq \mathrm{T}}] = (1-\delta)\int_t^T B(t,\tau)\mathbf{1}_{\tau\leq \mathrm{T}}d\tau = (1-\delta)$$

$$\int_t^T B(t,\tau)\frac{\partial F(\tau)}{\partial \tau}d\tau = (1-\delta)\int_t^T B(t,\tau)f(\tau|t)d\tau = (1-\delta)\int_t^T B(t,\tau)h(\tau|t)[1 - F(\tau|\tau > t)]d\tau$$

Equating the two equations, we have: $P(s,t) = B_D(h,t)$ which means that:

$$\int_t^T B(t,x)s(x)[1 - F_t(x)]dx = (1-\delta)\int_t^T B(t,\tau)h(\tau|t)[1 - F(\tau|\tau > t)]d\tau$$

As a result, we obtain:

$$B(t,\tau)s(\tau)[1 - F(\tau|\tau > t)] = (1-\delta)B(t,\tau)h(\tau|t)[1 - F(\tau|\tau > t)] \quad \text{or}$$
$$s(\tau) = (1-\delta)h(\tau|t)$$

For example, if the default (hazard) rate is constant (meaning that the default probability is of the exponential type), then: $h(\tau|t) = \bar{h}$ or $\bar{s} = (1-\delta)\bar{h}$ and therefore, $F(t) = 1 - e^{-[\bar{s}/(1-\delta)]t}$.

9.2 A Debt Multi-Agent CCAPM Model

A consumer whose wealth is W_0^k may also assume debt, D_0^k. While such debt may be used for current consumption, it can also be used a bet on future prices. For example, if a consumer expects the price of houses to increase he may assume more debt, while if prices are stable or deflationary, it will be less of an appealing to assume debt. Extending the CCAPM kernel pricing model of Chap. 8, and considering a borrower only, we have:

$$\underset{c_0^k, D_0^k}{\text{Max}}\, U_k = u_k(c_0^k) + \beta E\left(u_{k1}(c_1^k)\right)$$

$$\text{with } c_1^k = \frac{\left(W_0^k + D_0^k - \pi_0 c_0^k\right)\left(1 + \tilde{R}_1^k\right) - D_0^k\left(1 + R_D^k\right)}{\tilde{\pi}_1(\tilde{C})}, \quad \tilde{C} = \sum_{k=1} c_1^k$$

where R_D^k is the interest rate that a consumer with such wealth pays for debt. D_0^k used for consumption or savings (for future consumption), or both. Necessary conditions for the optimality of current and future consumption are then found by:

$$dU_k = \frac{\partial u_k(c_0^k)}{\partial c_0^k} dc_0^k + \beta E\left(\frac{\partial u_k(c_1^k)}{\partial c_1^k} dc_1^k + \frac{\partial u_k(c_1^k)}{\partial D_0^k} dD_0^k\right) = 0$$

$$0 = \frac{\partial u_k(c_0^k)}{\partial c_0^k}\frac{dc_0^k}{dc_0^k} + \beta E\left(\frac{\partial u_k(c_1^k)}{\partial c_1^k}\frac{dc_1^k}{dc_0^k} + \frac{\partial u_k(c_1^k)}{\partial D_0^k}\frac{dD_0^k}{dc_0^k}\right) = 0$$

$$\frac{\partial u_k(c_0^k)}{\partial c_0^k}\frac{dc_0^k}{dD_0^k} + \beta E\left(\frac{\partial u_k(c_1^k)}{\partial c_1^k}\frac{dc_1^k}{dD_0^k} + \frac{\partial u_k(c_1^k)}{\partial D_0^k}\right) = 0$$

which yields: $1 + E\left(\tilde{M}_k \frac{dc_1^k}{dc_0^k} + \tilde{M}_k \frac{\partial c_1^k}{\partial D_0^k}\frac{dD_0^k}{dc_0^k}\right) = 0, \tilde{M}_k = \beta \frac{\partial u_k(c_1^k)/\partial c_1^k}{\partial u_k(c_0^k)/\partial c_0^k}$. Noting that:

$$c_1^k = \frac{\left(W_0^k + D_0^k - \pi_0 c_0^k\right)\left(1 + \tilde{R}_1^k\right) - D_0^k\left(1 + R_D^k\right)}{\tilde{\pi}_1(\tilde{C})}, \quad \tilde{C} = \sum_{k=1} c_1^k,$$

By implicit differentiation:

$$0 = \frac{\left(W_0^k + D_0^k - \pi_0 c_0^k\right)\left(1+\tilde{R}_1^k\right) - D_0^k\left(1+R_D^k\right)}{\tilde{\pi}_1(\tilde{C})} - c_1^k,\ \tilde{C} = \sum_{j\neq k} c_1^j + c_1^k \quad \text{and}$$

$$\frac{dc_1^k}{dD_0^k} = -\frac{\partial\Omega/\partial D_0^k}{\partial\Omega/\partial c_1^k} = -\frac{\left(\tilde{R}_1^k - R_D^k\right)}{\tilde{\pi}_1(\tilde{C}) + \left[\left(W_0^k + D_0^k - \pi_0 c_0^k\right)\left(1+\tilde{R}_1^k\right) - D_0^k\left(1+R_D^k\right)\right]\frac{\partial \ln \tilde{\pi}_1(\tilde{C})}{\partial C}}$$

$$\frac{dD_0^k}{dc_0^k} = -\frac{\partial\Omega/\partial D_0^k}{\partial\Omega/\partial c_0^k} = \frac{\pi_0\left(1+\tilde{R}_1^k\right)}{\tilde{R}_1^k - R_D^k}$$

$$\frac{dc_1^k}{dc_0^k} = -\frac{\pi_0\left(1+\tilde{R}_1^k\right)}{\tilde{\pi}_1(\tilde{C}) + \left[\left(W_0^k + D_0^k - \pi_0 c_0^k\right)\left(1+\tilde{R}_1^k\right) - D_0^k\left(1+R_D^k\right)\right]\dfrac{\partial \ln \tilde{\pi}_1(\tilde{C})}{\partial C}}$$

As a result,

$$1 = -E\left(\tilde{M}_k \frac{dc_1^k}{dc_0^k}\right) \text{ and } \frac{dc_0^k}{dD_0^k} = -E\left(\tilde{M}_k \frac{dc_1^k}{dD_0^k}\right)$$

Since $\dfrac{\partial}{\partial \tilde{C}_1}\left(\dfrac{1}{\tilde{\pi}_1(\tilde{C}_1)}\right) = \left(\sum_{j=1}^{n}\left(W_0^j + D_0^j - \pi_0 c_0^j\right)\left(1+\tilde{R}_1^j\right) - D_0^j\left(1+R_D^j\right)\right)^{-1}$

we have:

$$1 = E\left(\tilde{M}_k \frac{\left(1+\tilde{R}_1^k\right)/\left(1+\tilde{\eta}_1(\tilde{C}_1)\right)}{1-\left[\dfrac{\tilde{c}_1^k}{\tilde{C}_1}\right]}\right),$$

$$\frac{\tilde{c}_1^k}{\tilde{C}_1} = \frac{\left(W_0^k - \pi_0 c_0^k\right)\left(1+\tilde{R}_1^k\right) + D_0^k\left(\tilde{R}_1^k - R_D^k\right)}{\sum_{j=1}^{n}\left\{\left(W_0^j - \pi_0 c_0^j\right)\left(1+\tilde{R}_1^j\right) + D_0^j\left(\tilde{R}_1^j - R_D^j\right)\right\}}$$

Set:

$$p_k = \frac{\left(W_0^k - \pi_0 c_0^k\right)}{\Lambda},\ \Lambda = \sum_{j=1}^{n}\left(W_0^j - \pi_0 c_0^j\right) \text{ and } q_k = \frac{D_0^k}{\Psi},\ \Psi = \sum_{j=1}^{n} D_0^j$$

Then,

$$\frac{\tilde{c}_1^k}{\tilde{C}_1} = \frac{p_k\left(1+\tilde{R}_1^k\right) + \Gamma q_k\left(\tilde{R}_1^k - R_D^k\right)}{\sum_{j=1}^{n}\left\{p_j\left(1+\tilde{R}_1^j\right) + \Gamma q_j\left(\tilde{R}_1^j - R_D^j\right)\right\}}$$

where Γ is the debt to equity ratio of all consumers while q_k is the consumer k individual share of total credit. Or

$$1 = E\left(\tilde{M}_k \frac{(1+\tilde{R}_1^k)}{1+\tilde{\eta}_1(\tilde{C}_1)}\left(1+\frac{p_k(1+\tilde{R}_1^k)+\Gamma q_k(\tilde{R}_1^k - R_D^k)}{\sum_{j\neq k}^{n}\left\{p_j\left(1+\tilde{R}_1^j\right)+\Gamma q_j\left(\tilde{R}_1^j - R_D^j\right)\right\}}\right)\right),$$

and therefore,

$$E\left(\tilde{M}_k\left(\frac{1+\tilde{R}_1^k}{1+\tilde{\eta}_1(\tilde{C}_1)}\right)\right) = 1 - E\left(\tilde{M}_k \frac{(1+\tilde{R}_1^k)}{1+\tilde{\eta}_1(\tilde{C}_1)}\frac{p_k(1+\tilde{R}_1^k)+\Gamma q_k(\tilde{R}_1^k - R_D^k)}{\sum_{j\neq k}^{n}\left\{p_j\left(1+\tilde{R}_1^j\right)+\Gamma q_j\left(\tilde{R}_1^j - R_D^j\right)\right\}}\right)$$

In other words, if markets consist of an infinite number of agents and markets are competitive (i.e. markets are complete), then the CCAPM price model would be $1 = E\left(\tilde{M}^*\left(\frac{1+\tilde{R}_1^k}{1+\tilde{\eta}_1(\tilde{C}_1)}\right)\right)$ while for multi agents, the price is necessarily smaller than one dollar. In other words in a multi-agent CCAPM, consumers will be willing to pay less for future returns. As a result, we can write:

$$E\left(\tilde{M}_k\left(\frac{1+\tilde{R}_1^k}{1+\tilde{\eta}_1(\tilde{C}_1)}\right)\right)$$
$$= \left\{E\left(\tilde{M}^*\left(\frac{1+\tilde{R}_1^k}{1+\tilde{\eta}_1(\tilde{C}_1)}\right)\right) - E\left(\tilde{M}_k \frac{(1+\tilde{R}_1^k)}{1+\tilde{\eta}_1(\tilde{C}_1)}\frac{p_k(1+\tilde{R}_1^k)+\Gamma q_k(\tilde{R}_1^k - R_D^k)}{\sum_{j\neq k}^{n}\left\{p_j\left(1+\tilde{R}_1^j\right)+\Gamma q_j\left(\tilde{R}_1^j - R_D^j\right)\right\}}\right)\right\}$$

Setting $\tilde{M}^{(k)} = \frac{\tilde{M}^*}{\tilde{M}_k}$ as well as

$$1+\ \bar{\Re}_k = E\left(\frac{1+\tilde{R}_1^k}{1+\tilde{\eta}_1(\tilde{C}_1)}\left(1+\frac{p_k(1+\tilde{R}_1^k)+\Gamma q_k(\tilde{R}_1^k - R_D^k)}{\sum_{j\neq k}^{n}\left\{p_j\left(1+\tilde{R}_1^j\right)+\Gamma q_j\left(\tilde{R}_1^j - R_D^j\right)\right\}}\right)\right),$$ we

have:

$$1 = \frac{1}{1+\ \bar{\Re}_k}E\left(\tilde{M}^{(k)}\left(\frac{1+\tilde{R}_1^k}{1+\tilde{\eta}_1(\tilde{C}_1)}\right)\right)$$

Note that $\bar{\Re}_k$ is defined in terms of an expectation of all individual agents rates of return $\tilde{R}_1^j, j = 1,\ldots,n$. Further $\tilde{M}^{(k)}$ denotes a ratio of consumer k marginal utilities of consumption under two modes—a first under a pricing model independent of other consumers (or equivalently, a pricing model where all consumers' share of disposable income are extremely small) and the multi-agent price model (with consumers' share of disposable income that is not negligible). As a result, $\tilde{M}^{(k)}$ is a relative kernel.

Next assume that all consumers-agents invest in inflation indexed risk free bonds, then:

$$\frac{1+\bar{\Re}_k^{ind}}{1+R_1^{ind}} = \left(1+\frac{p_k\left(1+R_1^{ind}\right)+\Gamma q_k\left(R_1^{ind}-R_D^k\right)}{\left(1-p_k\right)\left(1+R_1^{ind}\right)+\Gamma\sum_{j\neq k}^{n} q_j\left(R_1^{ind}-R_D^j\right)}\right) \text{ and}$$

$$\mathrm{E}\left(\tilde{M}^{(k)}\right) = \left(\frac{1+\bar{\Re}_k^{ind}}{1+R_1^{ind}}\right)$$

which provides the following implied pricing model for each agent k:

$$1=\frac{1}{\left(1+R_1^{ind}\right)}\frac{1+\bar{\Re}_k^{ind}}{1+\bar{\Re}_k}E^{\tilde{M}^{(k)}}\left(1+\tilde{R}_{real}^k\right),\ 1+\tilde{R}_{real}^k=\frac{1+\tilde{R}_1^k}{1+\tilde{\eta}_1(\tilde{C}_1)}\quad k=1,2,\ldots,n$$

Note that such a solution requires the simultaneous solution of k simultaneous and nonlinear equations. If all agents rate of return is defined by an investment strategy in bonds and equity, $\tilde{R}_{real}^k\left(a_k,\ \tilde{\Re}\right)$, then:

$$\begin{aligned}1&=\frac{1}{\left(1+R_1^{ind}\right)}\frac{1+\bar{\Re}_k^{ind}}{1+\bar{\Re}_k}E^{\tilde{M}^{(k)}}\left(1+\tilde{R}_{real}^k\left(a_k,\ \tilde{\Re}\right)\right),\ 1+\tilde{R}_{real}^k\left(a_k,\ \tilde{\Re}\right)\\&=\frac{1+\tilde{R}_1^k}{1+\tilde{\eta}_1(\tilde{C}_1)}\quad k=1,2,\ldots,n\end{aligned}$$

While,

$$1+\bar{\Re}_k=E_{\Re}\left(\frac{1+R_1^k\left(a_k,\ \tilde{\Re}\right)}{1+\tilde{\eta}_1(\tilde{C}_1)}\left(1+\frac{p_k\left(1+R_1^k\left(a_k,\ \tilde{\Re}\right)\right)+\Gamma q_k\left(R_1^k\left(a_k,\ \tilde{\Re}\right)-R_D^k\right)}{\sum_{j\neq k}^{n}\left\{p_j\left(1+R_1^j\left(a_j,\ \tilde{\Re}\right)\right)+\Gamma q_j\left(R_1^j\left(a_j,\ \tilde{\Re}\right)-R_D^j\right)\right\}}\right)\right)$$

which is a function of the equity rate of return $\tilde{\Re}$ assume in this case to be equally predicted by all agents.

Of course, if all agents use the same stochastic discount kernel, then:

$$E^{\tilde{M}^{(*)}}\left(\tilde{R}_{real}^k(a_k)\right)=\left(1+R_1^{ind}\right)\frac{1+\bar{\Re}_k(a_k)}{1+\bar{\Re}_k^{ind}}-1,\quad k=1,2,\ldots n$$

which provides a system of nonlinear equation in the investment strategy. Explicitly, if,

$$\tilde{R}_{real}^k(a_k)=a_kR_1^{ind}+(1-a_k)\,\tilde{\Re}_M^{ind}$$

where $\tilde{\Re}_M^{ind}$ is the inflation indexed market rate of return, then:

$$\left(\frac{1+R_1^{ind}}{1-a_k}\right)\left(\frac{1+\bar{\Re}_k(a_k)}{1+\bar{\Re}_k^{ind}}\right)-\left(\frac{1+a_kR_1^{ind}}{1-a_k}\right)=E^{\tilde{M}^{(*)}}\left(\tilde{\Re}_M^{ind}\right),\ k=1,2,\ldots n$$

where $E^{\tilde{M}^{(*)}}\left(\tilde{\Re}_M^{ind}\right)$ is the valuation of an index rate of return valued equally by all consumers-agents.

Also,

$$1+\bar{\Re}_k=E\left(\left(1+a_kR_1^{ind}+(1-a_k)\,\tilde{\Re}_M^{ind}\right)\left(1+\frac{\left\{\begin{array}{l}p_k\left(1+a_kR_1^{ind}+(1-a_k)\,\tilde{\Re}_M^{ind}\right)+\\+\Gamma q_k\left(a_kR_1^{ind}+(1-a_k)\,\tilde{\Re}_M^{ind}-R_D^k\right)\end{array}\right\}}{\sum_{j\neq k}^{n}\left\{\begin{array}{l}p_j\left(1+a_jR_1^{ind}+(1-a_j)\,\tilde{\Re}_M^{ind}\right)+\\+\Gamma q_j\left(a_kR_1^{ind}+(1-a_j)\,\tilde{\Re}_M^{ind}-R_D^j\right)\end{array}\right\}}\right)\right)$$

and

$$1+\ \bar{\Re}_k^{ind}=\left(1+R_1^{ind}\right)\left(1+\frac{p_k\left(1+R_1^{ind}\right)+\Gamma q_k\left(R_1^{ind}-R_D^k\right)}{(1-p_k)\left(1+R_1^{ind}\right)+\Gamma\sum_{j\neq k}^{n}q_j\left(R_1^{ind}-R_D^j\right)}\right)$$

A number of cases arise which we consider as examples below.

- *Aggregate consumption and approximate prices*

For a specific consumer, whose disposable wealth and debt are known, the proportional consumption of this consumer is a function of the market index and his risk investment preference that determines the portfolio's parameter a_k. Explicitly,

$$\frac{\tilde{c}_1^k}{\tilde{C}_1}=\frac{p_k\left(1+\tilde{R}_1^k\right)+\Gamma q_k\left(\tilde{R}_1^k-R_D^k\right)}{\sum_{j=1}^{n}\left\{p_j\left(1+\tilde{R}_1^j\right)+\Gamma q_j\left(\tilde{R}_1^j-R_D^j\right)\right\}}=\frac{\Phi_k\left(R_1^{ind},1+\ \tilde{\Re}_{real}^{ind}\right)}{\Phi\left(R_1^{ind},1+\ \tilde{\Re}_{real}^{ind}\right)}$$

where Φ_k and Φ are two functions, with $\Phi=\sum_{j=1}^{n}\Phi_j$. It is convenient to write:

$$\Phi_k\left(R_1^{ind},\ \tilde{\Re}_{real}^{ind}\right)=\alpha_k+\lambda_k\tilde{\Re}_{real}^{ind}\quad \text{and}$$

$$\Phi\left(R_1^{ind},\tilde{\Re}_{real}^{ind}\right)=\sum_{j=1}^{n}\alpha_j+\ \tilde{\Re}_{real}^{ind}\sum_{j=1}^{n}\lambda_j$$

and thereby,

$$\frac{\tilde{c}_1^k}{\tilde{C}_1} = \frac{\alpha_k + \lambda_k \tilde{\Re}_{real}^{ind}}{\sum_{j=1}^{n} \alpha_j + \tilde{\Re}_{real}^{ind} \sum_{j=1}^{n} \lambda_j}$$

As a result,

$$1 = E\left(\tilde{M}_k \left(1 + a_k R_1^{ind} + (1 - a_k)\, \tilde{\Re}_{real}^{ind}\right) \left(1 + \frac{\left(\alpha_k + \lambda_k \tilde{\Re}_{real}^{ind}\right)}{\sum_{j \neq k}^{n} \alpha_j + \tilde{\Re}_{real}^{ind} \sum_{j \neq k}^{n} \lambda_j} \right) \right)$$

Consider the term, $\tilde{x} = \sum_{j \neq k}^{n} \alpha_j + \tilde{\Re}_{real}^{ind} \sum_{j \neq k}^{n} \lambda_j$, a two terms Taylor series expansion about its mean is:

$$\frac{1}{\tilde{x}} \approx \frac{1}{\hat{x}} - \frac{1}{\hat{x}^2}(\tilde{x} - \hat{x}) = \frac{1}{\sum_{j \neq k}^{n} \alpha_j + \hat{\Re}_{real}^{ind} \sum_{j \neq k}^{n} \lambda_j} - \frac{1}{\left(\sum_{j \neq k}^{n} \alpha_j + \hat{\Re}_{real}^{ind} \sum_{j \neq k}^{n} \lambda_j\right)^2} \left(\left(\tilde{\Re}_{real}^{ind} - \hat{\Re}_{real}^{ind} \right) \sum_{j \neq k}^{n} \lambda_j \right)$$

and therefore, the term in parenthesis above is:

$$1 + \left(\frac{\left(\alpha_k + \lambda_k \tilde{\Re}_{real}^{ind}\right)}{\sum_{j \neq k}^{n} \alpha_j + \hat{\Re}_{real}^{ind} \sum_{j \neq k}^{n} \lambda_j} - \frac{\left(\alpha_k + \lambda_k \tilde{\Re}_{real}^{ind}\right) \left(\left(\tilde{\Re}_{real}^{ind} - \hat{\Re}_{real}^{ind} \right) \sum_{j \neq k}^{n} \lambda_j \right)}{\left(\sum_{j \neq k}^{n} \alpha_j + \hat{\Re}_{real}^{ind} \sum_{j \neq k}^{n} \lambda_j\right)^2} \right)$$

which leads to a pricing model in terms of the first three moments of the market rate return index. This model is left as an exercise

- *A Linear Kernel*

The linear multi-agent CCAPM assumes that the pricing kernel is a linear function of the index rate of returns $\tilde{M}_k = \alpha_k + b_k \tilde{\Re}_{real}^{ind}$ and therefore

$$1 = E\left(\left(\alpha_k + b_k \tilde{\Re}_{real}^{ind}\right) \left(1 + a_k R_1^{ind} + (1 - a_k)\, \tilde{\Re}_{real}^{ind}\right) \left(1 + \frac{\left(\alpha_k + \lambda_k \tilde{\Re}_{real}^{ind}\right)}{\sum_{j \neq k}^{n} \alpha_j + \tilde{\Re}_{real}^{ind} \sum_{j \neq k}^{n} \lambda_j} \right) \right)$$

which provides a pricing model in terms of the first 4 moments of the inflation indexed Market index. Of course, since this model is again a nonlinear function in consumers' agent investment.

- *Debt borrowing and lending*

Consider two parties—one borrowing and the other lending. In this case, both parties' prices are:

$$\tilde{c}_1^B = \frac{\left(W_0^B + D_0^{LB} - c_0^B\right)\left(1+\tilde{R}_1^B\right) - D_0^{LB}\left(1+R_D^{LB}\right)}{\tilde{\pi}_1\left(\tilde{C}\right)}$$

$$\tilde{c}_1^L = \frac{\left(W_0^L - D_0^{LB} - \pi_0 c_0^L\right)\left(1+\tilde{R}_1^L\right) + D_0^{LB}\left(1+R_D^{LB}\right)}{\tilde{\pi}_1\left(\tilde{C}\right)}$$

$$1 = E\left(\tilde{M}_B \frac{\pi_0\left(1+\tilde{R}_1^B\right)/\left(1+\tilde{\eta}_1(\tilde{C}_1)\right)}{1-\left[\dfrac{\tilde{c}_1^B}{\tilde{c}_1^B+\tilde{c}_1^L}\right]}\right) \text{ and } 1 = E\left(\tilde{M}_L \frac{\pi_0\left(1+\tilde{R}_1^L\right)/\left(1+\tilde{\eta}_1(\tilde{C}_1)\right)}{1-\left[\dfrac{\tilde{c}_2^L}{\tilde{c}_1^B+\tilde{c}_1^L}\right]}\right)$$

Thus,

$$1 = E\left(\tilde{M}_B \frac{\pi_0\left(1+\tilde{R}_1^B\right)/\left(1+\tilde{\eta}_1(\tilde{C}_1)\right)}{1-\left[\dfrac{\left(W_0^B + D_0^{LB} - \pi_0 c_0^B\right)\left(1+\tilde{R}_1^B\right) - D_0^{LB}\left(1+R_D^{LB}\right)}{\left(W_0^B + D_0^{LB} - \pi_0 c_0^B\right)\left(1+\tilde{R}_1^B\right) + \left(W_0^L - D_0^{LB} - \pi_0 c_0^L\right)\left(1+\tilde{R}_1^L\right)}\right]}\right)$$

$$1 = E\left(\tilde{M}_L \frac{\pi_0\left(1+\tilde{R}_1^L\right)/\left(1+\tilde{\eta}_1(\tilde{C}_1)\right)}{1-\left[\dfrac{\left(W_0^L - D_0^{LB} - \pi_0 c_0^L\right)\left(1+\tilde{R}_1^L\right) + D_0^{LB}\left(1+R_D^{LB}\right)}{\left(W_0^B + D_0^{LB} - \pi_0 c_0^B\right)\left(1+\tilde{R}_1^B\right) + \left(W_0^L - D_0^{LB} - \pi_0 c_0^L\right)\left(1+\tilde{R}_1^L\right)}\right]}\right)$$

where $\tilde{M}_B$ and $\tilde{M}_B$ are the kernel prices for each of the parties while the B consumer borrows D_0^{LB} paying the interest rate R_D^{LB}. Each of the parties has different wealth, consumption etc. but pay the same prices for consumption.

If both the borrower and the lender have identical kernels (i.e. valuing consumption equally) but of different wealth levels, to:

$$\frac{\left(W_0^B + D_0^{LB} - \pi_0 c_0^B\right)}{D_0^{LB}\left(1+R_D^{LB}\right)} E\left(1+\tilde{R}_1^B\right)^2 - \frac{\left(W_0^L - D_0^{LB} - \pi_0 c_0^L\right)}{D_0^{LB}\left(1+R_D^{LB}\right)} E\left(1+\tilde{R}_1^L\right)^2$$
$$= E\left(1+\tilde{R}_1^B\right) + E\left(1+\tilde{R}_1^L\right)$$

setting:

$$Q^* = \frac{(W_0^B + D_0^{LB} - \pi_0 c_0^B)}{D_0^{LB}(1+R_D^{LB})}\left(\text{var}(1+\tilde{R}_1^B) + (E(1+\tilde{R}_1^B))^2\right) - E(1+\tilde{R}_1^B)$$

$$Q^* = \frac{(W_0^L - D_0^{LB} - \pi_0 c_0^L)}{D_0^{LB}(1+R_D^{LB})}\left(\text{var}(1+\tilde{R}_1^L) + (E(1+\tilde{R}_1^L))^2\right) + E(1+\tilde{R}_1^L)$$

The consumer selects a debt D_0^{LB} while the lender demands an interest rate R_D^{LB}. If both seek to maximize their rate of return for future consumption, a game result, (Chap. 12) with the borrower selecting how much to borrow and the lender determines the borrowing rate.

$$D_0^{LB} = \frac{(W_0^B - \pi_0 c_0^B)\left(\text{var}(1+\tilde{R}_1^B) + (E(1+\tilde{R}_1^B))^2\right)}{(1+R_D^{LB})(Q^* + E(1+\tilde{R}_1^B)) - \left(\text{var}(1+\tilde{R}_1^B) + (E(1+\tilde{R}_1^B))^2\right)}$$

$$1 + R_D^{LB} = \frac{(W_0^L - D_0^{LB} - \pi_0 c_0^L)\left(\text{var}(1+\tilde{R}_1^L) + (E(1+\tilde{R}_1^L))^2\right)}{Q^* D_0^{LB} - D_0^{LB} E(1+\tilde{R}_1^L)}$$

which provides a schedule for the credit which is function of Q^*. To determine Q^*, D_0^{LB} and R_D^{LB} are selected such that Q^* is as defined above.

9.3 Global Finance and Risks

Global finance is embedded in terms of countries' macroeconomic factors, trade, political and social agendas, countries interdependence and their many derivatives. They provide both opportunities and risks to individual investors, albeit with greater complexity, uncertainty and consequences that may dwarf the traditional view of social agendas. People and capital migration, a growth of wealth and social inequalities, religions and geographies in collision on unprecedented scales are precursors of a global volatility. Foreign exchange (FX) markets meant to provide a price for converting one currency into another reached by financial parties, exporters and importers, global investors, central banks and the many institutions and financial parties that make up the global economy and intervene in such markets are also gradually altered from free exchange financial markets to markets in collision.

Questions and challenges abound. Is globalization sustainable? Is globalization contributing to economic growth in all countries or only to some? Does growth favor the few or the many? Is globalization contributing to a better world, to greater equality between countries? Is it contributing to greater inequalities within countries, making the rich, richer and leading to the disappearance of a middle economic class? How is globalization affecting the engineering risk of finance? Can traditional valuation theory capture the complexity of pricing international assets and their exchange? What are the implications of strategic economies (Asia, US, Europe, Latin America) on the dominance of a single exchange currency (the dollar)? Etc.

Emerging facts are pointing to a greater "equalization" of the global economy and therefore, the loss of traditional competitive advantage. This trend, in its historical perspective is countering a trend that has accelerated inter-nations inequalities ever since the beginnings of the first Industrial revolution. Greater "equalization" is assuming many faces. It occurs in technology diffusion, in industrial might (through outsourcing), A-national firms and investors seeking greater yield and less regulation and controls as well as National Champions or Sovereign States Firms expanding their might globally. Equalization is accelerated by open borders and a "comparative advantage"—whether in natural resources, in technology or know-how or based on financially engineered and repressions policies in trade and global finance and taxation. This underlies a Tsunami from West to East, with countries such as China, India and Asia in general assuming a far greater importance with increasingly large and expanding firms while the West assuming a greater debt to meet needs it has long grown to believe and taken their supply for granted. For example, (*The Economist*, Dec.-17–30,' 2011 issue, p. 109) State controlled firms account today for 80% of the market capitalization of the Chinese stock market, more than 60% of Russia's and 35% of Brazil. They also make up now more than 19 of the world 100 multinational biggest companies and 28 of the top 100 among emerging markets. These firm are now found everywhere from China, Russia, the Middle East. Trade repression is also pervasive, applied either directly or indirectly. Brazil for example imposed in December 2012 a 30 percentage point tax increase on cars with less than 65% local content, taking the tax on some imported models up to 55% on top of import tariffs. Such tax was imposed on the pretext of foreign imports creating a de-industrialization of the Brazilian economy.

The tenets of comparative advantage remain however only in natural endowments, most of which are in limited supply and with prices rising due to greater demands by a growing population (currently over seven billion and predicted to reach the ten billions) and to a global increases in consumption. While in other areas, no country can maintain long a comparative advantage. For example, commodities' prices are already rising dramatically due to a growth of demand in Asia with both economic and political repercussions all over the world. At the same time, countries endowed with large populations providing them cost advantages, are heralding a shift of industries to countries where costs are low and demand plenty. Symptoms to these emerging States and these states of affair abound. As stated above, it underlies a transfer of industrial and financial might with traditional rich countries assuming debt they cannot service. It leads to a gradual departure of paying jobs from the West to the East with the rich richer, a collapsing middle class and internal political challenges. It leads to the faltering of financial foundations with a rise of a global Knightian uncertainty (namely, a future where the futures are not defined, see Chap. 10). In fact globalization with its greater interdependence, complexity and transparency and non-transparency, combined with extreme weather patterns and pollution altering the face of the earth by desertification, migration etc. are also challenging appreciably future economics and finance, increasingly strategic and risk-strategic.

A summary of essential risk concerns amplified by the financial crisis of 2007–2008 include:

- **A multi-polar and multi-agent global finance world**: with competing financial markets and economic regions (US, ASIA, The Euro Zone, The nascent Latin American pole) with increased interventions of Sovereign States and large A-national firms combined with financial repression (taxation and regulation hindering globalization)
- **A volatile currencies world**: Currencies' volatility followed by the Lack faith in any specific agent (country). An increased "lack of faith" in a currency contributes to creditors' "credit risk." If there is no viable and safe alternative to a "viable and safe currency," the global financial system and global trade can falters. Such situations lead as well to a flight from risk as noted below.
- **Flight from risk** results from greater global instability and international uncertainty contributing to a "flight from globalization,". In such an economic environment, global risks "might not be diversifiable" or done easily away with by some investments portfolio. Global investments (in particular where they are needed the most) are then reduced.
- **The external global environment**: Creeping desertification, global warming and changing weather patterns, water shortage and quality and population and concentration growth determine the global supply and demand of commodities and thereby their prices. These elements, combined with contagious grass roots movements, a tyranny of minorities, ill-defined and networked through new networking technologies contribute to "re-evolutions" that pits political agendas internally and externally providing new sources of economic and financial instability.
- **Statistical and contagious dependence**: Statistical dependence arises when the events in one country affect the events in another. A contagions arises however when one event triggers a process of some sort in a series of States. For example, default can be contagious as a failure in one State may induce failure in another. States intervening to seek advantageous terms in trade through tariff and non-tariff barriers, quotas and taxes on imports and incentives for exports can also lead to contagious retaliation by other countries. By the same token, downgrading a whole banking sector subject to increased regulation may induce a tightening of its loan standards and value at risk requirements that dampen further their ability to lend and profit (and thus induces a greater probability of failure).
- **A persistent transformation of information technology** and the growth of a new brand of technology risks including the use of technology to leverage information, misinformation, social and cyber risks etc. with a global impact.

- **Information and power asymmetry** with moral hazard and adverse selection allowing the few to profit at the expense of the many. Global finance requires global information and controls a global scale, and therefore it tends to favor those financiers that have the means to engage in it. Such risks contribute as well to potential economic inequalities both within countries and globally.

These risks are amplified by political risks, external and natural risks as well as by the limits of traditional financial analysis based on assumptions that do not hold in a very complex, dependent and global finance. For example, from a fundamental finance technical viewpoint, pricing depends on three essential assumptions:

- Predictable asset's state preferences (i.e. that future state prices are assumed known)
- A references measure used to price assets (such as a risk free rate or a basket of currencies)
- Information needed to price assets

When the economy consists of heterogenous agents, the information needed to value and price assets is far greater. Lacking this information, markets are incomplete. A unique markets price for exchange cannot then be set. In other words, complete markets finance is not applicable simply. Financial firms recognize today that "no one can fully predict tomorrow's risks." But in today's fundamentally changed investment world, the risks are less foreseeable and more than ever—and can cause lasting damage.

9.3.1 *Pricing International Assets and Foreign Exchange Risk*

In this section, we pursue both the CCAPM approach and the fundamental Arrow-Debreu approach to pricing international financial assets and investments. In a global-international environment the assumptions made by pricing models may differ. There are a number of reasons to justify such differences:

- In a global setting, investors differ in their value, preferences, sophistication etc. and thus finding a typical "consumer-investor" might not be realistic. Rather, "consumers are heterogenous."
- Regulations, taxes on financial profits and trades, government controls etc. differ from country to country and therefore the risk-returns equivalence of international trades for investors in different countries may differ appreciably.
- In international financial markets, investors tend to consist of large institutions and funds. As a result, rather than having atomic investors, investors are confronted by "oligarchic" investors competing for returns on a global scale.
- FX rates' prices, their inter-temporal evolution, manipulations by Sovereign State macroeconomic and Trade policies add sources of uncertainty that may be difficult to integrate in theory of complete markets for assets' prices.

- Politics and conflicts normally neglected in assets prices such as nationalization, selective actions assumed by governments against or for foreign investors are also important and can lead to approaches that differ between assets prices for domestic investors and assets prices for foreign investors.

These and other factors may thus alter the pricing of international assets compared to domestic assets pricing. Below examples are considered. It is noteworthy however that pricing of FX has essentially been based on macroeconomic models based on a global equilibrium with unique and equivalent prices realigned by the FX between countries.

Example: Global Portfolios Consider an investor maximizing current and future consumption. To do so, part of the investor's wealth is consumed initially and its residual is invested in a portfolio providing a rate of return, defining his future wealth which is then consumed again partly or wholly and his investment preferences. Such investment can be diversified across national boundaries and thus assume a foreign exchange risk exposure. The following cases outline the basic mechanics to price such investments over two periods. Let a consumer-investor wealth be \$1, part of which, say $I_0 = 1 - c_0$, is invested for its consumption in the next period, with $\tilde{I}_1 = I_0(1 + \tilde{R}_1) = \tilde{c}_1$ where $\tilde{R}_1$ is the investor's portfolio rate of return. If the portfolio is invested solely in a risk free bond, then $I_1 = I_0(1 + R_f)$. For a portfolio invested in both an equity k and in a risk free asset, each of which has rate of return $\tilde{R}_k$ and R_f, we then have:

$$\tilde{I}_1 = I_0(1 + \tilde{R}_1) = \alpha I_0(1 + R_f) + (1 - \alpha)I_0(1 + \tilde{R}_k) \quad \text{or}$$
$$\tilde{R}_1 = R_f + (1 - \alpha)(\tilde{R}_k - R_f)$$

where α is the proportion of one's initial portfolio invested in the risk free asset. By the same token, consider a domestic investor investing locally and in a foreign country in both risk free and risk price assets. The portfolio price a period hence is then:

$$\tilde{I}_1 = I_0(1 + \tilde{R}_1) = \alpha_D I_0\left(1 + R_f^D\right) + \alpha_F I_0 \xi_{DF}^0 \tilde{\xi}_{FD}^1 \left(1 + R_f^F\right)$$
$$+ \beta_D I_0(1 + \tilde{R}_k^D) + \beta_F I_0 \xi_{DF}^0 \tilde{\xi}_{FD}^1 (1 + \tilde{R}_k^F)$$

where $\alpha_D + \alpha_F + \beta_D + \beta_F = 1$ denote the proportions of one investment in both domestic and foreign markets while ξ_{DF}^0, ξ_{FD}^0 are the foreign exchange rates for exchanging domestic currency into a foreign one and vice versa and therefore, $\xi_{DF}^0 = 1/\xi_{FD}^0$. Further, since:

$$\xi_{DF}^0 \tilde{\xi}_{FD}^1 = \xi_{DF}^0 (\xi_{FD}^0 + \Delta \tilde{\xi}_{FD}^1) = (\xi_{FD}^0 + \Delta \tilde{\xi}_{FD}^1)/\xi_{DF}^0 = 1 + \tilde{R}_{FD}^{\xi},$$

where $\tilde{R}^1_{FD}$ is the rate of return (or a loss) resulting from the foreign currency FX rate. In this case, the portfolio rate of return priced in the domestic currency:

$$\tilde{R}_1 = \alpha_D R^D_f + \beta_D \tilde{R}^D_k + \alpha_F \left(R^F_f \left(1 + \tilde{R}^\xi_{FD} \right) + \tilde{R}^\xi_{FD} \right) + \beta_F \left(\tilde{R}^F_k + \tilde{R}^\xi_{FD} + \tilde{R}^F_k \tilde{R}^\xi_{FD} \right)$$

Note in particular the effects of the FX rate of return on the portfolio rate of return. For example, while a bet on a foreign security may disappoint, it may be compensated by an appreciable rate of return on the currency, and vice versa. The portfolios expected returns and their standard deviation will also (most likely) not be normal. For example, the expected rate of return of such a portfolio would be also a function of the co-variations of the foreign equity asset price and the country FX rate since:

$$\hat{R}_1 = \alpha_D R^D_f + \beta_D \hat{R}^D_k + \alpha_F \left(R^F_f \left(1 + \hat{R}^\xi_{FD} \right) + \hat{R}^\xi_{FD} \right) + \beta_F \left(\hat{R}^F_k + \hat{R}^\xi_{FD} + E\left(\tilde{R}^F_k \tilde{R}^\xi_{FD} \right) \right)$$

For these reasons, a "summary" rate of return for a "global" investor ought to account as well for currencies co-variations.

The pricing of such a portfolio, assuming a kernel $\tilde{M}_t$, would then indicate that the price of the portfolio rates of returns is:

$$1 = \frac{1}{1 + R^D_{f,t+1}} E_t \left(\frac{\tilde{M}_t}{E(\tilde{M}_t)} \left\{ \begin{array}{l} \alpha_D R^D_{f,t+1} + \beta_D \tilde{R}^D_{k,t+1} + \alpha_F \left(R^F_{f,t+1} \left(1 + \tilde{R}^\xi_{FD,t+1} \right) + \tilde{R}^\xi_{FD,t+1} \right) \\ + \beta_F \left(\tilde{R}^F_{k,t+1} + \tilde{R}^\xi_{FD,t+1} + \tilde{R}^F_{k,t+1} \tilde{R}^\xi_{FD,t+1} \right) \end{array} \right\} \right)$$

Of course, since the future return is a random variable (and therefore returns are risk prone), the kernel price is a deflator that accounts for the risk that the investor assumes both in the domestic market and in the foreign one. It implies that under the probability measure M defined to maintain a risk neutral price estimate:

$$1 = \frac{1}{1 + R_{f,t+1}} E^M_t \left(1 + \tilde{R}_{t+1} \right) \quad \text{and explicitly,} \quad 1 = \frac{1}{1 + R_{f,t+1}} \sum_{i=1}^{n} p^M_i \left(1 + R_{i,t+1} \right)$$

If rates of returns are risk free, then $\tilde{R}_{t+1} = R_{f,t+1}$ and therefore $1 = E^M_t(1)$. For example, for a portfolio of two risk free assets, one in domestic and the other in a foreign country, we have after some elementary manipulations (with $\alpha_D + \alpha_F = 1$):

$$1 = \frac{\alpha_D R^D_{f,t+1} + (1 - \alpha_D) R^F_{f,t+1}}{1 + R^D_{f,t+1}} + \frac{(1 - \alpha_D)(1 + R^F_{f,t+1})}{1 + R^D_{f,t+1}} E^M_t \left(\tilde{R}^\xi_{FD,t+1} \right)$$

While for the international portfolio consisting of risk free and securities assets, we have:

$$1 = \frac{1}{1+R^D_{f,t+1}} \begin{pmatrix} \alpha_D R^D_{f,t+1} + \alpha_F R^F_{f,t+1} + \beta_D E^M_t\left(\tilde{R}^D_{k,t+1}\right) + \left(1+\alpha_F R^F_{f,t+1}\right) E^M_t\left(\tilde{R}^{\xi}_{FD,t+1}\right) \\ + \beta_F E^M_t\left(\tilde{R}^F_{k,t+1}\right) + \beta_F E^M_t\left(\tilde{R}^{\xi}_{FD,t+1}\right) + \beta_F E^M_t\left(\tilde{R}^F_{k,t+1}\tilde{R}^{\xi}_{FD,t+1}\right) \end{pmatrix}$$

Note in particular, the term $E^M_t\left(\tilde{R}^F_{k,t+1}\tilde{R}^{\xi}_{FD,t+1}\right)$ points out to the effects of co-variations of the foreign asset price and FX rate of return.

Example: Implied International Prices: An Arrow-Debreu Model International portfolio investments require some considerations including:

- Pricing and risk assessment of foreign investments
- Expected regime change inducing uncertainty regarding future investments
- Market frictions
- Anticipating natural future global change (desertification, global weather patterns change)
- Sovereign debt and credit risk of Sovereign States and rating of countries
- Price stickiness and overshooting
- Etc.

International portfolios can consist of both specific investments (such as mining firms in Australia in Australian dollars, in Chinese firms listed on the Hong Kong or Shanghai stock markets denoted therefore in HK dollars or in Yuan), etc. Alternatively, there are numerous ETF funds marketed to provide opportunities to investors to invest in a portfolio of commodities traded globally, on regional economic activity policies, on indexed investments to specific financial markets, etc. Such investments, while they may provide greater yields than those accessible in domestic markets may also have also greater risks as we shall see later on. This example considers first a portfolio consisting of fixed income investments (bonds). Subsequently, a number of cases are considered including a portfolio consisting of fixed income and equity as well as debt. Assume an investor in a domestic country whose initial wealth is W^D_0, maximizing his utility of consumption over two periods only. Part of this wealth is consumed now while the remaining part is invested in a domestic and foreign portfolio of risk free bonds. At the next and last period, the portfolio is liquidated and the proceeds are consumed. The current portfolio price is assumed for simplicity to be defined in terms of fixed income investments, thus:

$$c^D_0 = W^D_0 - P^D_0, \quad P^D_0 = B^D_D + B^D_F \quad \text{(in domestic currency)}$$

where P^D_0 is the investment consisting of two bonds, one domestic and the other foreign, given by $\left(B^D_D, B^D_F\right)$, each of which has a risk free rate of return $\left(r_d, r_f\right)$ and ξ^0_{DF}

is the current FX rate for buying the foreign currency and thus investing in the foreign bond market. A period later, the foreign investment is repatriated and the portfolio is liquidated and consumed. Its price (and therefore the amount consumed) is:

$$\begin{aligned}\tilde{P}_1^D = \tilde{c}_1^D &= B_D^D(1+r_d) + B_F^D(1+r_f)\xi_{DF}^0\tilde{\xi}_{FD}^1 \\ &= B_D^D(1+r_d) + B_F^D(1+r_f)\left(1+\tilde{R}_{FD}^{\xi}\right)\end{aligned}$$

while the price of the future consumption is:

$$P_0^D = B_D^D + B_F^D \frac{1+r_f}{1+r_d}\left(1+E^Q\left(\tilde{R}_{FD}^{\xi}\right)\right)$$

Consider next an international domestic investor in Sovereign Bonds, whose portfolio is as stated above: $P_0^D = B_D^D + B_F^D(1-\delta_{DF})$ where δ_{DF} is the investors' transaction cost when buying the foreign currency. A period later, the investment portfolio is worth (in the domestic currency):

$$\begin{aligned}\tilde{P}_1^D = \tilde{c}_1^D = (1-\nu_d)B_D^D(1+r_d) + (1-\nu_d)(1-\nu_f)B_F^D(1+r_f) \\ (1-\delta_{FD})(1-\delta_{DF})(1+\tilde{R}_{FD}^{\xi})\end{aligned}$$

where (ν_d, ν_f) are countries inflation rates, transaction costs are (δ_{DF}) and (δ_{FD}) for reconverting the foreign currency into a domestic one. Maximization of consumption over the two periods by, say a domestic investor is then:

$$U^D(.) = \underset{B_D^D, B_F^D}{\text{Max}}\, u_0^D(c_0) + \mu E^P u_1^D(\tilde{c}_1^D), \quad \left(c_0 = W_0^D - P_0^D; \tilde{c}_1^D = \tilde{P}_1^D\right) \quad \text{or}$$

$$U^D(.) = \left\{\begin{array}{l}\underset{B_D^D, B_F^D}{\text{Max}}\, u_0^D\left(W_0 - \left(B_D^D + B_F^D(1-\delta_{DF})\right)\right) \\ +\mu E u_1^D\left(\begin{array}{l}(1-\nu_d)B_D^D(1+r_d) \\ +(1-\nu_d)(1-\nu_f)B_F^D(1+r_f)(1-\delta_{FD}) \\ (1-\delta_{DF})(1+\tilde{R}_{FD}^{\xi})\end{array}\right)\end{array}\right\}$$

The necessary conditions for the portfolio optimality (with the utility of consumption defined equally for both periods) are:

$$\begin{aligned}\frac{\partial U^D(.)}{\partial B_D^D} &= -\frac{\partial u_0^D(.)}{\partial c_0^D} + \mu(1-\nu_d)(1+r_d)E^P\left(\frac{\partial u_0^D}{\partial \tilde{c}_1^D}\right) = 0 \\ \text{or}\ \frac{1}{(1-\nu_d)(1+r_d)} &= E^P\left(\mu\frac{\partial u_1^D(.)/\partial \tilde{c}_1^D}{\partial u_0^D(.)/\partial c_0^D}\right)\end{aligned}$$

$$\frac{\partial U^D(.)}{\partial B_F^D} = -\frac{\partial u_0^D(.)}{\partial c_0^D}(1-\delta_{DF}) + \mu(1-\nu_f)(1-\delta_{DF})(1-\delta_{FD})(1+r_f)$$
$$E^P\left(\mu\frac{\partial u_1^D(.)/\partial \tilde{c}_1^D}{\partial u_0^D(.)/\partial c_0^D}(1+\tilde{R}_{FD}^{\xi})\right) = 0$$

Setting $\tilde{M}^D = \left(\mu\frac{\partial u_1^D(.)/\partial \tilde{c}_1^D}{\partial u_0^D(.)/\partial c_0^D}\right)\frac{1}{(1-\nu_d)(1+r_d)} = E^P\left(\tilde{M}_0^D\right)$ and assuming a Q probability measure, we have:

$$\frac{(1-\nu_d)(1+r_d)}{(1-\nu_f)(1-\delta_{FD})(1+r_f)} = E^Q\left(1+\tilde{R}_{FD}^{\xi}\right) \quad \text{Or}$$
$$\xi_{FD}^0 = \frac{(1-\nu_f)(1-\delta_{FD})(1+r_f)}{(1-\nu_d)(1+r_d)}E^Q\left(\tilde{\xi}_{FD}^1\right)$$

where $E^Q\left(1+\tilde{R}_{FD}^{\xi}\right)$ expresses the domestic investor expectation with respect to a Q probability measure. Note that in this case, the current FX rate is deflated by a number of economic variables including the countries risk free rates, the transaction costs in FX and the countries inflation rates. Thus, under this measure, the rate of return in the foreign exchange currency (an asset class in this case) is equal to the relative risk free rates of returns between the domestic and the foreign countries. This valuation is the one that expresses the domestic valuation of the FX rate. For the foreign country (investor) we have instead:

$$\xi_{DF}^0 = \frac{(1-\nu_d)(1-\delta_{DF})(1+r_d)}{(1-\nu_f)(1+r_f)}E^V\left(\tilde{\xi}_{DF}^1\right), \quad \xi_{DF}^0 = \frac{1}{\xi_{FD}^0}$$

Again, ξ_{DF}^0 expresses the valuation of the FX rate by the foreign party which at time t $= 0$ must necessarily maintain the equality $\xi_{DF}^0 = \frac{1}{\xi_{FD}^0}$ (assuming no transaction cost). In this case:

$$\xi_{DF}^0 = \frac{(1-\nu_d)(1-\delta_{DF})(1+r_d)E^V\left(\tilde{\xi}_{DF}^1\right)}{(1-\nu_f)(1+r_f)} = \frac{(1-\nu_d)(1+r_d)}{(1-\nu_f)(1-\delta_{FD})(1+r_f)E^Q\left(\tilde{\xi}_{FD}^1\right)}$$

And therefore,

$$E^Q\left(\tilde{\xi}_{FD}^1\right)E^V\left(\tilde{\xi}_{DF}^1\right) = \frac{1}{(1-\delta_{DF})(1-\delta_{FD})}$$

What if in our calculations $\xi_{FD}^0 \neq \frac{1}{\xi_{DF}^0}$, this would mean that FX prices are not priced equally between countries which will provide opportunities for arbitrage as it will allow one party to transfer funds from one place to the other and make a riskless and instantaneous profit (presuming of course that there are no transaction and other costs to include). If there is no inflation and no transaction costs, these equations are reduced to our previous results:

$$1=\frac{1+r_f}{1+r_d}E^Q\left(1+\tilde{R}^{\xi}_{FD}\right) \quad \text{or} \quad \frac{r_d-r_f}{1+r_f}=E^Q\left(\tilde{R}^{\xi}_{FD}\right),\ \frac{r_f-r_d}{1+r_d}=E^V\left(\tilde{R}^{\xi}_{DF}\right)$$

If $(1-\nu_d)(1+r_d)=(1-\nu_f)(1-\delta_{FD})(1+r_f)$, then $E^Q\left(\tilde{R}^{\xi}_{FD}\right)=0$ and $E^V\left(\tilde{R}^{\xi}_{DF}\right)=0$ which points out to a FX stability, as FX rates will be maintained at a given level over time. Of course, if FX rates are maintained to a fixed level arbitrarily, while countries risk free rates, transaction costs and inflation differ, we will also have an expected zero rate of return in the currencies market $E\left(\tilde{R}^{\xi}_{FD}\right)$ $=E\left(\tilde{R}^{\xi}_{DF}\right)=0$ which will contradict the "market price" for the FX rates. Again, such a disparity opens up opportunities for betting for or against the currency in the expectation that eventually, the true FX rate market price will prevail.

Note that the countries risk free deflator implied by kernel pricing is a function of both countries discount rates as well as inflationary forces in both countries and currency transactions costs. For example, if foreign currencies FX rates of returns are null for the domestic investor, then $E^Q\left(\tilde{R}^{\xi}_{FD}\right)=0$ and thereby the interest rate on the foreign bond is a function of countries relative inflation rates and currencies transaction costs:

$$r_f=\frac{(1-\nu_d)(1+r_d)}{(1-\nu_f)(1-\delta_{FD})}-1 \text{ and } r_f=\frac{(1-\nu_f)(1+r_f)}{(1-\nu_d)(1-\delta_{DF})}-1$$

These interest rates are not of course equal and provide therefore an incentive to investors to move their income from one country to the other. For example, if for a domestic investor to repatriate his investment, the transaction cost is 10% while that of the foreign investor is 0%, then ceteris paribus (zero rate of returns in currency speculation), we have: $r_f=\frac{(1-\nu_d)(1+r_d)}{0.90(1-\nu_f)}-1$ and $r_d=\frac{(1-\nu_f)(1+r_f)}{(1-\nu_d)}-1$.

And interest rates differentials are necessarily not equal to zero but a function of the countries' inflation rates. Or,

$$r_f-r_d=\frac{(1-\nu_d)^2(1+r_d)-0.90(1-\nu_f)^2(1+r_f)}{0.90(1-\nu_f)(1-\nu_d)}\neq 0$$

If countries have equal inflation rates, we have: $r_f=0.0555+1.0555r_d$ which requires that the domestic Sovereign treasury rate be much higher. If this is not the case, then investment capital will flow to the foreign country.

This result is generalized next by recursion to indicate over n periods. Considering the case of no transaction costs and no inflation, then:

$$1 = \left(\frac{1+r_f}{1+r_d}\right)^n E^Q\left(\frac{\tilde{\xi}^n_{FD}}{\xi^0_{FD}}\right) \quad \text{and} \quad \xi^0_{FD} = \left(\frac{1+r_f}{1+r_d}\right)^n E^Q\left(\tilde{\xi}^n_{FD}\right)$$

Thus, at the limit, when n is large, $\lim_{n\to\infty}\left(\frac{1+r_f}{1+r_d}\right)^n = e^{-(r_d-r_f)n}$ and therefore, over a period of continuous time t, we have the well known pricing functions:

$$\xi^0_{FD} = e^{-(r_d-r_f)t} E^Q_0\left(\tilde{\xi}^t_{FD}\right) \text{ and also } \xi^t_{FD} = e^{-(r_d-r_f)(T-t)} E^Q_t\left(\tilde{\xi}^T_{FD}\right)$$

where E^Q_t indicates an expectation under the probability measure Q at time t and based on all the information commonly available and shared by all investors (also called the filtration at time t). For call and put options on a foreign currency, the following holds:

$$C^0_{FD}\left(\xi^0_{FD}\right) = e^{-(r_d-r_f)T} E^Q_0\left(C^0_{FD}\left(\tilde{\xi}^T_{FD}\right)\right) = e^{-(r_d-r_f)T} E^Q_0\left(\text{Max}\left(\tilde{\xi}^T_{FD} - \bar{\xi}^C_{FD}, 0\right)\right)$$
$$P^0_{FD}\left(\xi^0_{FD}\right) = e^{-(r_d-r_f)T} E^Q_0\left(P^0_{FD}\left(\tilde{\xi}^T_{FD}\right)\right) = e^{-(r_d-r_f)T} E^Q_0\left(\text{Max}\left(\bar{\xi}^P_{FD} - \tilde{\xi}^T_{FD}, 0\right)\right)$$

where $\bar{\xi}^C_{FD}$ is the strike of the call option and $\bar{\xi}^P_{FD}$ the strike of the put option with T, the options exercise time.

Example: The FX and the Consumer Price Index (CPI) Assume next that there is a purchasing power parity (PPP) between two countries CPI (consumer Price indexes), then $\xi^t_{FD} = \frac{CPI^t_F}{CPI^t_D}$ or:

$$\xi^t_{FD} = \frac{CPI^t_F}{CPI^t_D} \quad \text{and} \quad \tilde{\xi}^T_{FD} = \frac{(CPI)^T_F}{(CPI)^T_D} \quad \text{and} \quad \frac{CPI^t_F}{CPI^t_D} = e^{-(r_d-r_f)(T-t)} E^Q_t\left(\frac{(CPI)^T_F}{(CPI)^T_D}\right)$$

Note however that if the expected inflation rates in CPI indexes over a period of time T-t and in both countries are: $\upsilon^{T,t}_D$ and $\upsilon^{T,t}_F$, then at time t:

$$\frac{CPI^t_F + \Delta CPI^{T,t}_F}{CPI^T_F} 1 + \frac{\Delta CPI^{T,t}_F}{CPI^T_F} = 1 + \upsilon^{T,t}_F \quad \text{and} \quad \frac{CPI^t_D + \Delta CPI^{T,t}_D}{CPI^T_D} 1 + \frac{\Delta CPI^{T,t}_D}{CPI^T_D} = 1 + \upsilon^{T,t}_D$$

Or current expectation of future inflation is embedded in the current inflation rates υ^t_F, υ^t_D and the probability measure Q.

$$\frac{1+\upsilon^t_F}{1+\upsilon^t_D} = e^{-(r_d-r_f)(T-t)} E^Q_t\left(\frac{1+\upsilon^{T,t}_F}{1+\upsilon^{T,t}_D}\right)$$

The actual FX rate results however from interacting market forces reflecting the preferences of the parties (Sovereign States), their economic states, inflation, financial and trade repression and incentives and risk attitudes. In other words, FX rates are defined globally and simultaneously by all countries economic financial and risk characteristics. These are further fueled by investors and speculators in currencies markets who view FX rates as an asset that compounds the uncertainty of their real or financial international transactions. While we have considered above the domestic side of the FX rate, similar results are obtained for the foreign side. Their differences and manipulations of parameters Sovereign States can control thus contribute to a strategic evolution in currency markets.

Example: A Portfolio risk neutrality under a Q probability measure and currency options Let P_0 be a portfolio consisting of domestic and international asset valued at the FX rate $\tilde{\xi}^1_{DF}$. A period later, the portfolio price is $\tilde{P}_1\left(1+\tilde{R}^{\xi}_{FD}\right)$. Under the Q (risk neutral) probability measure, we then have for the portfolio:

$$P_0^D = \frac{1}{1+r_d} E^Q \tilde{P}_1\left(1+\tilde{R}^{\xi}_{FD}\right).$$

This result is verified by including as well the currencies transaction costs. Note that:

$$\begin{aligned} P_0^D &= B_D^D + B_F^D(1-\delta_{DF}) \quad \text{and} \\ \tilde{P}_1^D &= B_D^D(1+r_d) + B_F^D(1+r_f)(1-\delta_{DF})(1-\delta_{FD})(1+\tilde{R}^{\xi}_{FD}) \end{aligned}$$

with δ_{FD} a transaction cost when converting the foreign currency into a domestic one, then:

$$\begin{aligned} P_0^D\left(\xi^0_{DF}\right) &= \frac{1}{1+r_d} E^Q \tilde{P}_1\left(\tilde{\xi}^1_{FD}\right) \Rightarrow B_D^D + B_F^D(1-\delta_{DF}) \\ &= \frac{1}{1+r_d}\left(B_D^D(1+r_d) + B_F^D(1+r_f)(1-\delta_{DF})(1-\delta_{FD})E^Q(1+\tilde{R}^{\xi}_{FD})\right) \end{aligned}$$

However, in this case $(1-\delta_{FD})(1+r_f) = \frac{1}{(1+r_d)} E^Q\left(1+\tilde{R}^{\xi}_{FD}\right)$ and therefore it verifies the "risk neutrality of the portfolio price since":

$$\begin{aligned} & B_D^D + B_F^D(1-\delta_{DF}) \\ & = \frac{1}{1+r_d}\left(B_D^D(1+r_d) + B_F^D(1+r_f)(1-\delta_{DF})(1-\delta_{FD})\frac{(1+r_d)}{(1-\delta_{FD})(1+r_f)}\right) \end{aligned}$$

In this case, $\frac{(1-\delta_{FD})(1+r_f)(1+r_d)-1}{(1+r_d)} = \frac{1}{1+r_d}E^Q\left(\tilde{R}_{FD}^{\xi}\right)$ is the required rate of return for an investor in a currency transaction. Similarly, for a currency call option whose strike is $\bar{R}_{FD}^{\xi}$ and whose price is C_{FD}^0, we have:

$$C_{FD}^0 = \frac{(1-\delta_{FD})(1+r_f)}{(1+r_d)}E^Q\mathrm{Max}\left(1+\tilde{R}_{FD}^{\xi} - \left(1+\bar{R}_{FD}^{\xi}\right),0\right)$$

Such an option provides its owner a profit in case the FX currency price increases (requiring more foreign money to buy back the domestic currency to be repatriated). Now consider a Put option with the same strike, its price is then:

$$\Pi_{FD}^0 = \frac{(1-\delta_{FD})(1+r_f)}{(1+r_d)}E^Q\mathrm{Max}\left(1+\bar{R}_{FD}^{\xi} - (1+\tilde{R}_{FD}^{\xi}),0\right)$$

Such an option provides a speculative profit in case the FX rate declines (in other words, fewer units of the foreign currency will buy a unit of the domestic currency). Therefore their usefulness depends on the belief regarding future changes in FX markets. A portfolio of options can therefore be used to engineer financial hedging and speculative strategy. The investment problem treated above provides a perspective from the domestic investor. A similar situation arises in reverse for the foreign investor. We summarize these results below where for notational simplicity we use:

$$\frac{1}{1+\mathbf{R}^D} = \frac{(1-\nu_f)(1-\delta_{FD})(1+r_f)}{(1-\nu_d)(1+r_d)} \text{ and } \frac{1}{1+\mathbf{R}^F} = \frac{(1-\nu_d)(1-\delta_{DD})(1+r_d)}{(1-\nu_f)(1+r_f)}$$

And therefore for the two countries D and F,

$$\left\{\begin{array}{l} 1 = \left(\frac{1}{1+\mathbf{R}^D}\right)^T E^Q\left(1+\tilde{R}_{FD}^{\xi}(T)\right),\ 1 = \left(\frac{1}{1+\mathbf{R}^F}\right)^T E^V\left(1+\tilde{R}_{DF}^{\xi}(T)\right) \\ \frac{1+\nu_F^t}{1+\nu_D^t} = \left(\frac{1}{1+\mathbf{R}^D}\right)^{T-t} E_t^Q\left(\frac{1+\nu_F^T}{1+\nu_D^T}\right),\ \frac{1+\nu_D^t}{1+\nu_F^t} = \left(\frac{1}{1+\mathbf{R}^F}\right)^{T-t} E_t^V\left(\frac{1+\nu_D^T}{1+\nu_F^T}\right) \end{array}\right\}$$

Discussion: *A quanto option*: A quanto option is an option on a foreign currency denominated asset but the payoff is a domestic currency. In this situation, the holder of the option has a double exposure. On the one hand he is exposed to the asset price risk and to the FX risk. Payoffs can then be determined in several manners. Below are a few cases where ξ_T denotes the FX rate, S_T is the asset price in the foreign currency while K_T is the strike denominated in the foreign currency as well:

(a) The foreign asset is defined in a foreign currency: $C(S_T,\xi_T,T) = \xi_T\max\left(S_T - K_T^F,0\right)$

(b) The foreign asset is denominated in the domestic currency: $C(S_T, \xi_T, T) = \max(\xi_T S_T - K_T^D, 0)$
(c) Exchange rates are fixed at a given level: $C(S_T, \xi_T, T) = \xi_0 \max(S_T - K_T^D, 0)$
(d) Asset linked to the foreign exchange call: $C(S_T, \xi_T, T) = S_T \max(\xi_T - \bar{\xi}_T, 0)$, $\bar{\xi}$ a strike

Other versions can be constructed as well combining hedging (optional) transactions on the underlying foreign asset and on the effective foreign exchange rate at the option's maturity.

9.3.2 *International Credit, Debt Leverage and the Investment Portfolio*

A domestic investor in bonds in a foreign country is a creditor while the foreign country is now a debtor, and vice versa. In many cases, borrowing costs may be smaller in one country than in the other and therefore, investors may incur debt in one country to invest in the other. In such situations the domestic party is the credit holder while the foreign party is the debt holder and both may assume a currency risk in addition to credit risk. To consider the effects of the debt let the initial funds available for both consumption and investment, $\hat{W}_0^D$ with:

$$\hat{W}_0^D = W_0^D + D_D^D + \xi_{FD}^0 D_F^D (1 - \delta_{FD}) - D_D^F.$$

$$\textit{Available Funds} = \textit{Initial Wealth} + \textit{Dom. Debt} + \textit{Foreign Borrowing} - \textit{Foreign Lending}$$

This consists of the domestic investor initial wealth W_0^D, the debt contracted locally D_D^D and the debt contracted in the foreign country in its currency is D_F^D (and therefore subject to a FX rate risk and transaction-conversion cost $\xi_{FD}^0(1 - \delta_{FD})$ for the domestic country). Similarly, D_D^F is the debt contracted by the foreign party to the domestic one. Similarly, in the foreign country, we have:

$$\hat{W}_0^F = W_0^F + D_F^F + \xi_{DF}^0 D_D^F (1 - \delta_{DF}) - D_F^D.$$

Both $\hat{W}_0^D$ and $\hat{W}_0^F$ define the funds available to an investor (or a Sovereign State) which can be applied for three purposes: Consumption (for example, financing services, infrastructure investment, feeding the needy, etc.); invest in one's own country by buying bonds of various denominations (Treasury bills, corporate bonds, Municipalities bonds etc.) and finally investing in foreign countries by lending to the other country and of course investing in both fixed income projects, bonds, equity etc. While ones' own wealth is say W_0^D, the country leverage is:

$$\Lambda_D = \frac{D_D^D + \xi_{FD}^0 D_F^D(1-\delta_{FD})}{W_0^D + D_D^F} = \frac{D_D^D + \xi_{FD}^0 D_F^D(1-\delta_{FD})}{D_D^D + \xi_{FD}^0 D_F^D(1-\delta_{FD})};$$
$$\Lambda_F = \frac{D_F^F + \xi_{DF}^0 D_D^F(1-\delta_{DF})}{W_0^F + D_F^D} = \frac{D_F^F + \xi_{DF}^0 D_D^F(1-\delta_{DF})}{D_F^F + \xi_{DF}^0 D_D^F(1-\delta_{DF})}$$

where Λ_D, Λ_F denote the amount invested (whether risk free or risk assets and programs) divided by the disposable wealth of the party that can be either invested or loaned to another party for whatever purposes that the party will choose. International credit risk arises when transactions (such as a loan to a country's treasury bills) falters. Namely when the domestic country has loaned D_D^F to a foreign country with an expected reimbursement, say a year later, with an added interest, and the foreign country is unable to repay the debt it has acquired, there will then be a debtor's default. Potential defaults, partly or wholly defines the credit risk and thereby affect the FX rate of that country. When countries are debtors and default, these are called Sovereign Default. Although such events are rare, they tend to be extremely damaging when they occur. Potential defaults are more common however in Corporate paper and debt. The probability of default by a party can arise for many reasons which include (among many potential factors) external and internal factors, political events in a country and not least the debtor party leverage Λ_D. For example, a party which is highly leveraged might not be able to meet his commitments if returns forecasts turn out to be disappointing. For these reasons, creditors often demand from debt holder collaterals they may take possession of in case of default.

Say that the domestic party consumes c_0^D in the current period and therefore, the residual available for investment for future consumption is P_0^D and $c_0^D = \hat{W}_0^D - P_0^D$. For simplicity, say that this investment consists of investment in Sovereign State Treasury bonds only. In this case, local debt will be reimbursed locally at the local risk free rate, while a debt acquired in a foreign country will be repaid at the country's risk free rate for foreign debt. Such investments and their returns are however a function of the foreign exchange rates between these countries, transaction costs in FX markets and finally taxation imposed on foreign (and domestic revenues). To simplify, taxation issues will not be considered here. As a result, investments in fixed income are defined by the portfolios:

$$P_0^D = B_D^D + B_F^D(1-\delta_{DF}) \quad \text{and}$$
$$c_0^D = W_0^D + D_D^D + \xi_{FD}^0 D_F^D(1-\delta_{FD}) - D_D^F - B_D^D - B_F^D(1-\delta_{DF})$$

$$PP_0^F = B_F^F + B_D^F(1-\delta_{FD}) \quad \text{and}$$
$$c_0^F = W_0^F + D_F^F + \xi_{DF}^0 D_D^F(1-\delta_{DF}) - D_F^D - B_F^F - B_D^F(1-\delta_{FF})$$

A period later, the domestic portfolio is:

$$\begin{aligned}
\tilde{P}_1^D &= B_D^D(1+r_d) + B_F^D(1+r_f)(1-\delta_{DF})(1-\delta_{FD})(1+\tilde{R}_{FD}^{\xi}) \\
&\quad - D_D^D(1+\tilde{R}_{DD}) - \xi_{FD}^0 D_F^D(1+\tilde{R}_{DF})(1-\delta_{DF})\tilde{\xi}_{DF}^1 + D_D^F(1+\tilde{R}_{FD}) \\
\tilde{P}_1^F &= B_F^F(1+r_f) + B_D^F(1+r_d)(1-\delta_{FD})(1-\delta_{DF})(1+\tilde{R}_{DF}^{\xi}) \\
&\quad - D_F^F(1+\tilde{R}_{FF}) - \xi_{DF}^0 D_D^F(1+\tilde{R}_{FD})(1-\delta_{FD})\tilde{\xi}_{FD}^1 + D_F^D(1+\tilde{R}_{DF})
\end{aligned}$$

For the domestic party, $B_D^D(1+r_d)$ is the return in the domestic bond market while the domestic investment in the foreign market return is $B_F^D(1+r_f)(1-\delta_{DF})$ $(1-\delta_{FD})(1+\tilde{R}_{FD}^{\xi})$ consisting of the foreign rate of return on treasury bills, the rate of return in the foreign currency and the transaction costs in currencies in both directions—from and back to the domestic country. Debt repayments to the foreign country is $\xi_{FD}^0 D_F^D(1+\tilde{R}_{DF})(1-\delta_{DF})\tilde{\xi}_{DF}^1$ including the local (credit risk) interest rate $\tilde{R}_{DF}$ as well as its currency exchange effects (to be repaid in the foreign currency and including a transaction cost), $D_D^D(1+\tilde{R}_{DD})$ is the local debt reimbursed at a random borrowing rate. Finally, $D_D^F(1+\tilde{R}_{FD})$ is the debt contracted by the foreign party which is reimbursed with a rate of return $\tilde{R}_{FD}$ in the domestic currency. Both $\tilde{R}_{FD}$ and $\tilde{R}_{DF}$ express a counterparty risk that each of the party assumes when they become creditors. Of course, bankruptcy occurs in one or the other (or both) countries id the portfolios at the end of period are smaller than the absolutely required consumption needed by both countries' $\left(\bar{c}_1^D, \bar{c}_1^F\right)$. Then $P\left(\tilde{P}_1^F \leq \bar{c}_1^F\right)$ denote the probability of say, the foreign part, not repaying partly or wholly its debt.

Domestic and foreign parties' utilities' of consumption over two periods (with a reimbursement of the debt in the second period) are thus:

$$U^D(.) = \left\{ \begin{array}{l} \underset{B_D^D, B_F^D, D_D^D, D_D^F}{\text{Max}} \; u_0^D\left(W_0^D + D_D^D + \xi_{FD}^0 D_F^D(1-\delta_{FD}) - D_D^F - B_D^D - B_F^D(1-\delta_{DF})\right) \\ + \mu E u_1^D \begin{pmatrix} (1-\nu_d)B_D^D(1+r_d) + (1-\nu_d)B_F^D(1+r_f)(1-\delta_{DF}) \\ (1-\delta_{FD})(1+\tilde{R}_{FD}^{\xi}) - (1-\nu_d)D_D^D(1+\tilde{R}_{DD}) - (1-\nu_f) \\ \xi_{FD}^0 D_F^D(1+\tilde{R}_{DF})(1-\delta_{DF})\tilde{\xi}_{DF}^1 + (1-\nu_d)D_D^F(1+\tilde{R}_{FD}) \end{pmatrix} \end{array} \right\}$$

$$U^F(.) = \left\{ \begin{array}{l} \underset{B_F^F, B_D^F, D_F^F, D_F^D}{\text{Max}} \; u_0^D\left(W_0^F + D_F^F + \xi_{DF}^0 D_D^F(1-\delta_{DF}) - D_F^D - B_F^F - B_D^F(1-\delta_{FD})\right) \\ + \mu E u_1^D \begin{pmatrix} (1-\nu_f)B_F^F(1+r_f) + (1-\nu_f)B_D^F(1+r_d)(1-\delta_{FD}) \\ (1-\delta_{DF})(1+\tilde{R}_{DF}^{\xi}) - (1-\nu_f)D_F^F(1+\tilde{R}_{FF}) - (1-\nu_d) \\ \xi_{DF}^0 D_D^F(1+\tilde{R}_{FD})(1-\delta_{FD})\tilde{\xi}_{FD}^1 + (1-\nu_f)D_F^D(1+\tilde{R}_{DF}) \end{pmatrix} \end{array} \right\}$$

We consider the case of the domestic party first, then:

$$\frac{\partial U^D(.)}{\partial D_D^D} = \frac{\partial u_0^D(c_0^D)}{\partial c_0^D} - \mu E \frac{\partial u_1^D(c_1^D)}{\partial c_1^D}(1-\nu_d)(1+\tilde{R}_{DD}) = 0$$

$$\frac{\partial U^D(.)}{\partial D_D^F} = -\frac{\partial u_0^D(c_0^D)}{\partial c_0^D} + \mu E \frac{\partial u_1^D(c_1^D)}{\partial c_1^D}(1-\nu_d)(1+\tilde{R}_{FD}) = 0$$

$$\frac{\partial U^D(.)}{\partial B_D^D} = -\frac{\partial u_0^D(c_0^D)}{\partial c_0^D} + \mu E \frac{\partial u_1^D(c_1^D)}{\partial c_1^D}(1-\nu_d)(1+r_d) = 0$$

$$\frac{\partial U^D(.)}{\partial B_F^D} = -\frac{\partial u_0^D(c_0^D)}{\partial c_0^D}(1-\delta_{DF}) + \mu E \frac{\partial u_1^D(c_1^D)}{\partial c_1^D}(1-\nu_d)(1+r_f)$$
$$(1-\delta_{DF})(1-\delta_{FD})(1+\tilde{R}_{FD}^{\xi}) = 0$$

And therefore, setting $\tilde{M}^D = \mu \frac{\partial u_1^D(c_1^D)/\partial c_1^D}{\partial u_0^D(c_0^D)/\partial c_0^D}$ these reduce the kernel pricing equations to their Q probability measure:

$$E(\tilde{M}^D) = \frac{1}{(1-\nu_d)(1+r_d)}, \quad 1 = \frac{1}{(1+r_d)} E\left(\frac{\tilde{M}^D}{E(\tilde{M}^D)}(1+\tilde{R}_{DD})\right)$$
$$= \frac{1}{(1+r_d)} E^Q(1+\tilde{R}_{DD})$$
$$1 = \frac{1}{(1+r_d)} E\left(\frac{\tilde{M}^D}{E(\tilde{M}^D)}(1+\tilde{R}_{FD})\right)$$
$$= \frac{1}{(1+r_d)} E^Q(1+\tilde{R}_{FD})$$
$$1 = \frac{(1+r_f)(1-\delta_{FD})}{(1+r_d)} E\left(\frac{\tilde{M}^D}{E(\tilde{M}^D)}(1+\tilde{R}_{FD}^{\xi})\right)$$
$$= \frac{(1+r_f)(1-\delta_{FD})}{(1+r_d)} E^Q\left(1+\tilde{R}_{FD}^{\xi}\right)$$

By the same token for the foreign party, we have under a probability measure V:

$$E(\tilde{M}^F) = \frac{1}{(1-\nu_f)(1+r_f)} \quad \text{And therefore :}$$

$$1 = \frac{1}{(1+r_f)} E\left(\frac{\tilde{M}^F}{E(\tilde{M}^F)}(1+\tilde{R}_{FF})\right) = \frac{1}{(1+r_f)} E^V(1+\tilde{R}_{FF})$$
$$1 = \frac{1}{(1+r_f)} E\left(\frac{\tilde{M}^F}{E(\tilde{M}^F)}(1+\tilde{R}_{DF})\right) = \frac{1}{(1+r_f)} E^V(1+\tilde{R}_{DF})$$
$$1 = \frac{(1+r_d)(1-\delta_{FD})}{(1+r_f)} E\left(\frac{\tilde{M}^F}{E(\tilde{M}^F)}(1+\tilde{R}_{DF}^{\xi})\right) = \frac{(1+r_d)(1-\delta_{DF})}{(1+r_f)} E^V\left(1+\tilde{R}_{DF}^{\xi}\right)$$

These lead to the six pricing equations which we summarize below using their probability measure (alternatively, we could use their kernel prices):

$$1 = \frac{1}{(1+r_d)} E^Q(1+\tilde{R}_{DD}),\ 1 = \frac{1}{(1+r_d)} E^Q(1+\tilde{R}_{FD}),$$
$$1 = \frac{(1+r_f)(1-\delta_{FD})}{(1+r_d)} E^Q\left(1+\tilde{R}^{\xi}_{FD}\right)$$
$$1 = \frac{1}{(1+r_f)} E^V(1+\tilde{R}_{FF});$$
$$1 = \frac{1}{(1+r_f)} E^V(1+\tilde{R}_{DF}),\ 1 = \frac{(1+r_d)(1-\delta_{DF})}{(1+r_f)} E^V\left(1+\tilde{R}^{\xi}_{DF}\right)$$

Note that the probability measures Q and V are not similar, as they account for variety of factors that are specific to the countries. In addition, $E^Q(1+\tilde{R}_{DD}) = E^Q(1+\tilde{R}_{FD})$ which only means that the price of debt (if it is not constrained locally or externally) domestically or in the foreign country are equal. For example, set$\tilde{R}_{DD} = \bar{R}_{DD}$ is the price of debt in US dollars (assumed to be riskless), then the probability measure indicates that $\bar{R}_{DD} = E^Q\left(\tilde{R}_{FD}\right)$ and $\tilde{R}_{FD}$ equals some rate of return which includes a risk premium for the risk of this debt. Considering again the domestic party only, we have:

$$1 = \frac{(1+r_d)}{(1+r_f)(1-\delta_{FD})} E^Q\left(1+\tilde{R}^{\xi}_{FD}\right) \text{ or } \xi^0_{FD} = \frac{(1+r_d)}{(1+r_f)(1-\delta_{FD})} E^Q\left(\tilde{\xi}^1_{FD}\right)$$

which indicates that the FX rate, under the Q measure is debt free. This is of course not the case, since the effects of debt are embedded in the kernel price which define the probability measure Q. Similarly, for the foreign country,

$$1 = \frac{(1+r_f)}{(1+r_d)(1-\delta_{DF})} E^V\left(1+\tilde{R}^{\xi}_{DF}\right) \text{ or } \xi^0_{DF} = \frac{(1+r_f)}{(1+r_d)(1-\delta_{DF})} E^V\left(\tilde{\xi}^1_{DF}\right)$$

The two pricing equations below express what each party considers the current FX price ought to be. It is not therefore the market FX rate unless: $\xi^0_{FD} = 1/\xi^0_{DF}$:

$$\xi^0_{FD} = \frac{(1+r_d)}{(1+r_f)(1-\delta_{FD})} E^Q\left(\tilde{\xi}^1_{FD}\right) \text{ and } \xi^0_{DF} = \frac{(1+r_f)}{(1+r_d)(1-\delta_{DF})} E^V\left(\tilde{\xi}^1_{DF}\right)$$

and

$$\xi^0_{FD} = 1/\xi^0_{DF} \text{ iff } E^Q\left(\tilde{\xi}^1_{FD}\right) E^V\left(\tilde{\xi}^1_{DF}\right) = (1-\delta_{DF})(1-\delta_{FD})$$

For example, let there be options on the following FX rates

$$C_{FD}\left(\xi_{FD}^{0},K_{FD}\right)=\frac{(1+r_d)}{(1+r_f)(1-\delta_{FD})}E^{Q}Max\left(\tilde{\xi}_{FD}^{1}-K_{FD},0\right)$$

$$C_{DF}\left(\xi_{DF}^{0},K_{DF}\right)=\frac{(1+r_f)}{(1+r_d)(1-\delta_{DF})}E^{V}Max\left(\tilde{\xi}_{DF}^{1}-K_{DF},0\right)$$

And say that $\tilde{\xi}_{FD}^{1}=\left(\tilde{\xi}_{FD}^{+},\tilde{\xi}_{FD}^{-}\right)$, $\tilde{\xi}_{DF}^{1}=\left(\tilde{\xi}_{DF}^{+},\tilde{\xi}_{DF}^{-}\right)$, thus the implied probabilities are:

$$C_{FD}\left(\xi_{FD}^{0},K_{FD}\right)=p^{Q}\frac{(1+r_d)}{(1+r_f)(1-\delta_{FD})}\left(\tilde{\xi}_{FD}^{+}-K_{FD}\right)$$

$$\text{and } p^{Q}=\frac{C_{FD}\left(\xi_{FD}^{0},K_{FD}\right)(1+r_f)(1-\delta_{FD})}{(1+r_d)\left(\tilde{\xi}_{FD}^{+}-K_{FD}\right)}$$

$$C_{DF}\left(\xi_{DF}^{0},K_{DF}\right)=p^{V}\frac{(1+r_f)}{(1+r_d)(1-\delta_{DF})}\left(\tilde{\xi}_{DF}^{+}-K_{DF}\right)$$

$$\text{and } p^{V}=\frac{C_{DF}\left(\xi_{DF}^{0},K_{DF}\right)(1+r_d)(1-\delta_{DF})}{(1+r_f)\left(\tilde{\xi}_{DF}^{+}-K_{DF}\right)}$$

In which case,

$$\begin{aligned}\xi_{FD}^{0}&=\frac{(1+r_d)}{(1+r_f)(1-\delta_{FD})}\left(p^{Q}\tilde{\xi}_{FD}^{+}+(1-p^{Q})\tilde{\xi}_{FD}^{-}\right)\\&=\frac{(1+r_d)}{(1+r_f)(1-\delta_{FD})}\left(\left(\frac{C_{FD}\left(\xi_{FD}^{0},K_{FD}\right)(1+r_f)(1-\delta_{FD})}{(1+r_d)\left(\tilde{\xi}_{FD}^{+}-K_{FD}\right)}\right)\left(\tilde{\xi}_{FD}^{+}-\tilde{\xi}_{FD}^{-}\right)+\tilde{\xi}_{FD}^{-}\right)\\\xi_{DF}^{0}&=\frac{(1+r_f)}{(1+r_d)(1-\delta_{DF})}\left(p^{V}\tilde{\xi}_{DF}^{+}+(1-p^{V})\tilde{\xi}_{DF}^{-}\right)\\&=\frac{(1+r_f)}{(1+r_d)(1-\delta_{DF})}\left(\frac{C_{DF}\left(\xi_{DF}^{0},K_{DF}\right)(1+r_d)(1-\delta_{DF})}{(1+r_f)\left(\tilde{\xi}_{DF}^{+}-K_{DF}\right)}\left(\tilde{\xi}_{DF}^{+}-\tilde{\xi}_{DF}^{-}\right)+\tilde{\xi}_{DF}^{-}\right)\end{aligned}$$

And therefore, the market price for the FX rate is defined if: $\xi_{FD}^{0}=\frac{1}{\xi_{DF}^{0}}$ or:

$$\begin{aligned}&\frac{(1+r_d)}{(1+r_f)(1-\delta_{FD})}\left(\left(\frac{C_{FD}\left(\xi_{FD}^{0},K_{FD}\right)(1+r_f)(1-\delta_{FD})}{(1+r_d)\left(\tilde{\xi}_{FD}^{+}-K_{FD}\right)}\right)\left(\tilde{\xi}_{FD}^{+}-\tilde{\xi}_{FD}^{-}\right)+\tilde{\xi}_{FD}^{-}\right)\\&=\left\{\frac{(1+r_f)}{(1+r_d)(1-\delta_{DF})}\left(\frac{C_{DF}\left(\xi_{DF}^{0},K_{DF}\right)(1+r_d)(1-\delta_{DF})}{(1+r_f)\left(\tilde{\xi}_{DF}^{+}-K_{DF}\right)}\left(\tilde{\xi}_{DF}^{+}-\tilde{\xi}_{DF}^{-}\right)+\tilde{\xi}_{DF}^{-}\right)\right\}^{-1}\end{aligned}$$

These results allow us then to compare the prices of call options (as well put options) to test whether currency FX prices are in equilibrium or they provide arbitrage opportunities.

Example: Using the Kernel Prices Using the kernel prices for both countries, we have:

$$E(\tilde{M}^D) = \frac{1}{(1-\nu_d)(1+r_d)}, \quad E(\tilde{M}^F) = \frac{1}{(1-\nu_f)(1+r_f)} \quad \text{and}$$

$$\tilde{M}^{\bar{D}} = \frac{\tilde{M}^D}{E(\tilde{M}^D)}, \tilde{M}^{\bar{F}} = \frac{\tilde{M}^F}{E(\tilde{M}^F)}$$

where

$$\tilde{M}^D = \mu_D \frac{\partial u_1^D(c_1^D)/\partial c_1^D}{\partial u_0^D(c_0^D)/\partial c_0^D}, \tilde{M}^F = \mu_F \frac{\partial u_1^F(c_1^F)/\partial c_1^F}{\partial u_0^F(c_0^D)/\partial c_0^F}, \quad \text{with } u_1^i(x) = \frac{x^{1-\gamma_i}}{1-\gamma_i}$$

Then:

$$\tilde{M}^D = \mu_D \left(\frac{c_1^D}{c_0^D}\right)^{-\gamma_D}, \tilde{M}^F = \mu_F \left(\frac{c_1^F}{c_0^F}\right)^{-\gamma_F} \quad \text{and}$$

$$\tilde{M}^{\bar{D}} = \mu_D \left(\frac{\tilde{c}_1^D}{E(\tilde{c}_1^D)}\right)^{-\gamma_D}, \tilde{M}^{\bar{F}} = \mu_F \left(\frac{\tilde{c}_1^F}{E(\tilde{c}_1^F)}\right)^{-\gamma_F}$$

And therefore, as seen earlier:

$$1 = \frac{1}{(1-\nu_d)(1+r_d)} E\left(\mu_D \left(\frac{\tilde{c}_1^D}{E(\tilde{c}_1^D)}\right)^{-\gamma_D} (1+\tilde{R}_{DD})\right),$$

$$1 = \frac{1}{(1-\nu_f)(1+r_f)} E\left(\mu_F \left(\frac{\tilde{c}_1^F}{E(\tilde{c}_1^F)}\right)^{-\gamma_F} (1+\tilde{R}_{FF})\right)$$

$$1 = \frac{1}{(1-\nu_d)(1+r_d)} E\left(\mu_D \left(\frac{\tilde{c}_1^D}{E(\tilde{c}_1^D)}\right)^{-\gamma_D} (1+\tilde{R}_{FD})\right),$$

$$1 = \frac{1}{(1-\nu_f)(1+r_f)} E\left(\mu_F \left(\frac{\tilde{c}_1^F}{E(\tilde{c}_1^F)}\right)^{-\gamma_F} (1+\tilde{R}_{DF})\right)$$

$$1 = \frac{(1-\nu_f)(1+r_f)(1-\delta_{FD})}{(1-\nu_d)(1+r_d)} E\left(\mu_D \left(\frac{\tilde{c}_1^D}{E(\tilde{c}_1^D)}\right)^{-\gamma_D} (1+\tilde{R}_{FD}^{\xi})\right),$$

$$1 = \frac{(1-\nu_d)(1+r_d)(1-\delta_{FD})}{(1-\nu_f)(1+r_f)} E\left(\mu_F \left(\frac{\tilde{c}_1^F}{E(\tilde{c}_1^F)}\right)^{-\gamma_F} (1+\tilde{R}_{DF}^{\xi})\right)$$

Note that each of these equations valuations in both countries can be different. Using for each its own probability measure, we have:

$$1 = \frac{1}{(1-\nu_d)(1+r_d)} E^Q\left(1+\tilde{R}_{DD}\right), \quad 1 = \frac{1}{(1-\nu_f)\left(1+r_f\right)} E^V\left(1+\tilde{R}_{FF}\right)$$
$$1 = \frac{1}{(1-\nu_d)(1+r_d)} E^Q\left(1+\tilde{R}_{FD}\right), 1 = \frac{1}{(1-\nu_f)\left(1+r_f\right)} E^V\left(1+\tilde{R}_{DF}\right)$$
$$1 = \frac{(1-\nu_f)\left(1+r_f\right)(1-\delta_{FD})}{(1-\nu_d)(1+r_d)} E^Q\left(1+\tilde{R}^{\xi}_{FD}\right),$$
$$1 = \frac{(1-\nu_d)(1+r_d)(1-\delta_{FD})}{(1-\nu_f)\left(1+r_f\right)} E^V\left(1+\tilde{R}^{\xi}_{DF}\right)$$

A thoughts experiment on debt risk and its price: On Inflation and rate of returns (check this again): Assume that inflation is high, then the effect of the risk free rate is smaller since $(1-\nu_d)(1+r_d) < (1+r_d)$. For example, borrowing money in a foreign country in its currency when the inflation is extremely large (or forthcoming) is advantageous. In such cases, the required rate of return of the borrower will be smaller. Similar considerations are applied on FX rates. However, note that both transaction costs (in the domestic and in the foreign country) can be appreciable which requires a higher rate of return on foreign (currency) investments. To compensate these effects, Quantos options are needed to manage the potential loss in currency conversion.

9.3.3 FX Rates Risk, Bonds and Equity

Assume an investment portfolio consisting of domestic and foreign risk free bonds and equity (say an equity index in both countries). We also assume no debt. Then:

$$P_0^D = B_D^D + n_D^D S_D^D + B_F^D + n_F^D S_F^D$$

where $n_D S_D$ and $n_F S_F$ are the investment in two equity indexes—one domestic and the other foreign. Similarly, for the foreign country, we have (in the foreign country currency):

$$P_0^F = B_F^F + n_F^F S_F^F + B_D^F + n_D^F S_D^F$$

At the next period, the price of the portfolio is random with and equal to:

$$\tilde{P}_1^D = B_D^D(1+r_d) + n_D^D S_D^D(1+\tilde{R}_D) + \left(B_F^D(1+r_f) + n_F^D S_F^D(1+\tilde{R}_F)\right)(1+R^{\xi}_{FD})$$

with $\tilde{R}_D, \tilde{R}_F$, are the domestic and the foreign indexes rates of returns. For the foreign investor, we have:

$$\tilde{P}_1^F = B_F^F(1+r_f) + n_F^F S_F^F(1+\tilde{R}_F) + \left(B_D^F(1+r_d) + n_D^F S_D^F(1+\tilde{R}_D)\right)(1+R^{\xi}_{DF})$$

The investors expected utility of consumption in both countries over a two periods model are then:

$$U_D(c_0^D, \tilde{c}_{11}^D) = u_D(c_0^D) + \beta E_0 u_D(\tilde{c}_1^D) = u_D(W_0^D - P_0^D) + \beta E_0 u_D(\tilde{P}_1^D)$$
$$U_F(c_0^F, \tilde{c}_1^F) = u_F(c_0^F) + \beta E_0 u_F(\tilde{c}_1^F) = u_F(W_0^D - P_0^F) + \beta E_0 u_F(\tilde{P}_1^F)$$

We consider first the domestic investor problem is:

$$\underset{B_D^D, B_F^D, n_D^D, n_F^D}{\text{Max}} U_D(c_0^D, \tilde{c}_{11}^D) = u_D(W_0^D - P_0^D) + \beta E_0 u_D(\tilde{P}_1^D)$$

The necessary conditions for optimality are then:

$$\frac{\partial U_D}{\partial B_D^D} = -\frac{\partial u_D}{\partial P_0^D}\frac{\partial P_0^D}{\partial B_D^D} + \beta E_0\left(\frac{\partial u_D}{\partial P_1^D}\frac{\partial \tilde{P}_1^D}{\partial B_D^D}\right)$$
$$\text{and } \frac{\partial U_D}{\partial B_D^D} = -\frac{\partial u_D}{\partial P_0^D} + \beta(1+r_d)E_0\left(\frac{\partial u_D}{\partial P_1^D}\right) = 0$$
$$\frac{\partial U_D}{\partial B_F^D} = -\frac{\partial u_D}{\partial P_0^D}\frac{\partial P_0^D}{\partial B_F^D} + \beta E_0\left(\frac{\partial u_D}{\partial P_1}\frac{\partial \tilde{P}_1^D}{\partial B_F^D}\right)$$
$$\text{and } \frac{\partial U_D}{\partial B_F^D} = -\frac{\partial u_D}{\partial P_0^D} + \beta(1+r_f)E_0\left(\frac{\partial u_D}{\partial P_1^D}(1+\tilde{R}_{FD}^{\xi}\right) = 0$$
$$\frac{\partial U_D}{\partial n_D^D} = -\frac{\partial u_D}{\partial P_0^D}\frac{\partial P_0^D}{\partial n_D^D} + \beta E_0\left(\frac{\partial u_D}{\partial P_1}\frac{\partial \tilde{P}_1}{\partial n_D^D}\right)$$
$$\text{and } \frac{\partial U_D}{\partial n_D^D} = -S_D^D\frac{\partial u_D}{\partial P_0^D} + \beta E_0\left((1+\tilde{R}_D)S_D^D\frac{\partial u_D}{\partial P_1}\right) = 0$$
$$\frac{\partial U_D}{\partial n_F^D} = -\frac{\partial u_D}{\partial P_0^D}\frac{\partial P_0^D}{\partial n_F^D} + \beta E_0\left(\frac{\partial u_D}{\partial P_1}\frac{\partial \tilde{P}_1^D}{\partial n_F^D}\right)$$
$$\text{and } \frac{\partial U_D}{\partial n_F^D} = -S_F\frac{\partial u_D}{\partial P_0^D} + \beta S_F E_0\left(\frac{\partial u_D}{\partial P_1^D}(1+\tilde{R}_F)(1+\tilde{R}_{FD}^{\xi})\right) = 0$$

And therefore, setting $\tilde{M}_0^D = \left(\beta\frac{\partial u_D/\partial \tilde{P}_1^D}{\partial u_D/\partial P_0^D}\right)$ we have:

$$\frac{1}{1+r_d} = E_0^D(\tilde{M}_0^D), \quad \frac{1+r_d}{1+r_f} = E_0^D\left(\frac{\tilde{M}_0^D}{E(\tilde{M}_0^D)}\left(1+\tilde{R}_{FD}^{\xi}\right)\right)$$
$$1 = \frac{1}{1+r_d}E_0\left(\frac{\tilde{M}_0^D}{E(\tilde{M}_0^D)}(1+\tilde{R}_D)\right),$$
$$1 = \frac{1}{1+r_d}E_0\left(\frac{\tilde{M}_0^D}{E(\tilde{M}_0^D)}(1+\tilde{R}_F)\left(1+\tilde{R}_{FD}^{\xi}\right)\right)$$

Assuming the Q probability measure for the domestic investor, we have:

$$\frac{1}{(1+r_d)} = E_0^Q(1), \quad \frac{1+r_d}{1+r_f} = E_0^Q\left(1+\tilde{R}_{FD}^{\xi}\right)$$
$$1 = \frac{1}{1+r_d}E_0^Q\left(1+\tilde{R}_D\right), \quad 1 = \frac{1}{1+r_d}E_0^Q\left(\left(1+\tilde{R}_F\right)\left(1+\tilde{R}_{FD}^{\xi}\right)\right)$$

These equations can be written conveniently with respect to the probability measure Q as follows:

$$\frac{r_d - r_f}{1+r_f} = E_0^Q\left(\tilde{R}_{FD}^{\xi}\right), \quad r_d = E_0^Q\left(\tilde{R}_D\right), \quad r_d = E_0^Q\left(\tilde{R}_F\right) + E_0^Q\left(\tilde{R}_{FD}^{\xi}\right) + E_0^Q\left(\tilde{R}_F\tilde{R}_{FD}^{\xi}\right)$$

Replacing $E_0^Q\left(\tilde{R}_{FD}^{\xi}\right)$, we have,

$$\frac{r_d - r_f}{1+r_f} = E_0^Q\left(\tilde{R}_D\right) - E_0^Q\left(\tilde{R}_F\right) - E_0^Q\left(\tilde{R}_F\tilde{R}_{FD}^{\xi}\right)$$

Of course, if $r_d = r_f$ then $-E_0^Q\left(\tilde{R}_F\tilde{R}_{FD}^{\xi}\right) = E_0^Q\left(\tilde{R}_F\right) - E_0^Q\left(\tilde{R}_D\right)$. And therefore, the difference expected rates of returns from an investment in a domestic and foreign country (from the domestic investor perspective) is that the required additional rate of return $E_0^Q\left(\tilde{R}_F\right) - E_0^Q\left(\tilde{R}_D\right)$ depends on the co-variation of $\tilde{R}_F$ and $\tilde{R}_{FD}^{\xi}$. Explicitly, since $E_0^Q\left(\tilde{R}_F\tilde{R}_{FD}^{\xi}\right) = E_0^Q\left(\tilde{R}_F\right)E_0^Q\left(\tilde{R}_{FD}^{\xi}\right) + \rho_{F,FD}^Q\sigma_F^Q\sigma_{FD}^Q$ and:

$$0 = E_0^Q\left(\tilde{R}_F\right)\left[E_0^Q\left(1+\tilde{R}_{FD}^{\xi}\right)\right] - E_0^Q\left(\tilde{R}_D\right) + \rho_{F,FD}^Q\sigma_F^Q\sigma_{FD}^Q$$
$$= E_0^Q\left(\tilde{R}_F\right)\frac{1+r_d}{1+r_f} - E_0^Q\left(\tilde{R}_D\right) + \rho_{F,FD}^Q\sigma_F^Q\sigma_{FD}^Q = E_0^Q\left(\tilde{R}_F\right) - E_0^Q\left(\tilde{R}_D\right) + \rho_{F,FD}^Q\sigma_F^Q\sigma_{FD}^Q$$

As a result, if $\rho_{F,FD}^Q < 0$, meaning that the greater the FX equity market, the stronger its currency (in term of its exchange for the domestic currency) and therefore, $E_0^Q\left(\tilde{R}_F\right) = E_0^Q\left(\tilde{R}_D\right) + \left(\left|\rho_{F,FD}^Q\right|\sigma_F^Q\sigma_{FD}^Q\right)$ which define the required (FX) premium for a foreign investment. When the fixed income rates of returns are not equal, $r_d \neq r_f$ we have:

$$E_0^Q\left(\tilde{R}_F\right) = \frac{1+r_f}{1+r_d}\left(E_0^Q\left(\tilde{R}_D\right) - \rho_{F,FD}^Q\sigma_F^Q\sigma_{FD}^Q\right) - \frac{r_d - r_f}{1+r_d}, \quad \rho_{F,FD}^Q < 0$$

For example, if $r_d < r_f$, the required rate of return $E_0^Q\left(\tilde{R}_F\right)$ is increased due to a greater rate of return in its (competing) fixed income rate r_f. Using a CAPM-model notation, we set:

$$\beta^D_{F,FD} = -\rho^Q_{F,FD}\frac{\sigma^Q_F}{\sigma^Q_{FD}},\quad -\rho^Q_{F,FD}\sigma^Q_F\frac{\sigma^{Q\ 2}_{FD}}{\sigma^Q_{FD}} = \beta^D_{F,FD}\sigma^{Q\ 2}_{FD},\quad \beta^D_{F,FD} > 0$$

And therefore $E^Q_0\left(\tilde{R}_F\right) = \frac{1+r_f}{1+r_d}\left(E^Q_0\left(\tilde{R}_D\right) + \beta^D_{F,FD}\sigma^{Q\ 2}_{FD}\right) - \frac{r_d - r_f}{1+r_d},\ \rho^Q_{F,FD} < 0$

$$\frac{1+r_d}{1+r_f}\left(E^Q_0\left(\tilde{R}_F\right) + \frac{r_d - r_f}{1+r_d}\right) = E^Q_0\left(\tilde{R}_D\right) + \beta^D_{F,FD}\sigma^{Q\ 2}_{FD}$$

Again, if the fixed income rates of returns are equal, then:

$$E^Q_0\left(\tilde{R}_F\right) = E^Q_0\left(\tilde{R}_D\right) + \beta^D_{F,FD}\sigma^{Q\ 2}_{FD}$$

where $\beta^D_{F,FD}\sigma^{Q\ 2}_{FD}$ denote the required rate of returns premium. If this is not the case, then note that if $r_d > r_f$ the left hand side of the equation will be greater, requiring a greater risk premium for investing in the foreign country, and vice versa if $r_d < r_f$.

Note that $E^Q_0\left(\tilde{R}_F\right), E^Q_0\left(\tilde{R}^{\xi}_{FD}\right), E^Q_0\left(\tilde{R}_F\tilde{R}^{\xi}_{FD}\right)$ are expectations under a probability measure Q and can thus be estimated on the basis of options prices. This Q-Beta CAPM model differs from the standard CAPM model. Therefore, the $\beta^D_{F,FD}$ used here may be called a "forward beta." Say that:

$$E^Q_0\left(\tilde{R}_D\right) = 0.05,\ E^Q_0\left(\tilde{R}_F\right) = 0.12,$$
$$\text{then}:\ \frac{1+r_d}{1+r_f}\left(0.12 + \frac{r_d - r_f}{1+r_d}\right) = 0.05 + \beta^D_{F,FD}\sigma^{Q\ 2}_{FD}.$$

Thus for the foreign country to reduce the premium foreign investors require to invest in their country, they will have to increase their risk free rate. Say that $r_d = 0.04$ then:

$$\frac{1.04}{1+r_f}\left(0.12 - \frac{r_f - 0.04}{1.04}\right) = 0.05 + \beta^D_{F,FD}\sigma^{Q\ 2}_{FD} \quad \text{and} \quad r_f = \frac{0.11 - \beta^D_{F,FD}\sigma^{Q\ 2}_{FD}}{1.05 + \beta^D_{F,FD}\sigma^{Q\ 2}_{FD}}$$

This latter expression indicates the relationship between the required risk premium and the foreign risk free rate. If risk free rates are equal, $r_f = 0.04$, then $\beta^D_{F,FD}\sigma^{Q\ 2}_{FD} = \frac{0.15}{1.04} = 0.1428$ which is the risk required risk premium. Say that the foreign country currency is appreciating as a rate of 6% (in repatriating invested funds in the foreign country), then $E^Q_0\left(\tilde{R}^{\xi}_{FD}\right) = 0.06$, and $1.12E^Q_0\left(\tilde{R}^{\xi}_{FD}\right) = 0.0672$ then of course it is advantageous for the domestic investor to invest in the foreign country. However if the currency depreciates at the same rate we still have

$-0.0672 > -0.07$ which continues to justify investing in the foreign country. For a foreign investor, similar considerations are applied (but in reverse). In this case:

$$\beta^F_{D,DF} = \frac{1}{\left(\sigma^Q_{DF}\right)^2}\left\{\left(E^V_0\left(\tilde{R}_F\right) - \frac{r_f - r_d}{1 + r_d}\right) - E^V_0\left(\tilde{R}_D\right)\left(\frac{1 + r_f}{1 + r_d}\right)\right\}$$

or

$$\beta^F_{D,DF} = -\frac{1}{\left(\sigma^Q_{DF}\right)^2}\left\{E^V_0\left(\tilde{R}_F\right)\left(1 + E^V_0\left(\tilde{R}^\xi_{DF}\right)\right) - \left(E^V_0\left(\tilde{R}_D\right) - E^V\left(\tilde{R}^\xi_{DF}\right)\right)\right\}$$

Since for a foreign investor, an appreciation of his currency is necessarily the inverse of that of the domestic currency (it is not necessarily true in this case, as the expectations are taken with respect to two different probability measures), we have in case $E^Q_0\left(\tilde{R}^\xi_{FD}\right) = 0.06$ then $E^Q_0\left(\tilde{R}^\xi_{DF}\right) = -0.06$ in which case:

$$\beta^F_{D,DF} = -\frac{1}{\left(\sigma^Q_{DF}\right)^2}\{0.12(1 - 0.06) - (0.05 + 0.06)\}$$

$$= -\frac{1}{\left(\sigma^Q_{DF}\right)^2}\{0.1128 - (0.11)\} < 0$$

And therefore, the foreign investor would invest in the foreign country.

Example. Foreign exchange prices are as stated above, relative prices. A foreign exchange rate is then measured relative to say, the dollar—normally assumed to be a "strong" and a reference currency. Say that two countries (say the US and China) and consider:

$$\frac{1 + r_d}{1 + r_f}\left(E^Q_0\left(\tilde{R}_F\right) + \frac{r_d - r_f}{1 + r_d}\right) = E^Q_0\left(\tilde{R}_D\right) + \beta^D_{F,FD}{\sigma^Q_{FD}}^2;$$

$$\frac{1 + r_f}{1 + r_d}\left(E^V_0\left(\tilde{R}_D\right) + \frac{r_f - r_d}{1 + r_f}\right) = E^V_0\left(\tilde{R}_f\right) + \beta^F_{D,DF}{\sigma^Q_{DF}}^2$$

First note that the correlation between the rates of returns in the foreign equity market and the FX rate from repatriating revenues from the foreign country is negative. This is the case, since when the FX rate increases, the investor would have to use more of the foreign currency to repatriate the income he made in that country. And vice versa, when the foreign economy has a declining FX rate, the required rate of return is smaller, as it is compensated by the increased currency exchange rate of return. When there is no correlation, then $\beta^D_{F,FD} = \beta^F_{D,DF} = 0$ and therefore:

$$E^Q_0\left(\tilde{R}_F\right) = \frac{1 + r_f}{1 + r_d}E^Q_0\left(\tilde{R}_D\right) - \frac{r_d - r_f}{1 + r_d} \text{ and } E^V_0\left(\tilde{R}_D\right) = \frac{1 + r_d}{1 + r_f}E^V_0\left(\tilde{R}_f\right) - \frac{r_f - r_d}{1 + r_f}$$

In other words, say that the risk free rate in the domestic country is greater than in the foreign country. In this case, $r_d - r_f > 0$ and therefore the rate of return in the foreign equity market is greater than that in the domestic market.

$$E_0^Q\left(\tilde{R}_F\right) = \frac{E_0^Q\left(\tilde{R}_D\right)\left(1+r_f\right) - \left(r_d - r_f\right)}{\left(1+r_d\right)} \quad \text{or} \quad E_0^Q\left(\tilde{R}_F\right) > E_0^Q\left(\tilde{R}_D\right)$$

Of course, if the risk free rates of return in both countries are equal then $r_d = r_f$ and therefore,

$$E_0^Q\left(\tilde{R}_F\right) = E_0^Q\left(\tilde{R}_D\right) + \beta_{F,FD}^D {\sigma_{FD}^Q}^2 \text{ and } E_0^V\left(\tilde{R}_D\right) = E_0^V\left(\tilde{R}_f\right) + \beta_{D,DF}^F {\sigma_{DF}^Q}^2$$

Evidently both the FX rate volatility and the foreign market equity index increases the required rate of return on the foreign index. There are then two countervailing forces, on the one hand stability of the FX rate reduces the required foreign exchange risk premium and on the other, the strong (negative) correlation between the foreign equity index and FX lead a growth in the index required rate of return premium. These observations can be used to make the case for China's FX policy on the condition that there is no volatility in this exchange. Inversely, by maintaining equity prices in China's index to be far more in concordance with "world" rates of returns, the correlation between these returns and a free floating exchange rate would necessary be smaller but with a greater volatility in the China foreign exchange rate.

Thought experiments: Consider a present the risk premium for local investors to invest in their own country. Based on our conditions, for optimality, we have:

$$E_0^Q\left(\tilde{R}_D\right) - r_d = \left(\frac{1+r_d}{1+r_f} E_0^Q\left(\tilde{R}_F\right) - \frac{r_f(1-r_d)}{1+r_f} - \beta_{F,FD}^D {\sigma_{FD}^Q}^2\right)$$

$$E_0^V\left(\tilde{R}_f\right) - r_f = \left(\frac{1+r_f}{1+r_d} E_0^V\left(\tilde{R}_D\right) - \frac{r_d(1-r_f)}{1+r_d} - \beta_{D,DF}^F {\sigma_{DF}^Q}^2\right)$$

Since, if $E_0^Q\left(\tilde{R}_D\right) - r_d = E_0^V\left(\tilde{R}_f\right) - r_f = 0$, we have:

$$E_0^Q\left(\tilde{R}_F\right) = \frac{1}{\left(1+r_d\right)}\left(r_f(1-r_d) + \left(1+r_f\right)\beta_{F,FD}^D {\sigma_{FD}^Q}^2\right)$$

$$E_0^V\left(\tilde{R}_D\right) = \frac{1}{\left(1+r_f\right)}\left(r_d(1-r_f) + \left(1+r_d\right)\beta_{D,DF}^F {\sigma_{DF}^V}^2\right)$$

If equity markets have the same rate of return (priced by investors in their respective country), then: $E_0^Q\left(\tilde{R}_F\right) = E_0^V\left(\tilde{R}_D\right)$ and therefore the following relationship holds between the risk free rates and the risk premiums:

$$\frac{\left(1+r_f\right)r_f\left(1-r_d\right)}{\left(1+r_d\right)^2} - \frac{r_d\left(1-r_f\right)}{\left(1+r_d\right)} + \frac{\left(1+r_f\right)^2}{\left(1+r_d\right)^2}\beta_{F,FD}^D {\sigma_{FD}^Q}^2 = \beta_{D,DF}^F {\sigma_{DF}^V}^2$$

Say that China "stabilizes" its FX rate by not letting it fluctuate. In this case, the propensity of investors in China will be to accept a smaller spread between investment in China's equity and bond markets. The implications of such a policy for China is that such stability (assuming that the FX rate is a fair one rather than an artificially managed one) will contribute to foreign capital inflows, attracted by the risk free returns and their conversion into the domestic currency. In this sense, foreign investors will buy China's obligations (rather than risk investment in China's equity markets) which would require an increase in China's foreign currency. When country's that are not as large or as powerful as China are faced with such a situation, a potential for arbitrage arises with currency bets engineered to profit when the foreign country can no longer finance the stability of its currency. Such a situation arose when the Sterling was "attacked" by a group of hedge funds and resulting in a great loss to England's Central Bank decades ago. It assumes may forms today such as "a play" on Hong Kong with a position taken by a Hedge Fund in the hope that the Honk Kong dollar which is "indexed" to the US dollar, to be revalued (*The Economist*, September 26, 2011). If this were to happen, then Hedge Fund would realize and appreciable profit.

These thoughts exercise are of course biased by our considering relatively simple portfolios as well as our considering one country at a time. For two simultaneous countries, we have:

$$r_d = E_0^Q\left(\tilde{R}_D\right),\ \frac{r_d - r_f}{1 + r_f} = E_0^Q\left(\tilde{R}_{FD}^{\xi}\right),\ E_0^Q\left(\tilde{R}_F\right) = r_f - \frac{\beta_{F,FD}^{D}\left(\sigma_{FD}^{Q}\right)^2}{E_0^Q\left(1 + \tilde{R}_{DF}^{\xi}\right)}$$

$$r_f = E_0^V\left(\tilde{R}_F\right),\ \frac{r_f - r_d}{1 + r_d} = E_0^V\left(\tilde{R}_{DF}^{\xi}\right),\ E_0^V\left(\tilde{R}_D\right) = r_d - \frac{\beta_{D,DF}^{F}\left(\sigma_{DF}^{V}\right)^2}{E_0^V\left(1 + \tilde{R}_{FD}^{\xi}\right)}$$

with $\tilde{\xi}_{DF}^{1} = 1/\tilde{\xi}_{FD}^{1}$.

These equations are of course more difficult to analyze as both are defined with respect to two different probability measures.

9.4 Additional Applications

9.4.1 *Finance and Insurance: Pricing Contrasts and Similarities*

Deregulation has allowed both insurance and financial firms to enter into each other domains. These have led to financial innovation in insurance and credit derivatives as well as induced greater competition and a greater complexity of financial products. However, instead of banks and insurance firms' mergers, we have witnessed a growth of financial institutions, becoming Too Big To Fail (Taleb and Tapiero 2010), through an expansion of their business to "one stop financial-insurance shops."

These had led to systemic risks, depending on the actions that these TBTF—both insurance and finance firms TBTF, would do. It is leading to a demand for greater regulation to reduce both the excessive use of their power on the one hand and mitigate their systemic risks. Insurance and financial firms are not similar however, requiring thereby specific approaches to their control. Weber (2002), for example, indicates that non (firms)-convergence is due to many reasons including:

- Regulation—financial and bank holding companies are federally regulated while insurance firms are state regulated. . . . New insurance products need to have state approval before they are allowed to be marketed, which is a slow process.
- Technology—banks are able to offer interactive online services, while insurance products apparently don't lend themselves to it. Further, insurance are slower to adopt new technology.
- Financial reasons—Return on equity for insurers have tended to be lower than those of insurers. . . . Probably Citigroup spun off Traveler's because it did not make double digit growth, a norm for Citibank (at that time, 2002). In addition, Banks have two goals in mind, including: Revenue diversification and Product diversification for their customers while insurance firms' goals are for more financial sustainability with horizons that are much longer, etc.

These and other reasons, explain that while there is a partial convergence in insurance and financial products, there is also an "institutional competition" between these two financial institutions that has contributed to the growth of each based on "the turf of the other."

Insurance and banks have thus maintained some of their differences and their approaches to valuation, to pricing and to risk management. Specific elements include how they price the risk premium, their valuation of time and liquidity and the management of their financial assets. Selected issues and their differences pertaining to dealing with time, uncertainty and pricing, liquidity, insurance and risk management are outlined below:

- The complexity of risk models they use and their pricing;
- Their regulatory, accounting and tax rights and responsibilities (Insurance firms in particular are likely to face far greater regulated capital adequacy ratios than banks);
- Their access to liquidity, to capital funds (such as through the Federal Reserve window, financial deposits etc.).
- The economic maturities of risk assets held by insurers and banks. Banks have shorter risk maturities while insurance firms (life insurance contracts for example) have longer and random maturities;
- Differences in insurers and banks value added, their transaction costs, the liquidity they are required to maintain and their financial policies. In addition, to the moral hazard and information and power asymmetry risks they own or are subjected to.
- Insurers and banks sensitivities to both macroeconomic trends and "black swans."

9.4.2 *Insurance and Finance: Pricing Examples*

Embrechts et al. (1997, 2001), Goovaerts et al. (1984), Dacorogna (2004) have repeatedly pointed out to such differences between the insurance and the financial approach. They then point out that "The possibility of buying and selling at time t represents the possibility of a take-over of this policy." Such assumption implies a complete market, i.e. no arbitrage, complete liquidity and therefore existence of a risk neutral distribution measure such that the "tradable claims" can be defined by a Q-martingale. Such rationality has been applied to pricing credit derivatives and "structured insurance financial products." Questionable practices violating the underlying assumptions of financial markets such as moral hazard and information asymmetry, the transfer of non-transparent risk among many parties through complex and financial networks etc. has led to mispricing however. Failure of risk pricing, has thus led insurance and credit liquidity to be supplied by traditional insurance and "actuarial accounting."

9.4.3 *Contrasts of Actuarial and the Financial Approaches*

In insurance the demand-price for insurance is defined by an actuarial measure of historical risks, an insured risk attitude and the firm's ability to bear risks (summarized by both its utility and kernel pricing utility-rational approaches as indicated in the previous chapter). In other words, an actuarial state of knowledge and a risk attitude of the insured and/or insurer determine an implied market price. And vice versa, a market price and an actuarial state of knowledge imply a risk attitude. And finally, any two imply the third. In this framework, insurance risk pricing formulas abound due to the individual risk attitudes and actuarial calculations of insurance firms (and therefore the difficulties in setting a fair market price). In a financial approach, securitization of assets and market pricing are used. Failure of financial markets due to the risks and the conditions they are able to account for leads as well to their failure. In other words, in such an approach, prices are based on the implied price of insurance or the Arrow-Debreu state preference theory which requires the definition of a future well defined by its potential losses or gains and by the numerous conditions that require for financial markets to be complete. For example, in some circumstances, we can argue that portfolio securitization of multiple risk products (such as MBS, Life insurance and their like) imbeds in fact an incompleteness due to our misunderstanding of the structure of assets' probability dependence and the uncertainty these products may have. These can therefore contradict assumptions made when we use the fundamental theory (complete markets) approach to pricing. These two approaches are reconciled however by the inverse problem defining prices as implied in the information used as indicated in the previous chapter.

9.4.4 Franchises

Franchises are defined by a contractual relationship between a franchiser and franchisees, for the purpose of exploiting a particular concept or advantage provided by the franchiser. It is mostly an economic agreement based on an exchange between parties made for profit, each party expecting to draw some advantage from the agreement. This general principle underlies franchise contracts, outsourcing agreements, joint partnerships etc. Franchises are thus a contract between two legally independent economic entities establishing a long-term relationship where the franchiser grants to the franchisee the right to use the franchiser's trademark, the use of a specific (potentially patented) technology etc. In exchange, the franchisee pays a lump sum fee and annual royalties at an agreed percentage of sales.

A franchise may involve several other provisions as well as options that each of the parties may grant to the other. For example, risk sharing, exclusive territories with optional agreement appended to these agreements, promotional efforts sharing, buy-back provisions (Marvel 1982; Rey 1992; Rey and Tirole 1986; Tirole 1988; Mathewson and Winter 1986; Klein and Saft 1985). Over one third of all retail sales in the US occur through a franchise system which is also growing globally with the growth of global brands (such as Benetton, Mc Donald and many others). In many cases production can be centralized while distribution may be franchised (e.g. car selling, some food and department stores, fast food, clothing trademarks etc.). In some cases as well, image and advertising is centralized but production is decentralized, franchised to companies focused in manufacturing (as it is increasingly the case).

The economic rationale for franchises arises due to the increased dominance of consumption as the driving economic fuel to industry, the emergence of very large but extremely diffused markets, the very high set up costs in selling as well as to encountering problems in managing complex and diffused distribution systems. A franchiser may thus construct a franchise system where franchisees would invest parts, if not all, of the required local investment (both reduce the commitments of capital by the franchiser and to augment the franchisee stake in the economic performance of the franchise). Typically, such an agreement is made for definite or indefinite periods of time, which the owner of a protected trademark grants to franchisees, for some consideration, the right to operate under this trademark for the purpose of producing or distributing a product or service (Caves and Murphy 1976). Because the value and risks of such assets are defined by their use, these contracts involve difficult contractual relations. Franchisee fees assume then many variations such as royalties, or commission, resale price maintenance, exclusive territories, exclusive dealing as well exclusivity relationships of various sorts with reciprocal agreements for the conduct of mutual services. The study of franchises involves as a result many issues such as resource constraints (thus the franchise will grant access to financial capital, market expertise and managerial talent of franchisee); risk incentives to assure their performance and of course efficiency.

Traditionally, an expected utility framework based on the parties' utilities for money is used to value franchise contracts. This value is not a market price however since it is not traded. Application of a real CCAPM to be considered in Chap. 8 may provide an implied pricing approach based on the many optional characteristics that define the franchise contract.

A typical franchise contract consists of a lump sum payment which may or may not be refundable, involving optional choices maintained by one or the other party to the contract over a certain length of time (at which time the franchise contract can be renegotiated) as a way to commit the franchisee to entrepreneurial activity and safeguard from the misuse of the franchise. Similarly, an advantage (or disadvantage) can be gotten through a tax on current inputs, such as selling current input at prices larger than the franchisor's marginal cost. For some contracts the franchisor participates in fixed operating costs (when he leases land that he owns) combined (or not) with provisions to recapture the franchise (which alters the franchisee utility). Thus, even with the most stringent contract, franchises are subject to many risks. Risks of "milking" the franchise; asymmetry risks (in power and in information) and other risks resulting in sub-performing franchisees can harm the franchise brand as a whole. We will next consider a number of simple problems to highlight only some of these issues.

9.4.5 *Outsourcing and Risks*

Outsourcing is essentially defined as the transfer of previously in-house activities to a third party (see also, Gattorna 1988; La Londe and Cooper 1989). Such transfers may reduce fixed costs, allow firms to focus on high yield activities (such as retail sales, services etc. and do away with operations and production) and augment their flexibility. Outsourcing creates risks however embedded in the following questions: Should a firm strive to maintain its capacity to produce or turn to an external (and therefore hardly controllable) supplier? Will a firm's technological positioning (and therefore its knowledge base in the future) be reduced? What are the firm's strategic options and contingent plans if it loses its suppliers? Outsourcing could thus lead to numerous problems. For example, demand for product types may be volatile and require a perpetual innovation in product design and manufacture which can be lost when the firm has outsourced its production potential. Apple for example, has demonstrated that by maintaining an internal product design arm and an aggressive marketing policy and outsourcing its production, it has succeeded financially beyond anyone's expectations. Such outsourcing has risks however—risks that the producer may decide to compete with Apple. Apple's reckoning with the risks it has engineered through its global outsourcing is yet to be seen.

9.5 Subjective Kernel Distributions

The question we are concerned with is: How to reconcile the real-or-subjective and the financial model (risk neutral) distributions given observed prices. Risk attitudes are implied by the risks persons take based on their utility function and subjective estimate of future prices while a current price equals (under a risk neutral measure) its future discounted expectation. This was exploited by Jackwerth (1999, 2000), Ait-Sahalia (1996) and Ait-Sahalia and Lo (2000a, b) in extracting a measure of risk aversion in a standard dynamic exchange economy (Lucas 1978). The underlying rationale of this approach is that a trader (or investor) whose wealth is one dollar will maximize the utility of a portfolio with future values and with a current price of \$1. Explicitly, let $f_T^S(.), f_T^Q(.)$ be the subjective and the risk neutral probability distribution of future prices. A person maximizing his utility of a future expectation subject to its current price (and therefore the price he is paying now for the future prospect) is given by:

$$\underset{W_T}{\text{Max}}\, E_S(u(W_T)) = \int_{-\infty}^{\infty} u(W_T) f_T^S(W_T) d(W_T) \ \text{Subject to}:$$

$$1 = e^{-R_f(T-t)} \int_{-\infty}^{\infty} W_T f_T^Q(W_T) d(W_T)$$

Let λ be the constraint Lagrange multiplier of the optimization problem above, the first order condition is:

$$u'(W_T) = \lambda e^{-R_f(T-t)} \frac{f_T^Q(W_T)}{f_T^S(W_T)} \quad \text{or}$$

$$u''(W_T) = \lambda e^{-R_f(T-t)} \frac{f_T' Q(W_T) f_T^S(W_T) - f_T'^S(W_T) f_T^Q(W_T)}{[f_T^S(W_T)]^2}$$

and

$$-\frac{u''(W_T)}{u'(W_T)} = -\frac{f_T' Q(W_T)}{f_T^Q(W_T)} + \frac{f_T'^S(W_T)}{[f_T^S(W_T)]} \quad \text{and} \quad -A_P(W_T) = \frac{d}{dW_T} \ln\left(\frac{f_T^Q(W_T)}{f_T^S(W_T)}\right)$$

And therefore,

$$\frac{f_T^Q(W_T)}{f_T^S(W_T)} = e^{K - \int_0^{W_T} A_P(x)dx} \quad \text{or} \quad f_T^Q(W_T) e^{-K + \int_0^{W_T} A_P(x)dx} = f_T^S(W_T)$$

Since

$$1 = \int f_T^S(W_T) = e^{-K} \int e^{\int_0^{W_T} A_P(x)dx} f_T^Q(W_T) \quad \text{or} \quad e^K = \int e^{\int_0^{W_T} A_P(x)dx} f_T^Q(W_T)$$

and the following fundamental relationship is obtained:

$$f_T^S(W_T) = \frac{e^{\int_0^{W_T} A_P(x)dx} f_T^Q(W_T)}{\int e^{\int_0^{W_T} A_P(x)dx} f_T^Q(W_T)dW_T}$$

In particular, if the utility is of the exponential type with a constant risk aversion index μ then:

$$f_T^S(W_T) = \frac{e^{\mu W_T} f_T^Q(W_T)}{\int e^{\mu W_T} f_T^Q(W_T)dW_T}$$

Note that the relationship above is also the Esscher transform (1932) commonly used in actuarial science and insurance.

On the basis of these results, we can also calculate (as a function of the parameters the utility function parameters) the differences between the mean price forecasts of an investor and those believed by investors and speculators in financial markets. Extensive research, both theoretical and empirical has extended this approach and indicated a number of important results. For example, risk attitude is not only state varying but is time varying as well. Further, it clearly sets out the concept of risk attitude in terms of a distance between the subjective and the risk neutral (market) distributions. For notational simplicity, set:

$$g(W_T) = \ln\left(\frac{f^P(W_T)}{f^Q(W_T)}\right) \quad \text{then :} \quad \frac{g''(W_T)}{g'(W_T)} - g'(W_T) = A_\pi$$
$$\text{or} \quad g''(W_T) = (A_\pi + A_P)A_P$$

In this equation, we clearly see the relationship between the risk neutral, the subjective and the risk attitude of the investor both with respect to the index of absolute risk aversion and the investor's prudence.

9.5.1 The HARA Utility

Let the utility of an investor be HARA (Hyperbolic Absolute Risk Aversion) with an index of absolute risk aversion $A(W_T)$:

$$u(W_T) = \frac{1-\gamma}{\gamma}\left\{\frac{aW_T}{1-\gamma} + b\right\}^{\gamma}, A(W_T) = \frac{a}{\left\{\frac{aW_T}{1-\gamma} + b\right\}}$$

Then:

$$A(W_T) = \frac{a(1-\gamma)}{aW_T + b(1-\gamma)} = \frac{d}{dW_T} g(W_T)$$
$$\text{and} \quad f^S(W_T) = (aW_T + b(1-\gamma))^{(1-\gamma)} f^Q(W_T).$$

For example, say that both the subjective and the risk neutral distributions are normally distributed with means and variances given by: (μ_S, σ_S^2) and (μ_Q, σ_Q^2), then:

$$\frac{f^S(W_T)}{f^Q(W_T)} = \frac{\sigma_Q}{\sigma_S} \exp\left[-\frac{(W_T - \mu_S)^2}{2\sigma_S^2} + \frac{(W_T - \mu_Q)^2}{2\sigma_Q^2}\right]$$

And therefore, this implies that risk aversion is proportional to the prospect since:

$$A(W_T) = \frac{d}{dW_T}\left(\ln(\sigma_Q - \sigma_S) - \frac{(W_T - \mu_S)^2}{2\sigma_S^2} + \frac{(W_T - \mu_Q)^2}{2\sigma_Q^2}\right) = W_T\left(\frac{\mu_S}{\sigma_S^2} - \frac{\mu_Q}{\sigma_Q^2}\right)$$

Thus, if $\frac{\mu_S}{\sigma_S^2} < \frac{\mu_Q}{\sigma_Q^2}$, risk aversion decreases linearly in wealth and vice versa. Note that a risk aversion is then defined by the sign of $A(W_T)$, as defined in this artificial example. In this sense, a risk attitude can be revealed by observed behaviors.

Problem: The Information-Entropy Approach Let an information metric (see Chap. 6) and say that $\ln\left(\frac{f^S(r)}{f^Q(r)}\right)$ measures the information distance between a subjective and a risk neutral distribution. Such a distribution is determined by minimizing the expected ln metric:

$$\text{Min } E^S\left\{\ln\left\{\frac{f^S(r)}{f^Q(r)}\right\}\right\} \text{ Subject to} : r_0 = (1 + R_f)^{-1} E^Q(r) \text{ and } \int_{r\in\Re} f^Q(r) = 1$$

Additional Sources and References

Models of debt risk (credit risk) and their derivatives abound, emphasizing the many aspects of risk in borrowing and in lending. The issues span partial information (and thereby counterparty risk), default risk as well as pricing credit risk and credit derivatives. A number of additional references on these topics include,

Acharya and Johnson (2007) on insider trading in credit derivatives, Duffie (1999) on CDS valuation, Hull and White (2000, 2001, 2003, 2006), KMV on model definition of risk, Alejksiejuk and Holyst (2001), on a model of bank bankruptcies. These models include as well implied pricing techniques we have used in the text including implied probability estimates (for example, Rubinstein 1994).

Applications to IPO pricing, signaling and information have also pointed out to the risks of IPO. Allen and Faulhaber (1989) pointed out to the role of signaling in IPO pricing, Benveniste and Spindt (1989) on pricing of new issues by bankers, Drake and Vetsuypens (1993), Welch (1989) on seasoned offerings, imitation costs, and the underpricing of initial public offerings, Ritter (1991) on the long-run performance of initial public offerings, Chemmanur and He (2011), on IPO waves, Chemmanur and Yan (2009) on advertising new equity issues, Chemmanur and Fulghieri (1999) present a theoretical framework for the going-public decision, Ritter (1991) on IPO and risk (we referred as well in Chap. 8) and the long-run performance of initial public offerings.

Tirole (1988), *The Theory of Industrial Organization*, is an important reference book on the topic as well as a background appreciation of risks in such organizations. These problems are vast encompassing the economics of franchising (Rey 1992; Rey and Tirole 1986; Mathewson and Winter 1986; Klein and Saft 1985; Caves and Murphy 1976).

Environmental economic risks and their price provide a broad set of alternative approaches. These include the personal price of pollution, as well as its market (as with the CO_2) and of course the price of risk externalities. The number of references in these domains is large with a few papers referred throughout this text. Additional references include, Marcus and Willig (Editors) (1997) of *Moving Ahead with ISO 14000: Improving Environmental Management and Advancing Sustainable Development*, Laffont (1995) on regulation, moral hazard and insurance of environmental risks, Decanio (1997), on modeling false tradeoff between the environment and economics.

In the 1990s, infrasturture research has attracted much attention due to a growing realization that public expenditure in infrastrucutre are contributing appreciably to the productivity of the private sector. In particular, attention was focused on the multiplier effect of infrastructure investments. Econometric studies by Aschauer (1989a, b), raising the question and making a case for public expenditure productivity. These papers were extended further and econometric tests conducted in different countries and on different sectors (see the surbvey of Gramlich 1994). Batten, and Karlsson, edityed a book on infrastructure and the complexity of economic development, Berndt and Harrison (1991), measured the contribution of public infrastructure in Sweden, Nijkamp and Ubbel on reliability of estimates of infrastructure costs. Nijkamp and Blaas (1993) on transportation infrasturcutre.

Holtz-Eakin and Schwartz (1994, 1995) provided a number of important studies on infrastructure in a structural model of economic growth, Morrison and Schwartz (1996), on state infrastructure and productive performance. Munnell (1992), on infrastructure investment and economic growth *as well as* Sanchez-Robles (1998a, b).

Chapter 10
Uncertainty Economics

Abstract This chapter seeks to reconcile fundamental financial approaches with uncertainty. Uncertainty is defined by the unknown rather than the predictable, counted and accounted for. While financial decisions are reached based on what we know, what we can predict and what we can presume based on experience and the rationalities that financial agents assume. The uncertainty we consider is defined in a limited sense, namely, a partial knowledge of future state preferences and their quantification. There are many approaches to do so such as negligence of the unknown, human intentional rationalities as well behavioral and psychological approaches to confront the unknown. This chapter focuses its attention on the use of entropy for "non-extensive systems" (a term commonly used in physics with its parallel in finance, which we define as "incompleteness") based on a parametric generalization of the Boltzmann–Gibbs entropy (which assumes extensive systems). Optimization of the Tsallis parametric entropy for non-extensive systems is then used to derive implied power laws and standardized probability distributions that are both asymmetric and have fat tails. This approach provides a parametric definition of the "missing", namely the tail probabilities not accounted for in selecting an asset future price distribution.

Subsequently, the chapter outlines a number of approaches to robust decision models and ex-post risk management. It concludes with a discussion of risk externalities in financial and environmental regulation and draws a parallel between "banks' risks" for which they do not assume responsibility for and pollution risks of firms and consumers who consume and who do not assume their pollution consequences. Both cases, call for an efficient regulation and statistical controls which is the topic of Chap. 11.

(with OREN TAPIERO)

C.S. Tapiero, *Engineering Risk and Finance*, International Series in Operations Research & Management Science 188, DOI 10.1007/978-1-4614-6234-7_10,

10.1 Introduction

Throughout all professions, persons' lives and societies, business, safety, health and security can finance, uncertainty prevails. To deal with uncertainty we construct risk models while residual uncertainty is what remains and not accounted for. Uncertainty thus accounts for our neglect, a lack of prior experience or a lack of knowledge, or to a mis-specification or unknown consequences. When this is the case, uncertainty primes and risk models are incomplete. A distinction between risk and uncertainty regarding states countability was pointed out by Knight (1921) (see also Weston 1954; Friedman 1976; LeRoy and Singell 1987; Langlois and Cosgel 1993) emphasizing that risk is mostly associated to complete state preferences, uncertainty is associated to unknown state preferences and to everything else risk models seem to neglect. What is meant by an "incomplete set of state preferences" was and is still under debate. Technically, risk implies that future states are ex-ante well defined and that probabilities can be measured and used to define the expectations of such states. The modern theory of finance in particular is based on a risk model, competitive and fair markets and a market rationality that precludes situations that are not sustainable (and therefore off equilibrium or various conditions with no arbitrage). The fundamental theories of risk assets pricing assume that markets are complete, defined explicitly as stated in Chaps. 7 and 8 (as well as in Chaps. 1 and 2). Incomplete markets are then defined simply by "markets that are not complete", which may include a broad number of reasons. In other words, any departure from any of the assumptions that define theoretical complete market is a source of incompleteness and uncertainty. This chapter provides a number of exploratory approaches to uncertainty economics. While some explicit results are presented, this chapter is necessarily biased and therefore incomplete in the treatment of uncertainty and its valuation.

10.2 Risk and Uncertainty, Time and Pricing

For Buddhism, "time is timeless" and therefore there is only a "present". St Augustine (Confessions), also claims that there is only a present: A "present of the past" (reflecting our memories and its filter representing and interpreting of past events) and a "present of the future" (an expectation of future states embedded in the present). In this sense, there is "only a present" merging the past and the future into one current instant. Economic and financial theories have developed pricing models based on the past (mostly actuarial and insurance based models) and some on the future (fundamental theory of finance and the Arrow-Debreu framework, used mostly to price financial assets in complete markets and based on current information as noted in Chaps. 7, 8 and 9). Such distinctions underlie the basic quantitative insurance and financial approaches we use to price risk assets. For insurers, prices are determined by an actuarial past, while for finance it is justified by equilibrium price models of future state preferences (Arrow, Debreu, Lucas, Samuelson, Fisher Black, Merton and others). Explicitly, for insurance, modeling future uncertainty (i.e. constructing a risk model) is based on statistical (actuarial) analyses of past events and a present subjective evaluation of future states. While in finance, models are "virtual", resulting from theoretical and complex interactions of buyers and sellers in financial markets summarized by fundamental economic principles. In this sense, future prices are traded through forward markets and inversely, quantitative future prospects are implied in the prices of current trades. As a result, while for insurance, the present is defined in terms of past experiences, compounded by an attitude that can be used to define future states (of losses, of gains, etc. that an insured or insurer are aware of or not, but states they assume explicitly by definition of the insurance contract), for finance a present price is implied in a "rational expectations model of the future". For these reasons, "insurance prices" and "financial prices" for future risks differ. For one, prices are anchored in a growing state of knowledge while for the other it is based on the evolving exchange patterns of buyers and sellers operating in a complete market.

In practice, insurance firms set the insurance price to exceed (predicted actuarially) future losses with a time scale usually far greater (1 year to several decades) than the time scale used by financial firms where profits are generated far more by the services they provide and by trading. Reserves thresholds (such as VaR and CAR regulation) for insurance firms are also set to higher levels than banks. For these reasons, insurance firms are often seen as "risk warehouses that aggregate risks" and are thus required to be capital intensive (which justifies the demand by investors to earn a commensurate rate of return for the risks insurance firms assume). The price of an insurance contract is then generally defined by the expected loss to be paid in case of a claim, the cost of capital (also called risk "loading") for covering unexpected losses and the internal marketing and writing costs of policies. Technically, actuarial loss models, NPV models as well as RAROC (RORBC, RBC, etc.) models are used. Outstanding issues remain such as: "are risks insurable?" (i.e. can risks be perfectly hedged? "what risk premium to apply for a given risk?", "what are the effects of deductibles and collaterals on the risk premium and the demand and supply of insurance?", "what to exclude from insurance contracts?", "can risk be exported and how much it ought to be shared, exchanged, transferred or securitized", etc. Such considerations are also prevalent in financial products such as Bonds where covenants are used to eliminate a broad set of obligations that are considered either not "insurable" or not "countable" in a financial sense and therefore uncertain.

In contrast to insurance, pricing by financial firms is based on standardization and securitization of portfolios and exchange. Typical financial products include options, credit risk derivatives defined by their quality (for example, their implied probabilities of default), their liquidity and their structure. When a price is not evident (i.e the market for financial insurance is grossly incomplete), financial insurance may falter and provide opportunities for arbitrage (as they have during the financial crisis 2007–2008).

Insurance and finance have in the past used risk models differently. For insurance, the emphasis is to personal and corporate insurance valuation while for finance it is and remains a valuation of exchange prices reached in financial markets—either explicitly or implicitly through risks models such as the CCAPM and the AD theoretical (martingale) framework (see Chaps. 7, 8 and 9).

Departure from these models arises for many reasons. Below, we shall consider some specific case.

10.3 Assets Pricing with Countable and Non-countable States

Let $\left(S_1^1, S_1^2, \ldots, S_1^n\right)$ be n potential future stock prices and let the price of a current stock be S_0. The buyer of such a stock has assumed in fact all its future price consequences $\left(S_1^1, S_1^2, \ldots, S_1^n\right)$ by the current price. If the price he is willing to pay for each future state is $\pi_i,\ i = 1, 2, \ldots, n$ then in a Walrasian equilibrium sense (i.e. for deterministic and competing markets, see Chap. 8) we have:

$$S_0 \Leftarrow \left(\pi_1 S_1^1 + \pi_2 S_1^2 + \ldots + \pi_n S_1^n\right) \quad \text{or} \quad S_0 = \sum_{i=1}^n \pi_i S_1^i$$

A standardization of these prices, set in terms of "probabilities" provides then a "risk model". To do so, information regarding future states are used (see Chap. 7). In particular, if a future asset provides a risk free rate of return regardless of the future states S_1^i with $S_1^i = S_0(1 + R_f)$, then:

$$S_0 = \sum_{i=1}^n \pi_i S_1^i = \sum_{i=1}^n \pi_i S_0(1 + R_f) \quad \text{and therefore,} \quad \frac{1}{1 + R_f} = \sum_{i=1}^n \pi_i$$

An appropriate standardization is thus $p_i^n = \pi_i \Big/ \sum_{i=1}^n \pi_i$ which leads to "risk neutral" pricing:

$$1 = \frac{1}{1 + R_f} \sum_{i=1}^n p_i^n S_1^i, \quad 0 \le p_i^n \le 1 \quad \text{and} \quad \sum_{i=1}^n p_i^n = 1$$

Note that these states and "probabilities" are well defined with respect to the state preferences and with a probability distribution that sums to one. When these probabilities do not sum to one, there are events that are not accounted for that render these distributions incomplete. For example, if we assume m states when there are in fact $n > m$ such states, then:

$$1 \neq \sum_{i=1}^m p_i^m \left(\frac{S_1^i}{1 + R_f}\right), \quad 0 \le p_i^m \le 1 \quad \text{and} \quad \sum_{i=1}^m p_i^m < 1$$

The question remains, how to account for the residual $m - n$ states? And how to standardize these probabilities such that they can be "called" (in a technical sense) "probabilities". Rational, behavioral and psychological research in economic and finance have approached these problems from different vantage points. For example, the Laplace Principle of Insufficient Reason is based on the presumption that one has to assume the least when one has no supporting information. This principle in a quantitative setting is reduced to the maximization of the Boltzmann–Gibbs Entropy function as we shall see later on. In this case, a measure (metric) of information is used whose optimization in constrained by known information. This approach provides a probability distribution of states which is then concordant with Laplace's Principle. When no information is available, the Laplace Principle indicates that an equi-probable distribution is to be used (see Sect. 10.2). Such a result, is based on the presumption that uncertainty is defined by a random uniform distribution, expecting the best and the worst (or any other state) with an equal probability. Of course, such a conclusion might not be deemed appropriate.

Alternatively, an approach based on Tsallis entropy metric provides a characterization of uncertainty which might be more acceptable. This approach has the particular characteristic that it generalizes the Boltzmann–Gibbs metric and therefore, by definition is parametrically its generalization. It is defined as follows: Set $0 < q = 1 - \sum_{i=1}^{m} p_i^m < 1$ as the sum of probabilities not accounted for by the risk model based on m countable and assume of states. In this case, the probability $p_i^m \Big/ \sum_{i=1}^{m} p_i^m = p_i^m \Big/ (1-q)$ defines "weighted" probabilities summing to one since: $\sum_{i=1}^{m} \frac{p_i^m}{1-q} = 1$. Thus, if there are indeed m states then $q = 0$ and the entropy of such a distribution is the Boltzmann entropy, or $H_B(m) = E\left\{\ln\left(\frac{1}{p_i^m}\right)\right\}$. Define instead a Tsallis generalized entropy given by:

$$H_T(m,q) = \left(\frac{\sum_{i=1}^{m} \left(p_i^m\right)^q - 1}{1-q}\right) \quad \text{or}$$

$$H_T(m,q) = \left[\frac{E\left(p_i^m\right)^{q-1} - 1}{1-q}\right]$$

Note that, by L'Hopitale rule:

$$\operatorname*{Lim}_{q\to 1}\left(\frac{\sum_{i=1}^{m} \left(p_i^m\right)^q - 1}{1-q}\right) = \operatorname*{Lim}_{q\to 1}\left(\frac{\sum_{i=1}^{m} \left(p_i^m\right)^q \ln\left(p_i^m\right)}{-1}\right) = \sum_{i=1}^{m} p_i^m \ln\left(\frac{1}{p_i^m}\right)$$

and therefore the Tsallis entropy is a functional generalization of the BG entrophy, a function of the tail probability $1 - q$ which, as seen above is indicative to the number of states that were not accounted for. Note however that still, $\sum_{i=1} p_i^m(q) < 1$ and therefore it is a measure of the Boltzmann–Gibbs entropy only if $m = n$ and the system is extensive (complete) or alternatively, if $q = 0$. In this case, $\partial p_i^m(q)/\partial q$ provides a measure of sensitivity to the distribution incompleteness. For example, if we price an underlying asset using a distribution $p_i^m(q)$ instead than say a distribution $p_i^n(q = 0)$, then

$$S_0^q = \sum_{i=1}^{m} [p_i^m(q)] \left(\frac{S_1^i}{B_1}\right) \text{ while } S_0 = \sum_{i=1}^{n} [p_i^n(q)] \left(\frac{S_1^i}{B_1}\right)$$

And the difference $S_0 - S_0^q$ may represent an uncertainty spread—a price we pay for using an incomplete states probability distribution. In this vein, the smaller q the more states are accounted, for and at the limit: $\lim_{q \to 0} S_0^q = S_0$. These problems are analyzed technically below.

10.4 Maximization of Boltzmann Entropy

Laplace postulated the "Principle of Insufficient Reason" for decision problems under complete ignorance. This principle states that when we do not know what are the probabilities of the states of nature in a given problem, we can assume that all probabilities are equally likely, reflecting a state of utmost ignorance. Thus, lacking any expected information, all states are equally likely! The "idea" behind this principle is that we adopt a distribution which assumes the least which consists of equal probabilities. Such an approach assumes however that all states are known.

The Principle of Maximum Entropy in essence defines quantitatively a distribution "that assumes the least" as that distribution which has the largest entropy. To maximize the entropy of a probability distribution subject to a set of informational constraints results then in a distribution that assumes the least and at the same time has the informational properties specified a-priori. Of course, with no constraint, maximization of entropy with respect to a complete probability distribution yields a uniform distribution:

$$E^P\left(\ln\frac{1}{p_i}\right) = \sum_{i=1}^{n} p_i \ln\left(\frac{1}{p_i}\right); \quad \sum_{i=1}^{n} p_i = 1, \quad 0 \le p_i \le 1 \quad \text{yields} \quad p_i = 1/n$$

Jaynes (1957a, b) gave perhaps the best description for the maximum entropy formalism for estimating distributions. Jayne's states that maximum entropy answers the following question:

"What knowledge of $E[f_i(x)] = \int f_i(x)p(x)dx$ tells us about $p(x)$?"

In other words, say that an expected value of $f_i(x)$ is observed. How can one infer $p(x)$ from such knowledge? In particular, given the mean of a distribution what is the least committal distribution (in the sense of a Maximum entropy) that has such a mean? What if both the mean and the variance are given, etc. Borland and Bouchaud (2004), Brody et al. (2007) and others, indicate that the Boltzmann maximum entropy does not capture the power law tails property observed empirically however (and deemed a better characterization of financial risk models). Rather, the maximum Boltzmann entropy is appropriate for risk models which are mostly defined by thin tails distributions. Entrophy Metrics that do point out to for tail disributions have been suggested by Renyi (1961, 1970) and Tsallis (1988). These properties contribute to the attractiveness of the generalized entropy functions.

Example: The Maximum Entropy Distribution Given Its Moments Define a finite number of states n, and set $S_i, \ \ i \in [1, n]$ and the mean μ_S:

$$\underset{p_i, i=1,2,\ldots,n}{\mathrm{Max}} \ E^P\left(\ln\left(\frac{1}{p_i}\right)\right); \text{ Subject to} : \mu_S = \sum_{i=1}^{n} S_i p_i, \quad \sum_{i=1}^{n} p_i = 1, \ \ 0 \leq p_i \leq 1$$

The problem's Lagrangian is:

$$\underset{p_i, i=1,2,\ldots,n}{\mathrm{Max}} \ E^P\left(\sum_{i=1}^{n} p_i \ln\left(\frac{1}{p_i}\right)\right) + \lambda_1\left(\sum_{i=1}^{n} S_i p_i - \mu_S\right) + \lambda_2\left(\sum_{i=1}^{n} p_i - 1\right)$$

And therefore, necessary conditions for optimality are $-\ln(p_i) - 1 + \lambda_1 S_i + \lambda_2 = 0$ or $p_i = e^{-1+\lambda_1 S_i + \lambda_2}$ where Lagrange parameters are found by: $\sum_{i=1}^{n} S_i e^{-1+\lambda_1 S_i+\lambda_2} = \mu_S, \sum_{i=1}^{n} e^{-1+\lambda_1 S_i+\lambda_2} = 1$. As a result, the Lagrange multiplier λ_1 is solved for by the first equation below which is inserted in the second equation for individual probabilities:

$$\sum_{i=1}^{n} S_i\left(\frac{e^{\lambda_1 S_i}}{\sum_{i=1}^{n} e^{\lambda_1 S_i}}\right) = \mu_S, \quad p_i = \frac{e^{\lambda_1 S_i}}{\sum_{i=1}^{n} e^{\lambda_1 S_i}}$$

A generalization to multiple expected data points, a distribution variance and other probability moments is a straightforward technical exercise. For example, assuming that the following expectations are given (for example, option prices) $\mu_k = \sum_{i=1}^{n} C_k(S_i) p_i, \quad \sum_{i=1}^{n} p_i = 1$, then, the following distribution results:

$$p_i = \frac{e^{\sum_k \lambda_k C_k(S_i)}}{\sum_{i=1}^{n} e^{\sum_k \lambda_k C_k(S_i)}} \text{ with } \mu_k = \sum_{i=1}^{n} C_k(S_i) p_i = \sum_{i=1}^{n} \frac{C_k(S_i) e^{\sum_k \lambda_k C_k(S_i)}}{\sum_{i=1}^{n} e^{\sum_k \lambda_k C_k(S_i)}}, \quad k = 1, 2 \ldots$$

These equations provide a nonlinear system of equations to be solved simultaneously. "Informational assumptions" regarding an underlying probability distribution leads therefore to various probability distributions as indicated in Table 10.1 below. A particular mean–variance case, is shown below to imply (in a Laplace Principle, Boltzmann–Gibbs entropy sense), a normal probability distribution.

Example: A Continuous State Problem Let $f(\theta)$ be a probability distribution with known mean and known variance. First consider the maximum Boltzmann–Gibbs entropy distribution based on the mean only, then:

$$\text{Max } H = -\int_0^\infty f(\theta) \log[f(\theta)] d\theta \quad \text{Subject to : } \int_0^\infty f(\theta) d\theta = 1, \ \hat{\theta} = \int_0^\infty \theta f(\theta) d\theta,$$

The solution of this problem is straightforward, and yields an exponential probability distribution:

$$f(\theta) = \frac{1}{\hat{\theta}} e^{-\theta/\hat{\theta}}, \quad \theta \geq 0$$

Assume next that the variance is specified. The resulting optimization problem is:

$$\text{Max } H = -\int_0^\infty f(\theta) \log[f(\theta)] d\theta \quad \text{Subject to :}$$

$$\int_0^\infty f(\theta) d\theta = 1, \quad \hat{\theta} = \int_0^\infty \theta f(\theta) d\theta, \quad \sigma^2 = \int_0^\infty (\theta - \hat{\theta})^2 f(\theta) d\theta,$$

whose solution yields the normal probability distribution with mean and variance $(\hat{\theta}, \sigma^2)$ (the mathematical solution of this problem is left as an exercise).

Additional as well as various types of information such as boundaries on the distributions, expectation of functions of the probability states etc. will alter the resulting functional form of the implied distribution. For example, let the natural logarithm of security prices have a known mean and a known variance, then the maximum entropy security distribution turns out to be a log-normal probability distribution (see Table 10.2) and therefore the distribution rates of return will assume a normal probability distribution. Assuming the pertinence of the BG maximum entropy, it may be applied equally when the probability distribution is bounded, multivariate with specified marginal distributions, representing time

Table 10.1 Examples of distribution obtained from maximizing the Boltzmann Entropy

Name	Required constraints	Maximum Boltzmann entropy distribution	Functional form
Exp.	$\mu = \int xp(x)dx$	$p(x) = e^{-(1+\omega_0)-\omega_1 x}$	$\frac{1}{\mu}e^{\frac{-x}{\mu}}$
Norm.	$\mu = \int xp(x)dx, \quad \sigma^2 = \int (x-\mu)^2 p(x)dx$	$p(x) = e^{-(1+\omega_0)-\omega_1 x-\omega_2(x-\mu)^2}$	$\left(\sigma\sqrt{2\pi}\right)^{-1} e^{-\frac{(x-\mu)^2}{2\sigma^2}}$
LN	$\mu = \int \ln(x)p(x)dx, \quad \sigma^2 = \int (\ln(x)-\mu)^2 p(x)dx$	$p(x) = e^{-(1+\omega_0)-\omega_1 \ln(x)-\omega_2(\ln(x)-\mu)^2}$	$\left(\sigma\sqrt{2\pi}\right)^{-1} e^{-\frac{(\ln(x)-\mu)^2}{2\sigma^2}}$
Gam.	$\mu = \int xp(x)dx, \quad \frac{\Gamma'(\mu)}{\Gamma(\mu)} = \int \ln(x)\,p(x)dx$	$p(x) = e^{-(1+\omega_0)-\omega_1 \ln(x)-\omega_2 \ln(x)}$	$\frac{1}{\Gamma(\mu)}e^{-x}x^{\mu-1}$
Chi-2	$\mu = \int xp(x)dx, \quad \frac{\Gamma'(\frac{1}{2})}{\Gamma(\frac{1}{2})} + \ln 2 = \int \ln(x)\,p(x)dx$	$p(x) = e^{-(1+\omega_0)-\omega_1 x-\omega_2 \ln(x)}$	$\left(2^{\frac{1}{2}\mu}\Gamma(\mu)\right)^{-1} e^{-\frac{1}{2}x}x^{\frac{1}{2}\mu-1}$
Weib	$1 = \int x^m p(x)dx, \quad m > 0$	$p(x) = e^{-(1+\omega_0)-\omega_1 x^m}$	$mx^{m-1}e^{-x^m}$
GED	$c(v) = \|x\|^m p(x)dx$	$p(x) = e^{-(1+\omega_0)-\omega_1 \|x\|^m}$	$C(v)e^{\|x\|^m}$
Cauc	$\int \log(1+x^2)p(x)dx = 2\ln 2$	$p(x) = e^{-(1+\omega_0)-\omega_1 \ln(1+x^2)}$	$\frac{1}{\pi(1+x^2)}$

Table 10.2 Boltzmann and Tsallis Entropies vs. EU and generalizations of EU

	Boltzmann entropy	EU	Tsallis entropy	Generalizations of EU
Functional form	$-\int p(x)\ln p(x)dx$	$\int u(x)p(x)dx$	$-\int p^q(x)\ln_q p(x)dx$	$\int \omega(p(x))u(x)dx$
Implications	Does not account for information incompleteness, hidden correlations and nonlinear relations	Probability density function is not distorted	Takes into account information incompleteness and possible nonlinear relationships	Distorted subjective probability density function
Special case	–	–	$-\int_\Omega p^q(x)\ln_q p(x)dx \xrightarrow{q\to 1} -\int_\Omega p(x)\ln p(x)dx$	$\omega(p(x)) = p(x)$

variant probability distributions etc. For example, the maximum entropy of a multivariate distribution with specified mean and known variance–covariance matrix turns out also to be a multivariate normal probability distribution. When a data sample indicates both skewness and leptokurtosis, then the resulting distribution will assume a more complex form.

10.5 The Subjective, the Q Distributions and BG Entropy

Consider next two probability measures with, $p^S(i)$ and $p^M(i)$, one subjective based on prior or experimental data and the other theoretical, meeting the pricing Arrow–Debreu or the standardized CCAPM kernel risk pricing models:

$$\mu = E^M(i) = \frac{1}{1+R_f}\sum_{i=0}^{n} ip^M(i) \text{ and } 1 = \sum_{i=0}^{n} p^M(i)$$

Unlike the expected BG entropy $E^M \ln\left(\frac{1/n}{p^M(i)}\right)$ considered above we use $E^M \ln\left(\frac{p^S(i)}{p^M(i)}\right)$, a relative entropy. The maximum BG entropy M-distribution, is then defined by the following:

$$\underset{p^M(i)}{\text{Max}}\ E^M \ln\left(\frac{p^S(i)}{p^M(i)}\right) \text{ Subject to :}$$

$$\mu = E^M(i) = \frac{1}{1+R_f}\sum_{i=0}^{n} ip^M(i) \quad \text{and} \quad 1 = \sum_{i=0}^{n} p^M(i)$$

whose solution yields again an exponential probability distribution for the M-measure with parameters selected to be concordant with the information supplied. In particular, we use the following Lagrange optimization problem:

$$\underset{p^Q(i)}{\text{Max}} \sum_{i=1}^{n} p^Q(i) \ln\left(\frac{p^S(i)}{p^Q(i)}\right) + \lambda\left(\frac{1}{1+R_f}\sum_{i=0}^{n} i p^Q(i) - \mu\right) + \nu\left(\sum_{i=0}^{n} p^Q(i) - 1\right)$$

which yields:

$$p^Q(i) = \frac{p^S(i) e^{\left(\frac{\lambda}{1+R_f}\right)i}}{\sum_{i=0}^{n} p^S(i) e^{\left(\frac{\lambda}{1+R_f}\right)i}} \text{ with } \mu = \frac{1}{1+R_f} \frac{\sum_{i=0}^{n} i p^S(i) e^{\left(\frac{\lambda}{1+R_f}\right)i}}{\left(\sum_{i=0}^{n} p^S(i) e^{\left(\frac{\lambda}{1+R_f}\right)i}\right)}$$

Note that if $\mathrm{E}(\mathrm{e}^{iz})$ denote the moment generating function $M_S(z)$ of the probability distribution $p^S(i)$, $\sum_{i=0}^{n} p^S(i) \mathrm{e}^{\left(\frac{\lambda}{1+R_f}\right)i} = M_S\left(\frac{\lambda}{1+R_f}\right)$. Further $\frac{\partial M_S(z)}{\partial z} = \mathrm{E}(i\mathrm{e}^{iz}) \quad z = 0$, $\frac{\partial M_S(0)}{\partial z} = \mathrm{E}(\mathrm{i}) = \mu$.

And therefore, λ is solved using μ below:

$$\frac{p^M(i)}{p^S(i)} = \frac{\mathrm{e}^{\left(\frac{\lambda}{1+R_f}\right)i}}{M_S\left(\frac{\lambda}{1+R_f}\right)} \quad \text{with} \quad \mu = \frac{1}{1+R_f} \left.\frac{\partial \ln M_S(z)}{\partial z}\right|_{z=\frac{\lambda}{1+R_f}}$$

In other words, given our best forecast of market prices, $p^S(i)$ and given a mean market price, the M-measure probability is as defined above.

Example: The Continuous State Problem Consider the continuous state problem:

$$\underset{f^M(x)}{\text{Max}} \; E^M \ln\left(\frac{f^S(x)}{f^M(x)}\right) \text{ Subject to} : \mu = E^M(x) = \frac{1}{1+R_f}\int x f^M(x)dx \text{ and } 1 = \int f^M(x)dx$$

Or $\underset{f^M(x)}{\text{Max}} \int f^M(x) \ln\left(\frac{f^S(x)}{f^M(x)}\right) + \lambda\left(\frac{1}{1+R_f}\int x f^M(x)dx - \mu\right) + \nu\left(\int f^M(x)dx - 1\right)$, thus:

$$\ln\left(\frac{f^S(x)}{f^M(x)}\right) - 1 + \lambda\frac{x}{1+R_f} + \nu = 0 \text{ and therefore,}$$

$$f^M(x) = \frac{f^S(x) e^{\lambda\frac{x}{1+R_f}}}{\int f^S(x) e^{\lambda\frac{x}{1+R_f}} dx} = \frac{f^S(x) e^{\lambda\frac{x}{1+R_f}}}{M_S\left(\frac{\lambda}{1+R_f}\right)}$$

Since, $\frac{\partial}{\partial z}\int f^S(x)e^{xz}dx = \int xf^S(x)e^{xz}dx$ and if $v = \frac{\lambda}{1+R_f}$ then, $\mu = \frac{M_S^{-1}(v)}{1+R_f}\int xf^S(x)e^{vx}dx$ we have:

$$\mu = \frac{1}{1+R_f}\frac{\partial M_S(v)/\partial v}{M_S(v)} = \frac{1}{1+R_f}\frac{\partial \ln M_S(v)}{\partial v},\ v = \frac{\lambda}{1+R_f}$$

which provides one equation in the Lagrange multiplier λ. In particular, say that the S distribution has a moment generating function given by $M(u) = \frac{1}{(1-u\beta)^{\alpha}}$. In this case $\ln M(u) = -\alpha\ln(1-u\beta)$ and $\partial \ln M(u)/\partial u = \alpha\beta/(1-u\beta)$ and therefore

$$\mu = \frac{1}{1+R_f}\frac{\partial M_S(v)/\partial v}{M_S(v)} = \frac{\alpha\beta}{1+R_f-\lambda\beta} \quad \text{or} \quad \frac{\lambda}{1+R_f} = \frac{\alpha}{\mu(1+R_f)} + \frac{1}{\beta}$$

Thus,

$$f^M(x) = \frac{f^S(x)e^{\lambda\frac{x}{1+R_f}}}{\int f^S(x)e^{\lambda\frac{x}{1+R_f}}dx} = \frac{f^S(x)e^{x\left(\frac{\alpha}{\mu(1+R_f)}+\frac{1}{\beta}\right)}}{M_S\left(\frac{\alpha}{\mu(1+R_f)}+\frac{1}{\beta}\right)}$$

And finally, since $f^S(x) = \frac{x^{\alpha-1}e^{-x/\beta}}{\beta^{\alpha}\Gamma(\alpha)}$ we obtain an explicit M-Gamma probability distribution:

$$f^M(x) = \frac{x^{\alpha-1}e^{x\left(\frac{\alpha}{\mu(1+R_f)}+\frac{1}{\beta}-\beta\right)}}{\beta^{\alpha}\Gamma(\alpha)\left(1+\frac{1}{\beta}-\beta-\frac{\alpha}{\mu(1+R_f)}\right)^{-\alpha}}$$

This distribution has assumed complete state preferences and is complete by the technical definition of risk neutral pricing. If state preferences are incomplete then maximizing the entropy of Tsallis provides a more general approach. Below we consider such a problem.

10.6 The Tsallis Maximum Entropy and Incomplete States Preferences

In physics, incompleteness is defined by non-extensive systems, transformed parametrically to extensive ones (i.e. defined by a complete set of state preferences) based on some physical metric. Say that some losses are assessed to be more important than others; there may therefore be more than two ways to express such a fact, either by using a utility principle or by weighting their probabilities. The introduction of subjective weights to express a concern for "undetectability" or "non-insurability" of potential events expresses such concerns. For example, in industrial engineering FMECA ranking risk prospects (see Chap. 6) gives a greater weight to events that are more difficult to detect. The assumption of such a weighting is based on the premise that what can be predicted can be avoided while what is not predicted or is not seen cannot be avoided and could, therefore, have important consequences. In valuing such prospects, it is therefore natural to give a greater weight to these probabilities.

Allais' (1953) utility paradox showed that utility is not necessarily linear in probabilities (and therefore an expectation might be nonlinear in probabilities). Allais' original contribution (recognized by a Nobel Prize) as well as others' contributions based on behavioral observations and psychological interpretations have motivated extensive research to provide alternatives to the expected utility paradigm. In particular Kahneman and Tversky (1979) seminal papers on prospect theory (also recognized by a Nobel Prize) were conceived in more general terms accommodating both psychological biases like framing and Allais and Ellsberg paradoxes, thereby, providing an interpretation to behavioral anomalies within one simple model. As with most theories generalizing expected utility, prospect theory suggests a deformation of subjective probabilities such that smaller probabilities are relatively and individually overweight (as indicated by the Tsallis entropy metric). In addition, prospect theory defines utility in terms of returns rather than wealth. Thus it incorporates the notion of loss–aversion and risk seeking behavior when facing loss prospects. Prospect theory definition of subjective probabilities, in addition, violates stochastic dominance criteria. Levy and Levy (1991) pointed to another critical aspect of prospect theory pointing out that it values prospects associated with gains and losses while financial investment includes also the possibility of mixed prospects. The notion of a deformed subjective distribution remains valid in the presence of mixed prospects however.

Some consensus on methods to deform subjective probabilities applied globally rather than individually seem to be accepted. Tversky and Kahneman (1992) cumulative prospect theory can be seen as an endorsement of such probability weighting (or deformation). There are also some similarities on how expectations are computed in these models and how they are calculated in *non extensive* statistics via escort probabilities (as we shall see below).

Thus, while expected utility theory has some difficulties to account for uncertainty with an incomplete set of prospects, almost all generalizations of expected utility theory concentrate on the deformation of the subjective probability. The Tsallis entropy metric we shall use below provides a parametric and quantitative characterization of such weighting. Table 10.2 above highlights the effects of both the Boltzmann and the Tsallis entropy metrics on the expected valuation of utility functions.

10.6.1 Tsallis Entropy and the Power Law

Tsallis (1988) entropy, generalizes the BG entropy ln metric in two ways. First it is the expected value of a generalized logarithmic function (q-logarithm) and second, the probability distribution function is deformed by a power q which is used to define escort probabilities (as will be defined below).

Let the Tsallis entropy be defined by:

$$H_q^T(f^M) = \frac{1 - E^M\left(\left[f^M(.)\right]^{q-1}\right)}{q-1} = E^M\left(\ln_q f^M(.)\right), \quad q > 1$$

where $\ln_q f^M(.)$ defines a q-logarithmic function: $\ln_q f^Q(.) = \frac{\left[f^Q(.)\right]^{1-q} - 1}{1-q}$. Further,

$$-\int \left[f^M(x)\right]^q \ln_q f^M(x)dx = -\int \left[f^M(x)\right]^q \frac{\left[f^M(.)\right]^{1-q} - 1}{1-q} dx = \frac{1 - \int \left[f^M(x)\right]^q dx}{q-1},$$

Or,

$$H_q^T(f^Q) = -\int \left[f^M(x)\right]^q \ln_q f^M(x)dx$$

Note that when q tends to 1, then by applying l'Hopitale rule for derivatives, the Tsallis entropy converges to the Boltzmann–Gibbs entropy. In this sense, the Tsallis entropy is a generalization of Boltzmann entropy as stated earlier. This relationship between Tsallis and Boltzmann functions stems mathematically from the relation between ordinary and Jackson q-derivatives (see below) with the ordinary derivative recovered from Jackson's q-derivative when q converges to one (see also Oren Tapiero 2012c).

10.6.2 A Mathematical Note: (Abe 1997; Borges and Roditi 1998)

A large number of papers have suggested generalized entropy metrics. Abe (1997) proposes to generate entropy functionals based on the Jackson q-derivative operator. This is a discrete derivative which is defined as follows. Let $f(x)$ be a function and define its q-derivative by:

$$\frac{d_q f(x)}{d_q x} = \frac{f(qx) - f(x)}{qx - x}$$

Consider next a probability distribution $\{p_i\}, \quad p_i \geq 0, \quad \sum p_i = 1, i = 1, \ldots n$ and its deformed probability distribution $\{p_i^q\}, \quad p_i^q \geq 0, \quad g(q) = \sum_{i=1} p_i^q < 1$ where $g(1) = 1$. Deformation, as stated earlier is commonly practiced in both psychology and behavioral finance. A derivative of the sum $g(q)$ and its limit leads to the Boltzmann entropy while using a Jackson q-derivative of $g(q)$ leads at the limit $q = 1$ to the Tsallis entropy. Explicitly,

$$\operatorname*{Lim}_{q\to 1}\left\{-k\frac{dg(q)}{dq}\right\} = -k\sum_{i=1} p_i \ln\left(\frac{1}{p_i}\right) = \mathrm{H}^{BG}(p_i) : \text{ The Boltzmann-Gibbs entropy}$$

$$\operatorname*{Lim}_{\alpha\to 1}\left\{-k\frac{d_q g(\alpha)}{d_q\alpha}\right\}_{\alpha=1} = -k\frac{\sum_{i=1}(p_i)^{q\alpha} - \sum_{i=1} p_i{}^{\alpha}}{q-1} = k\frac{1-\sum_{i=1}(p_i)^q}{q-1} = S^T(p_i : q) : \text{ The Tsallis entropy}$$

Note that when $q \to 1$, we recover the BG entropy. Further, the Jackson q-derivative is a generalized version of the usual derivative and therefore it is to be expected that the Tsallis entropy obtained is in some sense a generalization of the principle of insufficient reason used to justify a probability distribution obtained by maximizing the entropy. In the general case, expressing a function by $E(w(p_i))$ provides a mechanism to express the relative usefulness of a probability, either due to subjective factors or due to the valuation of certain outcomes, it is therefore a "utility of probabilities". Of course, if to each event whose probability is denoted by p_i we also associate an outcome x_i, we may then value such outcomes by their utility $u(x_i)$ and as we saw earlier, its value defined by its expected utility. However, if in addition we provide a weighting of probabilities by $(p_i)^q$ then expectation of such events is $\sum_{i=1}(p_i)^q u(x_i)$ which introduces a double bias. On the one hand the utility bias (albeit rational) and the probability bias. Such issues were confronted by psychologists but are difficult to justify as the bias in both probabilities and outcomes will define each other. To circumvent such problems, a careful interpretation of deformed probabilities is essential. In our case, the parameter q was interpreted as expressing an incompleteness of the probability distribution (rendering its sum smaller than one) and therefore, the sum $\sum_{i=1}^{m<n}(p_i)^q u(x_i)$ is incomplete. However, since probabilities we use are presumably complete, we use probabilities coined escort probabilities that have the property that they sum to 1. To do so, we simply set:

$$p_i^e(q) = \frac{(p_i)^q}{\sum_{i=1}^{m}(p_i)^q} \quad \text{and therefore,} \quad 0 \le p_i^e(q) \le 1 \text{ and} \sum_{i=1}^{m} p_i^e(q) = 1$$

In which case, we can presume that the expected value of outcomes (or their utility) is given by:

$$\sum_{i=1}^{m} p_i^e(q)u(x_i)$$

The estimation of these parameters is however empirically difficult to verify although numerous behavioral experiments are made to evaluate the probabilities' parameters.

Generalizing further the Jackson q-derivative, Abe suggested the following derivative: $\frac{d_{A,A'}f(x)}{d_{A,A'}x} = \frac{f(Ax)-f(A'x)}{Ax-A'x}$, A, A' real. A derivative of $-\,kg(\alpha)$ at the limit $\alpha \to 1$ yields then a generalized entropy which can be expressed as a difference of two Tsallis entropies providing thereby a measurement of "relative" (Tsallis) entropy distances, or

$$S^A = k\frac{\left(\sum\limits_{i=1}(p_i)^{A'} - \sum\limits_{i=1}(p_i)^{A}\right)}{A-A'} = \frac{((1-A')S^T_{A'} - (1-A)S^T_{A})}{A-A'}$$

10.6.3 *The Maximum Tsallis Entropy and the Power Law Distribution*

Consider next the following M measures probability constraints:

$$\int f(x,T)dx = 1, \quad \int \left[f(x,T)\right]^q dx = e^{-R_f(T-t)} \text{ and } \int C^T_k(x,T)\left[f(x,T)\right]^q dx = C^T_k(t)$$

where $f(x,T)$ is a M probability measure at time T (and therefore a market price distribution), $[f(x,T)]^q$ is the "Tsallis kernel" (in other words, in the CCAPM model, we assume a kernel distribution defined by the deformed distribution). At time t, we therefore have $\int (f(x,T)^q dx = e^{-R_f(T-t)}$ while the expected price of call options $C^T_k(x,T)$ (for example $C^T_k() = (x(T)-K_k)^+$ at time t are $C^t_k(t) = \int C^T_k(x,T)\left(f^Q(x,T)\right)^q dx, \quad k = 1,2,\ldots$ The underlying assumption of this problem is that the risk neutral distribution is not known and therefore the deformed probability distribution $[f(x,T)]^q$ accounts for the distribution incompleteness. The distribution that maximizes the Tsallis entropy subject to the specific constraints stated above is then a Lagrangian optimization problem:

$$L = \int\limits_{\Omega}\left[f(x,T)\right]^q \ln_q f(x,T)dx + \lambda_0\int f(x,T)dx + \sum_{k=1}\lambda_k\int C^T_k(x,T)\left[f(x,T)\right]^q dx$$

A straightforward optimization yields a family of power law density functions, defined by the problem constraints above,

$$f(x,T)=\frac{\left\{(1-(1-q))\sum_{k=1}\lambda_k C_k^T(x,T)\right\}^{\frac{1}{1-q}}}{Z(q,\lambda_k)};$$

$$Z(q,\lambda_k)=\int\left\{(1-(1-q))\sum_{k=1}\lambda_k\int C_k^T(x,T)\right\}^{\frac{1}{1-q}}$$

where $Z(q,\lambda_k)$ is a normalizing function. The Lagrange multipliers are given by a simultaneous nonlinear system of equations:

$$C_k^T(t)=\int C_k^T(x,T)\left[\frac{\left\{(1-(1-q))\sum_{k=1}\lambda_k C_k^T(x,T)\right\}^{\frac{q}{1-q}}}{Z^q(q,\lambda_k)}\right]dx,\quad k=1,2,\ldots.$$

However, since $\left[f^Q(x,T)\right]^q$ is a kernel, and since for such a kernel

$$E(M(x))=\int M(x)f(x)dx=\int\left[f(x)\right]^q dx=e^{-R_f(T-t)}$$

We note that Tsallis entropy is in this case a function of the reference risk free asset and the q-parameter, reducing the entropy as q increases. In other words, as q becomes smaller, the number of states that are not accounted for is increasingly small while at q = 1, all states are accounted for and therefore the distribution is complete. Thus, $H_q^T=\frac{e^{-R_f(T-t)}}{q-1},\quad q>1$ where the index q is the sole parameter governing incomplete states information uncertainty. Since entropy is considered as a measure of expected information (where $-\ln_q f(x)$ measures the information of a corresponding state), then as the parameter q increases, the expected information decreases as indicated above. For this reason, we interpret the parameter q as discounting for possible incomplete states information.

10.6.4 The Tsallis Entropy and Subjective Estimate of the M-Distribution

Consider at present the use of the Tsallis entropy as a measure metric of a subjective (known or assumed) probability distribution $f^S(x)$ and a Q-complete market price distribution. We seek to maximize the relative Tsallis entropy where the estimate of current option prices as biased of a deformed probability:

$$\underset{f^M(x,T)}{\text{Max}} \int_{\Omega} \left[f^M(x)\right]^q \ln_q\left(\frac{f^M(x)}{f^S(x)}\right) dx \text{ Subject to} : 1 = \int f^M(x)dx \text{ and } C_k^T = \int C_k^T(x)\left[f^M(x)\right]^q dx$$

Since

$$\ln_q\left(\frac{f^M(x)}{f^S(x)}\right) = \frac{\left[f^M(x)/f^S(x)\right]^{1-q} - 1}{1-q},$$

The Lagrangian is:

$$\text{L} = \frac{\left[f^M(x)\right]\left[f^S(x)\right]^{q-1} - \left[f^M(x)\right]^q}{1-q} + \lambda_0\left[f^M(x) - 1\right] + \sum_{k=1} \lambda_k\left\{C_k^T(x)\left[f^M(x)\right]^q - C_k^t\right\}$$

And therefore, conditions for optimality are:

$$-\frac{\left[f^S(x)\right]^{q-1} - q\left[f^M(x)\right]^{q-1}}{q-1} + \lambda_0 + \sum_{k=1} \lambda_k q\left\{C_k^T(x)\left[f^M(x)\right]^{q-1}\right\} = 0$$

And therefore,

$$f^M(x) = \frac{f^S(x)}{(q-1)^{\frac{1}{q-1}}}\left\{\frac{q}{q-1} + \lambda_0 + \sum_{k=1} \lambda_k q\left\{C_k^T(x)\right\}\right\}^{-\frac{1}{q-1}}$$

Where the Lagrange multipliers are determined by the solution of the nonlinear system of equations:

$$1 = \int f^M(x)dx = \int \frac{\left[f^S(x)\right]}{(q-1)^{\frac{1}{q-1}}}\left\{\frac{q}{q-1} + \lambda_0 + \sum_{k=1} \lambda_k q\left\{C_k^T(x)\right\}\right\}^{-\frac{1}{q-1}} dx$$

$$C_k^t = \int C_k^T(x)\left[\frac{\left[f^S(x)\right]^q}{(q-1)^{\frac{q}{q-1}}}\left\{\frac{q}{q-1} + \lambda_0 + \sum_{k=1} \lambda_k q\left\{C_k^T(x)\right\}\right\}^{-\frac{q}{q-1}}\right] dx$$

Again, setting,

$$e^{-R_f T} = \int \left[f^M(x)\right]^q dx = \int \left[\frac{\left[f^S(x)\right]^q}{(q-1)^{\frac{q}{q-1}}}\left\{\frac{q}{q-1} + \lambda_0 + \sum_{k=1} \lambda_k q\left\{C_k^T(x)\right\}\right\}^{-\frac{q}{q-1}}\right] dx$$

which provides an estimate for q in terms of the risk free rate. Again setting the subjective estimate as a Gamma probability distribution, we have the nonlinear equation in q (with the previous nonlinear system of equations in moment expectations providing a system of nonlinear equations in the Lagrange parameters):

$$e^{-R_f T} = \frac{\int_0^\infty \left[x^{q(\alpha-1)} e^{-xq/\beta} \left\{ \frac{q}{q-1} + \lambda_0 + \sum_{k=1} \lambda_k q \{ C_k^T(x) \} \right\}^{-\frac{q}{q-1}} \right] dx}{\beta^{q\alpha} (\Gamma(\alpha))^q (q-1)^{\frac{q}{q-1}}}$$

10.6.5 Maximum Tsallis Entropy with Escort Probabilities

As noted previously, it is conceptually more natural to use escort probabilities:

$$p_i^e(q) = \frac{(p_i)^q}{\sum_{i=1}^m (p_i)^q}, \quad \sum_{i=1}^m p_i^e(q) = 1, \quad \Phi = \sum_{i=1}^m (p_i)^q < 1$$

And therefore, $p_i = \left(\Phi\, p_i^e(q)\right)^{\frac{1}{q}}$ which defines the escort probability relative to the true probability p_i. Such a standardization allows the transformation of the Tsallis entropy maximization problem using the escort probabilities. Below, we consider two examples,

Example: We consider first the maximization of a Tsallis entropy subject to a mean estimate (or known value) of the underlying distribution using escort probabilities. The optimization also indicates the properties of the deformed probabilities parameter to be consistent with the maximization of entropy. Consider the problem:

$$\text{Max } S_e^T = \frac{1 - E \sum \left(p_i^e(q)\right)^{g(q)}}{g(q)} = \frac{1 - \sum \left(p_i^e(q)\right)^{1+g(q)}}{g(q)} \text{ with } \sum_{i=1}^m p_i^e(q) = 1, \quad \sum_{i=1}^m C_i^+ p_i^e(q) = C_0$$

The Lagrangian function is:

$$L = \frac{1 - \sum \left(p_i^e(q)\right)^{1+g(q)}}{g(q)} + \lambda_0 \left(\sum_{i=1}^m p_i^e(q) - 1 \right) + \lambda_1 \left(\sum_{i=1}^m C_i^+ p_i^e(q) - C_0 \right)$$

And the first order necessary conditions for optimality yield:

$$p_i^e(q) = \frac{\left(1 - \frac{\lambda_1 C_i^+}{\lambda_0}\right)^{\frac{1}{g(q)}}}{\left(\frac{1+g(q)}{\lambda_0 g(q)}\right)^{\frac{1}{g(q)}}} \text{ with } Z = \left(\frac{1+g(q)}{\lambda_0 g(q)}\right)^{\frac{1}{g(q)}} = \sum_{i=1}^m \left(1 - \frac{\lambda_1 C_i^+}{\lambda_0}\right)^{\frac{1}{g(q)}}$$

Then, at the limit $g(q) \to 0, p_i^e(q) \sim e^{-\lambda_0 C^+}$, when setting $\lambda_0 = 1/g(q)$. A second derivative of the Lagrangian yields,

$$\frac{\partial^2 L}{\partial p_i^e(q)^2} = -(g(q)+1)p_i^e(q)^{g(q)-1}$$

which requires for maximization that $(g(q)+1) > 0$ and therefore, $g(q) > -1$. Further, since $\partial S_e^T/\partial q > 0$ leads to $\partial g(q)/\partial q > 0$. In particular, setting $g(q) = q$ we obtain a Tsallis entropy. In other words (with $\lambda_0 = 1/q$):

$$p_i^e(q) = (1+q)^{-\frac{1}{q}}(1-\lambda_1 q C_i^+)^{\frac{1}{q}} \text{ with}$$
$$C_0 = \sum_{i=1}^m C_i^+ p_i^e(q) = (1+q)^{-\frac{1}{q}} \sum_{i=1}^m C_i^+ (1-\lambda_1 q C_i^+)^{\frac{1}{q}}$$

which provides a solution to the Lagrange multiplier λ_1. Let the solution be $\lambda_1^*(C_0, q)$, thus:

$$p_i^e(q) = \frac{p_i^q}{\sum_{i=1}^m p_i^q} = (1+q)^{-\frac{1}{q}}(1-\lambda_1^*(C_0,q)qC_i^+)^{\frac{1}{q}} \text{ with}$$

Which provides a solution for p_i^q and therefore for $p_i = \left(p_i^q\right)^{\frac{1}{q}}$. Explicitly, say that m = 2, with $C_1^+ > 0$ and $C_2^+ = 0$ thus

$$C_0 = C_1^+ p_1^e(q) = (1+q)^{-\frac{1}{q}} C_1^+ (1-\lambda_1 q C_1^+)^{\frac{1}{q}} \quad \text{and}$$
$$\frac{C_0}{C_1^+} = p_1^e(q) = (1+q)^{-\frac{1}{q}}(1-\lambda_1 q C_1^+)^{\frac{1}{q}}$$

Or

$$\lambda_1 = \left\{1-(1+q)\left(\frac{C_0}{C_1^+}\right)^q\right\} \Big/ qC_1^+$$

As a result,

$$p_1^e(q) = \frac{p_1^q}{p_1^q + p_2^q} = (1+q)^{-\frac{1}{q}}\left(1-\left\{1-(1+q)\left(\frac{C_0}{C_1^+}\right)^q\right\}\right)^{\frac{1}{q}}$$
$$1 - p_1^e(q) = \frac{p_2^q}{p_1^q + p_2^q} = 1-(1+q)^{-\frac{1}{q}}\left(1-\left\{1-(1+q)\left(\frac{C_0}{C_1^+}\right)^q\right\}\right)^{\frac{1}{q}}$$

With

$$\left(\frac{p_1}{p_2}\right)^q = \left(\frac{p_1}{1-p_1}\right)^q = \left(\frac{(1+q)^{-\frac{1}{q}}\left(1-\left\{1-(1+q)\left(\frac{C_0}{C_1^+}\right)^q\right\}\right)^{\frac{1}{q}}}{1-(1+q)^{-\frac{1}{q}}\left(1-\left\{1-(1+q)\left(\frac{C_0}{C_1^+}\right)^q\right\}\right)^{\frac{1}{q}}}\right)$$

As a result, an estimate for q can be obtained by an estimate of a logit regression on the distribution of the odds $\ln\left(\frac{p_1}{1-p_1}\right)$. Given such an estimate, and given $\frac{C_0}{C_1^+}$, an estimate of the parameter q is obtained by equating the logit estimate with the equation above.

For example, say that $q = 1.3$ and $\frac{C_0}{C_1^+} = 0.8$ then:

$$\frac{p_1}{p_2} = \frac{p_1}{1-p_1} = \left(\frac{(2.3)^{-\frac{1}{1.3}}\left(1-\left\{1-(2.3)(0.8)^{1.3}\right\}\right)^{\frac{1}{1.3}}}{1-(2.3)^{-\frac{1}{1.3}}\left(1-\left\{1-2.3(0.8)^{1.3}\right\}\right)^{\frac{1}{1.3}}}\right)^{\frac{1}{1.3}}$$

The problem treated here raises a number of questions. First, are there other deformed probability systems that are coherent with the entropy maximization and second, what if q is to be estimated from sample and statistical data. How would the underlying maximum entropy distribution function be altered by its randomization.

First note that the deformed function $g(q)$ is greater than -1 and further its derivative is increasing. As seen above, $q > 1$ which recovers the Tsallis entropy is certainly one solution. Assume instead that g(q) assume the functional form of a beta distribution with $1 + q = 1 + \beta$ and let β have a Beta distribution with known parameters (a, b). In this case, the underlying distribution is a mixture distribution which is a function of parameters (a, b):

$$p_i^e(\beta|a,b) = \frac{\left(1-\frac{\lambda_1 C_i^+}{\lambda_0}\right)^{\frac{1}{1+\beta}}}{\left(\frac{2+\beta}{\lambda_0(1+\beta)}\right)^{\frac{1}{1+\beta}}} \text{ with } Z = \left(\frac{2+\beta}{\lambda_0(1+\beta)}\right)^{\frac{1}{1+\beta}} = \sum_{i=1}^{m}\left(1-\frac{\lambda_1 C_i^+}{\lambda_0}\right)^{\frac{1}{1+\beta}}$$

In particular, first note that a solution for the two parameters is given by:

$$\left(\frac{2+\beta}{\lambda_0(1+\beta)}\right)^{\frac{1}{1+\beta}} = \sum_{i=1}^{m}(1-\lambda C_i^+)^{\frac{1}{1+\beta}},\ \lambda = \frac{\lambda_1}{\lambda_0},$$

$$\sum_{i=1}^{m} C_i^+(1-\lambda C_i^+)^{\frac{1}{1+\beta}} = C_0\left(\frac{2+\beta}{\lambda_0(1+\beta)}\right)^{\frac{1}{1+\beta}} = C_0\sum_{i=1}^{m}(1-\lambda C_i^+)^{\frac{1}{1+\beta}}$$

Or,

$$f(\beta) = \sum_{i=1}^{m} \left(C_i^+ - C_0\right)\left(1 - \lambda C_i^+\right)^{\frac{1}{1+\beta}} = 0$$

Since β is small (and in the extensive case it equals 0), we may consider a first order Taylor series approximation. In this case,

$$f(\beta) = f(0) + \beta \frac{\partial f(0)}{\partial \beta} = 0 \quad \text{and} \quad \beta = \frac{f(0)}{\partial f(0)/\partial \beta} \quad \text{with}$$

$$f(0) = \sum_{i=1}^{m} \left(C_i^+ - C_0\right)\left(1 - \lambda C_i^+\right) \quad \text{and}$$

$$\frac{\partial f(0)}{\partial \beta} = -\sum_{i=1}^{m} \left(C_i^+ - C_0\right) \ln\left(1 - \lambda C_i^+\right) \frac{\left(1 - \lambda C_i^+\right)}{(1+\beta)^2}$$

$$= \sum_{i=1}^{m} \left(C_i^+ - C_0\right) \ln\left(1 - \lambda C_i^+\right)\left(1 - \lambda C_i^+\right)$$

And therefore,

$$\sum_{i=1}^{m} \left(C_i^+ - C_0\right)\left(1 - \lambda C_i^+\right) + \beta\left[\sum_{i=1}^{m} \left(C_i^+ - C_0\right) \ln\left(1 - \lambda C_i^+\right)\left(1 - \lambda C_i^+\right)\right] = 0$$

which is one equation in λ when β is known.

When β has a probability distribution, note that by introducing the change of variable $y = (1+\beta)^{-1}$ we obtain:

$$\int_0^1 (1-s)^{\frac{1}{1+\beta}} f_B(\beta|a,b)d\beta = -\int_1^{1/2} y^2(1-s)^y f_Y(y|a,b)dy$$

where $f_Y(y|a,b)$ is an inverted Beta probability distribution.

Example (Oren Tapiero 2012*)*: We consider next a relative optimization problem and consider:

$$\text{Max}_{p_i^e(q)} \Phi \sum_{i=1}^{m} p_i^e(q) \ln_q\left(\Phi^{\frac{1}{q}}\left[p_i^e(q)\right]^{\frac{1}{q}} \Big/ \left(p_i^S\right)^{\frac{1}{q}}\right) \quad \text{Subject to :}$$

$$1 = \sum_{i=1}^{m} p_i^e(q) \quad \text{and} \quad C_k^T = \Phi \sum_{i=1}^{m} C_k^T(i) p_i^e(q)$$

Since,

$$\ln_q p_i = \frac{[p_i]^{1-q} - 1}{1-q}, \quad \ln_q\left(\Phi^{\frac{1}{q}}\left[p_i^e(q)\right]^{\frac{1}{q}} \Big/ \left(p_i^S\right)^{\frac{1}{q}}\right) = \frac{\left[\Phi^{\frac{1-q}{q}}\left[p_i^e(q)\right]^{\frac{1-q}{q}}\left(p_i^S\right)^{-\frac{1-q}{q}}\right] - 1}{1-q}$$

And therefore,

$$\underset{f^{Q}(x,T)}{\text{Max}}\ \Phi \sum \left\{\Phi^{\frac{1-q}{q}}\left(p_i^S\right)^{-\frac{1-q}{q}}\left[p_i^e(q)\right]^{\frac{1}{q}} - 1\right\} \Big/ (1-q) \ \text{Subject to}:$$

$$1 = \sum_{i=1}^{m} p_i^e(q) \ \text{and} \ C_k^T = \Phi \sum_{i=1}^{m} C_k^T(i) p_i^e(q)$$

Whose Lagrangean is:

$$L = \frac{\Phi^{\frac{1-q}{q}}\left(p_i^S\right)^{-\frac{1-q}{q}}\left[p_i^e(q)\right]^{\frac{1}{q}}}{(1-q)} + \lambda_0 \sum_{i=1}^{m} p_i^e(q) + \lambda_k \Phi \sum_{i=1}^{m} C_k^T(i) p_i^e(q)$$

Thus

$$p_i^e(q) = \left(\frac{q(q-1)\left[p_i^S\right]^{\frac{1-q}{q}}}{\Phi^{\frac{1-q}{q}}}\left(\lambda_0 + \lambda_k \Phi\, C_k^T(i)\right)\right)^{\frac{q}{1-q}}$$

Inserting $[p_i]^q = \Phi\, p_i^e$ we have instead:

$$p_i = \left[p_i^S\right]^{\frac{1}{q}}\left(q(q-1)\left(\lambda_0 + \lambda_k \Phi\, C_k^T(i)\right)\right)^{\frac{1}{1-q}}$$

And the Lagrange parameters may be estimated using the known expected values.

In this framework, higher order moments can also be considered.

Example: Let the continuous probability-Tslallis entropy problem with escort probabilities:

$$\text{Max } S_q^T(f) = \frac{1 - E\left([f(x)]^{q-1}\right)}{q-1} \quad \text{Subject to}: \ 1 = \int_{-\infty}^{+\infty} f(x)dx, \quad \int_{-\infty}^{+\infty} x^2 [f^e(x)]^q dx = \sigma_q^2$$

where $f^e(x)$ are the escort probabilities: $f^e(x) = [f(x)]^q \Big/ \int_{-\infty}^{+\infty} [f(x)]^q dx$:

$$f(x) = Z_q^{-1}\left[\left(1 - \frac{q - 1x^2}{(3-q)\sigma_q^2}\right)_+\right]^{\frac{1}{1-q}}, \quad Z_q^{-1} = \left\{\frac{\Gamma\left(\frac{1}{2}\right)\Gamma\left(\frac{1}{q-1} - \frac{1}{2}\right)}{\sqrt{\frac{q-1}{(3-q)\sigma_q^2}}\Gamma\left(\frac{1}{q-1}\right)}\right\} \quad \text{where}$$

$$\Gamma(x) = \int_0^\infty t^{x-1}e^{-t}dt, \quad x > 0$$

Note that for $q \to 1$, we recover the standard normal distribution. It is called therefore the q-Gaussian distribution (Aki-Hiro Sato 2010). The parameter ‘q’ can be, estimated as follows. First note that:

$$f^q(x) = Z_q^{-q}\left[\left(1 - \frac{q - 1x^2}{(3-q)\sigma_q^2}\right)_+\right]^{\frac{q}{1-q}}, \quad Z_q^{-q} = \left\{\frac{\Gamma\left(\frac{1}{2}\right)\Gamma\left(\frac{1}{q-1} - \frac{1}{2}\right)}{\sqrt{\frac{q-1}{(3-q)\sigma_q^2}}\Gamma\left(\frac{1}{q-1}\right)}\right\}^q$$

Inserting this equation in the variance constraint, we have: $\sigma_q^2 \int f^q(x)dz = \int x^2 f^q(x)dz$ which is reduced to an integral equality to be solved for q.

$$0 = \int_{-\infty}^{\infty} \left(x^2 - \sigma_q^2\right)\left[\left(1 - \frac{(1-q)x^2}{(3-q)\sigma_q^2}\right)_+\right]^{\frac{q}{1-q}} dx$$

These problems are generally resolved by numerical means however.

These models, based on Tsallis generalized entropy provide therefore a kernel characterization of valuation and pricing models with fat tails and partial information.

10.7 Choice, Rationality, Bounded Rationality and Making Decision Under Uncertainty

Risk models of uncertainty are essentially a set of constraints and a rationality that we impose to bound uncertainty. To do so, we require (Ariel Rubinstein 1998):

- Knowledge of the problem
- Clear preferences
- An ability to optimize
- Indifference to equivalent logical descriptions of alternative and choice sets

The fact that decisions are not always rational does not mean that there are no other reasons for choosing a decision. A systematic approach to departures from our conventional definition of rationality has thus been the topic of intense interest. Herbert Simon for example, distinguished between a substantive and procedural rationality. By substantive rationality he refers to a behavior that is appropriate to the achievement of given goals within the limits imposed by given conditions and constraints. However, behavior is procedurally rational when it is the outcome of appropriate deliberations, i.e. it is the outcome of some reasoning process.

There is in fact and in practice ample evidence that decisions do not always appear to be rational. For example, decision makers often have a tendency to "throw good money after bad" (known sunk costs). Although it is irrational, it is often practiced. Here are a few examples,

- "having paid for the movie, I will stay with it, even though it is a dreadful and time consuming movie"
- "an investment in a stock, even if it has failed repeatedly, may for some irrational reason generate a loyalty factor"

The reason we are so biased in favor to bringing existing projects to fruition whatever the cost is that such behavior is built into our brains which is resisting conceptual changes even though the rational decision would be to change. The problem is psychological, once we have made an irreversible investment, we imbue it with extra value—the price of our emotional "ownership". There are many variations to this phenomenon. One is the "endowment effect" in which a person who is offered $10,000 for a painting he paid for only $1,000 and refuses such an apparent generous offer. The premium he refuses may then be accounted for by the pride he has in the belief of his exceptional judgment. Choice is therefore imbued in both personal biases that may be difficult to reconcile with basic axioms of rationality as they were stated in Chaps. 6, 7, 8 and 9. There are nonetheless a number of approaches one can use and make a choice which may be sensitive to the residual effects of uncertainty (when risk models by themselves cannot mitigate their real uncertain consequences). These approaches can be summarized by the following:

- Decision risk models insensitivity to models' parameters assumptions. Such models include robust models.
- Ex-post Decision and recovery risk models, providing a contingent ability to recover from events that were not accounted for.

These models are both embedded in the selection of their objective functions as was indicated in Chap. 6 for example (including regret and Savages's criteria for decision making). Below we shall consider some of the approaches.

10.7.1 Models Sensitivity and Robustness

Experimentation and simulation are important approaches to assess the sensitivity of designs to parametric variations. When a system is being operated on, the number of intervening variables (and parameters) may be too large, some of which may be uncontrollable, it is essential that the underlying model on the basis of which decisions are reached be insensitive to parameters variations. The following example based on a planned built-in insensitivity of a model parameters highlights such an approach. To simplify, of course, the example we consider is technically tractable.

Example: Sensitivity Analysis of an Optimal Design Consider for convenience a two variables system whose well defined objective is a function of a given and uncontrollable parameter c. For example, c may be a policy parameter such a regulation (a VaR), an interest rate, a risk and safety requirement set by an appropriate authority, etc. The resulting function to optimize is again defined for convenience by the continuous function $f(x_1, x_2, c)$ (it may be a portfolio expected future value, etc.). For a continuous function, an optimal design consists in selecting the controllable parameters (x_1, x_2) with first order necessary conditions for optimality:

$$f_1(x_1, x_2, c) = 0 \text{ and } f_2(x_1, x_2, c) = 0 \quad \text{or} \quad x_1^* = x_1(c) \text{ and } x_2^* = x_2(c)$$

Then, under certain conditions, for the Jacobian not to be null:

$$\begin{vmatrix} \partial h_1/\partial x_1 & \partial h_1/\partial x_2 \\ \partial h_2/\partial x_1 & \partial h_2/\partial x_2 \end{vmatrix} \neq 0 \quad \text{we have}$$

$$\begin{pmatrix} \partial h_1/\partial x_1 & \partial h_1/\partial x_2 \\ \partial h_2/\partial x_1 & \partial h_2/\partial x_2 \end{pmatrix} \begin{pmatrix} \partial x_1^*/\partial c \\ \partial x_2^*/\partial c \end{pmatrix} = \begin{pmatrix} -\partial h_1/\partial c \\ -\partial h_2/\partial c \end{pmatrix}$$

And therefore, sensitivity of the controllable variables (x_1, x_2) to the uncontrollable variable c is:

$$\begin{pmatrix} \partial x_1^*/\partial c \\ \partial x_2^*/\partial c \end{pmatrix} = \begin{pmatrix} \partial h_1/\partial x_1 & \partial h_1/\partial x_2 \\ \partial h_2/\partial x_1 & \partial h_2/\partial x_2 \end{pmatrix}^{-1} \begin{pmatrix} -\partial h_1/\partial c \\ -\partial h_2/\partial c \end{pmatrix}$$

A small variation in parameter c from a given predictable level is:

$$df(x_1{}^*, x_2{}^*, c) = \frac{\partial f}{\partial x_1{}^*} dx_1{}^* + \frac{\partial f}{\partial x_2{}^*} dx_2{}^*; \quad dx_1{}^* = \frac{\partial x_1{}^*}{\partial c} dc, \quad dx_2{}^* = \frac{\partial x_2{}^*}{\partial c} dc$$

And therefore,

$$df(x_1{}^*, x_2{}^*, c) = \left(\frac{\partial f}{\partial x_1{}^*} \frac{\partial x_1{}^*}{\partial c} + \frac{\partial f}{\partial x_2{}^*} \frac{\partial x_2{}^*}{\partial c} \right) dc$$

Note that the inverse matrix is:

$$\begin{pmatrix} \partial h_1/\partial x_1 & \partial h_1/\partial x_2 \\ \partial h_2/\partial x_1 & \partial h_2/\partial x_2 \end{pmatrix}^{-1} = \left(\frac{\partial h_1}{\partial x_1}\frac{\partial h_2}{\partial x_2} - \frac{\partial h_1}{\partial x_2}\frac{\partial h_2}{\partial x_1}\right)\begin{pmatrix} \partial h_2/\partial x_2 & -\partial h_1/\partial x_2 \\ -\partial h_2/\partial x_1 & \partial h_1/\partial x_1 \end{pmatrix}$$

And therefore,

$$\frac{\partial x_1^*}{\partial c} = -\left(\frac{\partial h_1}{\partial x_1}\frac{\partial h_2}{\partial x_2} - \frac{\partial h_1}{\partial x_2}\frac{\partial h_2}{\partial x_1}\right)^{-1}\left\{\frac{\partial h_2}{\partial x_2}\frac{\partial h_1}{\partial c} + \frac{\partial h_1}{\partial x_2}\frac{\partial h_2}{\partial c}\right\}$$
$$\frac{\partial x_2^*}{\partial c} = -\left(\frac{\partial h_1}{\partial x_1}\frac{\partial h_2}{\partial x_2} - \frac{\partial h_1}{\partial x_2}\frac{\partial h_2}{\partial x_1}\right)^{-1}\left\{\frac{\partial h_2}{\partial x_1}\frac{\partial h_1}{\partial c} + \frac{\partial h_1}{\partial x_1}\frac{\partial h_2}{\partial c}\right\}$$

which lead to an assessment of the effects on the objective function of a small variation in the uncontrollable parameter c.

$$\frac{df(x_1{}^*, x_2{}^*, c)}{dc} = \frac{\frac{\partial f}{\partial x_1{}^*}\left\{\frac{\partial h_2}{\partial x_2}\frac{\partial h_1}{\partial c} + \frac{\partial h_1}{\partial x_2}\frac{\partial h_2}{\partial c} + \frac{\partial f}{\partial x_2{}^*}\left\{\frac{\partial h_2}{\partial x_1}\frac{\partial h_1}{\partial c} + \frac{\partial h_1}{\partial x_1}\frac{\partial h_2}{\partial c}\right\}\right\}}{\left(\frac{\partial h_1}{\partial x_2}\frac{\partial h_2}{\partial x_1} - \frac{\partial h_1}{\partial x_1}\frac{\partial h_2}{\partial x_2}\right)}$$

In other words, a small change in c (=dc) induces a change df. A system is insensitive if a small change in c has relatively no impact on the function f.

Of course higher order sensitivity measures can be obtained (for example, if c is randomly distributed with known probability distribution). In such a case, a second order Taylor series expansion yields:

$$df(x_1{}^*, x_2{}^*, c) = \left(\frac{\partial f}{\partial x_1{}^*}\frac{\partial x_1{}^*}{\partial c} + \frac{\partial f}{\partial x_2{}^*}\frac{\partial x_2{}^*}{\partial c}\right)dc + \frac{1}{2}\begin{pmatrix} \frac{\partial^2 f}{\partial x_1{}^{*2}}\left(\frac{\partial x_1{}^*}{\partial c}\right)^2 + \left(\frac{\partial f}{\partial x_1{}^*}\frac{\partial^2 x_1{}^*}{\partial c^2}\right) + \\ \frac{\partial f^2}{\partial x_2{}^{*2}}\left(\frac{\partial x_2{}^*}{\partial c}\right)^2 + \left(\frac{\partial f}{\partial x_2{}^*}\frac{\partial^2 x_2{}^*}{\partial c^2}\right) \end{pmatrix}(dc)^2$$

which requires a more complex analysis (which is left as an exercise). Designing for "parameter insensitivity" requires then that a selection of controllable variables be designed such that $df(x_1{}^*, x_2{}^*, c)$ be appropriately small with variations in the uncontrollable c. While such problems may seem difficult to assess technically, they are usually simple to implement by simulation of the optimal objective valued at the optimal controllable variables as a function of c.

Example: Robust Decision Making Robust decisions are decisions that are relatively insensitive to parameters variations. For example, when future states may be so complex and many that we restrict ourselves to only a "vital few". Explicitly, let (a_1, a_2, … a_n) be a set of experiments (or design "concepts") some of which are a function of a number of controllable factors—either experimental or design parameters. Say that n experiments are made with each group of experiments based on another set of (potentially competing or environmental) parameters. For example, say that there are five risk factors, two of which are non-controllable. Each factor is assumed for simplicity to have two possible values each and let (r_{ij}) be the result corresponding to a decision (or to an experiment) i when the controllable factors are set to their jth value. If A, B, C are the controllable factor and D and E the non-controllable ones, the following set of experimental results $R = [r_{ij}]$ are obtained (see Table 10.3) where "+" and "−" correspond to the values assumed by each of the factors (for example, expressing a deviation from a risk model assumed to be complete). To evaluate these results, we may proceed in a number of manners as indicated in Chap. 6. Below a number of relevant objectives for decision making under uncertainty are summarized. These include Expectations, Maximin, Minimax, Regret etc. (see Chap. 6).

(a) If all $2^2 = 4$ combinations of uncontrollable factors (D, E) have each a probability $p(D,E)$ denoted by p_j, $j = 1,2,3,4$ the expected value corresponding to plan (or design i) is then: $\overline{R}_i = \sum_{J=1}^{m} r_{ij}p_j$ and therefore under an expected valuation objective, we select the alternative $Max\left[\overline{R}_i\right]$, $1 \le i \le n$

(b) The Maximin criterion consists in maximizing the minimum gains or

$$\text{Max}\{\{\text{Min } r_{ij}\},\quad 1 \le j \le m\},\quad 1 \le i \le n.\quad \text{And in expectation :}$$

$$\underset{1 \le i \le n}{\text{Max}}\left\{\text{Min} \sum_{j=1}^{m} p_j r_{ij},\right\}$$

(c) The Minimax Regret or the Savage Criterion, is based on a ex-post valuation. Namely let r_{ij} be the outcome corresponding to a plan (or decision) i while the uncontrollable factors are j, occurring with probability p_j. The "regret" ex-post is then the best plan (decision) given j or $\{\text{Max } r_{ik}\}$, $1 \le k \le m$ and $\rho_{ij} = \left\{\underset{1 \le k \le m}{\text{Max}} \; r_{ik} - r_{ij}\right\}$. Thus, ρ_{ij} corresponds to the difference between the observed experimental (or consequential) result and the outcome obtained had we selected the optimal decision. The least regret decision is therefore:

$$\underset{1 \le i \le n}{\text{Min}} \rho_{ij} = \left\{\underset{1 \le j \le m}{\text{Max}} \; \rho_{ij}\right\}$$

(d) A Robust decision is based on selecting a plan (a decision) that is relatively insensitive to the outcomes j.

Example: Robust and Experimental Design Experiments, unless not performed by computer simulations, are usually costly. For example, computer simulating the effectiveness of a missile launch, it is far more expensive to test an actual launch (although both are used and both are important). Experimental designs, properly used provide a consistent procedures and principles for collecting data the effects of controllable and uncontrollable factors on a particular design can be assessed effectively (i.e. by using far less experiments than would be needed if appropriate experimental design were to not to be used). Experimental designs are both extensively researched and practiced and are mostly used to assess the statistical response of a system (whether complex or not) to selected control and uncontrollable variables. For example, to test a process it might be possible to limit the number of variables (i.e. maintain them in control) and assess their effects on a product's or a process' performance by those variables (or factors), deemed to be more important. Such principles can be applied similarly to making decision under uncertainty. For example, when a system (or process) can perform appropriately over a large set of variations, it is deemed to be robust. Robust design consists then in selecting controllable parameters that yield a robust function For quality manager, a robust design implies "fitness to use or fitness to needs", reflecting the performance of a product or a system in actual conditions rather than in a laboratory. Similarly, we may design an investment portfolio strategy that would perform not only according to a risk model that assume market completeness but also perform reasonably well when assumptions of market completeness do not hold in fact. Robustness is therefore an essential and desirable characteristic feature of portfolios, of products, of processes etc. In other words, a robust system is a system better adapted to meeting contingent uncertain events when they occur.

A robust design requires first that:

- We define what we need in precise and operational terms
- We define the controllable variables-factors (also called an internal array in experimental design) and the uncontrollable ones (also called the outer array in experimental factors).
- Apply an efficient experimental design to gather data which will provide the statistically more meaningful, direct and multiplicative experimental responses.
- Estimate a relationship between the response and the experimental factors (the independent variables) in terms of the dependent-internal factors-array.
- Optimize the controllable parameters (i.e. the design factors) such that the system, the being designed conforms or performs as needed over a broad range of environmental and uncontrollable conditions (the external array).

Table 10.3 outlines an array with three factors A,B,C endogenous (design, control), each assuming two values, "+" or "−". There are also two external factors D and E, assuming as well two experimental results with rij are the outcome from an application of factors A,B,C when the external factors are set as well. A system design is then defined by parameters associated to (A,B,C) with results r_{ij} associated to the design ABC and to the scenario parameters DE.

Over all parameters for the ith design, the following results are obtained (r_{i1}, r_{i2}, r_{i3}, r_{i4}) which a sample of results we can use to value the systems performance. For example, mean performance, its variance, its range, its mean to standard deviation as well as a signal to noise ratio which is in fact equivalent to mean response (the signal) to standard deviation (the noise). etc. can be used.

Table 10.3 An experiment/design—internal array of three factors and two for an external one

			D	+	+	–	–
			E	+	–	+	–
				Responses/consequences			
	A	B	C	1	2	3	4
1	+	+	+	r_{11}	r_{12}	r_{13}	r_{14}
2	+	–	+	r_{21}	r_{22}	r_{23}	r_{24}
3	+	+	–	...		...	...
4	+	–	–	...			
5	–	+	+	...			
6	–	+	+	...			
7	–	–	–	...	...	...	...
8	–	+	–	r_{81}	r_{82}	r_{83}	r_{84}

10.7.2 Ex-Post Decisions and Recovery

Ex-post, unlike ex-ante decisions, is reached once information is revealed and uncertainty is resolved. Each decision has then a regret that a decision maker may seek to minimize. The cost of a decision's regret represents the difference between the ex-ante payoff that would be received with a given outcome compared to the maximum possible ex-post payoff that would be received compared to the maximum possible ex-post payoff. Savage, Bell as well as Loomes and Sugden 1982, 1987 have pointed out the relevance of this objective to decision making under uncertainty. A decision is then selected to minimize the regrets associated to that decision. Of course, this is also a risk model since it is based on the assumption of a number of future states (normally defined as a potential scenarios) with respect to which each decision is valued (see also Chap. 6 and the above). For example, say that we expect the demand for a product to grow significantly, and as a result we decide to expand the capacity of our plants. In fact, our expectation for demand growth does not materialize due to some future event that are not accounted for. What can we do then, except regret our decision! Similarly, assume that we expect peace to come on earth and decide to spend less on weapons development. Optimism, however wanted, may unfortunately be not justified and instead we find ourselves facing a war for which we may be ill-prepared. What can we do? Not much, except regret our decision. The *Minimax regret* (also called the *Savage regret*) criterion, then, seeks to minimize the regret we may have in adopting a decision. Basically it is defined by,

$$\textit{Regret} = \textit{Best payoff received a-posteriori less the payoff actually received}$$

which is the opportunity loss. In practice, we may be concerned mostly about extreme regrets that can occur with small probabilities that render their valuation far too important and dominant. For example, facing extreme weather with disastrous consequences, building up weapons capabilities to meet any future security contingencies, etc. In finance and insurance, as discussed earlier, we use insurance contracts as well as options and other derivatives to mitigate and reduce contingent risks. Their importance (and in particular the importance of financial contracts such as swaps, credit derivatives, and securitized financial products) are extremely useful as they allow an exchange at a price between parties who want to assume the future uncertainties at a price they extract from the parties who seek to avoid these risks and uncertainties.

Other models define regret as a disappointment. It is then defined as psychological reaction to an outcome that does not meet a decision maker expectation. The measure of disappointment is assumed to be proportional to the difference between expectation and the outcome below the expectation. Elation, inversely, occurs when the outcome is better than the expectation. A more general treatment of this approach based on risk-value theory is dealt with by Jia and Dyer 1996. When expectation is given, the regret-disappointment approach to stochastic optimization can be implemented easily using simulation. In other words, given the ex-ante optimal solution of the stochastic optimization, the regret-disappointment can be calculated based on repeated simulation of the stochastic optimization problem.

Example: Ex-Post and Scenario Optimization and Simulation Scenario approaches are based on models defining future states a priori in terms of countable events, potential processes etc. These scenarios can be complex, varied and numerous as their analysis are mostly based on computer simulation techniques. Again, variables-factors may be defined in terms of controllable factors (about which a decision has to be reached) and uncontrollable factors defined by the potential scenarios they can assume or scenarios resulting from the interactions of the uncontrollable factors themselves and interactions between controllable and uncontrollable factors. The breadth of potential models on may construct is therefore extremely large. Under any specific scenario (or simulation), a performance results while under numerous scenarios, numerous (a cloud) of performances result. Selecting then particular design or controllable factors is expressed as a data-analytic process of selecting a position (solution) in cloud of possibilities generated by the computer. A decision maker may then seek to minimize the regret based on the position of the performance selected in the cloud whose effects, risks and values can be assessed.

Example: Scenario Optimization Dembo (1989, 1993) suggested a particular two-phase approach. In the first phase, an optimal solution is sought for each selected (or simulated) scenario, defining a point in the cloud as a function of all uncontrollable factors. At a second stage, a tracking model is used to find an acceptable solution. Technically, the problem's principle is formulated as follows:

$$\text{Minimize } \tilde{c}^T x \quad \text{Subject to} : \tilde{A}x = \tilde{b}, \; Dx = d, \; x \geq 0$$

with (A, b, c) random variables while (D, d) deterministic. At a first phase, a finite number of problems based on scenarios (say defined by random parameters) are solved, or:

$$v_s = \text{Minimize } c_s^T x \text{ Subject to} : A_s x = b_s, \;\; x \geq 0$$

where A_s, c_s and b_s are scenario matrices and $s = 1, 2, \ldots m$ is the individual run. The optimal solution of each sub problem is denoted by ν_s^* and denote a uni-dimensional cloud (a multivariate cloud would define instead a number of solutions $\nu_{s,k}^*, \;\; k = 1, 2, \ldots$ associated to each scenario). The second phase consists (for a uni-variate cloud) in solving the tracking problem:

$$\text{Minimize } E\left\{ \left| c_s^T x - v_s \right|^2 + |A_s x - b_s|^2 \right\} \text{ Subject to} : \quad Dx = d, \;\; x \geq 0$$

The first objective ensures that the solution, x, is (from the tracking problem viewpoint) close to optimal. A number of generalizations are possible however, by allowing other objectives that the quadratic tracking problem and of course in considering the problem as a multivariate Cloud.

When the number of scenarios is large, the underlying optimization problem can potentially be excessively large and thus may be very difficult to solve in a reasonable amount of computer time. This is the case particularly of "deterministic equivalent" programs resulting from two-stage stochastic programs. The appeal of scenario based schemes lie in the fact that each scenario define independent problems and thus can be solved in parallel when available computing machinery allows it. In order to achieve similar advantages for the solution of the classic, two stage stochastic program, it is necessary to resort to decomposition algorithms (Rockafellar and Wets 1992). Such techniques have both commercial and non-commercial software available however.

10.8 Uncertainty Economics, Risk Externalities and Regulation

What we do may or may not affect others. When we assume responsibility for what we do, there is no risk externality. However when what we do affects the welfare of others without our assuming any responsibility for their consequences, we have then a risk externality. In these circumstances, regulation and controls are on the one hand needed to protect "the others" and on the other hand, regulation that stifles freedom and economic activities that benefits the "commons" may also be a risk externalities. Risk externalities are also important sources of uncertainty as they occur often in an environment where there are information and power asymmetries as well as uncertainties or rare consequences and therefore potentially negligible by the parties concerned. For example, a TBTF bank might exercise its power, its information or assume systemic risks with substantial and collective consequences. The Dodd–Frank regulation act and the creation of an OFR (The Office of Finance Research) voted by Congress is an important decision sought to prevent and mitigate the consequences of such systemic risks. However, over-regulation has also consequences such as the migration of financial services to ports where they are both welcome and regulated much less. Similarly, investors hindered by excess regulation and over-controls seek to migrate where controls are less stringent. In an era of economic globalization where markets and countries are more open to a free flow of resources and financial assets, regulation unbalanced can have substantial risk externalities. Risk externalities are notorious also in the environment. A manufacturer spills chemicals in a lake that other people use for the water they need and for their pleasure. Often such spills are hidden from the public who assumes nonetheless its consequences. Similarly, oil spills at sea have consequences that, unless detected, can induce costs and risks to unsuspected communities along the sea shore. These two cases are a risk externality that others assume while the perpetrator of such a risk may or may not assume. When the price the perpetrators pay for what they have done, then the market is said to efficient. If they do not, then it is inefficient. Regulation controls and regulation enforcements are then justified and efficient when they are balanced with their risk consequences (whether externalities or others).

An externality occurs because no one owns a lake or the river; no one owns the sea, so everyone can use it as he needs to without harming the others. A lake, the sea, the air we breathe, etc. are therefore owned by "the commons", which is all of us. Similar issues recur in a broad set of circumstances. A person smoking in a closed space may threaten the well-being of others in that space; banks pursuing unsustainable risks are also a threat to the sustainability of financial markets that are needed for the economy as a whole, and so on.

Externalities can of course be positive or negative and are reciprocal. For example, an invention that cures a certain disease by a University Professor is a positive externality. Even if the University and the Professor will profit from it, its greater good and a reduction in health costs will be assumed by society. Similarly, investing in education and in infrastructure is a positive externality as it contributes both the educated individual and to society by the good he may contribute to. Externalities are negatives when these are not the case, namely when the well being of one is at the expense of the commons.

When partial information and uncertainty to one or to the whole, or when complex and stealth actions are undertaken to evade compliance to regulation, these externalities are difficult to account for and value and controls—whether statistical or not, are then necessary. For example, to enforce environmental regulation, public awareness, the media, incentives to comply and controls by the responsible agencies is needed. For a bank, offshore transactions and stealth financial flows through global financial networks are often the rule than an exception and require therefore both complex and intelligence means to compensate their consequences and control financial flows and transactions to be sustainable.

Public goods are a special kind of externality where the full effects of the positive externality are felt by every person in the economy. In this sense, events leading to systemic risks are necessarily "public events" that affect every person in the economy. Thus, banks' systemic risks as well as environmental and related risks are necessarily to be controlled to protect the public.

To mitigate risk externalities, value and price them, extensive, systematic and continuous measurements are needed to both detect and support the regulation control and policy processes. These may include among others:

- The power of Laws to create rules, incentives, enforcement and punishment agencies to mitigate risk externalities and their consequences.
- Internalize the costs of risk externalities in firms and perpetrators accounting systems by creating incentives, both positive and negative
- Be informed, measure and track the operations and the performance of institutions and sources of these risks
- Develop the theoretical frameworks needed to rationalize and balance regulation and its risks

Increased global dependence have increased our attention to risks, financial risk externalities and to their regulation, due to a growing awareness that risk externalities are a common threat to long term survival and sustainability of the current financial systems and society as it currently exists. At the same time, the need regulate and mitigate risk externalities has to be balanced with its consequences. For example, over regulating banks may lead to their migrating to areas where they may not be regulated, it may also reduce liquidity and contribute to credit starvation, increased employment and other effects. On the one hand, the pursuit of self-interest embedded in economic pursuits may lead to unsustainable social and economic consequences without regulation. Selected issues, pertaining to risk externalities and uncertainties are considered next.

10.8.1 Risk Externalities, Industry and the Environmental: Energy and Pollution

Environmental economic problems are marred by two essential problems; what level of economic activity (product) to assume and what price to pay for its environmental consequences. For China for example, producing rare earths to meet global demand and at the same time face its environmental consequences. On the one hand, it consists in assessing the value of such an economic activity and on the other how to assess how much consumers are willing to pay for an environmental state. In this perspective, it is expected that for some cases, (at a higher cost or in the reduction of economic activity), an improvement in our environmental may generate greater economic well-being.

Consider the following case. Industrial and other energy consuming activities may, necessarily produce pollution as well. Examples abound, car driving, energy expanded on agricultural development, home heating etc. produce carbon dioxide (Ang 1999; Ang and Pandiyan 1997; DeCanio 1997). These problems are well documented although there seems to be few theoretical frameworks that capture the complexity of the energy–economic–activity–pollution relationships. Further, most theoretical approaches have emphasized deterministic models, neglecting thereby the special and probabilistic character of pollution when it is derived from economic activity models (Greaker 2003, 2006, for example).

The recognition that energy and pollution are intimately related, as well as their major effects on industrial firms and our societies has attracted a great deal of empirically based studies. For example, to determine energy consumption of input and output used in sugar beet production in a farm region in Turkey data was collected data from 146 sugar beet farms using a questionnaire and indicated that total energy consumption in sugar beet production was accounted for by 49.33% in fertilizer energy, and 24.16% in diesel energy. The output/input energy ratio was thus 25.75 with 12.82% in renewable (and pollution free) form only with remaining parts contributing of course to pollution. Although energy consumption increased beet sugar yield production, it also resulted in problems such as land degradation, nutrient loading, pesticide pollution and potential contribution to global warming. Similarly, Granovskii et al. (2006) used published data to perform economic and environmental comparisons of four types of vehicles: conventional, hybrid, electric and hydrogen fuel cell. Their comparisons included economic indicators (prices of vehicles and fuels during the vehicle life and driving range) and environmental indicators (greenhouse gas and air pollution emissions). They then indicate some observations regarding the economic and environmental impact of these vehicles and found that hybrid and electric car technologies exhibit (efficiency) advantages over the other types. In addition, recent papers to appear in Energy Economics, point out to a profusion of recent studies elucidating the energy–pollution relationships. Brannlund et al. (2006) examine how exogenous technological progress, in terms of energy efficiency, affects consumption choice by Swedish households and thereby emissions of carbon dioxide (CO_2) as well as other pollutants. Using national consumption and pollution data, they used a logarithmic model to calculate the emission shares of various household groups and their polluting effects, providing thereby an important econometric relationship between energy and pollution. Such relationships have previously been surveyed in Greening, et al. (2000). Their model presented subsequently is based on a family of Cobb–Douglas and translog production and cost functions. However while the product produced is given as a function of energy, capital and labor, we can also assume that the consumption of energy is an inherent parameter of the technology used in the production of the underlying economic activity (or product), while energy consumption and production combine to "produce" pollution (for a model based on a queue process see Tapiero 2005a).

10.8.2 Networks and Externalities

Network externalities arise when one member of the network depends on its other members, each accounting only for its consequences with their network aggregated adverse consequences assumed by none. These externalities can be positive or negative. For example, a firm part of a successful networked set of firms (a supply chain) is a positive externality. By the same token, over-charging an electric supply grid is a negative externality as its number of users will affect the quality and the reliability of electricity supply. Dependence of a networked financial system can also have risk externalities. For example, multinational global banks failing may trigger a contagion with systemic consequences. This was the case with Merril Lynch, heralding the 2007 financial and credit crisis that since has migrated to Europe and the rest of the world. In some cases, the greater the number of interacting elements in a network (whether financial or not) can contribute to its economic effectiveness (Fig. 10.1). Consider the following network and say that an enterprise has devised a technology to assure all communications from eavesdropping and other interferences within its grid. Such technology is intended to replace a 1–1 safe line by a 1ine with $n(n-1)/2$ safe lines for each member of the network. Explicitly, if there are two members only, we will then have a 1–1 line while for one additional member to a network of three members, each has two potential communications. For six members, we have instead $6*5/2 = 15$ potential safe communication lines. Thus, the greater the number of members, the much (much) greater the number of potential communications and therefore the "much greater the service provided to each member due to the size of the network".

Such a rationale applied profusely to social media and intra-net networking systems justifies at times the extremely large IPO prices of firms that provide such services (for example, Economides 1996). By the same token, the greater the number of Mobile phone subscription to say the Veryzon 917-phone lists, the greater the number of persons of such phone number holders can access freely (and therefore, it is a network, positive externality). At the same time, network externalities can create greater dependence and thereby greater risks. For example, Economides (1996) defines a network critical mass as the smallest size a network that can sustain equilibrium. They argue that when the critical mass is substantial, market coverage will not be achieved—either the market does not exist or it is of insufficient coverage. Accordingly, the willingness to adopt and use an Internet service is an increasing function of network size and the services it provides (both of which have increased immensely), fueling the ever growing demands of the network itself (see also Gary Madden, Network Externalities and the Internet,

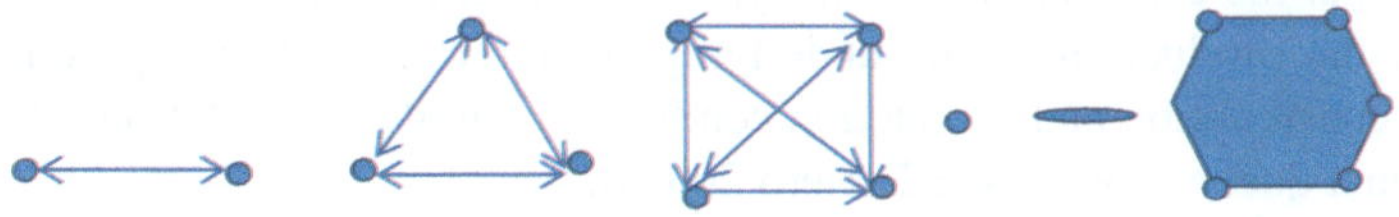

Fig. 10.1 The growth of complexity and interconnectedness

Russel Cooper School of Economics and Finance, University of Western Sydney, Gary Madden).

Network however can also contribute to stealth and non-regulation compliant financial flows rendering the detection and the control of such behaviors extremely difficult. To mitigate, networks sampling techniques (for example, Capobianco 1973; Tapiero, Capobianco and Lewin 1975) as well as assessing their economic consequences are sorely needed.

10.8.3 Infrastructure and Externalities

Aschauer (1989a, b) pointed out that infrastructure investments are underestimated, neglecting the productivity of public capital—a capital consumers do not pay for" but can enjoy its use or/and use it for profit. For example, a network of roads, bridges, etc. are such example. While in many cases, societies are taxing the use of some of its public capital (through taxes and usage costs) is mostly sustained by a collective non-profit entity whose purpose is to provide a service for the collective.

Positive externalities are embedded in public capital as pointed out to by econometric studies that have revealed an appreciable positive elasticity to infrastructure investments and GNP growth. It is explained for example by the appreciable economic contribution of highway construction (and other infrastructures), which lowers direct and indirect costs, increases the time value for individual drivers, etc. (Batten and Karlsson 1996; Berndt and Harrison 1991; Nijkamp and Blaas 1993). Isolating specific investments, such as a particular investment in a highway, in augmenting production and its efficiency or in a hub is more difficult to determine, however. For this reason specific investment in projects infrastructure positive externalities is more difficult to justify (see also Gramlich 1994; Morrison and Schwartz 1996) while their negative effects are instantly revealed by the complaints and court cases that seek to block such investments. A case of particular contentions is the fracking process across the US in search of Gas or the laying of pipelines to feed the US with Gas and Oil from the Canadian Tar Sand steppes. At the macroeconomic level, the evidence of a positive externality is profusely documented. Sanchez-Robles (1998a, b) points out to a broad number of analyses that have shown a wide dispersion but still high levels of elasticity. Munnell (1992) assessing a number of studies concluded that infrastructure research is still not conclusive for policy making in individual projects. (but not on a regional or macroeconomic scale).

The economic valuation of infrastructure investments is difficult to measure, but why? The reasons abound:

- Infrastructure Investments are means to ends . . . and therefore are therefore valued by their "derivatives".
- They are valued and subjugated to the many parties, different in their need, uses and valuation of such investments (for example, transports modes—train, highways, buses etc.), safety, etc.

- They are long term based; Less market sensitive than production and services and therefore valued economically over long periods of time. Thus, they require a strong commitment to the future
- They are market structure dependent;
- They involves both irreversible investments and service contracts, both of which can be risky and difficult to assess economically.

For example, say that a City is to build a port or an airport, who pays for it? How can we justify the billions of dollars needed for such an infrastructure enterprise? How much is a highway worth investing in? What is the relevant cash flow to apply in valuing these investments? What are the direct and the indirect positive and negative externalities and how can they be valued? How much is a particular project infrastructure needed and can it be justified? Who pays for it and how are costs shared in the present and in the future? What is the infrastructure yield? Should it be market assisted, self-financed or contracted? Can it be implemented without taking into consideration a potential for reneging current agreements? What are the optional values embedded in infrastructure investments and how do they mitigate the infrastructure economic risks?

In free economies, subject to market forces, public Development Policy for infrastructure is mostly out of fashion. Rather, privatization seems to be the order of the day just as outsourcing has been the order of day in the Western industrial world. Both have been fueled by an underlying reduction in set-up and fixed costs to economic activities based on cash flows. These comments imply that the misvaluation of infrastructures embeds important economic and other risks.

10.8.4 Economics and Externalities: Pigou and Coase

In general, an externality is an indication of market failure (an incomplete markets). Markets with externalities are no longer efficient, even if markets are competitive. For example, a gas fracking firm may contribute to energy independence on the one hand but it also produces a pollution which is harmful to neighboring farms. The question is then is fracking worthwhile after its externality costs are assessed and their economic and social costs are accounted for?

Risk externalities thus create a divergence between private costs and social costs. To compensate these costs Pigou suggested that we use a Tax (called the Pigouvian Tax). It consists essentially in taxing the perpetrator of such risks such that:

$$\text{Social cost} = \text{private cost} + \text{external cost or SC} = \text{PC} + \text{EC}$$

In marginal terms, we have:

$$\begin{aligned}\text{Marginal social cost} &= \text{Marginal private cost} + \text{Marginal external cost or MSC}\\ &= \text{MPC} + \text{MEC}\end{aligned}$$

The price of externality is thus equals its Marginal social cost less its private cost. In 1960, Ronald Coase pointed out however that externalities have a number of characteristics that will tend to invalidate the application of a Pigouvian tax. These consist of:

- Reciprocal
- Have an opportunity cost and
- Transaction cost

For example, in case of pollution, externalities may not be due to one party only but rather the result from two or more parties and therefore in some sense "both are at fault". For example, why would one or a firm pollute? Because the risk of being caught may be smaller than assuming the cost of pollution (in particular if there is regulation or regulation is not enforced). Or perhaps, the polluted, will find it acceptable to assume the pollution. For example, if a polluting firm in a particular settlement also pays local taxes and employs many people in this settlement, it is quite possible that local residents will accept such pollution in order not to lose these jobs. In this sense, risks and their consequences depend on more than the perpetrator of the polluting event.

Coase pointed out also to the role of opportunity cost in valuing externalities. When an externality cost or benefit is imposed on another party—that makes him a party to the transaction externality, whether he is protected by law or not, or whether he has a remedy to the cost that is imposed on him. In other words, persons subject to externality costs can find a remedy by paying to prevent it, which they do not.

Finally, say that a payment is made and is received by a party which is not part of a transaction. Examples include the lawyer fees, time and effort spent on negotiation, organizing the parties to the transaction, etc. Such costs, transaction costs, contribute to market inefficiencies and to inefficient contracts. In such situations, Coase suggests that the best legal rule depends on the facts of the situation. Efficiency tells us to pick the legal rule that places the burden on the party that is the lowest cost avoider. If a polluter is the least-cost avoider, then a legal rule allowing the injured parties to prevent pollution is efficient. The polluter will not be able to pay off the polluted (even if transaction costs are low), so it will change over instead. If the polluted are the least-cost avoiders, then a legal rule allowing the polluter to pollute is efficient. Coase "Theorem" articulates a number of principles to manage these risks externalities. These are summarized by three conditions hold:

- Property rights are well defined.
- Property rights are transferable.
- Transaction costs are sufficiently small.

In this sense, when elements of a risk model for property rights are defined (rather than an uncertain definition of rights) an economic approach to externalities can be defined. When this is not the case, we should expect externalities to be a problem. A similar situation arises in regulation. When regulation rules are well defined, responsibilities clearly set up and a market for deviations from regulating rules exists and is legal and if the transactions costs (controls) of regulation are sufficiently small, then regulation can be effective. When any of these conditions cannot be maintained, we then have a problem.

Property Rules and Liability Rules: Property rights are any and all rules that define people's control over the use of resources. While liability rules are means of enforcing property rights in the broader sense. When using the legal system, property rights are not always well defined, thus providing opportunities for risk externalities to occur. Similarly, if regulators are unable or do not assume their responsibilities risk externalities can also arise. Examples are numerous including emission standards for automobiles, detailed regulations relating to the disposal of toxic chemicals.

Of course, externalities are not only associated to pollution only. They can be safety-related, jobs-related and due to technological innovations, regulation related, etc. Below a n umber of examples are outlined.

Example of Negative Externality: Overproduction of goods generating negative externalities; Undersupply of goods generating positive externalities; Overproduction of negative externality-generating goods With externalities, the industry's supply curve will not reflect marginal social costs, only marginal private costs

Examples of production externalities: Air pollution from burning coal; Ground water pollution from fertilizer use; Food contamination and farm worker exposure to toxic chemicals from pesticide use; Irrigation water and consequential decline of waterfowl population in nearby wildlife refuge; Production of refrigerators using CFC's; Health issues resulting from gold mining; Health issues relating to asbestos uses in construction and other processes (Lloyd was charged with about 15 billion of euros only to refund damages related asbestos in about 20 years, with many court cases still outstanding).

Examples: Internalize the externality: By forming economic units of sufficient size that most of the consequence of any action occur within the unit Example: neighborhood the households may collectively decide that maintenance of the facilities that affect them all should be undertaken collectively or there must exist some way of enforcing the collective agreement

Measuring and valuing the economic effects of risk externalities is both complex and very challenging. These may include: Tail Value at Risk Techniques based on tail risks (Taleb and Tapiero 2010; Oren Tapiero 2012c); Type II statistical risks (see Chap. 3 and the Chap. 11).; Networks Risks (Contagions risks for example, with an application in Chap. 11, etc.); Strategic Regulation Risks which are considered explicitly in the next chapter.

References and Additional Reading

Uncertainty and risks are issues that remain debatable. Brock (1986) and Brock and Dechert (1988) distinguish between random and deterministic systems. In Chap. 5 these issues were both discussed and referred to, each approach seeking some way to formalize our lack of knowledge in some rational and quantified form. In particular, we have used incomplete state preferences (Knight 1921) to define uncertainty. It is however a specific type of uncertainty. Other approaches seek to define alternatives to rational decision making based on a weighting of probabilities. For example, Machina 1982, 1987; Samuelson 1963, 1965 (on the limits of the explanatory power of "numbers"), Machina and Munier 1999; Abdellaoui and Munier 1997, 1999 on the belief and probability assignments to rare events (see also Munier 1986, 1995, 1991; Munier et al. 1999; Rabin 1998 on economics psychology), Simon 1979, 1982 on bounded rationality, Munier 2012 (on volatility and uncertainty supported by an extensive empirical analysis of agricultural commodities), Ahlbrecht and Weber 1997 on the gradual resolution of uncertainty, etc.

The literature in this domain is thus extremely large and applied to a broad set of issues and problems. The references below define various formalizations to decision making with behavioral and psychological factors: Dyer and Jia 1997, Jia and Dyer 1996, Jia et al. 2001 on perceived risks, Eeckoudt and Kimball 1991, Eeckhoudt et al. 1996, Gollier and Pratt 1996 (on background risks), Ellsberg 1961 on ambiguity and Savage Axioms (regret), Evans on Psychological Reasoning and rationality, Gilboa and Schmeidler 1995, Kahneman and Tversky 1979 on Prospect Theory, Thaler et al. 1997; Wakker 1994, 2001, 2010; Wakker and Tversky 1993; Taleb 2007, 2008, 2009.

Implications to economics include seminal papers by Arrow (1963b) on welfare of medical care, Arrow (1982) on the risk perception and psychology in economics, Arrow et al. (1996) on Rational Foundations for decision making, Debreu (1953) on the economics of uncertainty, Friedman and savage (1948, 1952) on utility analysis, Kindleberger (1978) on Mania and panics, Laffont (1989, 1995) on the economics of uncertainty, regulation insurance and environmental risks, Loomes and Sugden (1982, 1987), Mulvey et al. (1991) on scenario based uncertainty and Magill and Quinzii (1996) on markets incompleteness.

Risk externalities are a special form of uncertainty (at least for the public at large, unaware of the actions that parties take with consequences they do not assume). In this sense, Coase (1937, 1960), important contributions discussed in this chapter provide a guideline to assess their effects. For applications, references are given in the text. We refer in particular to Xepapadeas (1994, 1995) on risk externalities and environmental issues.

Chaos, complexity or "the lack of order" are some of the more important sources of uncertainty. Chaos is a field originating in forms of "unpredictability". For example, a process that cannot converge to any particular end, bifurcating continuously in an unpredictable manner is chaotic. This leads to statements, that a "sneeze" may lead one catch a cold in New York! Namely, that anything and everything are unpredictable and therefore everything may occur. Lorenz (1966) for example, studied and defined chaos in atmospheric data, May (1974) defined dynamic systems in biological systems that are "chaotic", Gleick (1987) and Peter Edgar (1995) have written popular books on chaos, etc.

In physics, uncertainty in its economic sense, is characterize by non-extensive systems (parallel to incomplete financial markets) and extensive systems (complete financial markets).The characterization of such systems are difficult however. Statistical physics following Gibbs (1961), Boltzmann, Shannon and others, have used entropy as a measure of disorder. The greater the entropy, the greater "disorder". In this sense entropy has become a "probability" measure to define "disorder" in terms of particular metrics. Stanley et al. (1996) for example, have made immense contributions to such approaches based statistical, economic and financial physics. Numerous attempts have been made to generalize both the measures and the metrics to define and estimate "non-extensiveness". The results outlined in this chapter were based on such studies. In particular, Jackson (1909) provided an approach to calculate for "jump" derivatives. Jackson's formulation has shown in this chapter, that Tsallis entropy (Tapiero 1995a, b) and the BG entropy differ only in the metric used to define "disorder". Sato (2010) uses the Jackson derivative under mean–variance assumptions to obtain a q-Gaussian distributions (based on Tsallis entropy). Other references (not referred to in the text) include Kapur and Kesavan (1992) entropy maximization, Renyi (1961), Dempster et al. (2007) applied empirical copulas to CDO tranche pricing using relative entropy, Akturk et al. (2007) on Sharma-Mittal entropy, Nau (2011) and Naudts (2007) on statistical dynamics and entropy. And Fraundorf (2007) on complexity, Kullback (1959, 1987), Kullback and Leibler (1951), Good (1965, 1968) on information and utility, have provided however a statistical approach to entropy as a metric—both of information and a metric.

Fractal models defined in Chap. 5 (Mandelbrot 1997; Mandelbrot and Wallis 1968 on fractal models for example), provide also another form of uncertainty, modeled by the nonlinear growth of volatility. While information asymmetry and financial intermediation based on contractual agreements, unexpected credit losses, etc. are also sources of uncertainty we have considered in Chaps. 8 and 9 (see also Leland and Pyle 1977 on Riordan 1984; Amato and Remolona 2005).

Finally, networks, the intricate relationships of networked agents, firms, their exchange etc. are a source of uncertainty. Even though, there are some research papers that recognize specific characteristics of networks and networking in economic and financial systems, it remains a challenging topic (Economides 1996). Katz and Shapiro (1985, 1994) provide an outline of networks externalities and their effects on system competition, Liebowitz and Margolis (1994), emphasize as well the uncommon tragedy of network externality.

Chapter 11
Strategic Risk Control and Regulation

Abstract The purpose of this chapter is to outline elements of risk and strategic control. These include detection and inspection programs, tracking (through control charts) as well as their uses in risk management and financial regulation. The statistical sampling and control literature pertaining to these problems is immense and we therefore refer at the end of the chapter to standard references on such topics as well as to specific research papers relating to strategic controls and regulation. Strategic risk control is increasingly important but has received relatively little attention. For this reason, I chose to focus greater attention on such problems that are far more appropriate. Examples are used to highlight the approaches that this chapter outlines. A number of industrial control cases are outlined applicable as well in the financial sector.

11.1 Introduction

The controls we use are varied. Controls and statistical inspections are used to detect risk factors; to inspect whether they conform to specified standards or to expectations; to provide a signal and incentive to comply to say, regulation; to control variations; to preempt blames for a lack of controls; to provide a preventive defensive response in case a disaster occurs, etc. Controls are applied both ex-ante and ex-post—on the one hand to detect suspicious threats and on the other to learn and prevent their recurrence as well as learn to recover when risk events have nonetheless occurred. They may be used to inspect bridges and buildings; to inspect and control the safety of nuclear facilities; to inspect products quality prior to their delivery, to inspect these products prior to their reception, to inspect banks and hedge funds for non-complying policies and acts (such as insiders' trading, the stealth transfers of illicit funds) etc. When controls are made to assure that firms conform to expectations, to meet a set of required standards, to be compliant to regulation laws, to assure that their actions are legal etc., then such controls may be

C.S. Tapiero, *Engineering Risk and Finance*, International Series in Operations Research & Management Science 188, DOI 10.1007/978-1-4614-6234-7_11,

strategic—one party controlled and the other controlling, both at cross purposes. Below we consider a number of such control problems.

Example: Control and Statistical Tests Statistical tests are based on a prior specification of the risk probabilities that tests are assuming. As indicated in Chap. 3, these may be type I risks and type II risks. Type 1 risks are defined when an inspection is made and rejects a tested of "good standing", e.g. pursue a tax evader who is legally compliant, reject a production lot when it is conforming to accepted standards, misdiagnosed a healthy patient (and therefore requiring more tests), inspecting in-depth a passenger on line at the airport when the passenger is no threat, etc. Type II risks are defined when a faulty lot is accepted or letting a terrorist pass through airport controls, giving a "clean bill of health" to a deviant bank, misdiagnosing a patient. When these tests are made to control parties that have different and conflicting objectives, these tests are strategic introducing a family of risks such as information and power asymmetry risks, risks associated to not knowing what the counter-party may choose to do, when, etc. Controlling (inspecting) a regulated bank, inspecting the emission of pollution by a chemical plant and their like; a control by the IRS are such examples, where the controlled often seeks to evade the controller.

Controlling Risks are thus both specific and comprehensive activities, with a broad and diversified set of tools used by various professions to meet their well-defined needs.

Example: Operations Risks and Controls in Financial Services Operations Risks and Controls or OpRisk in financial services are typically defined with probabilities of defaulting due to operational causes such as a person involved in a transaction (for example misleading the buyer of the terms of a mortgage, making mistakes, etc.). Risks can result from mishaps resulting from the following: Transaction errors; Mis-or-partial information—either purposeful or not; Loss of/or damage to assets; Theft, fraud, and unauthorized activities; Legal liabilities. Etc. Controlling OpRisk are therefore both complex and costly (their risk consequences may have increased appreciably with the growth of complex banking systems into a global network of financial services with TBTF financial institutions). Nowadays most retail banks have built or acquired by extensive data analysis expertise providing a clear view of the distribution of the number of mishandling events (i.e., the number of transaction cancellations and amendments) per day as well as the mishandling loss severity distribution. For example, some research indicate that the number of events mishandled as a percentage of daily volume is the highest on Mondays, the lowest on Wednesday and high on Fridays. What are the consequences of these events and attitudes and what is their price? These are some of the issues that both banks and regulators are concerned with.

11.2 Statistical Risk Control: Inspection and Acceptance Sampling

In January 1990, ATT was struck by a system error that stopped for 9h all communications in the North East of the USA. This was due to a software error that generated a massive communication system breakdown. An investigation indicated that the culprit was a programming error that could have been avoided had prior inspection implemented. Knight Capital, a broker, similarly, lost hundreds of Millions of dollars due to a software glitch in August 2012. NASA discovers at its first launching that the Hubble telescope, whose cost was estimated then at four billion dollars, did not supply clear pictures because of a fault in design. These faults could have been detected had the manufacturer of the lenses tested them prior to their delivery. An ex-post investigation pointed out that the mirror was never tested in or out of its spec. Only a computer simulation (itself a model) was performed. A patient is found to have a colon cancer that could have been prevented by a regular colonoscopy. These as well as other and less consequential risks (such as the control of supplies deliveries, in process control of manufacture, etc.) justify the use of control tests, audits and inspections.

Financial regulators use controls to both provide an incentive to banks to comply to regulation (since when caught they may be subject to appreciable penalties) and mitigate the systemic risks that may occur due to contagious default of the banking system. A policy for regulators to finance their needs by the penalties they may collect for non-compliance may render such controls self serving however. In some cases they may even turn to be dubious, requiring that the « regulators be regulated and controlled as well for their excesses and the risks they may contribute to. In an imperfect world, in incomplete financial markets, in an environment of power and information asymmetries, controls are thus needed and may be used for a broad number of purposes.

Inspections are also conducted to detect potential events as well as monitor processes or persons. For example, when a Principal delegates a task to an agent who is better informed and their interests diverge, the agent may use his position to an unwarranted advantage (for example, a trader that turns rogue, a CEO short changing stockholders). Additional examples abound. A government (IRS) monitors the income of tax due, prevent and detect misreporting and tax evasion. The scope of inspections-controls-audits-tracking-monitoring of businesses, financial institutions, the environment, health, government, etc. are therefore varied, defined by their:

- Organization and relationships between the "inspector and the inspected", the power and information each has about its own processes, the others, their dependence or their independence
- The underlying inspection process, whether it is statistical (i.e. based on statistical sampling), are inspections complete or partial, what are the control consequences, are inspections triggered by a signal, private and/or public information etc.

The inspection process has over the years grown into a complex set of tools designed to deal with the increased complexity of the systems to control and their ability to evade these controls or their consequences. At the onset of the industrial revolution, production process inspections were introduced due to inherent technological evolutions of manufacturing processes and their organization. For example, producing parts to meet explicit specifications and assembled into complete products through an elaborate assembly process of men and machines. As products and processes became more complex, manufacturing standards have become more stringent requiring as well stringent control. By the same token, the emergence of supply chains and networked organizations of stakeholders have also contributed to strategic audits needed to assure the supply chain integrity and that members of the "network" conform to acceptable standards. Audits are then used as both monitoring and preventive tools. Seeking on the one hand to detect problems and understand why such problems occur, their structural causes—both internal and external, their remedies and their future prevention. Similarly, in the financial sector, size, collusion, products complexity, globalization and stealth networks of associated firms servicing the financial sector have rendered financial regulation extremely important and challenging. Products such as MBS (Mortgage Backed Securities), CDO, CLO, and their like (Credit Derivatives) have also rendered their risk properties stealth—even to some financial experts, and thus lead to their misuse and mispricing.

11.2.1 Elements Statistical Sampling

Statistical controls by inspection sampling are designed as a function of three essential considerations: The statistical risks of the control procedure; Its' value and costs and finally its complexity (see Fig. 11.1). Inspections are performed for many reasons. Acceptance sampling is used to provide to the user the assurance that the risks associated to accepting a prospect (a supply lot, a bank compliance to regulation), is within specified limits. Tests are implemented based on statistical criteria such as tests performed by attributes (qualitative characteristics to control), testing quantitative properties; tests implemented at scheduled, random times or triggered by internal or external information. All these tests, lead to special statistical test procedures that are specific to the problems they seek to control, with specific statistical and stringent properties. They are generally based on an approach that seeks to balance statistical risks and their costs-value.

A typical inspection test is performed as follows. First, risks we are willing to assume are specified (for example, type I or II risks, see Chap. 3), second, costs are assumed for both the costs of control and the costs of non-detection and finally we select the sampling technique to apply which allows us to reach a decision in a "statistically acceptable manner" (i.e. balancing the precision of the tests and its costs and other consequences). For example, let α and β be defined type I and II risks. This means that the probability of rejecting wrongly a test outcome has to be

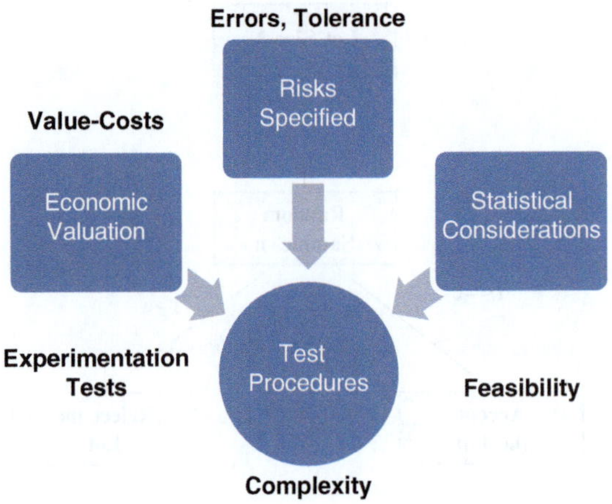

Fig. 11.1 Statistical testing: value and risks

smaller than α, while a type II risk, β, is the probability of accepting a test wrongly. These are given by:

$$1 - P_I(TestValue|Sampling\ Technique) \leq \alpha \quad \text{and}$$
$$P_{II}(Test\ Value|Sampling\ Technique) \leq \beta$$

Cost criteria may include inspection costs, the cost of a type I error, the cost of a type II error (leading potentially to risk externalities) as well as other preventive and other costs used to design risk control systems and inspection programs.

Consider for example and (n,c) control test where n is a sample size of regulatory directives and c is a Test Value (as indicated in Fig. 11.2) in a set of N directives to be controlled. Say that there are D non-compliance. The probability of observing r non-compliances in a set of n tests has then a hyper-geometric probability distribution given by:

$$f_{HG}(r|n,N,D) = \binom{D}{r}\binom{N-D}{n-r}\Big/\binom{N}{n}$$

If N is large and if we let the proportion of non-compliance be p = D/N, then an approximate probability distribution is given by the binomial distribution with mean and variance $E(r) = np; var(r) = np(1-p)$:

$$f_{HG}(r|n,N,D) = \binom{D}{r}\binom{N-D}{n-r}/\binom{N}{n} \Rightarrow f_B(r|n,p)$$
$$= \binom{n}{r}p^r(1-p)^{n-r}, \ r = 0,1,2,\ldots,n$$

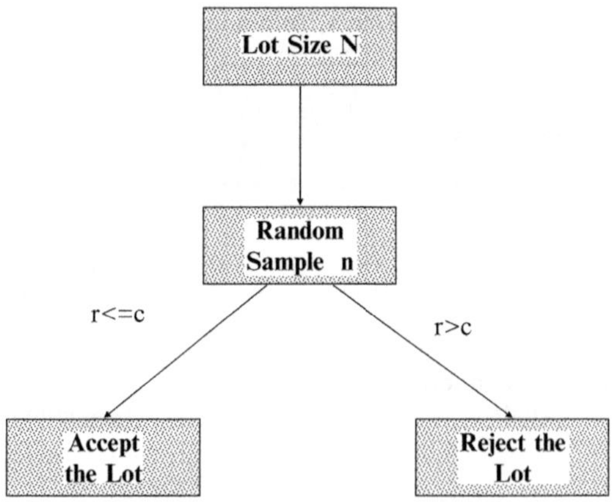

Fig. 11.2 An (n,c) risk control test

In this case, the cumulative distribution is (see also Chap. 3):

$$F_B(k) = \sum_{i=0}^{k} f_B(i) = \sum_{i=0}^{k} \binom{n}{i} p^i (1-p)^{n-i} \quad \text{or}$$

$$F_B(k) = n\binom{n-1}{k-1} \int_0^{\theta} t^{k-1}(1-t)^{n-k} dt$$

The control test is (see Fig. 11.2):

$$r \leq c \text{ accept compliance} \quad \text{and} \quad r > c \text{ reject compliance}$$

That is, if a test of a compliant bank has a proportion p_1 non-compliant while a non-compliant bank has a proportion $p_2 > p_1$, then the cumulative distribution (measuring the risks of reaching a wrong decision) under each of these alternatives are:

$$F_I(c) = \sum_{i=0}^{c} P(i) = \sum_{i=0}^{c} \binom{n}{i} p_1^i (1-p_1)^{n-i} \quad \text{and}$$

$$F_{II}(c) = \sum_{i=0}^{c} P(i) = \sum_{i=0}^{c} \binom{n}{i} p_2^i (1-p_2)^{n-i}$$

Setting the test probabilities to, $1 - F_I(c) \leq \alpha$ and $F_{II}(c) \leq \beta$ we obtain (as indicated in Chap. 3):

$$1 - F_I(c|n,p_1) \leq \alpha \quad \text{and} \quad F_{II}(c|n,p_2) < \beta$$

For example, let $n = 100, \quad c = 5, \quad p_1 = 0.05, \quad p_2 = 0.15$ and therefore,

$$F_I(10) = \sum_{i=0}^{5} \binom{100}{i} 0.05^i (1 - 0.05)^{100-i} \quad \text{and}$$

$$F_{II}(5) = \sum_{i=0}^{5} \binom{100}{i} 0.15^i (1 - 0.15)^{100-i}$$

Thus, given the risk constraints α, β as well as the proportions p_1 and p_2, the control test is found by a solution of the equations above, or finding the control sample and decision parameter (n, c), or: $(\alpha, \beta | p_1, p_2) \Rightarrow (n, c)$

In some cases, to facilitate calculations, Poisson and Normal approximations to the binomial model are used. For the Poisson approximation we have (see also Chap. 3),

$$P(R = r) = \frac{\lambda^r e^{-\lambda}}{r!}, \quad \lambda = np, \quad n \text{ large}, \quad p \text{ small}, \quad r = 0, 1, 2, \ldots. \quad \text{and}$$

$$F_I(c) = \sum_{r=0}^{c} \frac{(np_1)^r e^{-np_1}}{r!} \quad \text{and} \quad F_{II}(c) = \sum_{r=0}^{c} \frac{(np_2)^r e^{-np_2}}{r!}$$

where the cumulative distribution of the Poisson is tabulated by chi-square distributions,

$$F(m, c) = \sum_{r=0}^{c} \frac{(m)^r e^{-m}}{r!} = \frac{1}{c} \int_{m}^{\infty} t^c e^{-t} dt \quad \text{and}$$

$$\sum_{r=0}^{c} \frac{(m)^r e^{-m}}{r!} = P\left[\chi^2 | 2(c+1)df\right] = e^{-m}\left\{1 + \sum_{j=0}^{c} \frac{(2m)^j}{2.4.8 \ldots (2j)!}\right\}$$

Similarly, for a normal approximation we have:

$$P(R = r) = \binom{n}{r} p^r (1 - p)^{n-r} \rightarrow f(R = x)$$

$$= \frac{1}{\sqrt{2\pi\sigma^2}} e^{-\frac{1}{2}\left(\frac{x-\mu}{\sigma}\right)^2}; \quad \mu = np, \ \sigma^2 = np(1 - p)$$

whose cumulative distribution is used to calculate the risk under both hypotheses. In some cases, control costs are accounted for in defining the control procedure and its stringency. Some of these costs include:

- The cost of statistical controls (sampling) as well as other costs incurred by the financial regulator and the controlled the regulated who incurs costs when controlled.

- Consequential costs, such as the costs of being found non-compliant or the costs of non-detecting non-compliance.
- Ex-post (external costs, occurring when non-compliance was not detected)
- Derived costs (costs incurred through contagion, related factors affected by non-compliance)
- Externality costs (costs that are not assumed by their perpetrators, such as pollution, financial systemic risks).

For example, consider the cost of sampling, the costs of making an error of the first kind and the cost of making an error of the second kind (which may also be interpreted as an externality when this risk is assumed by a consumer and with no penalty for a controller, as we will consider when the control of a banks is considered subsequently). In this case, we have:

$$\mathrm{Min}E\left\{c_I n + \tilde{c}_\alpha(N-n) + \tilde{c}_\beta(N-n)\right\} \text{ Subject to :}$$
$$\tilde{c}_\alpha = \begin{cases} C_\alpha & 1-F_I(c) \\ 0 & F_I(c) \end{cases} \quad \text{and} \quad \tilde{c}_\beta = \begin{cases} C_\beta & F_{II}(c) \\ 0 & 1-F_{II}(c) \end{cases}$$

whose expectation is:

$$\underset{n,c}{\mathrm{Min}} \quad c_I n + \left\{C_\alpha\left[1-F_{AQL}(c)\right] + C_\beta F_{LTFD}(c)\right\}(N-n)$$

These lead to the solution of a system of two simultaneous equations:

$$c_I + \left\{C_\beta \frac{\partial F_{II}(c,n)}{\partial n} - C_\alpha \frac{\partial F_I(c,n)}{\partial n}\right\}(N-n) = C_\alpha[1-F_I(c,n)] + C_\beta F_{II}(c,n)$$
$$-C_\alpha \frac{\partial F_I(c,n)}{\partial c} + C_\beta \frac{\partial F_{II}(c,n)}{\partial c} = 0$$

Note that $\frac{C_\alpha}{C_\beta} = \frac{\partial F_{II}(c,n)/\partial c}{\partial F_I(c,n)/\partial c}$ which provides a unique equation in $\frac{c_I}{C_\beta}$, the inspection cost to the external cost in terms of the control errors and their derivatives.

Other test can of course be constructed (see Tapiero 1996). Some are based on simple (single) sampling and some are based on double sampling as Fig. 11.3 indicates. Sampling techniques are thus based on the risk one is willing to assume and the complexity and the costs of the procedure set in place to control risks.

11.2.2 Bayesian Controls—A Medical Care Case

Bayesian controls are subjective approaches based on prior beliefs of future outcomes. As evidence accrues, prior beliefs are updated using Bayes rules for conditional probabilities (see Chap. 3). For example, say that the future is defined

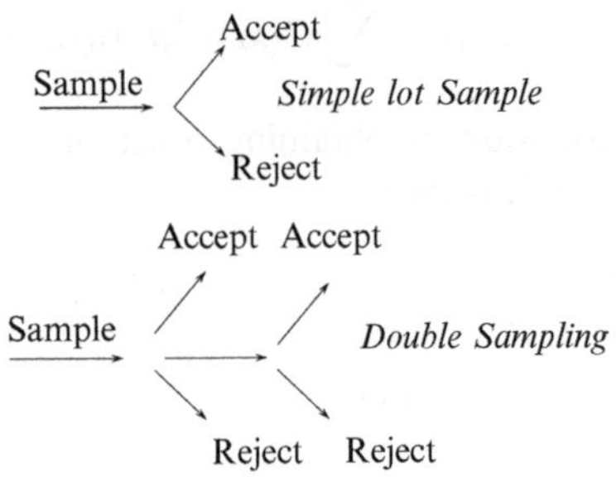

Fig. 11.3 Single and double sampling

by m states and let an attribute associated to each state (a probability, a cost, a patient, compliance, to a specific regulation, health state or disease etc.) be defined by a random variable $x_j,\ j = 1, 2, \ldots, m$. Say that we associate a prior (belief) probability to each of these states, denoted by $0 \le p(x_j) \le 1, \sum_{j=1}^{m} p(x_j) = 1$. Now say that a number of tests or experiments are made. For example, imagine a patient is being tested for potential m ailments. The Doctors recommends n tests, each of which has n potential outcomes $y_i,\ i = 1, 2, \ldots, n$ with the following (conditional) probability consequence $p(y_i|x_j)$ where test i whose outcome is a probability y_i of an ailment i given the prior belief of an ailment j. Ex-post such a test provides a posterior reassessment by the Doctor, leading him to the following conditional probability (see also Chap. 3 on Bayesian calculus):

$$(p(x_j), p(y_i|x_j)) \Rightarrow p(x_j|y_i) = \frac{p(y_i|x_j)p(x_j)}{\sum_{j=1}^{m} p(y_i|x_j)p(x_j)}$$

Assuming n tests, each with outcomes $\{y_i\},\ i = 1, \ldots, n$, we have:

$$p(x_j|y_i, y_2, \ldots, y_n) = \frac{p(y_i, y_2, \ldots, y_n|x_j)p(x_j)}{\sum_{j=1}^{m} p(y_i, y_2, \ldots, y_n|x_j)p(x_j)}$$

where $p(x_j|y_i, y_2, \ldots, y_n)$ is the posterior probability of the patient's ailment j. These probabilities may be used and valued accordingly for each of the states' (ailments) and their consequences. For example, say that the Doctor tests recommend a medical treatment for ailment k. And let $C(x_k|x_j)$ be "a cost" when treating ailment k when in fact it is j and a zero cost, $C(x_k|x_k) = 0$ when it is in fact a k ailment. The expected cost of a recommended treatment is then:

$$C_k(y_i, y_2, \ldots, y_n) = \sum C(x_k|x_j)p(x_j|y_i, y_2, \ldots, y_n)$$

The probability distribution of obtaining a set of test results when the true ailment is i, $f_i(y_i, y_2, \ldots, y_n)$, is then:

$$C_k(y_1, y_2, \ldots, y_n) = \sum_{j=1}^{m} C(x_k|x_j) \frac{p(x_j|y_1, y_2, \ldots, y_n)}{\sum\limits_{y_1 \in Y_1} \cdots \sum\limits_{y_1 \in Y_n} p(x_j|y_1, y_2, \ldots, y_n) f_j(y_i, y_2, \ldots, y_n)}$$

For example say that $C(x_k|x_j) = 1$ and $C(x_k|x_k) = 0$ and consider two states (potential ailments), j = 1,2, then

$$C_1(y_1, y_2, \ldots, y_n) = \sum_{j=1}^{m} C(x_k|x_j) \frac{p(x_2|y_1, y_2, \ldots, y_n)}{\sum\limits_{y_1 \in Y_1} \cdots \sum\limits_{y_1 \in Y_n} p(x_2|y_1, y_2, \ldots, y_n) f_j(y_i, y_2, \ldots, y_n)},$$

$$C(x_1|x_2) = 1, \quad C(x_2|x_1) = 1, \quad C(x_1|x_1) = 0, \quad C(x_2|x_2) = 0$$

Again, for simplicity, consider two tests with outcomes $\{y_{j1}; y_{j2}\} \equiv \{1, 0; 1, 0\}$ associated to a series of tests when the true ailment is j while the diagnostic was set to $k \neq j$. And let probability distributions $f_j(y_i, y_2)$ be bivariate bernoulli (see Chap. 4). Explicitly, we set:

$$f_j(y_{j1}, y_{j2}|\alpha_{j1}, \alpha_{j2}, \rho_j) = \frac{(\alpha_{j1})^{y_{j1}}(1-\alpha_{j1})^{1-y_{j1}}(\alpha_{j2})^{y_{j2}}(1-\alpha_{j2})^{1-y_{j2}}\rho^{y_{j1}y_{j2}}}{1+\alpha_{j1}\alpha_{j2}(\rho_j - 1)}, \quad j = 1, 2$$

And therefore, since a cost is incurred only if the Doctor has made the wrong diagnostic, its cost is:

$$C_1(y_{21}, y_{22}) = \frac{p(x_2|y_{21}, y_{22})}{\sum\limits_{y_{21}=0}^{1}\sum\limits_{y_{22}=0}^{1} p(x_2|y_{21}, y_{22}) \frac{(\alpha_{21})^{y_{21}}(1-\alpha_{21})^{1-y_{21}}(\alpha_{22})^{y_{22}}(1-\alpha_{22})^{1-y_{22}}\rho^{y_{21}y_{22}}}{1+\alpha_{21}\alpha_{22}(\rho_2 - 1)}}$$

Say that outcome $(y_{21}, y_{22}) = (1, 0)$ occurs, then, the denominator consists of the sums of:

$$p(x_2|0,0)\frac{(1-\alpha_{21})(1-\alpha_{22})}{1+\alpha_{21}\alpha_{22}(\rho_2 - 1)} + p(x_2|0,1)\frac{(1-\alpha_{21})(\alpha_{22})}{1+\alpha_{21}\alpha_{22}(\rho_2 - 1)}$$
$$+ p(x_2|1,0)\frac{(\alpha_{21})(1-\alpha_{22})}{1+\alpha_{21}\alpha_{22}(\rho_2 - 1)} + p(x_2|1,1)\frac{(\alpha_{21})(\alpha_{22})\rho_2}{1+\alpha_{21}\alpha_{22}(\rho_2 - 1)}$$

And therefore

$$C_1(1,0) = \frac{(1 + \alpha_{21}\alpha_{22}(\rho_2 - 1))p(x_2|1,0)}{\left\{\begin{array}{l} p(x_2|0,0)(1-\alpha_{21})(1-\alpha_{22}) + p(x_2|0,1)(1-\alpha_{21})(\alpha_{22}) + \\ p(x_2|1,0)(\alpha_{21})(1-\alpha_{22}) + p(x_2|1,0)(\alpha_{21})(\alpha_{22})\rho_2 \end{array}\right\}}$$

Next, say that based on prior experience (and numerous similar tests), we have the following estimates:

$$p(x_2|0,0) = 0.15, \quad p(x_2|0,1) = 0.5, \quad p(x_2|1,0) = 0.3, \quad p(x_2|1,1) = 0.05$$

And therefore,

$$C_1(1,0) = \frac{(1+\alpha_{21}\alpha_{22}(\rho_2-1))0.3}{\{0.15(1-\alpha_{21})(1-\alpha_{22})+0.5(1-\alpha_{21})(\alpha_{22})+0.3(\alpha_{21})(1-\alpha_{22})+0.05(\alpha_{21})(\alpha_{22})\rho_2\}}$$
$$= \frac{(1+\alpha_{21}\alpha_{22}(\rho_2-1))0.3}{(1-\alpha_{21})[0.15+0.35\alpha_{22}]+\alpha_{21}[0.3-(0.3-0.05\rho_2)\alpha_{22}]}$$

which are a function of the parameters α_{21}, α_{22} and ρ_2 that characterize the bivariate bernoulli distribution of the two tests outcomes when the true ailment is the second one.To complete this problem, we calculate $C_1(1,0)$, $C_2(1,0)$ and select the least cost decision. Generally, all costs $C_j(y_{j1}, y_{j2})$ can be calculated and a diagnostic selected as a function of the tests results.

Bayesian control problems recur in many situations in industry, in banking, in business and of control in general. In financial regulation, prior estimates of specific regulatory laws may be detected from historical and data pointing out to a systemic propensity to comply or not to comply to specific regulations.

11.2.3 Temporal Bayesian Controls

Say that we want to test competing hypotheses when our underlying information is a time series (for example, a financial time series to test competing financial processes, etc.). The evidence for a given hypothesis is revealed by the data associated to that hypothesis. For example, let $Y^t \equiv \{y(t), t \geq 0\}$ be a recorded time series, and define $H_j, j = 0, 1$ be two hypotheses—the Null and its alternative, defined by their prior probabilities. Let the prior probability for each alternative be:

$$\pi_0 : \text{ Prior Null Hypothesis;} \quad \pi_1 = 1 - \pi_0 : \text{ Prior Alternative Hypothesis}$$

The conditional probabilities for each of these hypotheses based on the time series are thus: $P[H_j|Y^t], \quad j = 0, 1$ with $\sum_{j=1}^{2} P[H_j|Y^t] = 1$. Define the likelihood $\Lambda(t)$ for the two hypotheses by:

$$\Lambda(t) = \frac{\pi_0 P[H_1|Y^t]}{(1-\pi_0)P[H_0|Y^t]} \quad \text{and} \quad Z(t) = \ln \Lambda(t)$$

where $Z(t)$ is the Log-Likelihood. A test for accepting or rejecting a hypothesis-model may be constructed by the following:

$$\text{If } Z(T) > F \Rightarrow \text{ Accept } H_1; \quad \text{If } Z(T) < F \Rightarrow \text{ Accept } H_0$$

To construct such a test, we specify again the risks we are willing to assume when we reach a decision. These are:

$$P_1(Z(T) > F) = \beta \text{ Under } H_1; \quad 1 - P_0(Z(T) \leq F) = \alpha \text{ Under } H_0$$

For example, say that under the alternative hypotheses j = 0,1, the underlying probability distribution is a Poisson distribution with parameters $\mu_j, u_j, q_j(0)$ denoting unknown constants, but about which we hypothesize a set of competing values under both hypotheses (j = 0,1) with:

$$P_j[S(t) = k] = e^{-q_j(t)} q_j^k(t)/k! \text{ with } \frac{dq_j(t)}{dt} = -\mu_j q_j + u_j, \quad q_j(0) \geq 0$$

The likelihood of the time series, each under a hypothesis, is thus:

$$\Lambda_j(t) = \prod_{i=0}^{t} P_j[S(t) = k_i] = \prod_{i=0}^{t} e^{-q_j(t)} q_j^{k_i}(t) \Big/ k_i!$$

The Ln Likelihood of each of these hypotheses are therefore:

$$Ln(\Lambda_j(t)) = \sum_{i=0}^{t} Ln\left(e^{-q_j(i)} q_j^{k_i}(i)/k_i!\right) = \sum_{i=0}^{t} -q_j(i) + k_i(\ln q_j(i)) - \ln(k_i!)$$

with $q_j(i)$ a solution of the differential equation above, expressed in terms of the unknown (hypothetical) parameters. An estimate for these parameters based on the Maximum Ln-Likelihood can be reached. In this case, we have: $Ln\left(\frac{\Lambda_1(t)}{\Lambda_0(t)}\right) = Ln(\Lambda_1(t)) - Ln(\Lambda_0(t))$ which is reduced to:

$$Ln(\Lambda_1(t)) - Ln(\Lambda_0(t)) = \sum_{i=0}^{t} \left\{-q_1(i) - q_0(i) + k_i \ln\left(\frac{q_1(i)}{q_0(i)}\right)\right\}$$

In our case, the statistical test is then:

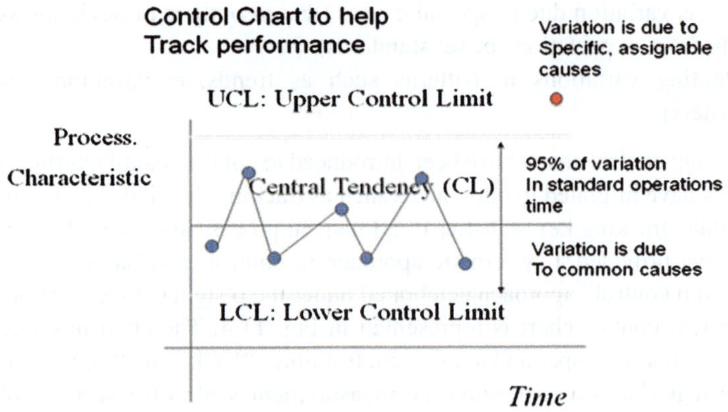

Fig. 11.4 A control chart

$$\text{If } Z(t) = \sum_{i=0}^{t} \left\{ -q_1(i) - q_0(i) + k_i \ln\left(\frac{q_1(i)}{q_0(i)}\right) \right\} > F \quad \text{accept } H_1$$

$$\text{If } Z(t) = \sum_{i=0}^{t} \left\{ -q_1(i) - q_0(i) + k_i \ln\left(\frac{q_1(i)}{q_0(i)}\right) \right\} \leq F \quad \text{accept } H_0$$

This test can then be used to accept or reject the alternative hypotheses. For simplicity, consider the decision threshold F, then

$$\textit{If } Z(T) > F \text{ accept } H_1; \quad \textit{If } Z(T) \leq F \text{ accept } H_0$$

Where $\alpha(T) = \text{Prob}[z(T) \geq F|H_0|$ and $\beta(T) = \text{Prob}[z(T) \leq F|H_1|$ denote type I and type II errors. To determine the threshold F, we can use Chernoff bounds as estimates of the cumulative distributions (see also Tapiero 1975a and Chap. 3 on probability bounds) for such an application. Of course, various risk models can also be used (e.g. lognormal processes, etc.) to test models hypotheses. Control may therefore be applied not only to test compliance or non-compliance to a set of regulated rules or to standards of performance but also to processes and their evolution.

11.3 Risk Control with Control Charts

Control charts are used to monitor and control repetitive processes (Fig. 11.4). They were introduced in 1931 by Shewart to control the variability of large volume parts manufacturing. Today they are used extensively to control various sources of variation including:

(a) Process variation from normal and expected standards (also called "common causes")

(b) Process variation due to special causes (due to outliers or deviating systematically from expectations or set standards)
(c) Detecting variations in patterns such as trends, co-variations and jumps (outliers).

Although control charts have been introduced to control manufacturing processes, their uses have migrated to other areas such as tracking the vital statistics of patients at hospitals, tracking key statistics of a financial process and so on. Further, control charts have provided a systematic approach to control risks based on a combined "design and control" approach celebrated under the 6 sigma model (see below).

A typical control chart is represented in Fig. 11.4. The chart has a center line "CL" as well as an upper and a lower control limit, "UCL" and "LCL" respectively. The vertical abscissa represents the measurement scale of a variable of interest (which can be quantitative or qualitative) while the horizontal abscissa stands for time. Measurements are recorded by star "*", which provide the visual record of sample results (positioned with respect to the CL, UCL and LCL). The distribution of measurements (points) on the control chart can be due to chance variation (common causes) or due to (special causes) assignable causes. Chance variations include the sum of all chance causes acting simultaneously on a process. Any chance cause has an individual and indistinguishable effect on the process however. For this reason, these are called common causes, and reflect inherent "normal" characteristics of the process. Such variations can be influenced only by changing the process or by introducing a structural change. For a process with normally distributed samples, control charts can be used to detect unexpected states (namely, events or series of events that defy in probability the assumption of their normality embedded in the control chart design). When surprises occur, these are assigned to special causes and causes are sought to explain these "surprising" observations.

Control charts are basically of two types: (1) Variable control charts and (2) Attribute control charts. Variable control charts use quantitative measures, including x-bar (x) charts whose purposes are to control a process average; the R-chart for the control of the range; the s-chart for the control of standard deviations; the s-squared-charts for the control of the variance and the X-chart for the control of individual values. In practice, the x-bar chart and the R chart are used together. The former to monitor samples average pattern while the latter providing a statistical measurement of the x-bar meaningfulness.

Attribute charts measure characteristics which need not be quantitative. These may include fraction defectives in a sample, the number of non-conformities etc. They include the p-chart for the control of fraction defectives, the np-chart for the number of nonconforming units, u-chart for the control of the number of nonconformities per unit produced and, finally, the c-chart for the number of non-conformities. Generally, we can apply control charts to monitor a broad set of indicators such as business performance, productivity indices and other measures of performance. All these charts are based however on a statistical definition of what may be called "normal" or "compliant". In this sense, control charts are merely a statistical measurement of variations.

When sample measurements are cumulated they are called CUSUM charts (cumulative sum charts). These charts are important for the detection of small but persistent shifts in a process and therefore complement the use of ordinary Shewart charts (that can detect larger shifts in a process). They were developed in the early 1950s (see Page 1954) and are useful for the rapid detection of small changes (Fellner 1961).

Control charts have proved over and over again to be simple and useful, highlighting non-conforming performances, detecting process instabilities. Their importance is both formal and informal (as an inducing, curative tool) as well as educational. Control charts do not solve problems however but provide information on the basis of which we can reasonably (in a statistical sense) presume that a process is "in control" or "out of control" i.e. compliant or not compliant. They can also indicate "false alarms" (errors that the system is out of control when it is not, the type I risk) as well as not detect situations which are out of control (i.e. type II risk).

There are essentially three approaches to construct control charts. First, use specified statistical characteristics (such as standards), second use historical data of a process operating over some periods of time in normal operating conditions (when, it is known a-priori that there are no special causes). Finally, use economic models and criteria to construct charts. The construction of a control chart and its analysis based on a working system is called the process capability study (see below).

11.3.1 Interpreting Charts

There are two essential ways to interpret control charts: First on the basis of the probability of observed unlikely events and second by observing patterns that depart from what may be defined as a "normal" expectation (see Figure 11.5). For example, cyclical patterns, stratification of samples results and other non-random patterns of data may suggest behaviors which are not random (and therefore due to some special recurrent causes). To determine whether points are randomly distributed, we can apply a number of tests for randomness (such a run tests) which provide a statistical verification of a process behavior over time. Figure 11.6 outlines a number of situations to consider when interpreting charts.

In practice, interpretations of control charts are based on detection of the unlikely in a probabilistic sense. These situations are a warning that there might be some special causes, or special situations, that warrant a concern to assess and explain their occurrence. When such events recur too often, or there is some pattern exhibited by the data, then, there are reasonable grounds to doubt both the processes and their statistical assumptions. Criteria should be carefully selected however. When selecting too many criteria, it may lead to erring in both our judgment and our decision regarding the process suitability as the probability of finding a "fault" may in such conditions be greater than warranted. Some criteria which are often used in practice include:

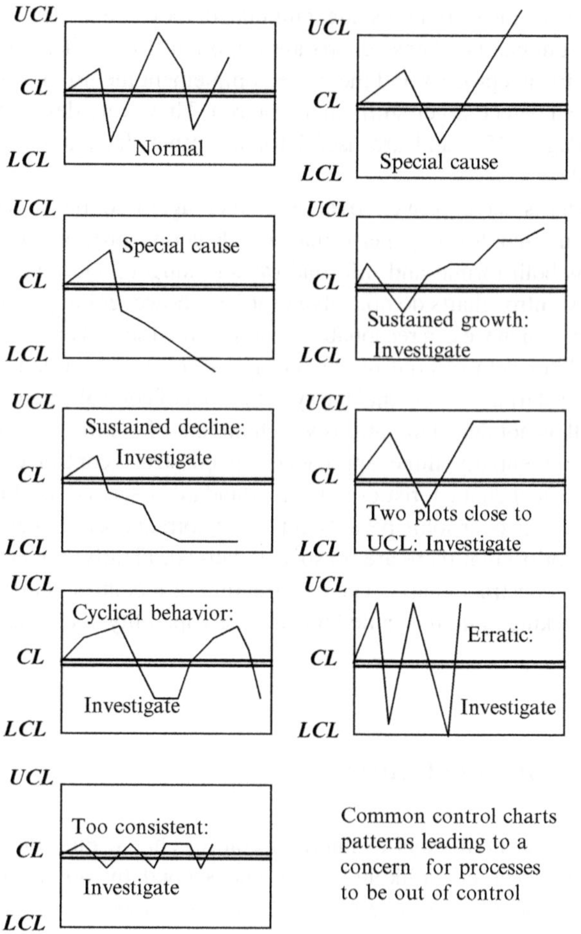

Fig. 11.5 Controlling charts

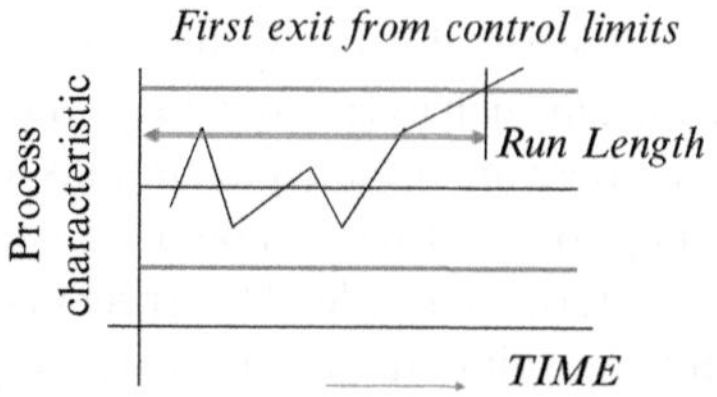

Fig. 11.6 Run lengths and control charts

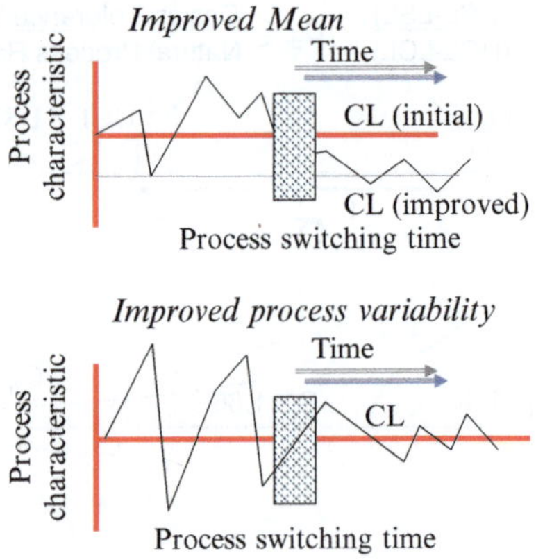

Fig. 11.7 Control charts for improvement mean tracking

- One or more points outside the control limits UCL, LCL
- Two or more points near the warning upper and lower limits
- Periodic patterns which highlight non stationary data behavior
- Points that are too close to the center line, implying that the warning limits were improperly drawn
- Run tests, which examine points positioned successively above or below the center line.

For example, a run of 7 points, a run of 10 in a string of 11 points, a run of 12 in 14 points, a run of 16 in 20 points have such small occurrence probabilities that they may be used to point out potential departures from a stable (controlled) process (Fig. 11.7). Other run tests, which use the number of points and their distance from the center lines, include a run of two or three points outside two deviations, or a run of four or five points outside one standard deviation limits. Say that a series of eight consecutive points remain above or below the center line (CL). The probability of obtaining such an event (0.49865) is 0.0039 where 0.49865 is the probability that a sample average is either above or below the line but inside the control limits. Thus, the probability of obtaining such an event is so small that when it occurs, it warrants further attention. It is a simple exercise to calculate the probabilities for each of these criteria. Clearly, they are all very small and therefore deemed unlikely small.

Various firms use, however, their own criteria. In practice, the simultaneous use of too many criteria can, as indicated above, be misleading, leading to too many false alarms and needless costs. For these reasons, a judicious choice of criteria to use requires a careful managerial judgment based on the risks we are willing to sustain and the costs they imply.

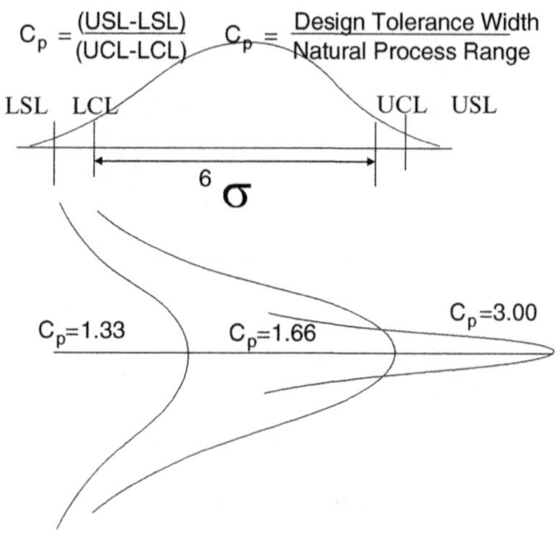

Fig. 11.8 Process capability and process improvement

11.3.2 6 Sigma and Process Capability

Process capabilities studies are conducted to compare the performance of a controlled process and its standard design requirements. The control limits of control charts, as stated above define that "95% of what we do" falls within these control limits, specifying the control chart UCL and LCL (upper and lower control limits). The error of rejecting a sample statistic outside the control limits has then a (type I) risk of $1 - 0.95 = 5\%$. However, say that a process has a standard specification of USL, and LSL (upper and lower specification limits). If design standards (USL and LSL) are greater than the control standards (UCL and LSL), that is USL > UCL and LSL < LCL, then what may seem to be uncontrolled from a manufacturing viewpoint (i.e. a sample statistic outside the UCL and LCL) would in fact be acceptable under the design specification standards. Therefore, if a control process has a standard deviation σ, then an index defining a "process capability" can be defined by a ratio of the control and the specification ranges as indicated in Fig. 11.8 above or:

$$C_p = \frac{USL - LSL}{UCL - LCL}$$

This index defines a "process potential", measuring the potential to meet the requirements, if operating properly. A control strategy can therefore be conceived as a simultaneous design-fault tolerant policy (i.e. expanding the USL and LSL limits and tightening the production process variability). It also implies that "zero errors" can be reached by a combined design/process strategy, consisting in expanding the design tolerance (as well as robustness) and the process stringency. Such an approach has managerial overtones which often is summarized by a 6 sigma "philosophy". In particular, set:

$$C_p = \frac{USL - LSL}{6\sigma}$$

Then, if $C_p = 2$, it will implies a tolerant process since it will almost never-never produce a non-acceptable part or product. For a process controlled with such a chart, the defect rate is then at most 3.4 ppm (parts per million!). For example, say that the design specification of a jet engine is 100 h flight time ($\pm 6hr$). Therefore the nominal spec limit is 100 while the USL and LSL are 106 and 94. If the part's life parameters have a mean of 100 and standard deviation 2, then $(\bar{x}, \sigma) = (100, 2)$. And the capability index is:

$$C_p = \frac{USL - LSL}{6\sigma} = \frac{106 - 94}{6 * 2} = 1$$

And therefore the capability index equals 1.0. The larger the index the better the performance.

The "evolution of a 6 sigma" approach to control occurred because of an increased awareness that non-conformance even in small probabilities can have important consequences. Thus instead of deviating more or less from the mean at $\pm 3\sigma$, one may want to increase the tolerance to $\pm 6\sigma$. Such tolerance is in fact and practically a "zero default" in everything we do. It has been used as a foundation for continuous an unabated improvement in everything we do—or the 6 sigma approach. Its migration to areas of TQM (total Quality Management), a focus on customers wants and satisfaction combined with data analysis, innovation, and financial performance etc. underlies the concept of 6 sigma, a means fuelling an evolution towards ever greater performance. The term 6 sigma was originally coined by William Smith an engineer at Motorola and is a registered trademark of Motorola. However, it was inspired by a fierce competition in the 1960s and 1970s with technology intensive Japanese manufacturing firms that were able to demonstrate that their fault production rate was very much smaller than that fault rates in the Western world. This approach, partly real, partly motivational and partly "philosophical" has an immense importance in fields other than manufacturing (for example health care) as they are all concerned with the increased cost of faults and the complexity of both products and processes (see also Tapiero 1996 on the Management and the Control of Quality).

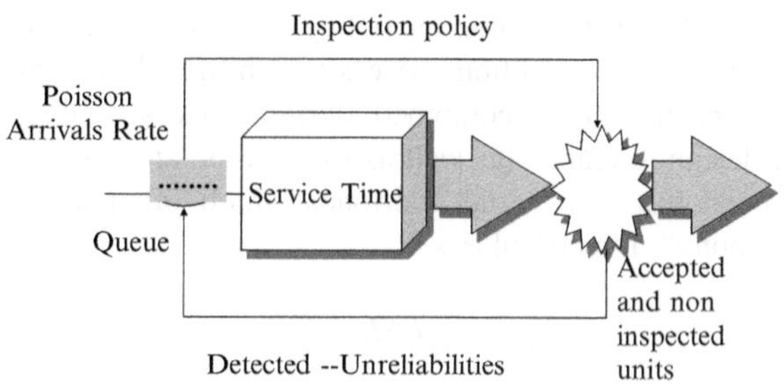

Fig. 11.9 Queue control

11.4 Queue Control

Queue models are generic stochastic systems applied broadly to many problems. The input or arrival process to a queue (defining the waiting line) may denote customers arriving to a bank for service, persons arriving by appointment to the dentist or to the hospital, an aircraft landing at the airport, incoming guests to a restaurant, random pollution emissions etc. When a service system is busy, an input process joins a capacitated (or un-capacitated) waiting line, seizing an available server following some priority rule, etc. The literature on stochastic and queue models is immense and varied (see for example, Gross and Harris 1985). Some problems such as servers' breakdown (also called queues with vacations, or queuing systems where servers may default and thus stop servicing) have been developed. However, when risk arises due to a qualitative (rather than a quantitative) malfunction allowing elements in the queue's system to pursue their route (even though they may be conforming to some standards or be non-compliant) to other servers, costs of time, resources, process, product and control costs can be meaningful. In a queuing network, this may assume processing and reprocessing on items that were already non-compliant. For example, banking transactions treated through queue network models of the banking system that render the process of non-compliance to be more complaint. Concerns for such problems have been attended to in numerous papers by Hsu and Tapiero (1987a, b, 1988a, b, 1992, 1994). Queue models consist essentially of (see Fig. 11.9):

- An input process consisting of arrivals joining a waiting line, occurring at deterministic or random times with a known distribution
- The waiting facility may be capacitated or not, providing a waiting space for incoming arrivals to wait for an available server. Seizing a server may be random, prioritized, scheduled, or follow defined disciplines (such as FIFO, LIFO, etc.)
- A service process defined by demands for a deterministic or by a random time, multiple or single needs, etc.

Each of these is subject to performance variations, and to default. These may include: a server breakdown, a service not fulfilled properly, a substandard part produced, a patient administered the wrong medicine or misdiagnosed, a bank faulty bank transaction, a stealth transfer of funds, a pollution emission, etc. For these reasons, inputs, outputs and intra-process detection, inspection and control of queues provide a model for certain risk control problems. For example, modeling the arrival of passengers to the airport, some of which may be ill intended, may inflict harm to both passengers and the air traffic system. Similarly, regulators are increasingly concerned with global banking where transfers are made to evade controls, illicit cash transfers and regulation. Further, intensely interconnnected banks that fail may also Contribute to systemic risks to the financial system as a whole etc. These situations may be modeled as networks of queuing systems. In Fig. 11.9 above we summarize some of these elements with an inspection-control policy added at the queue's output. When, a "serviced unit" is tested and found non-compliant then, it is redirected to the waiting queue (or to some other queue in a network of queues) for reprocessing. A simple quantitative model is considered a well below to highlight how such controls may be reached (for further studies see the numerous references of Hsu and Tapiero) as well as Tapiero and Hsu 1988).

11.4.1 The Simple M/M/1 Queue

Consider a queue model with a Poisson arrival rate λ (i.e. arrivals occur one at a time are independent and the time between arrivals have an exponential probability distribution). When a unit seizes the server, the mean time to be serviced and thus leaving the queue (and be controlled) has also an exponential probability distribution with mean $\mu > \lambda$. This latter condition assures the existence of a long run equilibrium for the queue model. Explicitly, say that at a given time, we have n units in the queuing system. This state arises from three possible events.

- There were n − 1 units and one arrives at a rate λ
- There were n + 1 units and one is serviced at a rate $\mu > \lambda$
- There were already n units and none arrived or departed.

These situations can be defined by the following:

$$P_n \Leftarrow \begin{cases} n-1 & P_{n-1} & \lambda \\ n & P_n & (1-\lambda)(1-\mu) \approx 1-\lambda-\mu \\ n+1 & P_{n+1} & \mu \end{cases}$$

For example, when there are n − 1 units in the system, with probability P_{n-1}, arrivals occur at a probability rate λ. Further, since all events—arrivals and service, occur one at the time, we have $\lambda\mu \approx 0$. Therefore, the following probability process results:

$$P_n = \lambda P_{n-1} + (1 - \lambda - \mu)P_n + \mu P_{n+1} \quad \text{and} \quad 0 = \mu P_1 - \lambda P_0$$

which we write as follows, with the added condition that $\sum_{n=0}^{\infty} P_n = 1$:

$$(\lambda + \mu)P_n = \lambda P_{n-1} + \mu P_{n+1}, n = 1, 2, \ldots. \quad \text{and} \quad \sum_{n=0}^{\infty} P_n = 1$$

The solution of this equation is straightforward and yields the probability distribution:

$$P_n = \rho^n(1 - \rho), \quad \rho = \lambda/\mu < 1$$

Note that the condition $\rho = \lambda/\mu < 1$ is necessary to assure the existence of such a geometric probability distribution and therefore all the statistical properties of the queue can be calculated.

11.4.2 *The Simple M/M/1 Queue and Non-compliance*

Assume next that each incoming unit (a production unit in process, a mistreatment in a medical facility, a financial transaction, etc.) may be misprocessed by the server with a probability θ, defining the server "unreliability". A unit then exits the queue "mistreated" or "non-compliant". A statistical control made at the queue's exit may however detect (or not) such a "mistreatment". Controls may assume then a number of approaches—attending at a pre-queue stage for the compliance of incoming units, an in-process stage for a process compliance and at a post treatment stage, for a finished unit compliance. Consider for example, a post-stage phase and let $W_i = 1, 0$ define the decision control a unit serviced ($W_i = 1$) when there are i units in the queuing system and let the probability of the unit serviced is "service non-compliant" be θ. In this case, the long run equilibrium probability distribution of the queuing system is (see Hsu and Tapiero 1987a and below for a proof) is:

$$P_n = \frac{\rho^n \beta_n}{\sum_{j=0}^{\infty} \rho^j \beta_j}, \quad \beta_n = \frac{1}{\prod_{i=0}^{n} [1 - (1 - \theta)W_i]}, \quad W_i = 1, 0; \quad W_0 = 0$$

Proof of the Probability Distribution Assume that ξ is the probability that a service process is non-compliant to a given standard (or say, faulty) and let W_i be the decision to control a unit in process when there are i units in the system (namely, there are n $-$ 1 in the queue). It is then simple to note that for n $=$ 0,1,2,. . .we have:

$$0 = \mu P_1 - \lambda P_0 \quad \text{and}$$
$$0 = \lambda P_{n-1} + \mu(1 - (1-\theta)W_{n+1})P_{n+1} - (\lambda + \mu(1 - (1-\theta)W_n))P_n$$

whose solution is:

$$P_n = \frac{(\lambda/\mu)^n \beta_n}{\sum_{j=0}^{\infty} (\lambda/\mu)^j \beta_j}, \quad \beta_n = \left(\prod_{i=1}^{n} (1 - (1-\theta)W_i)\right)^{-1}$$

This distribution indicates important properties of controls we have referred to previously, namely when control are applied, the controls do alter the performance of the system it controls. In other words, compliance and (the physical) efficiency of the system it controls are altered. First note that when all units (transactions) are service compliant then $\theta = 1$ in which case $\beta_n = 1$ and $P_n = (1-\rho)\rho^n$. In other words, the (physical) properties of the queue are not altered by controls since these controls are external to the queue-service and require no additional attention by the server (since all controls were external and none found to be non-compliant). Next assume that no controls are made, then $W_i = 0$ and therefore (again) $P_n = (1-\rho)\rho^n$ which, of course, has no effect on the queuing system but may have a series of exiting units to be "mistreated". If the service is in probability non-compliant and controls are exercised on random units (or financial transactions) with probability ξ, $\xi = \text{Prob}(W_i = 1)$ the probability distribution that n units are in the system (1 in service and n − 1 waiting to be serviced) is also a geometric distribution, albeit with parameters that account for the server "unreliability" and the controls for compliance that are exercised with $\beta_n = [1 - (1-\theta)\xi]^{-n}$, or:

$$P_n = \frac{\rho^n/[1-(1-\theta)\xi]^n}{\sum_{j=0}^{\infty} (\rho/[1-(1-\theta)\xi])^j} = \left(\frac{\rho}{1-(1-\theta)\xi}\right)^n \left(1 - \frac{\rho}{1-(1-\theta)\xi}\right)$$

$$P_n = \left(\breve{\rho}\right)^n \left(1 - \breve{\rho}\right), \; \breve{\rho} = \frac{\rho}{1-(1-\theta)\xi}$$

Note that in such situations, there is a required growth in the capacity to maintain such a system since:

$$\breve{\rho} < 1 \Rightarrow \rho < 1 - (1-\theta)\xi.$$

These equations provide of course the ability to measure the effects of a queue server "unreliability" θ and the effects of control for compliance on the queue operational characteristics. Such a parameter implies as well additional system costs due to a process-service needed to attend to reprocess "mistreated and detected" units. For example, consider the number of units in the queue, then:

$$L = E(n) = \frac{\rho}{1-\rho},\ \breve{L} = E(\breve{n}) = \frac{\breve{\rho}}{1-\breve{\rho}} \quad \text{or}$$

$$\frac{L}{\breve{L}} = \frac{\rho(1-\breve{\rho})}{(1-\rho)(\breve{\rho})} = 1 - \frac{(1-\theta)\xi}{1-\rho} < 1$$

If $\rho = 0.8$, $\theta = 0.1$, $\xi = 0.3$ then $\breve{\rho} > 1$ and such an unreliable system will diverge to an infinite queue. If instead, the server reliability is increased to $\theta = 0.05$ then $\breve{\rho} = 0.812$.

There are then three alternatives we may follow: (1) increase the service capacity (at a cost) to a reasonable level $\breve{\rho} < 1$; (2) Augment the reliability of service, i.e. reduce further θ (also at a cost); (3) Reduce the controls applied, at a risk of increasing the number of non-compliant services that are not detected and therefore may incur appreciable external costs.

Although a great many problems such problems recur in industrial systems, they also occur in health care systems, in banking and in financial services in general. For example, banks treating an extremely large number of transactions (some automatically), may have non-compliant treatments with appreciable risks of non-compliance to regulation (for example, money laundering, errors in transfer of funds, uncontrolled traders, processing illicit trades etc.). Controls of these transactions are needed to compensate these risks of loss and non-compliance as well as customers' costs due to banks' errors. Say that a bank considers that control costs are too costly, or the probability of the statistical control technique they apply is not effective and say, as a result, $\xi = 0$. In such circumstances, while "transaction" flows, would appear to be "normal", there will be a continuous flow of non-compliant transactions, each with a probability θ of non-compliance, that when detected externally (by a regulator or by unhappy customers or whistle blowers) can lead to potentially important costs as well as penalties.

11.4.3 The Continuous CSP-1 Control of Queues and Banking

The problem we have considered is of course a simplification of banks' performing continuously millions of transactions. Controls may then be applied not only to secure the reliability of the service output but both the reliability of the service system set in place and the good faith of the controlled (or the regulated). Random controls by the controlled (say the bank) may be applied even if a system is believed to be compliant. Full controls however may be applied when there are good reasons to believe that the service system is basically non-compliant or producing far more errors than would be expected. A control procedure may then be applied that is sensitive to such situations. In such cases, a $(CSP(m,\xi) - 1)$ procedures may be applied. This procedure consists in controlling continuously m units (transactions). If all are compliant, then a random control procedure with probability ξ is applied until one unit is found non-compliant—at which time, m units are controlled again and if none is found to be non-compliant, then control reverts to a random control. In this control, the resulting queue process turns out to be more complex (although and for practical ends, it will be simple to obtain simulated results). In our case, we have a system which is defined as follows:

$$0 = \mu P_1 - \lambda P_0 \quad \text{and}$$
$$0 = \lambda P_{n-1} + \mu\left(1 - (1-\theta)\tilde{W}_{n+1}\right)P_{n+1} - \left(\lambda + \mu\left(1 - (1-\theta)\tilde{W}_n\right)\right)P_n$$

where $\tilde{W}_n$ is the probability of a control when a $CSP(m, \xi) - 1$ control procedure is applied. As indicated in the previous section, we then have a probability distribution:

$$P_n(n|W_i) = \frac{\rho^n \beta(n|W_i)}{\sum_{j=0}^{\infty} \rho^j \beta(j|W_i)}, \quad \beta_n = \left(\prod_{i=1}^{n} (1 - (1-\theta)W_i)\right)^{-1}$$

We assume further (and for simplicity) that controls are queue's length insensitive such that $W_i = \tilde{W}$. The probability of a control can be calculated simply based on the long run average probability of a control under the CSP-1 control procedure. In other words, for a one detection cycle, i.e. one non-compliance detected, the number of units controlled is $m + \tilde{r}$ which is given by:

$$\tilde{z} = \begin{cases} m & w.p. \ \ (1-\theta)^m \\ \tilde{r} & \text{with } w.p. \ (\theta\xi)(1-\theta\xi)^{r-1}, \ r = 1, 2, \ldots m-1 \end{cases}$$

where $\tilde{r}$ is the number of units controlled for one found to be non-compliant. Since the probability of control is ξ and if the probability of a unit to be non-compliant is θ, we obtain a (truncated) geometric probability distribution for $\tilde{r}$ given by $(\theta\xi)(1-\theta\xi)^{r-1}, \quad r \leq m$ and therefore,

$$E(\tilde{z}) = m(1-\theta)^m + E^m(\tilde{r})$$

where $E^m(\tilde{r}) = \sum_{r=1}^{m-1} r(\theta\xi)(1-\theta\xi)^{r-1} = (\theta\xi)\sum_{r=0}^{m-1} r(1-\theta\xi)^{r-1} = \theta\xi \frac{\partial}{\partial x}\frac{1-x^m}{1-x}$, $x = (1-\theta\xi)$ and therefore,

$$E^m(\tilde{r}) = \theta\xi \frac{\partial}{\partial x}\left(\frac{1-x^m}{1-x}\right) = \left(\frac{1 - (1+\theta\xi(m-1))(1-\theta\xi)^{m-1}}{\theta\xi}\right)$$

and

$$E(\tilde{z}) = m(1-\theta)^m + \frac{1-(1+\theta\xi(m-1))(1-\theta\xi)^{m-1}}{\theta\xi}$$

Since only one non-compliant event is detected, the probability of detecting of a non-complying event is in the long run (based on renewal arguments)

$$p_D(m,\theta,\xi) = \frac{1}{E(\tilde{z})} = \frac{\theta\xi}{1+\theta\xi m(1-\theta)^m - (1+\theta\xi(m-1))(1-\theta\xi)^{m-1}}$$

Say that $\theta = p_C, p_{NC}$, are the probabilities that a system is compliant (p_C small) and non-compliant (p_{NC} larger), then:

$$p_D(m,p_C,\xi) = \frac{p_C\xi}{1+p_C\xi m(1-p_C)^m - (1+p_C\xi(m-1))(1-p_C\xi)^{m-1}} \quad \text{and}$$

$$p_D(m,p_{NC},\xi) = \frac{p_{NC}\xi}{1+p_{NC}\xi m(1-p_{NC})^m - (1+p_{NC}\xi(m-1))(1-p_{NC}\xi)^{m-1}}$$

Type I and II risks can then be defined by the length of the number of controls made. Explicitly, if a system is non-compliant we use m controls until we find the system to be compliant, then its probability $1 - p_D(m,p_{NC},\xi)$ which is the type II risk. For example, let $p_{NC} = 0.15, \quad m = 20, \quad \xi = 0.20$, then:

$$p_D(NC) = \frac{0.03}{1+0.6(0.85)^{20} - 1.57(0.97)^{19}}$$

And therefore the probability of a type II risk is fairly high as it equals $1 - p_D(NC)$.

11.4.4 *Networks and Queues*

Networks of queues are used intensively to model digital communication systems, electricity grids, robotized manufacturing systems, networked healthcare facilities etc. They may equally be applied to a network of interconnected banks. They have been applied to unreliable industrial systems where controls are applied at all levels of industrial queues (see Hsu et al. 1993 for a survey, Prabhu 1965, 1980 and numerous papers by Hsu and Tapiero and Tapiero and Hsu listed in the references). Most networks of queues are difficult to analyze theoretically (except for a type of network called Jackson queues), albeit Monte Carlo and queuing systems simulation software are profusely available.

Many results in queuing networks are based on decomposition results published by Jackson (1957, 1963) (see also Buzacott and Yao 1986; Ho and Cao 1983). Extensions and approximations to more complex queuing models abound (Whitt 1983). In this section, we consider "unreliable" queues in the spirit of Hsu's and Tapiero's research papers (although applied in a financial context).

The Jackson Open Network Queue (ON):

A ON Jackson queue consists of the following:

- N connected queues with $N < \infty$.
- All incoming units (financial transactions, or units to be processed) arrive at a known Poisson rate at some of the queues are then transiting in probability to other queues once their service has been concluded.
- Each queue has an unlimited waiting capacity
- Service is FCFS (First Come, First Served) in each of the queues and Service times are and arrival statistically independent of one another. Service may consist of one or more servers (for example, one or several machines that treat incoming transactions and then send them on for further treatment or let them exits the network).

Such situations may represent partly a network of interconnected banks with both internal and external flows (some across national boundaries) as well as a financial supply chain. Say that each queue has $[n_i],\ i = 1, 2, \ldots N$ units in process (either waiting to be processed or currently in service). The probability distribution $f(n_1, n_2, \ldots, n_N)$ then defines the "queue network states" of $(n_1, n_2, \ldots, n_n)$ units (transactions) in each of the queues. The essential result implied by Jackson queue networks is that such a distribution can be expressed by a product of individual queue distributions, or (for proofs and extensive developments we refer to Jackson as well as papers we refer to at the end of the chapter):

$$f(n_1, n_2, \ldots, n_N) = \prod_{i=1}^{N} f_i(n_i)$$

This decomposition is based on the statistical independence of each of the N queues in the network and the network Markov routing probabilities from queue i to queue j, given by $[\mathbf{P}] = [p_{ij}];\ i, j = 1, 2, \ldots, N$. External Poisson arrival rate are $[\lambda] = [\lambda_i];\ i = 1, 2, \ldots, N$. When service times are exponentially distributed a queue's exit rate has also a Poisson rate. A linear combination of Poisson rates is also a Poisson rate however which justifies the decomposition of the network of queues. An intuitive proof is as follows. Let Λ be the Poisson rate of arrivals to an N network and let p_{0i} be the probability that an incoming arrival joint the ith queue. Namely, the arrival rate to the ith queue has a Poisson rate Λp_{0i}. Assume that following the service provided by the queue, the unit is directed in probability to another queue, say j, for another service (or exits the queue) with probability p_{ij}. The net probability for an incoming flow to the ith queue is thus given by the probability of an external arrival plus the sum of the probabilities that has been directed from all other queues times the probability of their exiting from the jth queue, or:

$$e_i = p_{0i} + \sum_{j=1}^{N} e_j p_{ji}$$

Setting $\alpha_i = e_i \Lambda$ we have the following vector equation $[\alpha] = [\alpha_i];\ i = 1, 2, \ldots, N$. Explicitly, the matrix equation: $[\alpha]^T = [\lambda]^T + [\alpha]^T [\mathbf{P}]$ results which is explicitly written as:

$$\alpha_i = \lambda_i + \sum_{j=1}^{n} \alpha_j p_{ji}$$

In other words, the queuing network has been decomposed to N queues, each with its arrival α_i (and thus, in equilibrium, an exit rate α_i as well). Let α_i be a solution of this linear system of equations. Thus, in a network of simple queues each with one server, we have a service rate for each queue $\mu_i\left(1 - (1 - \theta_i)W_{i,n_i+1}\right)$ and an arrival rate α_i and of course the service provided by the queue. Assuming that service is unreliable, the probability distribution of the number of units in queue i is given by:

$$P_{n_i} = \frac{(\alpha_i/\mu_i)^{n_i} \beta_{i,n_i}}{\sum_{j=0}^{\infty} (\alpha_i/\mu_i)^j \beta_{i,j}}, \quad \beta_{i,n_i} = \left(\prod_{k=1}^{n_i} \left(1 - (1 - \theta_i)W_{i,k}\right)\right)^{-1}$$

The probability distribution of the number of units (transactions) in the queuing network is thus:

$$f(n_1, n_2, \ldots, n_N) = \prod_{i=1}^{N} P_{n_i} = \prod_{i=1}^{N} \left(\frac{\left(\alpha_i/\left[\mu_i\left(1 - (1 - \theta_i)W_{i,n_i+1}\right)\right]\right)^{n_i}}{\sum_{j=0}^{\infty} \left(\alpha_i/\left[\mu_i\left(1 - (1 - \theta_i)W_{i,n_i+1}\right)\right]\right)^j} \right)$$

For example, letting q_i be the probability of an output control in queue i, or $q_i = W_{i,k}$ then: $\beta_{i,n_i} = (1 - (1 - \theta_i)q_i)^{-n_i}$ and:

$$P_{n_i} = \left(\frac{\alpha_i}{\mu_i(1 - (1 - \theta_i)q_i)}\right)^{n_i} \left(1 - \frac{\alpha_i}{\mu_i(1 - (1 - \theta_i)q_i)}\right);\ \alpha_i = \lambda_i + \sum_{j=1}^{n} \alpha_j p_{ji}$$

which can be used to study the mutual effects of controls on all queues. Other properties as well as more complex networks may be assessed as well, either analytically (when these can be defined by Jacksonian networks) or by approximations. For example, open multi server Jackson networks, closed queuing networks (with a constant number of units in the queue network) and other queue networks have this decomposition property. When this is not the case, approximations can be made.

11.5 Strategic Inspections and Controls (See Also Chap. 12 for a Review of Game Theory)

Statistical approaches have, for the most part, presumed that an underlying risk is neutral! In other words, risks are not motivated by the agenda of a particular party. In reality, this may not be the case. For example, multiple parties interacting with broadly varying motivations; information asymmetries; power and asymmetric relationships between interacting agents; a conflicting environment etc. have contributed to risks arising from such strategic relationships.

To control risks in such an environment, game theory (Chap. 12) can be useful. For references and applications, see Reyniers (1992), Reyniers and Tapiero (1995a, b) as well as Tapiero (1995b, 1996, 2004a, 2005a, c). Unlike risk neutral and exogenous risk models, the game theoretic framework considers parties intentionality and their risk attitudes. The applications we consider below will use traditional Type I and Type II errors when outcomes are both random and depend on the (motivated) choices each of the parties make. This approach thus reduces random payoff games to "risk-analysis" games which simplifies their solution.

The relationship between statistics, conflict and control as well as the role of statistical sampling in improving controls when multiple parties interact has to a large measure been neglected (Avenhaus 1994; Avenhaus et al. 1995, 1996; Avenhaus and Kilgour 2004; Canty et al. 2001). Statistic's failure to deal with conflict arose from its presumption that "risk" is not motivated. Interpreting risk and reducing its effects is then based on the presumption that our measurements and our acts are independent of their origins and their motivations (although there is an extensive economic literature on information asymmetries and its risks—adverse selection and moral hazard (Akerlof 1970; Holmstrom 1979, 1982; Holmstrom and Myerson 1983). Information asymmetry risks in a strategic environment in particular in the financial sector, thus induces a need for regulation to assure that outcomes are "fair" or "compliant" to some agreed on contracts and rules that are needed to maintain the fair and integrity of financial exchanges. For example, strategic audits (such as banks' audits for financial compliance to regulation rules, nuclear inspections for compliance to safety standards, and their like) always convey a number of messages: a control and a signal to the audited.

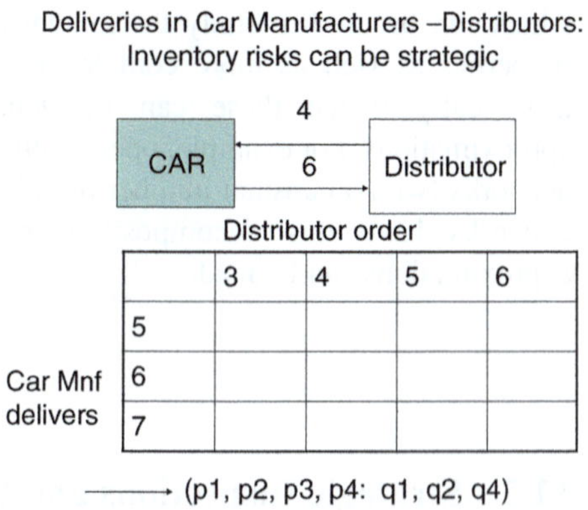

Fig. 11.10 A producer distribution game

Applications to strategic controls are outlined and solved to indicate how such problems may be handled.

Example: The Strategic Inventory Orders Consider a car manufacturer and a cars' distributor-agency. Traditionally, car manufacturers owned their distribution and sales point as well. It is today rarely the case as distributors are now agents to more than one manufacturer. Thus, when an agency orders only three cars, the car manufacturer may in fact ship 3,4,5 or 6 cars to transfer its excess car inventory as well as provide an incentive to the agency to sell its cars. This result in a strategic game with the distributor, ordering a particular number of cars, aware that he may receive, in probability, another number as Fig. 11.10 indicates. How much to order depend of course on the consequences of ordering "one thing" and receiving "another" for both the agency and the manufacturer. This results in a game whose solution may provide a particular number to order (and a particular number to supply) which is called a pure startegy. In some cases there is no such solution. Instead a "randomized solution" results. That is, each strategy of say the agency (ordering 3,4,5,6 cars) may be reached by choosing any one of these strategies in probability defined by the game solution.

11.5.1 Yield and Control in a Supplier–Customer Relationship

Consider a supplier of parts whose manufacturing yield, p, defines the propensity to produce a part that conforms to acceptable standards and whose cost is $C(p)$, $\partial C/\partial p > 0, \quad \partial^2 C/\partial p^2 > 0, \quad C(1) = \infty$. The yield can be improved either by investments in technologies and education of its workforce. A supplier's strategy is then a yield defined by what the supplier does. A customer's strategy consists instead in either accepting the parts' deliveries without any control, or with controls to assure that parts conform (statistically) as specified by the contract. Nonconformance or delivery of substandard parts entail costs which are defined by the contract terms for penalties agreed on a-priori between the supplier and the customer. To assess some of the relationships between the customer and the supplier inspection-control policies, the manufacturing yield and the parameters of the contract; a sensitivity analysis is performed. Existence of a number of potential solution to such a game can arise depending on the cost function $C(p)$ and on the contract terms. Using a game solution, we may show that a production technology (resulting in the yield p) and contract agreements generate various yields and a number of potential control policies.

A quantitative expression to this problem follows. Assume that a contract for the supply of parts has been negotiated and signed by two parties, the supplier and the buyer. Suppose that this contract stipulates that a payment π is made to the supplier while a penalty cost T is incurred by the supplier if parts are detected and found non-compliant by the buyer. Further, let p be the yield and assume that once a unit is inspected by the supplier, and found defective, then all subsequent units are inspected until a unit is demonstrably non defective. That is, the supplier uses a corrective sampling technique. Let ϕ be the buyer of parts selling price. If the buyer inspects parts with probability y and the supplier with probability x, the following bi-matrix (random payoff) game summarized in Table 11.1 results:

In this game, note that the buyer has two alternatives: control or not for parts compliance. The supplier has however, two decisions to reach, one regarding the manufacturing yield and the other regarding the controls defined by (x, p). Once a control solution is found as a function of the yield, the supplier may select the yield that optimizes the value of this game. Since the cost of consistently producing a good part, (equivalently a "zero-defects" technology of infinite costs $C(1) = \infty$), the yield is necessarily given by a probability $p < 1$. Both the "yield" and the "control" strategies of the supplier and therefore the control by the buyer are dependent, as the choice of one affects the other. For example, if p is very large (close to one), then it is possible that a 0% control policy may be optimal.

The bi-matrix game entries in Table 11.1 are evident. For example, say that the supplier produces a non-compliant part (with probability $1 - p$) and inspect it with probability x, then, whether the buyer inspects materials or not, the expected payoff is equal to the unit price π less the production cost $C(p)$ and its inspection cost c_i. Since parts are found defective, it must be replaced by other parts which are also inspected until found to be non-defective. Since p is the yield, the number of parts tested until one is non-defective is a random variable $\tilde{k}_1$ whose probability distribution is geometric and given by $p(1-p)^j$, $j = 1, 2, ..$ whose mean is $(1-p)/p$. Although only one part is sold at a price of π, the expected number processed and tested equals $E(1 + \tilde{k}_1) = 1/p$. The buyer's payoff is thus equal to $\phi - \pi - c_b$ if inspected and $\phi - \pi$ otherwise.

If the supplier does not inspect its parts and it is defective and detected by the buyer, the supplier's payoff is reduced by the contracted penalty cost T, paid to the buyer (in fact, it is a price break). With no inspection a total loss of $-\pi$ is incurred. Other costs such as post sales failure costs, warranty and related costs are not

Table 11.1 The (supplier, manufacturer) payoff matrix

Sampling policies	Control by the manufacturer With probability y	No control by the manufacturer With probability $1-y$
Control by supplier with probability x	$\begin{pmatrix} \pi - [C(p)+c_i] & w.p.\ p \\ \pi - [C(p)+c_i](1+\tilde{k}_1) & w.p.\ 1-p \end{pmatrix};$ $\phi - \pi - c_b$	$\begin{pmatrix} \pi - [C(p)+c_i] & w.p.\ p \\ \pi - [C(p)+c_i](1+\tilde{k}_1) & w.p.\ 1-p \end{pmatrix}; \phi - \pi$
No control by the supplier with probability $1-x$	$\begin{pmatrix} \pi - C(p) & w.p.\ p \\ \pi - C(p) - [C(p)+c_i](\tilde{k}_1) - T & w.p.\ 1-p \end{pmatrix};$ $\begin{pmatrix} \phi \dot{-} \pi - c_b & w.p.p \\ \phi - \pi - c_b + T & w.p.\ 1-p \end{pmatrix}$	$\pi - C(p);$ $-\pi$

considered for simplicity. Using the game defined above, its solution yields the following strategic control policies:.

Proposition: *For a risk neutral, maximizing expected payoff supplier, using a yield p, it is never optimal to fully sample. Further,*

(i) *If* $T \leq C(p) + c_i/(1-p)$ *it is never optimal to sample, i.e.* $x^* = 0$.
(ii) *If* $T \geq C(p) + c_i/(1-p)$ *it is optimal to sample, i.e.* $x^* > 0$.
(iii) *For a risk neutral manufacturer, the supplier inspection probability under condition (ii) above is given by:*

$$x^* = 1 - \frac{c_b}{\phi + T(1-p)}$$

Proof (See Also Chap. 12 for the Solution of Two Persons Non-zero Sum Games) For each of the entries of the supplier's game, we consider the expected utilities game, where the utility function $u(.)$, $u'(.) > 0$, $u''(.) \leq 0$,

$$\begin{aligned}
\hat{u}_{11} &= pu(\pi - [C(p) + c_i]) + (1-p)Eu(\pi - C(p) - [C(p) + c_i](1 + \tilde{k}_1));\\
\hat{u}_{12} &= \hat{u}_{11}\\
\hat{u}_{21} &= pu(\pi - C(p)) + (1-p)Eu(\pi - C(p) - [C(p) + c_i](\tilde{k}_1) - T);\\
\hat{u}_{22} &= u(\pi - C(p))
\end{aligned}$$

These are reduced to simple expectations for a risk neutral supplier. Note that $\hat{u}_{22} > \hat{u}_{11}(= \hat{u}_{12})$. Further, $\hat{u}_{21} > \hat{u}_{11}$ if $T \leq C(p) + c_i/(1-p)$. Thus the strategy to never sample by the supplier is always optimal as stated in (i) in the proposition. When this is not the case, both full sampling and no sampling strategies are not dominating and therefore the only remaining possibility is for a mixture of these alternatives, which lead to a sampling probability of $0 < x^* < 1$ as stated in (ii) above. Further, these two cases cover the proposition's statement that it is never optimal to fully sample (control).

When the game solution results in a randomized strategy, we have (as indicated in Chap. 12):

$$x^* = \frac{v_{21} - v_{22}}{v_{12} - v_{11} + v_{21} - v_{22}}$$

where the v_{ij}, $i = 1, 2;\ j = 1, 2$ are the expected values corresponding to the entries in the bi-matrix game.

$$v_{11} = \phi - \pi - c_b;\ v_{12} = \phi - \pi \text{ and } v_{21} = \phi - \pi - c_b + T(1-p);\ v_{22} = -\pi$$

In this case, we have:

$$x^* = \frac{\varphi - \pi - c_b + T(1-p) + \pi}{\varphi - c_b + T(1-p) + \pi + \varphi - \pi - c_b + T(1-p) + \pi}$$

which is reduced to the proposition statement. Q.E.D

The implications of this proposition is that if the contracted transfer cost T, is smaller than the expected cost of production plus the expected cost of control-sampling a compliant part, then it is not optimal to control. However, when it is larger, it is optimal to incur the control cost to avoid such a payment. In this case, the control probability is expressed in terms of the contract parameters and the buyer's control cost. Note that the larger the control cost c_b, the smaller the amount of control (since the buyer will control less). Further the larger the selling price and the larger the penalty cost, the more the supplier will sample-control. In this sense, the terms of the contract determine the amount of control of the supplier. We turn next to the buyer and prove the following proposition.

Proposition: *The manufacturer–customer has no dominating no-sampling strategy. Further, If $T \leq C(p) + c_i/(1-p)$, the supplier–customer optimal policy is to sample fully If $T \geq C(p) + c_i/(1-p)$ the optimal sampling-control strategy is to sample in probability:*

$$0 < y^* = \frac{1}{1 + \frac{p(T(1-p)-c_i-(1-p)C(p))}{(1-p(1-p))c_i+(1-p)C(p)}} < 1$$

Proof: Consider again the following entries $v_{11} = \phi - \pi - c_b$; $v_{12} = \phi - \pi$; $v_{21} = \phi - \pi - c_b + T(1-p)$; $v_{22} = -\pi$

Note that: $v_{11} < v_{12}$ while $v_{21} > v_{22}$ which rules out a dominating sampling strategy by the customer. As a result, the sampling strategy is randomized and given by solving for y,

$$\underset{y}{\text{Max}}\,\{v_{11}xy + v_{12}x(1-y) + v_{21}(1-x)y + v_{22}(1-x)(1-y)\}$$

As we saw in the first proposition, two situations arise. First, $x = 0$ which occurs if $T \leq C(p) + c_i/(1-p)$. In this case,

$$\underset{y}{\text{Max}}\,\{v_{21}y + v_{22}(1-y)\} = \underset{y}{\text{Max}}\,v_{22} + y\{v_{21} - v_{22}\}$$

Since, $v_{21} > v_{22}$, the optimal control policy is $y^* = 1$. Now consider the case, $T \geq C(p) + c_i/(1-p)$ where the supplier policy is to sample. In this case, we have:

$$\underset{0<y<1}{\text{Max}}\,\{v_{11}xy + v_{12}x(1-y) + v_{21}(1-x)y + v_{22}(1-x)(1-y)\}$$
$$\underset{0<x<1}{\text{Max}}\,\{\hat{u}_{11}xy + \hat{u}_{12}x(1-y) + \hat{u}_{21}(1-x)y + \hat{u}_{22}(1-x)(1-y)\}$$

And

$$0 = \{v_{11}x - v_{12}x + v_{21}(1-x) - v_{22}(1-x)\} \quad \text{and}$$
$$0 = \{\hat{u}_{11}y + \hat{u}_{12}(1-y) - \hat{u}_{21}y - \hat{u}_{22}(1-y)\}$$

which leads to:

$$0 = v_{21} - v_{22} + \{v_{11} - v_{12} - v_{21} + v_{22}\}x \quad \text{and}$$
$$0 = \hat{u}_{12} - \hat{u}_{22}\{\hat{u}_{11} - \hat{u}_{12} - \hat{u}_{21} + \hat{u}_{22}\}y$$

and therefore,

$$y^* = \frac{\hat{u}_{22} - \hat{u}_{12}}{\{\hat{u}_{11} - \hat{u}_{12} - \hat{u}_{21} + \hat{u}_{22}\}}$$

Inserting the relevant terms of the bi-matrix, we have:

$$\hat{u}_{11} = pu(\pi - [C(p) + c_i]) + (1-p)Eu(\pi - C(p) - [C(p) + c_i](1 + \tilde{k}_1));$$
$$\hat{u}_{12} = \hat{u}_{11}$$

$$\hat{u}_{21} = pu(\pi - C(p)) + (1-p)Eu(\pi - C(p) - [C(p) + c_i](\tilde{k}_1) - T);$$
$$\hat{u}_{22} = u(\pi - C(p))$$

$$y^* = \frac{\frac{(1-p(1-p))c_i + (1-p)C(p)}{p}}{\{T(1-p) - c_i - (1-p)C(p)\} + \frac{(1-p(1-p))c_i + (1-p)C(p)}{p}}$$

which lead to the proposition result of a randomized strategy when $T \geq C(p) + c_i/(1-p)$

$$0 < y^* = \frac{1}{1 + \frac{((T - C(p))(1-p) - c_i)}{(1-p(1-p))c_i + (1-p)C(p)}} < 1 \qquad \text{Q.E.D}$$

The implications of this proposition are again demonstrating the dependence of the buyer control policy as a function of the yield p, its cost function and of course the contract parameters.

We note from both propositions, that for the case $T \geq C(p) + c_i/(1-p)$, there is one unique equilibrium in mixed strategies for both the supplier and the buyer while when $T \leq C(p) + c_i/(1-p)$, there is also a unique equilibrium consisting of no sampling by the supplier and full sampling by the buyer.

To determine the optimal yield, the supplier will consider his value of the game accounting for the optimal control strategies. In this case, the optimal yield is found by solving and comparing the value of the game under the following two optimization problems representing the relative values of penalties:

$$\underset{0<p<1}{\text{Max}} \{\hat{u}_{11}x^*y^* + \hat{u}_{12}x^*(1-y^*) + \hat{u}_{21}(1-x^*)y^* + \hat{u}_{22}(1-x^*)(1-y^*)\}$$

$$\text{Subject to :} \quad T \geq C(p) + c_i/(1-p)$$

which is equivalent (in utility terms) to:

$$\underset{0<p<1}{\text{Max}} \frac{\hat{u}_{11}\hat{u}_{22} - \hat{u}_{12}\hat{u}_{21}}{\hat{u}_{11} + \hat{u}_{22} - \hat{u}_{12} - \hat{u}_{21}} = \underset{0<p<1}{\text{Max}} \frac{\hat{u}_{11}(\hat{u}_{22} - \hat{u}_{21})}{\hat{u}_{22} - \hat{u}_{21}} = \underset{0<p<1}{\text{Max}}\ \hat{u}_{11}$$

$$\text{Subject to : } T \geq C(p) + c_i/(1-p),$$

where (in expectations)

$$\hat{u}_{11} = \pi - \left(\frac{1}{p}\right)C(p) - \left(\frac{1-p(1-p)}{p}\right)c_i$$

As a result, if there is an interior solution, we have:

$$c_i = \left(\frac{pC'(p) - C(p)}{1-p^2}\right)$$

Note that, this requires as well that the derivative $C'(p)$ be: $C'(p) > C(p)/p$. For example, if $C(p) = A/(1-p)$, then,

$$\frac{c_i}{A} = \left(\frac{2p-1}{(1+p)(1-p)^3}\right) \quad \text{With } p > \frac{1}{2}$$

Alternatively, under the second equilibrium, we have:

$$\underset{0<p<1}{\text{Max}} \{\hat{u}_{21}\} = \underset{0<p<1}{\text{Max}}\ \pi - \left(\frac{1-p(1-p)}{p}\right)C(p) - c_i\frac{(1-p)^2}{p} + T(1-p)$$

$$\text{Subject to :} \quad T \leq C(p) + c_i/(1-p)$$

which yields a solution of the following equation in terms of the marginal cost of yield.

$$C'(p) = \frac{(c_i + C(p))\left(\frac{(1-p)(1+p)}{p}\right) - Tp}{1-p(1-p)}$$

Let (p_1^*, p_2^*) be the optimal solutions under both solutions, then, obviously, the supplier will adopt the solution that leads to the largest expected payoff of the game.

By changing the assumptions regarding the relative power each of the parties has, we will obtain, of course, other solutions. These are discussed below.

Each of the solutions considered here can be altered by an appropriate selection of contract parameters which can lead to a pre-posterior game analysis evaluated in terms of the yield and the penalty. If the supplier and the buyer do not cooperate, it is possible to create an incentive for the supplier to supply compliant parts by the selection of contract parameters. A sensitivity analysis of some of these solutions to highlight their effects on strategic controls follows. For convenience, consider only the case with interior solutions and study the effects of the penalty T on the propensity to sample-control. Obviously, the larger T, the less the buyer will control since

$$\frac{\partial y}{\partial T} < 0 \text{ Further, } \quad \frac{\partial x}{\partial T} = \frac{c_b}{(1-p)(\phi+T)^2} > 0$$

Similarly, relationships can be found by treating other parameters. The implications are that increasing penalties T provide an incentive for the supplier to control while for the buyer to control less. When either the buyer or the supplier is a leader and the other a follower, a Stackleberg game is defined (see Chap. 12). For example, say that the supplier is a leader and the buyer is a follower. Then, for a given (x, p), the buyer's problem is:

$$\operatorname{Max}_{y} \; V(y, x; p) = [\phi(1-x)(1-p) + T(1-x)(1-p) - c_b]y$$

And therefore the buyer control policy is either to fully control or none at all. Of course:

$$y = \begin{cases} 1 & \text{if} \quad \phi(1-x)(1-p) + T(1-x)(1-p) \geq c_b \\ 0 & \text{if} \quad \phi(1-x)(1-p) + T(1-x)(1-p) < c_b \end{cases}$$

The supplier's problem consists then in selecting a strategy (x,p) based on the buyer's response $y(x, p)$ given above. Namely,

$$\operatorname{Max}_{x,p} \; U(y; x; p) = \pi - C(p) - c_i x - \left[(C(p) + c_i)\frac{(1-p)^2}{p}\right](x + y(1-x)) - T(1-x)(1-p)y$$

Subject to $y(x, p) = 1, 0$ as sated above. In this case, we note that the control decision is always an all or nothing sampling policy. For example, say that y = 1, then the supplier turns to full sampling if $T(1-p) \geq c_i$, otherwise the supplier will not sample at all. However, if the buyer does not sample, then we note that the supplier does not sample either, since

$$\text{Max}_{x,p} \quad U(y;x;p) = \pi - C(p) - x\left(c_i + \left[(C(p) + c_i)\frac{(1-p)^2}{p}\right]\right)$$

has always a solution $x = 0$. Of course, the supplier yield will then be minimal. In this sense, when one of the parties has power over the other the yield will be lower (as it is the case in Stackleberg games but which does not hold true in Nash conflict games).

The problem we considered, although framed in a supplier–buyer context can be applied as well (with some modifications) to a Bank and a Regulator. For example, a bank compliance may be defined a probability p (the supplier yield) which can be increased at a cost $(C(p))$. In addition, the banks may institute statistical controls (for example, control of traders to prevent rogue trading) while a regulator would merely control banks statistically to prevent non-compliance and penalize banks when they are non-compliant. Regulation of banks is however more challenging as it involves both risk externalities and systemic risks that both banks and regulators are concerned by. Some of these issues are considered next.

11.6 Financial Regulation and Controls

11.6.1 Financial Regulation in a Post Crisis World

The need for the regulation of financial institutions has been reaffirmed following the 2007–2009 financial crisis and its contagious aftermath. The growth of a global and strategic (gaming) economic environment combined with global financial institutions has challenged regulation—both in theory and practice. Regulation seeks to protect "the social contract" between financial institutions, consumers of financial institutions and the public at large. Regulation and the control of compliance are challenged however on two fronts. First, increased globalization and too big to fail (TBTF) and global financial institutions have rendered financial regulation far more complex and costly. The growth of complexity, financial non-transparency combined with increased costs of regulation and shrinking Sovereign States funding, have challenged regulation further.

An extensive literature on what regulation does and its consequences has focused essentially on the effects of regulation on risk taking by regulated firms. Following the Dodd–Frank legislation, greater concern was given to the systemic risks that

TBTF banks and the financial systems may create (through contagion, a lack of global controls, etc.). These concerns are an add-on to the an extensive economic literature on the theory of the regulated firm that has provided an economic framework to study the policies of regulated firms, the pricing of their services and their propensity to assume risks. The current collision of TBTF and A-national banking systems and regulation provide important and strategic challenges, both in theory and in practice to regulation. On the one hand, an unregulated banking system, left to its own, may surely lead to another crisis while, over-regulating the banking systems may have appreciable effects on the supply of funds to the real economy. In this sense, both excessive regulation and a lack of regulation have seeds for future financial crises.

Some regulation risks have been pointed out for example in The Economist (February 2012). In particular they point out to the limits of regulation and the costs of complexity and of over-regulation. Examples abound, including among others: The migration of economic agents to areas where regulation is less pronounced (at a cost of a loss of jobs, downsizing of firms, outsourcing); The rise of stealth and virtual financial entities with a global economic activity; Counter-cyclicality leading to the accelerated demise of corporate entities in a down economic cycle; The complexity effects of regulation and their costs to regulated firms; The effects of a self-financing regulation (by penalties imposed, and leading to control and "out of court settlements"), etc. This latter element may further amplify if in the future micro-regulation may be the focus of regulators (namely the regulation of individual firms and economic entities) at the expense of their macro-economic and systemic effects. In other words, while regulation is needed to protect investors and the sustainability of financial markets, its tools and practice ought to recognize both the increased complexity and intricacies of a changing financial economic environment and their systemic macro-economic implications.

The purpose of the examples we shall treat here is to evoke a greater awareness of the complexity of these problems and in particular provides a risk analysis approach to regulation and control games. Controls are today both far more extensive and costly, expanding from a mere few principles regulated have to comply with to extremely complex and numerous regulations. The Dodd–Frank regulation initiated in 2011 has for example expanded into a complex regulation book consisting of 5,000 pages. This regulation is challenging to both banks and regulators alike. For example, given 100 regulations (there are in fact far more, rather in the thousands) there are 100! alternative controls a regulator or a regulated may pursue. If a regulator were to select only 10 of the 100 regulations, the number of alternative controls is then $\binom{100}{10}$ and therefore practically unwieldy. To circumvent the size, the complexity of controls and the information and power asymmetries of both the banking system and regulators, we shall consider a strategic and statistical approach. Such an approach is demonstrated using relatively simple examples.

11.6.2 Statistical Controls and Regulation

Say that there are N regulations and say that for D_1 (small) non-compliances, a bank may be considered compliant. While for $D_2 > D_1$ (D_2 large), the bank is deemed non-compliant. Given the large number of regulation articles, the bank selects to control a random subset of n_i regulations on the basis of which a decision is reached regarding the statistical compliance of the bank. Say that a regulator chooses n regulations to control and let r regulations found to be non-compliant. Define the control parameter q as follows:

If $r \geq q$ the bank is deemed non-compliant
If $r < q$ the bank is deemed compliant.

The random number r obtained depends however on both the assumptions made regarding the selection of the regulations to control, the total numner of regulations the control sample is selected from and of course the number of non-compliant regulations $D = D_1, D_2$. The probability distribution of obtaining r non-compliant articles given $D = D_1, D_2$ has then a hyper-geometric probability distribution (see Chap. 3):

$$f_{HG}(r|n,N,D) = \binom{D}{r}\binom{N-D}{n-r}\Big/\binom{N}{n}, \quad D = D_1, D_2$$

When N is large, the non-compliance proportion $p_k = D_k/N,\ k = 1,2$ can be used and the hypergeometric distribution approximated by the binomial distribution:

$$f_B(r_k|n,p_k) = \binom{n}{r_k} p_k^{r_k}(1-p_k)^{n-r_k}, \quad r_k = 0,1,2,\ldots,n$$

The cumulative distribution is thus:

$$F_B(q|n,p_k) = \sum_{r_k=0}^{q} \binom{n}{r_k} p_k^{r_k}(1-p_k)^{n-r_k} \quad \text{or}$$

$$F_B(q) = n\binom{n-1}{q-1}\int_0^{p_k} t^{q-1}(1-t)^{n-q}dt$$

And therefore, decision errors (risks) are:

If $r \leq q$ the bank is rightly deemed compliant with probability $F_B(q|n,p_1)$
If $r > q$ the bank is wrongly deemed compliant with probability $1 - F_B(q|n,p_1)$

If $r \leq q$ the bank is wrongly deemed compliant with probability $F_B(q|n,p_2)$
If $r > q$ the bank is rightly deemed compliant with probability $1 - F_B(q|n,p_2)$

The risks of such decisions define the type I and II risks the bank assumes:

$$\alpha_B(n,p_1) = 1 - F_B(q|n,p_1) \text{ and } \beta_B(n,p_2) = F_B(q|n,p_2)$$

Similar considerations are reached by the regulator. Thus, when each selects (due to partial information) his control sample, prior estimates of non-compliance and reaches decisions based on its criteria, we have a set of 4 possible errors from the regulator and from the bank viewpoints. These are defined below where R denotes the regulator and B the bank:

$$\alpha_B\left(n_B,p_1^B\right) = 1 - F_B(q_1^B|n_B,p_1^B) \text{ and } \beta_B\left(n_B,p_2^B\right) = F_B(q_2^B|n_B,p_2^B)$$
$$\alpha_R\left(n_R,p_1^R\right) = 1 - F_R(q_1^R|n_R,p_1^R) \text{ and } \beta_R\left(n_R,p_2^R\right) = F_R(q_2^R|n_R,p_2^R)$$

Type I and II errors depend of course on what actions both the bank and the regulator take. For example, if the bank adopts a strategy (n_B, q^B) while the regulator adopts a strategy (n_R, q^R) then the type I and II risk assumed by the public at large is given by both the bank and by the regulator who erred in claiming that the bank is compliant when it is not, or:

$$\beta\left(n_B,n_R;p_2^R,p_2^B\right) = \beta_R\left(n_R,p_2^R\right)\beta_B\left(n_B,p_2^B\right) = F_B\left(q_2^B|n_B,p_2^B\right)F_R\left(q_2^R|n_R,p_2^R\right)$$

This probability does not mean that an external and harmful event occurs but that its likelihood of occurring is greater. When its consequences are not assumed by the bank but by the public at large, it is a risk externality. The risk of non-compliances becomes systemic when two situations arise: First when the bank is TBTF (Too Big To Fail) and a catastrophic bank failure leads to an important and contagious market failure or when a bank risk externality, is spreading due to its connectivity and dependence with other banks. In this case, indexing a bank by i, $i = 1,2,\ldots,\mathbb{N}$, with type II individual risks for each individual bank $\beta\left(n_{B(i)},n_{R(i)};p_2^{R(i)},p_2^{B(i)}\right)$ we have an external type II risk arising from banks' contagion defined by: $\beta^{(i)}\left(n_{B(i)},n_{R(i)};p_2^{R(i)},p_2^{B(i)}\right)$, defined by a system of linear equations where μ_{ji} denotes the increasing effects of con-compliance of bank j on bank i, or:

$$\beta^{(i)}\left(n_{B(i)},n_{R(i)};p_2^{R(i)},p_2^{B(i)}\right) = \beta\left(n_{B(i)},n_{R(i)};p_2^{R(i)},p_2^{B(i)}\right) + \sum_{j\neq i}^{M}\mu_{ji}\beta^{(j)}\left(n_{B(j)},n_{R(j)};p_2^{R(j)},p_2^{B(j)}\right)$$

For example, for two banks, we have the following solution:

$$\beta^{(1)}\left(n_{B(1)}, n_{R(1)}; p_2^{R(1)}, p_2^{B(1)}\right) = \frac{\beta\left(n_{B(1)}, n_{R(1)}; p_2^{R(1)}, p_2^{B(1)}\right) + \mu_{21}\beta\left(n_{B(2)}, n_{R(2)}; p_2^{R(2)}, p_2^{B(2)}\right)}{1 - (\mu_{12} \cdot \mu_{21})}$$

$$\beta^{(2)}\left(n_{B(2)}, n_{R(2)}; p_2^{R(2)}, p_2^{B(2)}\right) = \frac{\beta\left(n_{B(2)}, n_{R(2)}; p_2^{R(2)}, p_2^{B(2)}\right) + \mu_{12}\beta\left(n_{B(1)}, n_{R(1)}; p_2^{R(1)}, p_2^{B(1)}\right)}{1 - (\mu_{12} \cdot \mu_{21})}$$

However, since a type II risk occurs only if both the regulator and banks fail to detect non-compliance, we have as indicated above.

$$\beta\left(n_{B(1)}, n_{R(1)}; p_2^{R(1)}, p_2^{B(1)}\right) = F_B\left(q_2^{B(1)} \middle| n_{B(1)}, p_2^{B(1)}\right) F_{R(1)}\left(q_2^{R(1)} \middle| n_{R(1)}, p_2^{R(1)}\right)$$

$$\beta\left(n_{B(2)}, n_{R(2)}; p_2^{R(2)}, p_2^{B(2)}\right) = F_B\left(q_2^{B(2)} \middle| n_{B(2)}, p_2^{B(2)}\right) F_{R(2)}\left(q_2^{R(2)} \middle| n_{R(2)}, p_2^{R(2)}\right)$$

which yields to a growth in bank risk externality due to banks' connectivity and dependence:

$$\beta^{(1)}\left(n_{B(1)}, n_{R(1)}; p_2^{R(1)}, p_2^{B(1)}\right) = \frac{\left\{\begin{array}{l} F_{B(1)}\left(q_2^{B(1)} \middle| n_{B(1)}, p_2^{B(1)}\right) F_{R(1)}\left(q_2^{R(1)} \middle| n_{R(1)}, p_2^{R(1)}\right) + \\ \mu_{21} F_{B(2)}\left(q_2^{B(2)} \middle| n_{B(2)}, p_2^{B(2)}\right) F_{R(2)}\left(q_2^{R(2)} \middle| n_{R(2)}, p_2^{R(2)}\right) \end{array}\right\}}{1 - \mu_{12}\mu_{21}}$$

$$\beta^{(2)}\left(n_{B(2)}, n_{R(2)}; p_2^{R(2)}, p_2^{B(2)}\right) = \frac{\left\{\begin{array}{l} F_{B(2)}\left(q_2^{B(2)} \middle| n_{B(2)}, p_2^{B(2)}\right) F_{R(2)}\left(q_2^{R(2)} \middle| n_{R(2)}, p_2^{R(2)}\right) + \\ \mu_{12} F_{B(2)}\left(q_2^{B(1)} \middle| n_{B(1)}, p_2^{B(1)}\right) F_{R(1)}\left(q_2^{R(1)} \middle| n_{R(1)}, p_2^{R(1)}\right) \end{array}\right\}}{1 - \mu_{12}\mu_{21}}$$

Their sum, define the increase of risk such a connectivity due to one or the other while their product indicates the probability that both lead to a systemic type II risk $\beta^{(\mathbf{1,2})}$, or:

$$\beta^{(1)}(..) + \beta^{(2)}(..) = \frac{\left\{\begin{array}{l} (1 + \mu_{12}) F_{B(1)}\left(q_2^{B(1)} \middle| n_{B(1)}, p_2^{B(1)}\right) F_{R(1)}\left(q_2^{R(1)} \middle| n_{R(1)}, p_2^{R(1)}\right) \\ (1 + \mu_{21}) F_{B(2)}\left(q_2^{B(2)} \middle| n_{B(2)}, p_2^{B(2)}\right) F_{R(2)}\left(q_2^{R(2)} \middle| n_{R(2)}, p_2^{R(2)}\right) \end{array}\right\}}{1 - \mu_{12}\mu_{21}}$$

and

$$\beta^{(\mathbf{1,2})} = \beta^{(1)}\left(n_{B(1)}, n_{R(1)}; p_2^{R(1)}, p_2^{B(1)}\right) \beta^{(2)}\left(n_{B(2)}, n_{R(2)}; p_2^{R(2)}, p_2^{B(2)}\right)$$

In other words, contagion, which is both spreading to other banks and fed back to the original bank increases dramatically the type II risks the public at large is confronted with. For example, say that:

$$F_{B(1)}\left(q_2^{B(1)}\middle|n_{B(1)},p_2^{B(1)}\right) = 0.15, F_{R(1)}\left(q_2^{R(1)}\middle|n_{R(1)},p_2^{R(1)}\right) = 0.20 \text{ or} \beta_{\mathrm{B}(1)}\ \beta_{\mathrm{R}(1)} = 0.30$$

$$F_{B(1)}\left(q_2^{B(1)}\middle|n_{B(1)},p_2^{B(1)}\right) = 0.10, F_{R(1)}\left(q_2^{R(1)}\middle|n_{R(1)},p_2^{R(1)}\right) = 0.20 \text{ or} \beta_{\mathrm{B}(1)}\ \beta_{\mathrm{R}(1)} = 0.20$$

With their sum with no contagion, equal to 0.05 while with contagion, we have:

$$\beta^{(1)}(..) + \beta^{(2)}(..) = \frac{(1+\mu_{12})0.15(0.20 + (1+\mu_{21})0.10(0.20)}{1-\mu_{12}\mu_{21}} = \frac{0.05 + \mu_{12}0.03 + \mu_{21}0.02}{1-\mu_{12}\mu_{21}} > 0.05$$

Thus, if the two banks are interconnected, with $\mu_{12} = 0.25,\ \mu_{21} = 0.3$, we have

$$\beta^{(1)}(..) + \beta^{(2)}(..) = \frac{0.05 + (0.25)0.03 + (0.3)0.02}{0.925} = \frac{0.05 + 0.075 + 0.006}{0.925} = 0.1416$$

which is an appreciable growth in the external risk. Its systemic risk externality is then $\beta^{(1)}(.)\cdot p^{(2)}(.)$ For these reasons, an inter-dependent banking system increases systemic risks. This example highlights a number of essential factors to reckon with:

(1) Bank inter-connectedeness and dependence induces external risks
(2) Internal bank controls is important and has to be regulated to induce banks to keep control of their external risks
(3) The control of compliance to regulation is essential both as an incentive for banks to control their own activities and their compliance to regulation and protect the public from risk externalities.

Controls are however costly, both to the banks and to regulators and therefore, the selection of efficient statistical control procedures and the inducements of penalties are important ingredients required to create an environment where risk externalities are implicitly mitigated.

Below we shall consider an additional example that highlights these elements. In such cases, both the bank and the regulator apply various private control strategies, where control is a randomized strategy that each the bank and the regulator apply. Randomization and the statistical control of compliance is thus applied to circumvent the extremely large costs of controls that full and continuous controls require.

Example: The Statistical Control of Compliance Control of compliance by either a bank (based on the internal information it has) or a regulator (based on past and statistical information, whistle blowers, or media's attention to banks' misshaps and practices, such as manipulating the LIBOR rate to their advantage) are based on both private and public information. These elements set an environment where each, the bank and the regulator, design and pursue their control strategy on the basis of their own information and of course, both fully aware of the consequences on compliance and non-compliance. To highlight their dependence, gaming and the effects of applying various control strategies, we shall consider a number of such approaches. The purpose of this example is to emphasize the substitution effects between a regulator and a bank in their control compliance and the risks they are willing to assume. Explicitly, the following cases are discussed:

(1) The regulator (or the bank), selects a-priori a number of regulations to control, say $N_{R,i}$, where $i, = 1, 2, \ldots$ denote an alternative set of regulations. A binomial control test, would require their full control to assess the number of non-compliances, theoretically given by a random variable $k = 1, 2, \ldots, N_{R,i}$. A control test is then set, say q with the following controls: If $k \geq q$, the bank is deemed (in a statistical sense) to be non-compliant. Otherwise, it is deemed "compliant" (although it may be required to attend to a detected non-compliance with potentially a smaller penalty). When $k \geq q$, the bank is assessed as non-compliant and therefore is penalized accordingly. In such a case, type I and type II risks can be calculated as indicated earlier.

Prior to such controls however, the bank may turn to its own internal controls based on its information and proceed to statistical controls using the parameters set defining both the costs of internal controls and the statistical stringency of the bank compliance to regulation. These situations lead to a random payoff game with controls and penalties that depend on the regulator and bank controls and their statistiscal stringency. Penalities are used to provide incentive to banks to control their compliance while at the same time compensate regulators for the costs they incur in controlling banks. These are however partial information games with the regulator and the bank having both private and common information (i.e. statistical industry and historical data based on historical performance—by type, size, location, etc., of all banks).

(2) Alternatively, to reduce the number of controls, the regulator and the bank may turn to a CSP-1 continuous control approach, selecting a sample of $N_{R,i}$ regulations and stopping controls as soon as one regulation is violated, or controlling the full sample $N_{R,i}$ if no regulation in the sample is violated. In this latter case, the bank is deemed compliant by the regulator.

These controls structured as single stage or multi-stage, under a broad set of informational and probability assumptions, define a plethora of control alternatives and approaches, all of which have cost and statistical implications. The problems we consider below are simple demonstrations.

Case: The Persistent Regulator Say that a "persistent" regulator controls a bank whose control strategy is j until one non-compliance is detected. The number of controlled regulations are then a random variable $\tilde{n}_{R,j}(1)$. If the bank non-compliance probability (when the bank control strategy j is applied) is $p_{2,B}^{j}$ and its estimate by the regulator is $p_{2,R}^{j}$, (either based on private information or based on industry data), the probability distribution of the number of controls $\tilde{n}_{R,ij}(1)$ by the regulator is $p_{2,R}^{j}(1-p_{2,R}^{j})^{\tilde{n}_{R,j}(1)-1}$. The expected number of control is thus $p_{2,R}^{j}/(1-p_{2,R}^{j})$ while the penalty is $P_B(1)$, applied when the bank is at last detected to be non-compliant, then the regulator control cost is a random variable, $C_R\left(\tilde{n}_{R,j}(1)\right)$. To cover its expected cost, the penalty is: $EC_R\left(\tilde{n}_{R,j}(1)\right) = P_B(1)$. For example, if $C_R\left(\tilde{n}_{R,j}(1)\right) = c_R\left(\tilde{n}_{R,j}(1)\right)^{\gamma}$, $\gamma < 1$, then:

$$E\left(\tilde{n}_{R,j}(1)\right)^{\gamma} = p_{2,R}^{j}\left(1 + 2^{\gamma}\left(1-p_{2,R}^{j}\right)^{1} + 3^{\gamma}\left(1-p_{2,R}^{j}\right)^{2} + 4^{\gamma}\left(1-p_{2,R}^{j}\right)^{3} + \ldots\right)$$

Of course, if $\gamma = 1$, then $E\left(\tilde{n}_{R,j}(1)\right) = p_{2,R}^{j}/(1-p_{2,R}^{j})$ and therefore $P_B(1) = c_R p_{2,R}^{j}/(1-p_{2,R}^{j})$. Type II risks are irrelevant as a bank will be controlled till it is found "non-compliant". Of course, if a bank is compliant and therefore with an extremely small non-compliance probability (i.e. $p_{2,B}^{j}$ very small), the persistent regulator will have to impose either extremely high penalties to cover his control costs, or assume increasingly greater regulation costs when banks become increasingly compliant. This is the case for some financial institutions (some hedge funds) that were investigated for years by regulators with no clear discovery of non-compliance in sight but a persistent regulation seeking to prove the fund non-performance due to insider trading. When a persistent regulator has a random prior estimate of a bank non-compliance, these results are altered. Explicitly, let the regulator estimate of the probability of non compliance be $\tilde{p}_R\left(= p_{2,R}^{j}\right)$, with a probability distribution $f_R(\tilde{p}_R)$. The number of controls until one non-compliance is detected is therefore a mixture of the geometric distribution:

$$\tilde{n}_{R,j}(1) \sim \int \tilde{p}_R(1-\tilde{p}_R)^{\tilde{n}_{R,j}(1)-1} f_R(\tilde{p}_R)\,d\tilde{p}_R$$

It is convenient to assume a Beta probability distribution for $f_R(\tilde{p}_R) = \frac{\tilde{p}_R{}^{a-1}(1-\tilde{p}_R)^{b-1}}{B(a,b)}$, $B(a,b) = \frac{\Gamma(a)\Gamma(b)}{\Gamma(a+b)}$ whose mean is (see Chap. 3) $p_{2,B}^{j} = a/(a+b)$ denoting the true mean of non-compliance of the bank and its variance, $ab/\left[a(a+b)^2(a+b+1)\right]$. In this case,

$$\tilde{n}_{R,j}(1) \sim \frac{1}{B(a,b)}\int_0^1 \tilde{p}_R{}^{a}(1-\tilde{p}_R)^{\tilde{n}_{R,j}(1)+b-2}\,d\tilde{p}_R = \frac{\Gamma(a+b)}{\Gamma(a)\Gamma(b)}\int_0^1 \tilde{p}_R{}^{a}(1-\tilde{p}_R)^{\tilde{n}_{R,j}(1)+b-2}\,d\tilde{p}_R$$

$$\frac{\Gamma(a+b)}{\Gamma(a)\Gamma(b)}\int_0^1 \tilde{p}_R{}^a(1-\tilde{p}_R)^{\tilde{n}_{R,j}(1)+b-2}d\tilde{p}_R = \frac{\Gamma(a+b)}{\Gamma(a)\Gamma(b)}\frac{\Gamma(a+1)\Gamma(b+\tilde{n}_{R,j}(1)-2)}{\Gamma(a+b+\tilde{n}_{R,j}(1))}$$

And therefore,

$$\tilde{n}_{R,j}(1) \sim \frac{\Gamma(a+b)}{\Gamma(b)}\frac{(a+1)\Gamma(b+\tilde{n}_{R,j}(1)-2)}{\Gamma(a+b+\tilde{n}_{R,j}(1))}$$

which can be used to calculate the expected number of controls the regulator will have until one non-compliance is detected. In this case,

$$E\left(\tilde{n}_{R,j}(1)\right) \sim \frac{(a+1)\Gamma(a+b)}{\Gamma(b)}\sum_{z=1}^{\infty} z\frac{\Gamma(b+z-2)}{\Gamma(a+b+z)}$$

Of course, if the regulator has no prior knowledge of the bank non-compliance and assumes that it is uniformly distributed in $[0, 1]$, then:

$$\tilde{n}_{R,j}(1) \sim \int_0^1 \tilde{p}_R(1-\tilde{p}_R)^{\tilde{n}_{R,j}(1)-1}d\tilde{p}_R = \frac{B(2,\tilde{n}_{R,j}(1)-1)}{B(2,\tilde{n}_{R,j}(1))}\int_0^1 \tilde{p}_R{}^{a-1}(1-\tilde{p}_R)^{\tilde{n}_{R,j}(1)-1}d\tilde{p}_R$$
$$= B(2,\tilde{n}_{R,j}(1)-1)$$

And therefore,

$$f_{\tilde{n}_{R,j}}\left(\tilde{n}_{R,j}(1)\right) \sim \frac{\Gamma(2)\Gamma(\tilde{n}_{R,j}(1))}{\Gamma(2+\tilde{n}_{R,j}(1))} = \frac{2}{\left(2+\tilde{n}_{R,j}(1)\right)\left(1+\tilde{n}_{R,j}(1)\right)}, \quad \tilde{n}_{R,j}(1) = 1, 2, \cdots$$

For example, calculating the probability of detecting a non-compliant regulation after 1, 2, 3, 4, etc. controls, we find that the probability of detection decreases with the number of controls, equalling 1/6 for the first controls, 1/10 and 1/15 for the third and fourth controls etc.

$$f_{\tilde{n}_{R,j}}\left(\tilde{n}_{R,j}(1)\right) \sim \left(1:\frac{1}{6}\right), \left(2:\frac{1}{6}\right), \left(3:\frac{1}{10}\right), \left(4:\frac{1}{15}\right), \ldots.$$

This implies that a persisten regulator would have less and less chance to detect a non-compliance. Such a result also observed for a persistent and "learning" regulator with a prior beta probability distribution, with controls that have not detected a non-compliance, updating the probability estimate of non-compliance (which will obviously result in a lower probability estimates). For example, if the prior mean non-compliance probability is: a/(a + b), then after one control such an estimate will be a/(a + b + 1) and after m controls with no result, the updated mean of non-compliance is a/(a + b + m). While these results, may be practically of limited usefulness, they point out to the challenges regulators are faced with.

Case: The Persistent Regulator and CSP-1 Limited Controls A persistent regulator may be confronted with control costs that are too great to sustain. This is certainly the case, given the complexity and the number of regulations and steps taken by banks to both evade and control their activities to be compliant. For this reason, and in order to mitigate the regulator (as well as the bank) costs, we shall consider a CSP-1 control strategy. It consists as indicated earlier, in controlling at most $N_{R,i}$ regulations (a regulator's alternative controls, $N_{R,i},\ i = 1, 2, \ldots, m$) and stop as soon as 1 non-compliance is detected ($\tilde{r}_i \le N_{R,i}$). The number of controls the regulator may have then, is:

$$\tilde{n}_{R,i}(1) = \begin{cases} N_{R,i} & \text{w.p. } \left(1 - p^j_{2,R}\right)^{N_{R,i}} \\ \tilde{r}_i & \text{w.p. } p^j_{2,R}\left(1 - p^j_{2,R}\right)^{\tilde{r}_i - 1},\ \tilde{r}_i = 1, 2, \ldots N_{R,i} \end{cases}$$

The expected number of controls is then equal to the probability of not detecting a non-compliance in $N_{R,i}$ controls or $\tilde{r}_i \le N_{R,i}$ if non-compliance is detected. As a result, the expected number of controls is:

$$E\left(\tilde{n}_{R,i} \middle| N_{R,i}\right) = N_{R,i}\left(1 - p^j_{2,R}\right)^{N_{R,i}} + p^j_{2,R} \sum_{\tilde{r}_i = 1}^{N_{R,i} - 1} \tilde{r}_i p^j_{2,R}\left(1 - p^j_{2,R}\right)^{\tilde{r}_i - 1}$$

Or

$$E\left(\tilde{n}_{R,i} \middle| N_{R,i}\right) = N_{R,i} p^j_{2,R}\left(1 - p^j_{2,R}\right)^{N_{R,i} - 1} + \frac{\left(1 - p^j_{2,R}\right)\left(1 - \left(1 - p^j_{2,R}\right)^{N_{R,i}}\right)}{p^j_{2,R}}$$

And therefore, the probability of detecting a non-compliance in a control cycle of $\min(\tilde{r}, N_{R,i})$ by the regulator, is (using the renewal theorem):

$$p_{D,R_i}\left(N_{R,i}\right) = \frac{1}{E\left(\tilde{n}_{R,i} \middle| N_{R,i}\right)}$$
$$= \frac{1}{N_{R,i} p^j_{2,R}\left(1 - p^j_{2,R}\right)^{N_{R,i} - 1} + \left(\left(1 - p^j_{2,R}\right)/p^j_{2,R}\right)\left(1 - \left(1 - p^j_{2,R}\right)^{N_{R,i}}\right)}$$

The probability of not detecting any non-compliance defines then the type II risk associated to the regulator strategy, or $\beta_{R,i} = 1 - p_{D,R_i}\left(N_{R,i}\right)$. The regulator cost of control are then twofold. On the one hand control costs $C_R\left(\tilde{n}_{R,i} \middle| N_{R,i}, p^j_{2,R_i}\right)$ and revenue from the bank's penalty if non-compliance is detected and given by $p_{D,R_i}(N_{R,i})P_B(1)$. On the other, the probability of a risk externality cost is incurred due to a default (type II risks) to detect non-compliance by both the bank and the regulator. A Sovereign State may then impose a risk constraint, to assure that the probability of not detecting non-compliance is bounded. The intent of such a constraint would be to balance on the one hand the risk to the banking system and on the other, not to overly hinder the consequences of over-regulation stiffling a bank ability to function. For a bank's internal controls of compliance, we have the following (and equivalent) results:

$$\tilde{n}_{B,j}(1) = \begin{cases} N_{B,j}\text{w.p. } \left(1-p^j_{2,B}\right)^{N_{B,j}} \\ \tilde{r}_j\text{w.p. } p^j_{2,B}\left(1-p^j_{2,B}\right)^{\tilde{r}_j-1}, \ \tilde{r}_j = 1,2,\ldots N_{B,j} \end{cases} ;\tilde{r}_j \le N_{B,j}$$

With an expected number of controls given by $E\left(\tilde{n}_{B,j}\middle|N_{B,j}\right) = N_{B,j}\left(1-p^j_{2,B}\right)^{N_{B,j}}$ $+\sum_{\tilde{r}_j=1}^{N_{B,j}} \tilde{r}_j p^j_{2,B}\left(1-p^j_{2,B}\right)^{\tilde{r}_j-1}$, or:

$$E\left(\tilde{n}_{B,j}\middle|N_{B,j}\right) = N_{B,j}p^j_{2,B}\left(1-p^j_{2,B}\right)^{N_{B,j}-1} + \frac{\left(1-p^j_{2,B}\right)\left(1-\left(1-p^j_{2,B}\right)^{N_{B,j}-1}\right)}{p^j_{2,B}}$$

And therefore, the probability of detecting a non-compliance in a bank control cycle where $N_{B,j}$ are the greatest number of controls if the bank does not detect non-compliance to regulation. The probability of detection is then:

$$p_{D,B_j}\left(N_{B,j}\right) = \frac{1}{E\left(\tilde{n}_{B,j}\middle|N_{B,j}\right)}$$
$$= \frac{1}{N_{B,j}p^j_{2,B}\left(1-p^j_{2,B}\right)^{N_{B,j}-1} + \left(\left(1-p^j_{2,B}\right)/p^j_{2,B}\right)\left(1-\left(1-p^j_{2,B}\right)^{N_{B,j}}\right)}$$

In this particular case, we shall assume that the bank will attend to its non-compliance. The probability of not detecting any non-compliance by the bank however, defines a type II risk associated to the bank's control policy, denote by: $\beta_{B,j} = 1 - p_{D,B_j}\left(N_{B,j}\right)$. As a result, assuming that both the bank and the regulator fail to detect non-compliance, the type II risk externality risk associated to the control strategies (i,j) is:

$$\beta_{ij} = \beta_{R,i}\beta_{B,j} = \left(1 - p_{D,R_i}\left(N_{R,i}\right)\right)\left(1 - p_{D,B_j}\left(N_{B,j}\right)\right)$$

Control costs of the bank are then twofold. On the one hand control the bank costs $C_B\left(\tilde{n}_{B,j}\middle|N_{B,j},p^j_{2,B}\right)$, and on the other avoid penalty costs in case the bank is detected by the regulator. Of course, if the bank detects that it is non-compliant and attends to it, being then compliant, the regulator may not detect non-compliance (although, a type I error may be made by the regulator, which I have chosen to ignor for simplicity's sake).

The regulation game between the regulator and bank, each adopting strategies (i,j) thus defines a random payoff game whose entries are:

$$\begin{cases} \Omega_R(ij) = C_R\left(\tilde{n}_{R,i}\middle|N_{R,i},\tilde{p}^j_{2,R_i}\right) + \begin{cases} P_B(1) & \text{wp } (1-\beta_{R,i})\beta_{B,j} \\ 0 & \text{wp } \beta_{R,i}\beta_{B,j}+1-\beta_{B,j} \end{cases} \\ \Omega_B(ij) = C_B\left(\tilde{n}_{B,j}\middle|N_{B,j},\tilde{p}^j_{2,B}\right) - \begin{cases} P_B(1) & \text{wp } (1-\beta_{R,i})\beta_{B,j} \\ 0 & \text{wp } \beta_{R,i}\beta_{B,j}+1-\beta_{B,j} \end{cases} + \begin{cases} C_{BC} & \text{wp } 1-\beta_{B,j} \\ 0 & \text{wp } \beta_{B,j} \end{cases} \end{cases}$$

Note that we have considered for simplicity only the random control costs and the penalties paid by the bank in case non-compliance is detected. For the regulator, an estimate of the banks non-compliance is in fact assumed random, depending on the probability of the bank not detecting non-compliance (its type II risk). Further, we assume that even if non-compliance is detected, the bank will never be able to eliminate all non-compliances but would reduce it to a probability $p^j_{1,B}$ deemed compliant. To do so, a cost C_{BC} incurred with probability $1-\beta_{B,j}$. In this case, the random probability of non-compliance estimated by an informed regulator is:

$$\tilde{p}^j_{2,R} = \tilde{p}^j_{2,B} = \begin{cases} p^j_{1,B} & wp\ 1-\beta_{B,j} \\ p^j_{2,B} & wp\ \beta_{B,j} \end{cases} \quad \text{or} \quad \hat{p}^j_{2,R} = p^j_{1,B}\left(1-\beta_{B,j}\right) + p^j_{2,B}\beta_{B,j} \text{ while}$$

$$\hat{p}^j_{2,B} = p^j_{2,B}\beta_{B,j}$$

While the bank compliance is merely equal to the probability of non-compliance times the probability of not detecting compliance. Note that in this case, type I errors (i.e. detecting a non-compliance when the bank is compliant, is assumed very small and therefore negligible). The regulation/bank game can thus be reduced to an expectation game subject to a type II risk externality constraint. This is formulated generally as follows. Let (m_R, m_B) be the number of alternatives each the regulator and the bank have. Let (x_i, y_j) be the probabilities that the regulator select control strategy i while the bank selects control strategy j. The matrix of expected cost for each is thus:

$$\Omega_R = \sum_{i=1}^{m_R}\sum_{j=1}^{m_B} x_i y_j\left(EC_R\left(\tilde{n}_{R,i}|N_{R,i},\tilde{p}^j_{2,R_i}\right)\right) - P_B(1)(1-\beta_{R,i})\beta_{B,j}\Big)$$

$$\Omega_B = \sum_{i=1}^{m_R}\sum_{j=1}^{m_B} x_i y_j\left(EC_B\left(\tilde{n}_{B,j}|N_{B,j},\tilde{p}^j_{2,B}\right)\right) + P_B(1)(1-\beta_{R,i})\beta_{B,j} + C_{BC}(1-\beta_{B,j})\Big)$$

$$\text{Subject to}: \sum_{i=1}^{m_R}\sum_{j=1}^{m_B} x_i y_j \beta_{ij} = \sum_{i=1}^{m_R}\sum_{j=1}^{m_B} x_i y_j \left(1 - p_{D,R_i}(N_{R,i})\right)\left(1 - p_{D,B_j}(N_{B,j})\right) \le \bar{\beta}$$

where $\bar{\beta}$ is a Sovereign state constraint imposed on the regulator to ascertain that on the average a type II risk externality is smaller than a given level and where:

$$\hat{\beta}_{ij} = \hat{\beta}_{R,i}\hat{\beta}_{B,j} = \left(\frac{1 - N_{B,j}\hat{p}^j_{2,B}(1-\hat{p}^j_{2,B})^{N_{B,j}-1} - \left((1-\hat{p}^j_{2,B})/\hat{p}^j_{2,B}\right)\left(1-(1-\hat{p}^j_{2,B})^{N_{B,j}}\right)}{N_{B,j}\hat{p}^j_{2,B}(1-\hat{p}^j_{2,B})^{N_{B,j}-1} + \left((1-\hat{p}^j_{2,B})/\hat{p}^j_{2,B}\right)\left(1-(1-\hat{p}^j_{2,B})^{N_{B,j}}\right)}\right)$$
$$* \left(\frac{1 - N_{R,i}\hat{p}^j_{2,R}(1-\hat{p}^j_{2,R})^{N_{R,i}-1} - \left((1-\hat{p}^j_{2,R})/\hat{p}^j_{2,R}\right)\left(1-(1-\hat{p}^j_{2,R})^{N_{R,i}}\right)}{N_{R,i}\hat{p}^j_{2,R}(1-\hat{p}^j_{2,R})^{N_{R,i}-1} + \left((1-\hat{p}^j_{2\hat{p},R})/\hat{p}^j_{2,R}\right)\left(1-(1-\hat{p}^j_{2,R})^{N_{R,i}}\right)}\right)$$

For a non-profit regulator whose costs are to be bourn by the penalties it extracts for non-compliance, we have:

$$\sum_{i=1}^{m_R}\sum_{j=1}^{m_B} x_i y_j\left(EC_R\left(\tilde{n}_{R,i}|N_{R,i},\tilde{p}^j_{2,R_i}\right)\right)\Big) - P_B(1)\sum_{i=1}^{m_R}\sum_{j=1}^{m_B} x_i y_j\left((1-\beta_{R,i})\beta_{B,j}\right) = 0$$

Or

$$P_B(1) = \frac{\sum_{i=1}^{m_R}\sum_{j=1}^{m_B} x_i y_j\left(EC_R\left(\tilde{n}_{R,i}|N_{R,i},\tilde{p}^j_{2,R_i}\right)\right)\Big)}{\sum_{i=1}^{m_R}\sum_{j=1}^{m_B} x_i y_j\left((1-\beta_{R,i})\beta_{B,j}\right)}$$

While controls are applied to maintain the Sovereign risk constraint $\bar{\beta}$. In general the game of regulation is extremely complex. For this reason, we shall consider simple a two strategies example one to control, the other not to control.

Example: A Two Control/No Control Strategies Let the regulator and the bank be: control 1, and no control 2. In this case, (with $\beta_{R,2} = \beta_{B,2} = 1$,) or:

$$\begin{aligned}\Omega_R &= xy\left(EC_R\left(\tilde{n}_{R,1}\middle|N_{R,1},\hat{p}^1_{2,R}\right)\right) - P_B(1)(1-\beta_{R,1})\beta_{B,1}\Big) \\ &\quad + x(1-y)\left(EC_R\left(\tilde{n}_{R,1}\middle|N_{R,1},\hat{p}^1_{2,R}\right) - P_B(1)(1-\beta_{R,1})\right) \\ \Omega_B &= xy\left(EC_B\left(\tilde{n}_{B,1}\middle|N_{B,1},\hat{p}^1_{2,B}\right) + P_B(1)(1-\beta_{R,1})\beta_{B,1} + C_{BC}(1-\beta_{B,1})\right) \\ &\quad + x(1-y)\left(P_B(1)(1-\beta_{R,1})\right) \\ &\quad + (1-x)y\left(EC_B\left(\tilde{n}_{B,1}\middle|N_{B,1},\hat{p}^1_{2,B}\right) + C_{BC}(1-\beta_{B,1})\right) \\ &\text{Subject to} : E\beta_{ij} = xy\beta_{R,1}\beta_{B,1} + x(1-y)\beta_{R,1} + (1-x)y\beta_{B,1} + (1-x)(1-y) \le \bar{\beta}\end{aligned}$$

Assume for simplicity that, $xy\beta_{R,1}\beta_{B,1} + x(1-y)\beta_{R,1} + (1-x)y\beta_{B,1} + (1-x)$ $(1-y) = \bar{\beta}$ which requires that:

$$0 \le x = \frac{1-\bar{\beta}-y(1-\beta_{B,1})}{(1-\beta_{R,1})(1-y(1-\beta_{B,1}))} \le 1$$

As a result, the solution of the game defined above is reduced to a constraint game, with:

$$\begin{aligned}\underset{0\le x\le 1}{\text{Max}}\ \Omega_R &= xy\left(EC_R\left(\tilde{n}_{R,1}\middle|N_{R,1},\hat{p}^1_{2,R}\right) - P_B(1)(1-\beta_{R,1})\beta_{B,1}\right) \\ &\quad + x(1-y)\left(EC_R\left(\tilde{n}_{R,1}\middle|N_{R,1},\hat{p}^1_{2,R}\right) - P_B(1)(1-\beta_{R,1})\right) \\ &\quad \text{Subject to} : x \ge \frac{1-\bar{\beta}-y(1-\beta_{B,1})}{(1-\beta_{R,1})(1-y(1-\beta_{B,1}))} \\ \underset{0\le y\le 1}{\text{Max}}\ \Omega_R\Omega_B &= xy\left(EC_B\left(\tilde{n}_{B,1}\middle|N_{B,1},\hat{p}^1_{2,B}\right) + P_B(1)(1-\beta_{R,1})\beta_{B,1} + C_{BC}(1-\beta_{B,1})\right) \\ &\quad + x(1-y)\left(P_B(1)(1-\beta_{R,1})\right) + (1-x)y \\ &\quad \left(EC_B\left(\tilde{n}_{B,1}\middle|N_{B,1},\hat{p}^1_{2,B}\right) + C_{BC}(1-\beta_{B,1})\right)\end{aligned}$$

And therefore, letting λ be the Sovereign state Lagrange multiplier, we have:

$$\frac{\partial\Omega_R}{\partial x} = 0 \Rightarrow y = \frac{EC_R\left(\tilde{n}_{R,1}\middle|N_{R,1},\hat{p}^1_{2,R}\right) - \lambda}{P_B(1)(1-\beta_{R,1})(1-\beta_{B,1})}, \quad x = \frac{1-\bar{\beta}-y(1-\beta_{B,1})}{(1-\beta_{R,1})(1-y(1-\beta_{B,1}))}$$

While

$$0 = \frac{\partial\Omega_B}{\partial y} \Rightarrow x = \frac{EC_B\left(\tilde{n}_{B,1}\middle|N_{B,1},\hat{p}^1_{2,B}\right) + C_{BC}(1-\beta_{B,1})}{P_B(1)(1-\beta_{R,1})(1-\beta_{B,1})}$$

Note that,

$$x^* = \frac{1 - \bar{\beta} - y(1 - \beta_{B,1})}{(1 - \beta_{R,1})(1 - y(1 - \beta_{B,1}))} = \frac{EC_B\left(\tilde{n}_{B,1} \middle| N_{B,1}, \hat{p}^1_{2,B}\right) + C_{BC}(1 - \beta_{B,1})}{P_B(1)(1 - \beta_{R,1})(1 - \beta_{B,1})}$$

which provides a solution for y which we equate to the first equation in the Lagrange multiplier, providing a solution for y and thereby x. This leads to:

$$y^* = \frac{1 - \frac{1-\bar{\beta}}{x^*(1-\beta_{R,1})}}{(1 - \beta_{B,1})\left(1 - \frac{1}{x^*(1-\beta_{R,1})}\right)} = \frac{EC_R\left(\tilde{n}_{R,1} \middle| N_{R,1}, \hat{p}^1_{2,R}\right) - \lambda}{P_B(1)(1 - \beta_{R,1})(1 - \beta_{B,1})}$$

Thus, $\lambda = EC_R\left(\tilde{n}_{R,1} \middle| N_{R,1}, \hat{p}^1_{2,R}\right) - P_B(1)(1 - \beta_{R,1})\left(1 - \frac{\bar{\beta}}{x^*(1-\beta_{R,1})-1}\right)$ and finally,

$$\lambda = P_B(1)(1 - \beta_{R,1}) \times \left\{ \frac{EC_R\left(\tilde{n}_{R,1} \middle| N_{R,1}, \hat{p}^1_{2,R}\right)}{P_B(1)(1 - \beta_{R,1})} + \frac{\bar{\beta}(P_B(1))}{\frac{EC_B\left(\tilde{n}_{B,1} \middle| N_{B,1}, \hat{p}^1_{2,B}\right)}{(P_B(1))(1-\beta_{B,1})} - \left(1 - \frac{C_{BC}}{P_B(1)}\right)} - 1 \right\}$$

Example: Choosing Banks to Control Regulators have in general a large number of banks to control and chose from. The question is then which bank to control at a particular time and what regulation control to apply from an extremely large number of regulations that may be violated. In a period of apparent stability (and a belief that banks are generally compliant), banks may be selected at random using a bernoulli probability, say q. When one bank is found to be non-compliant (using a well defined criterion on the severity of non-compliance), m banks are selected and all are fully controlled. If all are compliant, then the regulator reverts to the random (bernoulli) selection of a bank to control. However, as soon as 1 is found to be non-compliant, then an additional m banks are selected, re-starting a control cycle. Of course, in this approach, we assume that banks are selected at random, although, selection of banks can be based using prior information (such as media attention, targeting certain financial sectors, etc.) or based on the statistical analysis of banks with various attributes signalling to non-compliance.

For example, say that a bank group G has a non-compliance probability estimate $\hat{P}^G_{2,R}$ and a compliance probability $\hat{P}^G_{1,R}$. Say that a bank is non-compliant and define $\mathbb{Z}$ to be a particular event (of detecting a non compliant bank), when a bernoulli control is applied while $\bar{\mathbb{Z}}$ is the event when the regulator applies a full control of m banks. Over a control cycle, the number of banks controlled is then $m + \tilde{r}$ where $\tilde{r}$ is the number of banks controlled until one is found to be non-compliant. As noted above $\tilde{r}$ has a geometric distribution given by $\hat{P}^G_{2,R}(1-\hat{P}^G_{2,R})^{\tilde{r}-1}$, $\tilde{r}=1,2,....$ and therefore, $E(\tilde{r})=(1-\hat{P}^G_{2,R})/\hat{P}^G_{2,R}$ as a result, the expected number of banks controlled in a cycle is $m(1-\hat{P}^G_{2,R})^m+(1-\hat{P}^G_{2,R})/\hat{P}^G_{2,R}$. Since only one bank is found non-compliant, the probability of detecting a non-compliant banks is:

$$p_D^{G,NC}=\frac{1}{m(1-\hat{P}^G_{2,R})^m+(1-\hat{P}^G_{2,R})/\hat{P}^G_{2,R}}$$

And therefore $1-p_D^{G,NC}$ is the probability of detecting a non-compliant bank when it is (statistically compliant), or

$$p_D^{G,C}=\frac{1}{m(1-\hat{P}^G_{1,R})^m+(1-\hat{P}^G_{1,R})/\hat{P}^G_{1,R}}$$

In this case, the type II errors made by the banks regulator are:

$$\beta_R=1-p_D^{G,NC}\text{ and }\alpha_R=p_D^{G,C}$$

If the regulator cost of controlling a bank is C^G_R while the penalty a bank pays when detected as non compliant is Π^G, the controlled bank cycle cost is:

$$\tilde{C}^G=\begin{cases}C^G_R-\frac{\Pi^G_2}{m(1-\hat{P}^G_{2,R})^m+(1-\hat{P}^G_{2,R})/\hat{P}^G_{2,R}} & with\ \hat{P}^G_{2,R}\\ C^G_R-\frac{\Pi^G_1}{m(1-\hat{P}^G_{1,R})^m+(1-\hat{P}^G_{1,R})/\hat{P}^G_{1,R}} & with\ \hat{P}^G_{1,R}\end{cases};\ \Pi^G_2>\Pi^G_1$$

If the regulatgor prior probability of a bank being non-compliant is ξ, then the expected cost is:

$$\hat{C}^G=C^G_R-\frac{\xi\hat{P}^G_{2,R}\Pi^G_2}{m(1-\hat{P}^G_{2,R})^m+(1-\hat{P}^G_{2,R})/\hat{P}^G_{2,R}}-\frac{(1-\xi)\hat{P}^G_{1,R}\Pi^G_1}{m(1-\hat{P}^G_{1,R})^m+(1-\hat{P}^G_{1,R})/\hat{P}^G_{1,R}}$$

If $\hat{C}^G=0$ and if Π^G_1 is negligible, then $\frac{C^G_R}{\Pi^G_2}=\frac{\xi\hat{P}^G_{2,R}}{m(1-\hat{P}^G_{2,R})^m+(1-\hat{P}^G_{2,R})/\hat{P}^G_{2,R}}$. Regulation falure is further given by its type II risk $1-p_D^{G,NC}$.

The fraction of banks controlled FI^G until one bank is found non-compliant, and its average are:

$$FI^G=\begin{cases}q & z\in\mathbb{Z}\\ 1 & z\in\bar{\mathbb{Z}}\end{cases}\quad\text{and}\quad AFI^G=q\text{Prob}(\mathbb{Z})+1\text{Prob}(\bar{\mathbb{Z}})$$

Based on a renewal theorem, the probabilities can be computed easily (in particular see Dodge 1943 and Duncan 1974, pp. 360–361 as well as Tapiero and Hsu 1988) we have:

$$\text{Prob}(\mathbb{Z}) = \frac{v}{v+u} \quad \text{where} \quad v = \frac{1}{q\hat{P}^G_{2,R}} \quad \text{and} \quad u = \frac{1-\left(1-\hat{P}^G_{2,R}\right)^m}{\hat{P}^G_{2,R}\left(1-\hat{P}^G_{2,R}\right)^m}$$

And thereby, the average fraction inspected and finding one bank non-compliant is:

$$AFI^G = 1-(1-q)\text{Prob}(\mathbb{Z}) = 1-\frac{(1-q)}{1+\frac{q\left(1-\left(1-\hat{P}^G_{2,R}\right)^m\right)}{\left(1-\hat{P}^G_{2,R}\right)^m}}$$

11.6.3 *Private Information, Type I and II Risks and Externality Risks*

For simplicity, consider again a single bank and its regulator. The bank is subject to a multitude of regulation articles, which it can control internally by investments, defining the probability for the bank to be compliant in case of a control. We let q_B be the probability of compliance of the bank (its reliability). In other words, whether the bank is compliant or not, it is defined by the following:

$$\tilde{p}_B = \begin{cases} p_B^C & wp\ q_B \\ p_B^{NC} & wp\ 1-q_B \end{cases}$$

Where p_B^C and p_B^{NC} are used to denote the bank compliance or non-compliance according to its own standards of controls. The bank, thus uses statistical controls internally to assure that its intended investments in compliance is maintained by its bankers, officers and employers. These controls cannot be fail-safe however. Let j = 1,2,. . .n be the alternative control programs it can use and let $\left(\alpha_{B,ij}(p_B^C), \beta_{B,ij}(p_B^{NC})\right)$ be the type I and the type II risks the bank has when it uses a control strategy *j* while the regulator assumes a control strategy *i*. Note that these risk probabilities are a function of the banks probability of compliance resulting from its investments other than statistical controls. Consider next, the regulator with m alternative control programs, i = 1,2,. . ., m. Each of these alternatives pertain to the articles of regulation (or their combination) it chooses to control. To each such alternative there are two risks, type I and II denoted by be the alternative sampling programs used by the supplier. Correspondingly, we denote by $\left(\alpha_{R,ij}(p_R^C), \beta_{B,ij}(p_R^{NC})\right)$ and by q_R the reliability of regulator estimate of the bank compliance, or:

$$\tilde{p}_R = \begin{cases} p_R^C & wp\ q_R \\ p_R^{NC} & wp\ 1 - q_R \end{cases}$$

These type I and II risks are therefore summarized in the risks bi-matrix below.

$$\left[\left(\alpha_{R,ij}(p_R^C), \beta_{R,ij}(p_R^{NC})\right); \ \left(\alpha_{B,ij}(p_B^C), \beta_{B,ij}(p_B^{NC})\right)\right]$$

Each of these probabilities may be calculated according to the control tests associated to each of the bank and the regulator alternatives and the statistical assumptions made regarding these controls. Assume that the regulator selects strategy i with probability x_i while the regulator selects strategy j with probability y_j. Then, the expected risks for the regulator and the bank are for a joint strategy (i,j):

$$\hat{\alpha}_{R,i} = \sum_{j=1}^{n} \alpha_{R,ij}(p_R^C)y_j; \quad \hat{\beta}_{R,i} = \sum_{j=1}^{n} \beta_{R,ij}(p_R^{NC})y_j \quad \text{and}$$

$$\hat{\alpha}_{B,j} = \sum_{i=1}^{m} \alpha_{B,ij}(p_B^C)x_i; \quad \hat{\beta}_{B,j} = \sum_{i=1}^{m} \beta_{R,ij}(p_B^{NC})x_i$$

However, since the regulator controls a banks once the bank has already selected and implemented its control strategy, their resulting type I and II risks for a combined choice of strategies (i,j) are necessarily dependent. Explicit calaculations are made below.

We consider the type II risks as these are the more important for regulation, i.e. the probability that the banks does not detect a regulation non-compliance as well as that of the regulator non-detecting a bank non-compliance. For any strategic pair, selected with probabilities $(x_i y_j)$, a number of situations arise as stated below:

1. $\alpha_{B,j}\alpha_{R,i}$ with costs $C_{ij}^B = C_{RB} + C_j^B + 2C_{j,C}^{\alpha}$; $C_{ij}^R = C_i^R$
2. $\alpha_{B,j}(1-\alpha_{R,i})$ with costs $C_{ij}^B = C_{RB} + C_j^B + C_{j,C}^{\alpha}$; $C_{ij}^R = C_i^R$
3. $(1-\alpha_{B,j})\alpha_{R,i}$ with costs $C_{ij}^B = C_{RB} + C_j^B + C_{j,C}^{\alpha}$; $C_{ij}^R = C_i^R$
4. $(1-\alpha_{B,j})(1-\alpha_{R,i})$ with costs $C_{ij}^B = C_{RB} + C_j^B$; $C_{ij}^R = C_i^R$
5. $\left(1-\beta_{B,j}\right)\alpha_{R,i}$ with costs $C_{ij}^B = C_{RB} + C_j^B + C_{j,NC}^{\beta} + C_{j,C}^{\alpha}$; $C_{ij}^R = C_i^R$
6. $\left(1-\beta_{B,j}\right)(1-\alpha_{R,i})$ with costs $C_{ij}^B = C_{RB} + C_j^B + C_{j,C}^{\beta}$; $C_{ij}^R = C_i^R$
7. $\beta_{B,j}(1-\beta_{R,i})$ with costs $C_{ij}^B = C_{RB} + C_j^B + C_{j,NC}^{\beta} + P_i^B$; $C_{ij}^R = C_i^R - P_i^B$
8. $\beta_{B,j}\beta_{R,i}$ with costs $C_{ij}^B = C_{RB} + C_j^B + \mu_B\left(\beta_{B,j}\beta_{R,i}\right)$; $C_{ij}^R = C_i^R + \mu_R\left(\beta_{B,j}\beta_{R,i}\right)$

Note that in these cases, (1) + (2) equals $\alpha_{B,j}$, (3) + (4) equals $\alpha_{R,i}$, (5) + (6) equals $1-\beta_{B,j}$ while (7) + (8) equals $\beta_{B,j}$. In case 1 for example, both the bank and the regulator have made a type I error. However in this case, the bank incurs fixed costs due to controls by the regulator, plus its own internal control costs and costs of attending to assess why it is non-compliant when in fact it is compliant. Since the error is made twice, the bank is paying this cost twice. A similar rationality is applied to the other cases. Considering case 5, we note that the bank has detected internally non-compliance and thus attended to its non-compliance by incurring a cost $C_{j,NC}^{\beta}$. However since thereafter the regulator made a type I error, the bank is required to reassess its compliance (when it is already compliant). Finally, in case 8, both the bank and the regulator have made a type II error, and thus externality costs are incurred with consequences defined as a function of their external (type II) risks. Say that the total external cost is $\mu\left(\beta_{B,j}\beta_{R,i}\right)$ and assume that it is accounted for by the bank and the regulator. This particular cost will be discussed subsequently, as it might not be accounted for by neither the bank or the regulator but by the "public at large" (and thus it is an externality cost). In this conditions, we note that the expected costs incurrend by the bank and the regulator are:

$$\begin{aligned} C_{ij}^{B} &= C_{RB} + C_{j}^{B} + C_{j,C}^{\alpha}\left(\alpha_{B,j} + \alpha_{R,i} + \left(\alpha_{R,i} - \alpha_{R,i}\beta_{B,j}\right)\right) \\ &\quad + \left(C_{j,NC}^{\beta}\right)\left((1-\beta_{B,j}) + \beta_{B,j}(1-\beta_{R,i})\right) \\ &\quad + P_{i}^{B}\beta_{B,j}(1-\beta_{R,i}) + \beta_{B,j}\beta_{R,i}\mu_{B}\left(\beta_{B,j}\beta_{R,i}\right) \\ C_{ij}^{R} &= C_{i}^{R} - P_{i}^{B}\beta_{B,j}(1-\beta_{R,i}) + \beta_{B,j}\beta_{R,i}\mu_{R}\left(\beta_{B,j}\beta_{R,i}\right) \end{aligned}$$

Consider for example 2 alternative strategies for both the bank and the regulator. One consisting of a control and the other none. Then, the bi-matrix of expected values is:

$$y = 1,\ y = 0$$

$$x = 1 \quad \begin{bmatrix} C_{11}^{R} = C^{R} - P^{B}\beta_{B}(1-\beta_{R}) + \beta_{B}\beta_{R}\mu_{R}(\beta_{B}\beta_{R}); \\ \left\{ \begin{array}{l} C_{11}^{B} = C_{RB} + C^{B} + C_{C}^{\alpha}(\alpha_{B} + \alpha_{R} + \alpha_{R}(1-\beta_{B})) + \\ C_{,NC}^{\beta}((1-\beta_{B}) + \beta_{B}(1-\beta_{R})) + \\ P^{B}\beta_{B}(1-\beta_{R}) + \beta_{B}\beta_{R}\mu_{B}(\beta_{B}\beta_{R}) \end{array} \right\} \end{bmatrix} \begin{bmatrix} C_{10}^{R} = C^{R} - P^{B}(1-\beta_{R}) + \beta_{R}\mu_{R}(\beta_{R}); \\ \left\{ \begin{array}{l} C_{10}^{B} = C_{C}^{\alpha}\alpha_{R} + C_{NC}^{\beta}(1-\beta_{R}) + \\ P^{B}(1-\beta_{R}) + \beta_{R}\mu_{B}(\beta_{R}) \end{array} \right\} \end{bmatrix}$$

$$x = 0 \quad \begin{bmatrix} C_{01}^{R} = \beta_{B}\mu_{R}(\beta_{B}); \\ \left\{ \begin{array}{l} C_{01}^{B} = C_{RB} + C^{B} + C_{C}^{\alpha}\alpha_{B} + C_{NC}^{\beta}(1-\beta_{B}) \\ + P^{B}\beta_{B}(1-\beta_{R}) + \beta_{B}\beta_{R}\mu_{B}(\beta_{B}\beta_{R}) \end{array} \right\} \end{bmatrix} \begin{bmatrix} C_{00}^{R} = \mu_{R}(1); \\ C_{00}^{B} = \mu_{B}(1) \end{bmatrix}$$

Assuming a constraint on the externality risks $\beta_{R,i}\beta_{B,j}$ we have the folllowing problem:

$$\text{Min}\sum_{i=1,0}\sum_{j=1,0}\left\{C_{ij}^{B}\right\}x_i y_j;\ \text{Min}\sum_{i=1,0}\sum_{j=1,0}\left\{C_{ij}^{R}\right\}x_i y_j\ \text{Subject to}:\sum_{i=1,0}\sum_{j=1,0}\beta_{R,i}\beta_{B,j}x_i y_j \le \bar{\beta}$$

Each of these risk probabilities are defined of course by the control strategies each the bank and the regulator assume.

Consider first that the externality constraint is not binding. In this case, the Nash non-zero sum game yields a randomized strategy for each of the parties—or:

$$\text{Min}\left\{C_{11}^{B}xy + C_{10}^{B}x(1-y) + C_{01}^{B}(1-x)y + C_{ij}^{B}(1-x)(1-y)\right\};$$
$$\text{Min}\left\{C_{11}^{R}xy + C_{10}^{R}x(1-y) + C_{01}^{R}(1-x)y + C_{ij}^{R}(1-x)(1-y)\right\}$$

$$x^* = \frac{C_{01}^{B} - C_{00}^{B}}{C_{01}^{B} - C_{00}^{B} - C_{11}^{B} + C_{10}^{B}} = \frac{\left\{\begin{array}{r} C_{RB} + C^{B} + C_{C}^{\alpha}\alpha_{B,j} + C_{NC}^{\beta}(1-\beta_B) + P^{B}\beta_B(1-\beta_R)+ \\ \beta_B\beta_R\mu_B(\beta_B\beta_R) - \mu_B(1) \end{array}\right\}}{\left\{\begin{array}{r} -C_{C}^{\alpha}(\alpha_R(1-\beta_{B,j})) + C_{NC}^{\beta}(1-\beta_R)(1-\beta_B) + P^{B}(1-\beta_R) \\ \beta_R\mu_B(\beta_R) - \mu_B(1) \end{array}\right\}}$$

$$y^* = \frac{C_{00}^{R} - C_{10}^{R}}{C_{11}^{R} - C_{01}^{R} - C_{10}^{R} + C_{00}^{R}} = \frac{P^{B}(1-\beta_R) - C^{R} - \beta_R\mu_R(\beta_R) + \mu_R(1)}{\left\{\begin{array}{r} C^{R} + P^{B}(1-\beta_B)(1-\beta_R)+ \\ \beta_B\beta_R\mu_R(\beta_B\beta_R) - \beta_B\mu_R(\beta_B) - \beta_R\mu_R(\beta_R) + \mu_R(1) \end{array}\right\}}$$

Note that if both the regulator is not accounting for the risk externalities and if type I errors are not penalized, then:

$$x^* = \frac{C_{RB} + C^{B} + C_{NC}^{\beta}(1-\beta_B) + P^{B}\beta_B(1-\beta_R)}{C_{NC}^{\beta}(1-\beta_R)(1-\beta_B) + P^{B}(1-\beta_R)},\quad y^* = \frac{P_i^{B}(1-\beta_{R,i}) - C_i^{R}}{C^{R} + P^{B}(1-\beta_B)(1-\beta_R)}$$

And $\frac{\partial x^*}{\partial P^B} < 0$ and $\frac{\partial y^*}{\partial P^B} > 0$ which are always the case. However,

$$\frac{\partial x^*}{\partial C^B} > 0,\quad \frac{\partial x^*}{\partial C_{RB}} > 0,\quad \frac{\partial x^*}{\partial C_{NC}^{\beta}} > 0 \ \text{ if } \ P^B > \frac{C_{RB} + C^B}{1 - \beta_B(1-\beta_R)}$$

Namely, when the regulator will exercise more controls the greater the cost of control by the bank. Note however that this contributes to greater costs to the bank because they are controlled (C_{RB}) and the regulator does not assume the bank costs. Finally, a regulator will exercise more control if compliance costs of the bank are less than their expected penalty costs plus the fixed costs imposed on the bank when it is controlled. In the expression above, note that $(1 - \beta_B(1 - \beta_R))$ is the probability of the bank paying a penalty which denotes the probability of the bank not detecting non-compliance and the regulator detecting it.

By the same token, for the bank, we have:

$$\frac{\partial y^*}{\partial P^B} > 0, \quad \frac{\partial y^*}{\partial C^R} < 0 \quad \text{if} \quad 2C^R B(1 - \beta_R)\beta_B > 0$$

And therefore, the probability of the bank controlling its regulation compliance increases when the penalties for non-compliance increases, while it will control less if the costs of controls (inspection) by the regulator are less than half has than half the penalty times the probability that the bank detects non-compliance while the regulator does not. For example, let $\beta_B = 0.10$ and $\beta_R = 0.20$, then $\frac{\partial y^*}{\partial C^R} < 0$ if $C^R < 0.04P^B$. In other words, if penalties are small and there is a good chance that the regulator will not detect non-compliance, then the banks will control less. However note that in this case, a regulator whose costs of control are increasing and large, will increase importantly the penalties to impose for non-compliance to provide an incentive for the bank to control its non-compliance.

External costs are not accounted for by either the regulator and the bank as in the previous special case. Although they are difficult to assess, they are nevertheless important and underlie appreciable systemic costs. For this reason, regulation targetting the probability of not detecting non-compliance may be limited but provide at least an index that relates to systemic risks. In this case, a Sovereign State may impose on the regulator the responsibility to meet a type II risk constraint. These are then given by the following matrix:

$\hat{p}_R = p_R^{NC}$	y_B	$1 - y_B$
x_R	$\beta_B\beta_R$	β_R
$1 - x_R$	β_B	1

With an expected (constraint) external risk:

$$\beta_B\beta_R xy + \beta_R x(1 - y) + \beta_B(1 - x)y + (1 - x)(1 - y) \leq \bar{\beta}$$

In such a case, the least amount of control that the regulator will apply is dictated by both what the bank does and the regulation risk contraint of thee Sovereign State. Explicitly,

$$x \geq \frac{(1 - \bar{\beta}) + (1 - y)\beta_B}{(1 - \beta_B\beta_R + \beta_R + \beta_B)y + (1 - \beta_R)}$$

In this case the game that regulator and the bank face is defined by the risk constraint game:

$$\text{Min}\left\{C^B_{11}xy + C^B_{10}x(1-y) + C^B_{01}(1-x)y + C^B_{ij}(1-x)(1-y)\right\};$$

and

$$\text{Min}\left\{C^R_{11}xy + C^R_{10}x(1-y) + C^R_{01}(1-x)y + C^R_{ij}(1-x)(1-y)\right\}$$

Subject to :

$$x \geq \frac{(1-\bar{\beta}) + (1-y)\beta_B}{(1-\beta_B\beta_R + \beta_R + \beta_B)y + (1-\beta_R)}$$

Two cases arise. First the regulator constraint holds at equality (the overeign constraint is binding) and second the Sovereign constraint is not binding. The second case was solved above. When the the constraint is binding then:

$$x = \frac{(1-\bar{\beta}) + (1-y)\beta_B}{(1-\beta_B\beta_R + \beta_R + \beta_B)y + (1-\beta_R)}$$

Since the bank is unconstrained, but only by the controls and the penalties of the regulator, we have as calculated above:

$$y^* = \frac{P^B(1-\beta_R) - C^R}{C^R + P^B(1-\beta_B)(1-\beta_R)}$$

And, following elementary manipulations, we have:

$$\begin{aligned} x^{**} &= \frac{(1-\bar{\beta}) + (1-y)\beta_B}{(1-\beta_B\beta_R + \beta_R + \beta_B)y + (1-\beta_R)} \\ &= \frac{C^R(3-\bar{\beta}) + P^B(1-\beta_R)\left((1-\bar{\beta})(1-\beta_B) - {\beta_B}^2\right)}{2P^B(1-\beta_R) - (2\beta_R + \beta_B(1-\beta_R))C^R} \end{aligned}$$

Evidently, we note that the more stringent the Sovereign State externality constraint (i.e. $\bar{\beta}$ smaller), then the more the regulator will have to control. Full control occurring when the ratio of the regulator's control cost to te penalties imposed equal the following:

$$x^{**} = 1 \Leftarrow \frac{C^R}{P^B} = \frac{(1-\beta_R)\left(1 + {\beta_B}^2 - (1-\bar{\beta})(1-\beta_B)\right)}{3 - \bar{\beta} + 2\beta_R + \beta_B(1-\beta_R)}$$

For example, say that $\bar{\beta} = 0.10$ while $\beta_B = 0.10,\ \beta_B = 0.20$, then : $x^{**} = 1 \Leftarrow \frac{C^R}{P^B} = 0.047$ which implies an extremely large penalty to induce the regulator to a full control (given the statistical errors embedded in the type II risks of the bank and the regulator).

A “Regulation Game” is therefore complex, embedding a broad number of issues to be accounted for. In practice, the Sovereign State risk constraint and the cost and risk consequences for both the regulator and the bank define the game they may be both confronted by. Of course, more information and greater statistical sophistication (as well as the extensive development of a public of whistle blowers may also reduce the cost of controls), leading eventually to banks reducing the external risks they assume and thus augmenting further penalties to finance regulation. These are evidently political issues debated in both political circles and the media. For example, say that a bank increases its propensity to control, then of course, $\frac{\partial x^{**}}{\partial y} < 0$ which will reduce the intake of regulators from penalties as they will have to control less. In other words, to compensate a loss of income from controlling less will require that the penalties be increased further. Explicitly, since $0 < \frac{\partial y^{*}}{\partial P^{B}} \Leftarrow C^{R}(1-\beta_{R})(1+(1-\beta_{B})) > 0$, will control more when the penalties it extract to compensate for a loss of income.

Both the Sovereign State expressing its concern for a global equity, namely seeking both to maintain banks operating and at the same time protecting the public from systemic risks and power and information asymmetry risks, lead necessarily to the intervention of regulation. A full control, namely controlling continuously and forever every aspect of a bank (or other institutions) is clearly not feasible and therefore the only solutions are either statistical controls or a population of “whistle blowers and spies”, or both. For these reason, the unique and feasible solutions to the game considered are randomized control strategy for both the bank and the regulator.

Additional Reading and References

Control, statistical, strategic and other types of controls are broadly applied. There are numerous books and papers on statistical and quality control as well as experimental design. Some references consulted include Cochran and Cox (1957), Coleman and Montgomery (1993), Crosby (1984), Daudin and Tapiero (1996), Godfrey and Kenett (2007), Grant and Leavenworth (1988), Juran (1974, 1980), Kenett and Zacks (2004), Page (1954), Phadke (1986), Shewart (1931); Taguchi and Clausing (1990) as well as my book Tapiero (1996).

This chapter has been particularly concerned with inspection games or strategic controls. These problems have been studied by Maschler (1966), Reyniers and Tapiero (1995a, b), Ruckle (1992), Rothenstein and Zamir (2002), Von Stengel (1991) (on military, time), Avenhaus and Kilgour (2004), Avenhaus (1994), Avenhaus et al. (1996), Avenhaus et al. (1996), Avenhaus et al. (1995), Baston and Bostock (1991), Bird and Kortanek (1974), Canty et al. (2001), Cariou et al. (2008), Rothenstein and Zamir (2002) as well as others. Some applications include inspection games for industrial products, contracts compliance and quality controls (Reyniers and Tapiero 1995a, b; Tapiero 1995a,b Ruckle 1992; Tapiero 1996, 2005a, b, c).

While strategic enforcement and monitoring applications include Russel (1990), Canty et al. (2001), Gawande and Bohara (2005), Innes (1999), Antelman and Savage (1965), Benveniste et al. (1987), Chow et al. (1971a, b), Moustakides (1986), Pierskalla and Voelker (1976), Tolley et al. (1989) (monitoring insurance), Willsky (1976, 1986), Willsky and Jones (1976), Gawande and Bohara (2005).

Inspection and control of queues has essentially been developed initially by my own and Hsu research in her doctoral dissertation. These have included Hsu and Tapiero (1987a, b, 1994, 1988a, b, 1992, 1993, 1994, 2000a), Posner and Tapiero (1988), Tapiero (1990).

For a theoretical and applied studies of queues and in particular network of queues, we refer to Jackson (1957, 1963), Gordon and Newell (1967), Buzacott and Yao (1986), Ho and Cao (1983), Whitt (1983), Solberg (1977) and Goyal (2007) on economics of networks.

Environmental control is both important and is the subject of continuous research. Some references in clude both my own (Tapiero 1995a, b, 1996; Tapiero 2005a) as well Avenhaus (1994), Bird and Kortanek (1974), Florens and Foucher (1999), Harford (1987), Florens and Foucher (1999), Grau-Viladrich (2003), Harford (1987), Innes (1996, 1999), Malik (1993), Xepapadeas (1992, 1994, 1995).

Other applications include: On control food, food, we refer to Antle (1996, 2000), while for the control of web we have refer to Bay and Pazzani (1999) as well some of my own papers (Tapiero 1997). Lee and Tapiero (1986, 1989) have applied quality control in marketing while Reisman et al. (1987a, b) have used control on services (see also Tapiero 1994a on economic quality)

For the control of ports, we refer to Knapp and Franses (2007), Li and Zheng (2008), Cariou et al. (2008), Grau-Viladrich (2003), Hamada et al. (2002), Hare (1997), IMO (2002), Hamada et al. (2002) on ship inspections, Hare (1997), Knapp and Franses (2007), Li and Cullinane (2003).

Control in finance has essentially focused on the VaR (Basel Committee 2004, 2008, Cumperayot et al. 2001) which was considered and referred to in previous chapters

Finally, there is currently and intensive set of studies to apply control in regulation (for example, Innes 1996, 1999 on automobiles). This chapter has provided a number of original problems and has raised some outstanding issues to confront in strategic regulation.

Chapter 12
Games, Risk and Uncertainty

Abstract This chapter provides an intuitive review of game theory models. These models provide a logical foundation to solve decision problems when "what we do and what others do" depend and affect all parties—each party with its own agenda. Game theory is thus a logical expression of a more complex rationality than the one we used when we consider only one decision maker. This chapter provides and elementary overview of games and their solutions, emphasizing issues of strategic risks. Examples are used to illustrate the solution of some games.

12.1 Introduction

Game theory is used to formulate, to assess and make strategic decisions—decisions involving more than one party, each with its own information, a common and shared information and agenda. Games are mostly complex, extending the rationalities we have used in expected utility valuation or when applying the Arrow-Debreu framework to price market risk. Games are also strategic models, seeking to reconcile what is "good for us" when what we do and others do determine the consequences of our decisions. There are numerous types, solutions and approaches to modeling and solving games. For this reason, we shall concentrate our attention on merely few games. Games are defined broadly in terms of:

- The game participants; individual persons, firms, etc.
- The rules of the game, such as simultaneous moves, leader-led games, constraints games etc.
- The private and shared information each of the parties has, both about the environment within which the game is played and between the parties—such as prior information about each of the parties' intent, risk attitudes and propensity to choose a particular or a set of potential decisions.

C.S. Tapiero, *Engineering Risk and Finance*, International Series in Operations Research & Management Science 188, DOI 10.1007/978-1-4614-6234-7_12,

- The game conditional consequences for each of the parties resulting from their alternative choice set. These can be deterministic or random. In the latter case, games are called random payoff games.
- Game completeness. There are two forms of "uncertainty games" including: information asymmetries regarding the consequences of games and incompleteness regarding each of the parties' strategic alternatives (i.e. not knowing completely or partially what are a counter-party alternative decisions).

Game theory differs from traditional decision theory (DT) that seeks to reach decisions under risk or uncertainty, in an environment based on the specific preferences of an individual party. Decisions in such cases are reached based on rational preferences and external (rather than endogenous) risks specifically defined by the decision model. Game models add an endogenous risk defined by what we and other do as indicated in chap. 8.

These problems are difficult to analyze and require both principles of conduct and a rationality that can be applied to resolve the conflict between the parties to the game—namely, seeking decisions that will provide a "stable" solution or agreement to sustain parties' strategic choices.

There are many games and therefore models that lead to a broad set of potential techniques and solution concepts. Although the Nash conjecture (Nash 1951) for solving non-zero sum games are mostly used. References to such models abound. Noteworthy personalities that have contributed to game theory include Emile Borel, Von Neumann and Morgenstern (1944), Nash (1950), Shubik (1955, 2002), Harsanyi (1967–1968) and Harsanyi and Selten (1987), Luce and Raiffa (1958). Games may be mathematical, expressing a logical approach to conflict resolution or behavioral, based on a psychological understanding of human behaviors or on observations of persons or animals that can provide clues to the solution of the conflicts.

The formal mathematical theory of games seeks to define a (logical) rationality that explains and prescribes "how to make decisions, negotiate, and compete in a strategic setting." "Behavioral approaches", instead, seek to express how parties behave in particular situations and how they proceed to reach a decision in a given gaming environment.

Applications of game theory span many fields such as contract economics, competitive finance, business, industry, health care, environmental science, political and social science, behavioral science, trading, security, safety, risk management etc. Contributions by economists and mathematicians have also been recognized by a number of Nobel prizes to academics that have contributed to the development of this field and to its application. In this chapter we shall limit ourselves to review a set of games which we relate and apply to risk problems. These include Nash's equilibrium conjecture for the solution of "non-zero sum games," Stackelberg (1934, 1952) approach to Leader-Led games, to random payoff games and Bayesian (and partial information) games. These problems require some mathematical background which can be complemented by referring to the books and papers in the references. For this reason, they are presented here intuitively. A number of examples and applications pertaining to strategic risk problems are considered as well. These games will be used to better appreciate the complexity of strategic risks. we have defined in chaps. 8 and 9. This chapter does not provide a complete study of game theory however but only provides a cursory view of games and control.

12.1.1 Games, Risk and Uncertainty

Risk and Uncertainty in games result essentially from two sources:

- The actions (the choice set) taken by the "other" (or "others") player. When a party's potential set of actions regarding the other are completely defined, we shall assume that "his preferences are complete" while when they are not, we shall assume that his "preferences are incomplete." Such a distinction differentiates also between risk and uncertainty in a Knightian sense (see Chap. 1).
- The game conditional payoffs. When conditional consequences are unknown or known partially such games are incomplete in the sense of Harsanyi (also called Bayesian games).

Games' risks are strategic, and thus differ from the risks of "randomness" we assumed to be risks that are not motivated. For example, say that a firm has negotiated the delivery of a product of a certain (well defined and contracted) quality. Once the contract has been drafted and implemented, what are the conditional consequences if the supplying firm does not meet the quality contracted? Similarly, how would a creditor confront a potential debtor's default due to both external events (namely, independent of the debtor intents)? or purposeful default? (and therefore a "strategic default"). These intents assume many forms. For example, a debtor might decide purposely not to pay his debt or seek to erode the terms of a contract. These situations can lead to conflicts and to risks that are "endogenous" as they relate by what we do and other does. These problems recur for example in Global Finance and Trade. Sovereign States may for example, sign an agreement and then, for "political reasons," rescind it explicitly or implicitly. To better understand and study such and other strategic problems, the logic of games is useful.

12.2 Concepts of Games and Risk

Game theory was first proposed and defined by the French mathematician Emil Borel in 1921. The famed mathematician John von Neumann provided an analysis of games in 1928. In 1944, von Neumann, together with the economist Oskar Morgenstern, published a first fundamental book on game theory, "The Theory of

Games and Economic Behavior." This book was published about the same time that Dantzig developed the simplex algorithm in linear programming with zero-sum games (where the gains to one party is a loss to the other) solved by such programming problems. Applications to strategic alternatives and decisions problems in the military, in economic and financial problems and many others, have provided an extensive foundation to game theory and its uses.

Mathematically, game models can be categorized in several ways. Some examples include:

- **Discrete or continuous strategic games**: Each party selects a strategy from a discrete set or a continuous set of alternatives.
- **Two parties' games**: Games are defined between two specific parties with well defined preferences known completely or partially to each of the parties.
- **Zero-sum games**: Games with conditional payoffs for one party being conditional losses for the other. Their sum is therefore equal to zero or to a constant (constant sum games).
- **Non-zero-sum games**: Games with conditional payoffs to parties not summing to zero. The solution of such games is based on Nash conjecture (to be seen later).
- **Stackelberg games or leader-led games**: Games with one of the parties able to observe the strategy of the other and based on such an observation select his own strategy. In these games, the "Leader" announces his move first, fully cognizant of the response function of the led.
- **Information asymmetry games**: Games with some player having information the other does not possess.
- **Random payoff games**: Games with conditional uncertain payoffs to at least one of the two parties.
- **Coalition games**: Games with parties negotiating a strategy to collaborate and distribute the spoils of collaboration according to agreed-on rules or contracts.
- **Repeated games**: These are sequential games, repeated with learning and information acquired as the game is played. These include also some Bayesian games, consisting of prior estimates of players' strategic selections by parties, updated by the game being repeated with outcomes observed when a game is played.
- **Constrained games**. These are games with parties subject to individual or to common constraints. For example, risk constraints, regulation constraints (VaR or Value at Risk Regulation imposed on Banks), trade constraints, exchange constraints, as well as risk-constrained game as introduced in Chap. 11.
- **Inter-temporal games or differential games**: Games repeated continuously (or discretely) in time. These games consist typically of underlying processes (discrete or continuous time difference deterministic (or stochastic) differential equations with each party pursuing his own inter-temporal preference defined over a given time span—finite or infinite. These games are notoriously difficult to solve although numerous applications can be found in Kogan and Tapiero (2009) on Supply Chains Games and Risk.

- **Bayesian incomplete (risk) games**. Unlike Nash Games that assume that all parties have complete information, Bayesian games assume that parties have incomplete information. Incomplete game that use learning mechanisms based on prior estimates of the choice set a party may have, their payoffs and its posterior revision once experience is accumulated.

Applications abound. For example, to assess strategic international pricing, outsourcing, Co-development, co-production, international trade and foreign exchange, revenues and risks sharing, etc.

Games can be subject to many risks which we summarize by:

- **Ex-ante risks defined by a countable and accountable choice set of one's and others' parties. This may arise due to partial information**, a prior assessment of choice sets and their consequences.
- **Ex-ante strategic risks** arising from parties' preferences and their power. These are counter-party risks prevalent in the exchange between financial agents (e.g. creditors and debtors).
- **Ex-ante risks arising from environmental randomness**, arise from external factors none of which depend on the parties to the game. These factors may be a third party to the game or random events that alter both the choice sets and the consequences of the game.
- **Ex-post risks**, resulting from negotiated agreements that are not maintained by the parties. And therefore requiring both contracts' design and their controls.
- **Incomplete state preferences risks or stealth strategic risks**. Such situations arise when one of parties does not know all the choice sets, if or the others have. In such cases, predictability of a party's actions or counter-actions of a party cannot be accounted for.

Partial information games are particularly challenging. They include strategic risks often denoted by: One-sided moral hazard (for example, with a single party, profiting from an information edge over its counter-party) and two-sided moral hazard (where both parties have an information edge that affect the parties' conditional outcomes). Incomplete states, stealth and partial information games although encountered in current practice are theoretically and practically difficult to solve.

Formally, the solution of a game depends on the assumptions we make, including:

- Their information (whether shared, partial or common);
- Their power and Lead-Led relationships (if any);
- The randomness of outcomes and party's information processing
- The rationality of the game layers.

Treating information in games is not simple. Typically, in a game theoretic format, the value of information is measured by improvement of the game solution to each of the parties. In some case, all parties may profit from better or more information. In other cases, one may profit but not the other or neither may profit. For example, if more information contributes to a greater awareness of an irremediable conflict, each may choose a decision which is most conservative, precluding decisions that would benefit both parties had hey be less aware of their conflict. Hirshleifer and Riley (1979, 1992), noted such a phenomenon in exchange economies. Others have also indicated that in oligopolistic models, more information can either improve or be harmful to the quality of decisions. For example, if players' objectives in a duopoly are convex, then more information improves the (Nash) equilibrium. Other cases are also pointed out by Bassan (2003) and Bassan et al. (2003) who seek to bridge fundamental rational expectations and strategic games.

Some games have also acquired distinctive names. These include among many others:

- **Cournot games** (1838): These are non-coordinating (non-zero sum) games when both parties have full information of their choice sets and reach decisions simultaneously and independently. These lead to a simultaneous "game play" with each party's strategy, a function of the other. Cournot games are associated to quantity competition games (Vives 1984). Originally, Cournot did consider both price and quantity games however. Cournot games are based on the Nash conjecture, defining a stable equilibrium resulting from an unspecified learning process. This is in contradiction to incomplete games (see Bayesian game) where a Nash solution results from an iterative process. By comparison, the Cournot equilibrium is defined uniquely, while incomplete games can result in games in multiple equilibria (i.e. multiple solutions and therefore are incomplete in a financial sense).
- **Bertrand games**: These games are similar to Cournot's but with parties competing instead on price. In these games, firms are price setters rather than quantity setters (with price sensitive consumers determining the quantity).
- **The prisoner dilemma**: Imagine two parties (criminals) are arrested and proof is sought for the conviction of one or the other or both. One may squeal or may decide to cooperate. Four situations arise, each situation with separate consequences that define the prisoner dilemma. The solution of this game seeks to predict whether the prisoners will cooperate or not and how can a stable cooperation be reached.
- **Battle of the sexes**: These are game between two parties, say a loving husband and a loving wife, one wanting to go to see a movie (the wife), the other to a Football game (the husband). There are again four set of consequences for each of the parties that define a game. The problem is how to reconcile their wants?
- **The Hawk-Dove game**: Say that two countries are in conflict, each with a choice set consisting of two postures. One is that of a Hawk consisting in escalating the conflict (say, between these countries) until the opponent retreats while the other, a Dove Strategy, consists in displaying hostility but retreating prior to being injured in case the other party chooses to be a Hawk. Again, there are four conditional outcomes and a "solution": (or "solutions") defined by the parties conditional outcomes. The problem is predicting what each party ought to do? Such games may be applied to regulation, threatening or cajoling potential non-complying financial institutions.

- **Inspection (or statistical control) games**: These are statistical games constructed to determine how best to control ex-ante agreements and their expected consequences. For example, what if an international contract of nuclear non-proliferation is agreed between two parties. The agreement may call for both mutual (and potentially random) inspections to ensure that the contract is maintained. Inspection games are important in regulation games as note in Chap. 11. In these models, an inspector verifies that another party complies to well defined rules. For example, a bank not complying to legally regulated rules. Typically, a inspector's resources are limited so that verification can only be statistical.

These and other games are logical constructs or tools that one may use to better understand and manage strategic alternatives in a conflicting environment. Some of these games are formulated and solved simply to provide guidelines for decision making when there are multiple and dependent decision makers.

12.3 Two-Persons Zero-Sum and Non-zero Sum Games

Two-persons zero-sum games involve two players. Each has only one move (decision) to take and both make their moves simultaneously. Each player has a set of alternatives, defining the strategies it may follow, say $\mathbf{A} = (A_1, A_2, A_3, \ldots\ldots, A_n)$ for the first party and $\mathbf{B} = (B_1, B_2, B_3, \ldots\ldots, B_m)$ for the second party. When both players make their move, outcomes (a_{ij}, b_{ij}) are obtained, corresponding to their moves (A_i, B_j). These games assume:

1. $a_{ij} + b_{ij} = 0$, the gain of one party a_{ij} is the loss to the other $b_{ij} = -a_{ij}$ for zero-sum games.
2. In a non-zero-sum game, $a_{ij} + b_{ij} \neq 0$.
3. Consequences to both players, are given by a bi-matrix $(\mathbf{A}, \mathbf{B}) = [a_{ij}, b_{ij}]$.
4. Parties do not know what strategy will be taken by the other party.
5. Each party is rational, with an ordered set of preferences and a well-defined choice set.

12.3.1 Terms and Solution Concepts

The solution of a game depends on the rationality of conflict resolution. A number of approaches include:

- *Minimax strategies*: Consists in parties' minimizing their maximum losses
- *Maximin strategies*: Consists in parties' maximizing their minimum gain
- *Saddle points* are an equilibrium and unique solution
- *Pure (dominant) strategies* define specific alternatives to be optimal

- *Randomized strategies* arise when there are no unique decisions that can be optimal. In this case, equilibrium consists of a (randomized) linear combination over the set of alternatives the parties have.
- *Negotiated solution* are based on parties prior agreements
- *Cooperative solutions* are based on a prior agreement to collaborate (both negotiated & co-operative solutions may have side payments to compensate parties to agree to a strategic agreement).

Other solutions are based on by Behavioral approaches to game theory (Wakker and Tversky 1993; Wakker 1994, 2001, 2010; Thaler 1997, etc.).

Example: A zero-sum game is defined below with entries denoting a gain to one party (and therefore a loss to the other). Let the rivals be two firms, A and B. The firm A has three courses of action A_1, A_2, A_3 while firm B has four courses of actions B_1, B_2, B_3, B_4. The consequences to each of the firms are given by the payoff (or loss) corresponding to decision pairs as given in the following matrix:

	B_1	B_2	B_3	B_4
A_1	.6	−.3	1.5	−1.1
A_2	.7	.1	.9	.5
A_3	−.3	0	−.5	.8

This problem has a solution, called a *saddle-point*, because the "least greatest loss" to B equals the "greatest minimum" gain to A. When this is the case, the game is said to be *stable*, defined by the game *saddle-point*. This saddle-point is also called the *value of the game*, which is the least entry in its row, and the greatest entry in the columns. Not all games can have a pure, *single strategy*, *saddle-point* solution for each player. When a game has no saddle point, a solution to the game can be devised by adopting a *mixed strategy*. Such strategies result from a combination of pure strategies, each selected with a probability calculated by the (optimized) solution of the game. Such a mixed strategy results then in a solution which is stable, in the sense that player 1's *Maximin* strategy equals player 2's *Minimax* strategy. Mixed strategies, express the probabilities that each party selects any of his available strategies with a given probability. These probabilities define strategic risks, embedded in a party's not knowing for sure what the counter party will do. On the other hand, mixed strategies are a mathematical artifact that provides a solution to the game by defining a game strategy as a "planned" statistical event.

12.3.2 The Nash Conjecture

The Nash Conjecture is a "logical" conclusion to situations defined by non-zero-sum non-cooperating games. Consider a game with N players, each of which has a set of choice set, say y_i, $i = 1, \ldots, N$. The payoff to a player, say, i, is a function of other players feasible decisions given by the function $J_i(y_1, y_2, \ldots, y_n)$ with, $y_i \in Y_i$. Assume complete information to all players. Since two-player games can be extended straightforwardly to multiple players, we consider for simplicity two players A and B only. Each player's goal is to maximize his own payoff with the Nash conjecture providing a solution to both players and is defined as follows (Nash 1950):

The Nash Conjecture: A pair of strategies (y_A*, y_B*) *is said to constitute a Nash equilibrium if the following pair of inequalities is satisfied for all* $y_A \in Y_A$, *and* $y_B \in Y_A$

$$J_A(y_A{}^*, y_B{}^*) \geq J_A(y_A, y_B{}^*) \quad \text{and} \quad J_B(y_A{}^*, y_B{}^*) \geq J_B(y_A{}^*, y_B).$$

The definition implies that the Nash solution is

$$y_A{}^* = \underset{y_A \in Y_A}{\arg\max}\,\{J_A(y_A, y_B{}^*)\} \quad \text{and} \quad y_B{}^* = \underset{y_B \in Y_B}{\arg\max}\,\{J_B(y_A{}^*, y_B)\},$$

with any unilateral deviation from this solution resulting in a loss for both parties.

If this is a one period "play," the problem is static. If the choice sets are not constrained and if the payoff functions are continuously differentiable, a first-order (necessary) optimality condition for a Nash equilibrium results in the following system of two equations in two unknowns $y_A{}^*$, $y_B{}^*$:

$$\frac{\partial J_A(y_A, y_B{}^*)}{\partial y_A}\Big|_{y_A = y_A{}^*} = 0 \quad \text{and} \quad \frac{\partial J_B(y_A{}^*, y_B)}{\partial y_B}\Big|_{y_B = y_B{}^*} = 0.$$

Second order (sufficient) optimality conditions, ensure that the payoff is maximized, or:

$$\frac{\partial^2 J_A(y_A, y_B^*)}{\partial y_A{}^2}\Big|_{y_A = y_A{}^*} < 0 \quad \text{and} \quad \frac{\partial^2 J_B(y_A{}^*, y_B)}{\partial y_B{}^2}\Big|_{y_B = y_B{}^*} < 0.$$

Equivalently, one may determine $y_A^R(y_B) = \underset{y_A \in Y_A}{\arg\max}\,\{J_A(y_A, y_B)\}$ for each $y_B \in Y_B$ to find the best response function, $y_A = y_A^R(y_B)$, of player A and of player B, $y_B = y_B^R(y_A)$ which constitute a system of two equations in two unknowns.

Example: A Two parties Non-Zero Sum Game: The Randomized Nash Solution - Consider the bimatrix game $(\mathbf{A}, \mathbf{B}) = (a_{ij}, b_{ij})$. Let $\mathbf{x}$ and $\mathbf{y}$ be the vector of mixed strategies with elements x_i and y_j, and such that $\sum_{i=1}^{n} x_i = 1,\ 0 \leq x_i \leq 1,\ \sum_{j=1}^{m} y_j = 1,\ 0 \leq y_j \leq 1$. The value of the game for each of the players is given by:

$$V_a = \mathbf{x}\mathbf{A}\mathbf{y}^{\mathbf{T}}, V_b = \mathbf{x}\mathbf{B}\mathbf{y}^{\mathbf{T}} \quad \text{or} \quad V_a = \sum_{j=1}^{m}\sum_{i=1}^{n} x_i y_j a_{ij}, V_b = \sum_{j=1}^{m}\sum_{i=1}^{n} x_i y_j b_{ij}$$

The equilibrium is defined for each strategy selected by one or the other player if the following conditions hold:

$$\mathbf{Ay} \le V_a, \mathbf{xB} \le V_b$$

Explicitly, consider the 2*2 bi-matrix game. The value of the game to each of the parties are:

$$V_a = xy(a_{11} - a_{21} - a_{12} + a_{22}) - y(a_{22} - a_{21}) - x(a_{22} - a_{12}) + a_{22}$$
$$V_b = xy(b_{11} - b_{21} - b_{12} + b_{22}) - y(b_{22} - b_{21}) - x(b_{22} - b_{12}) + b_{22}$$

For an admissible solution, the first party requires that:

$$\begin{cases} x = 0 & \text{if} \quad Ay - a \le 0 \\ x = 1 & \text{if} \quad Ay - a \ge 0. \\ 0 < 1 & \text{if} \quad Ay - a = 0 \end{cases}$$

where $A = (a_{11} - a_{21} - a_{12} + a_{22})$ and $B = (b_{11} - b_{21} - b_{12} + b_{22}) : V_a(1, y) \le V_a(x, y)$; $V_a(0, y) \le V_a(x, y)$ which is equivalent to $A(1 - x)y - a(1 - x) \le 0$; $Axy - ax \ge 0$ where, $A = (a_{11} - a_{12} - a_{21} + a_{22}); a = (a_{22} - a_{12})$.

In this sense there can be three solutions $(0, y); (x, y); (1, y)$. Similarly for the second party using parameters B and b. Say that $A \ne 0$ and $B \ne 0$, a solution for x and y satisfies the following conditions:

$$\begin{cases} y \le a/A & \text{if} \quad A > 0 \\ y \ge a/A & \text{if} \quad A < 0 \\ x \le b/B & \text{if} \quad B > 0 \\ x \ge b/B & \text{if} \quad B < 0 \end{cases}$$

A simultaneous interior solution leads to the following equations for (x, y), which is found by setting:

$$0 = \frac{\partial V_a}{\partial x} \text{ and } 0 = \frac{\partial V_b}{\partial y} = 0$$

or

$$x^* = \frac{(b_{22} - b_{12})}{(b_{11} - b_{12} - b_{21} + b_{22})}, \; y^* = \frac{a_{22} - a_{21}}{a_{11} - a_{12} - a_{21} + a_{22}}$$

Example: An Inspection Games

Regulator Controls

		(N_1, R_1)	(N_2, R_2)	. . .	(N_m, R_m)
	(S_1, C_1)	**Consequences**			
Bank controls	(S_2, C_2)				
	.				
	.				
	.				
	(S_n, C_n)				

In this model both a regulator and a bank apply statistical controls consisting of (S_n, C_n) denoting the number of controls and the critical factor defining the stringency of the statistical test applied. Similarly, for the regulator we have (N_m, R_m). Note that if there exists a random control strategy, this means that inspection are strategic providing an incentive or a threat to the other party.

Example: The Prisoner Dilemma and the problems of collaboration Say that the joint venture of two firms in some country provides a gain to be shared by the two firms. This gain is subject to taxation in the home country. Two cases arise, if the two firms cooperate and share agreeably the spoils of their venture, then the profit to each will be (p, p) If both firms reports to tax authorities, both are penalized and their gain will be (q, q). When one firm reports and the other does not, then gain for the one and the other are (r, s). The 2×2 game matrix is a non-zero sum game which is given below:

Firm1/Firm 2	Collaborate (say nothing)	Reports to IRS (squeal)
Collaborate (say nothing)	p1,p2	s1,r2
Report to IRS (squeal)	r1,s2	q1,q2

For example:

Firm1/Firm 2	Collaborate (say nothing)	Reports to IRS (squeal)
Collaborate (say nothing)	(2.5,2.5)	(1,−3)
Report to IRS (squeal)	(1,−3)	(−1,−1)

The solution of the game depends of course on the parties' conditional consequences. For this game, the general (randomized) solution we have defined above is:

$$x^* = \frac{(s2 - q2)}{(p2 - s2 - r2 + q2)}, \ y^* = \frac{(s1 - q1)}{(p1 - s1 - r1 + q1)}$$

Say that $p = p1 = p2, \;\; q = q1 = q2, r = r1 = r2$ and $s = s1 = s2$. In this case, both parties will select the same strategy with an equal payoff. In other words,

$$x^* = y^* = \frac{(s-q)}{(p-s-r+q)} = \frac{1-(-1)}{2.5-1+1-0} = \frac{2}{1.5-1+3} = \frac{2}{3.3} = 0.571$$

Should they collaborate? Of course, as in this case each would collect 2.5 which is larger than other payoffs. However, if these firms do not know what the other will do, they may be tempted to squeal. For example, if a firm squeals and the other says nothing, then the firm squealing will collect 1 and the other will be punished, collecting -3. Since there is no dominant strategy, the randomized equilibrium strategy will be for each to collaborate with a probability of 0.571. This particular problem indicates therefore the importance of collaboration if the parties are assured that their collaboration will pay. Such problems belong of course to a family of games we have called "Prisoners' Dilemma."

Can collaboration be sustained? What if the game is repeated, would any of the players change his strategy? For example, cooperating first and subsequently squealing? Both will collaborate if the collaborative strategy is dominant, $(p,p) > (s,r)$ and $(p,p) > (q,q)$. If this is not the case, i.e. for the first firm (p) but (r) for the second then conflict arises whose solution is a randomized strategy.

The Prisoner dilemma game is important and has generated a great deal of research to define a stable solution where parties will find that it is to their advantage to cooperate. For example, can parties to the game exchange information first? (this explains why prisoners are isolated, to prevent an information exchange!). If this were the case, they might reach an agreement and split in some ways the additional payoff each can receive. In finance, there are many approaches to create incentives for collaborations. Payoffs are always greater if we can collaborate for the good of the whole, at the same time rogue and short sighted parties as well as a lack of trust might lead to counter party risks with a pursuit of one's own interests at the expense of the other.

Example: The battle of the Sexes Consider the following bi-matrix game:

Man/woman	Man choice	Woman choice
Man choice	(2,1)	(0,0)
Woman choice	(0,0)	(1,2)

Note that the randomized solution exists and is (2/3, 2/3), since:

$$x^* = \frac{(2-0)}{(1-0-0+2)} > 0, \; y^* = \frac{(2-0)}{(2-0-0+1)} > 0$$

The value of the game is then:

$$V_X = V_Y = \left(\frac{2}{3}\right)\left(\frac{1}{3}\right)(2) + \left(\frac{1}{3}\right)\left(\frac{2}{3}\right)(1) = \frac{6}{9} = \frac{2}{3} < 1$$

Obviously, there ought to be a better solution. If the woman complies, then each will collect 2 and 1 and vice versa if the man complies then each collects 1 and 2. A negotiation might then be a better alternative (say, better than "divorcing," each going his/her way and collecting nothing!). For these reasons, games are not only mechanisms to seek "a solution" but also to induce "an exchange" based on negotiated settlements. For example, say that the woman, very much in love with her man would yield to his wants. The gains are then 2 and 1 with an excess of $2 - 1 = 1$ to the man. What if he were to compensate her and buy her a jewel for ½? In such a case, they would both enjoy a gain of 1.5 and therefore both would feel they have gained by pleasing the other. Such problems recur in international exchanges as well. Would it be better to go to war to impose one's will or to find areas of accommodations where exchanges are contributing to all parties—even if this is not in an equal manner?

As with the prisoner's dilemma, reaching a cooperative solution is not simple. Strategies such as side payments, briberies, information exchange etc. are often needed and used as means to induce a cooperative solution. When a game is repeated, the information acquired by the parties might then build the trust (or a lack of it) to collaborate and select the "win-win" solution rather the "I win, you lose" solution. It might also reveal that such collaboration in not feasible due to parties' interests and attitudes and therefore seek some mechanism to render a hot conflict, a cold one.

Consider again the battle of sexes. If the man declares his strategy first, say with a gain of 2, the woman has a choice of gaining nothing (in which case both have a zero gain) or adopting the man's choice and gaining at least 1. Inversely, if the woman declares her strategy first then she will gain 2 and the man either 1 or both get nothing. In such a case, both are better off than adopting a randomized strategy. Such a situation is a special case. There are strategies that are better than the randomized equilibrium (which in some cases, it might be difficult to apply in practice) but such solutions might not be stable.

12.3.3 The Numerical Solution of Two Persons-Games: The Lemke-Howson Algorithm

The Nash conjecture defines the conditions for a Nash equilibrium to exist. To calculate it when the matrix is greater than say 3 strategies available to choose from (or select a randomized strategy), the computation of a solution is more difficult that the 2×2 matrix case considered here. Lemke and Howson (1964) have developed a technique which uses quadratic programming for finding the randomized strategies in a general (A,B) matrix. Let $(\mathbf{x}, \mathbf{y})$ denote the mixed strategies of both players. Namely, by definition of these probabilities:

$$(\mathbf{x}e_m, \mathbf{y}e_n) = (1, 1) \text{ with } \mathbf{x} \geq 0, \ \mathbf{y} \geq 0$$

where m is the number of choices given to the first player and n is the number of choices given to the second one. The parties expected outcome for the bi-matrix game are thus: $(\mathbf{xAy}^T, \mathbf{xBy}^T)$. Say that a Nash Solution is a pair of probability vectors $(\bar{\mathbf{x}}, \bar{\mathbf{y}})$ that satisfy the following conditions:

$$\left(\mathbf{xA\bar{y}}^T \leq \mathbf{\bar{x}Ay}^T\right) \text{ and } \left(\mathbf{\bar{x}By}^T \leq \mathbf{xB\bar{y}}^T\right)$$

But this implies also

$$\left(\mathbf{A\bar{y}}^T \leq \mathbf{\bar{x}Ay}^T e_m\right) \text{ and } \left(\mathbf{\bar{x}B} \leq \mathbf{xB\bar{y}}e_n{}^T\right)$$

Conversely, we suppose that $(\mathbf{A\bar{y}}^T \leq \mathbf{\bar{x}Ay}^T e_m)$ and$(\mathbf{\bar{x}B} \leq \mathbf{xB\bar{y}}^T e_n{}^T)$ holds for $(\bar{\mathbf{x}}, \bar{\mathbf{y}})$. Multiplying the first expression on the left by $\mathbf{x}$ and the second on the right by $\mathbf{y}^T$ we note that these expressions are equivalent to $(\mathbf{x}e_m, \mathbf{y}e_n) = (1, 1)$ with $\mathbf{x} \geq 0,\ \mathbf{y} \geq 0$ and $(\mathbf{xA\bar{y}}^T \leq \mathbf{\bar{x}Ay}^T)$ and $(\mathbf{\bar{x}By}^T \leq \mathbf{xB\bar{y}}^T)$. These equivalence are then used to construct the following quadratic programming problem (for a proof see Lemke and Howson 1964, pp. 413–423):

The Lemke–Howson Quadratic programming Problem:

$$\max_{x,y,p,q} \left(\mathbf{xA\bar{y}}^T + \mathbf{\bar{x}}B\mathbf{y}^T - \mathbf{p} - \mathbf{q}\right) \text{ Subject to : } \mathbf{A\bar{y}}^T \leq \mathbf{p}e_m \text{ and } \mathbf{B}^T\mathbf{x}^T \leq \mathbf{q}e_n$$

$$x_i \geq 0, i = 1, \ldots, m; y_j \geq 0, j = 1, \ldots, n \text{ and} \sum_{i=1}^{m} x_i = 1 \text{ and} \sum_{j=1}^{n} y_j = 1$$

12.3.4 *Negotiated Solution and the Nash Equilibrium*

In some cases, the solution of the two person game can be reached by maximizing one's outcome subject to a least outcome each can obtain. For example, let v^* and w^* be the minimax solution (value) each of the players can get. Let the probability that each party uses be p and q respectively. For a Nash solution, the value of the game is defined by $V(p,q) \leq v^*$ and $W(p,q) \leq w^*$. Now assume that the following conditions hold:

- Individual rationality, meaning that the Nash solution is always better than the minimax (zero sum game) solution, or $(V^*, W^*) \leq (v^*, w^*)$.
- Existence of a feasible solution (V^*, W^*)
- Pareto optimality. This means that there exists a feasible solution (V', W') and if $(V', W') \leq (V^*, W^*)$ then $(V^*, W^*) = (V', W')$
- Independence of irrelevant alternatives. This means that solutions over a smaller set of feasible solutions that contains the solution is also a solution.
- Independence of linear transformations. In other words the following have the same solution: $(V, W) \leq (\alpha_1 V + \beta_1, \alpha_2 W + \beta_2)$

- Symmetry. If the feasible set is symmetric and the values of both games are the same, then solutions are equal too, or $(V_1 = V_2)$.

When these conditions hold, Luce and Raiffa (1957) have shown that the Nash equilibrium solution is found by the constraint optimization problem:

$$\begin{aligned}&\mathrm{Max}(\mathrm{v} - \mathrm{V}(\mathrm{p},\mathrm{q}))(\mathrm{w} - \mathrm{W}(\mathrm{p},\mathrm{q}))\\ &\text{Subject to :}\\ &(V(p,q) \leq v),\ \ (W(p,q) \leq w),\ p \in [0,1],\ q \in [0,1]\end{aligned}$$

The solution of this problem is then solved by the usual Kuhn-Tucker conditions for optimality. Explicitly, define the general objective:

$$\operatorname*{Max}_{p,q}\,(v - V(p,q))(w - W(p,q)) + \lambda(V(p,q) - v) + \mu(W(p,q) - w)$$

The Necessary KT (Kuhn-Tucker) conditions for optimality are defined in terms of the following set of equations:

$$\begin{aligned}&-\frac{\partial V}{\partial p}(w - W(p,q)) - \frac{\partial W}{\partial p}(v - V(p,q)) + \frac{\partial V}{\partial p}\lambda + \frac{\partial W}{\partial p}\mu\\ &-\frac{\partial V}{\partial q}(w - W(p,q)) - \frac{\partial W}{\partial q}(v - V(p,q)) + \frac{\partial V}{\partial q}\lambda + \frac{\partial W}{\partial q}\mu\\ &\lambda = \begin{cases} 0 & \text{if } v - V(p,q) < 0 \\ > 0 & v = V(p,q) \end{cases};\ \mu = \begin{cases} 0 & \text{if } w - W(p,q) < 0 \\ > 0 & w - W(p,q) = 0 \end{cases}\end{aligned}$$

Again, such problems can be solved numerically. Note in particular that four problems are to be tested consisting of the following cases: $(\lambda > 0, \mu > 0)$, $(\lambda = 0, \mu > 0)$, $(\lambda > 0, \mu = 0)$, $(\lambda = 0, \mu = 0)$.

12.4 The Stackelberg Strategy

In a Stackelberg game, moves are sequential rather than simultaneous. The player who first announces his strategy is then the leader. The follower then chooses his best response to the leader's move. The leader has an advantage because he is able to optimize his objective function subject to the follower's best theoretical response. Formally this implies that if player A is the leader, then $y_B = y_B^R(y_A)$ is the best response for player B as determined by its Nash equilibrium. Since the leader is aware of this response, he then optimizes his objective function subject to $y_A = y_A^R(y_B) = y_A^R(y_B^R(y_A))$. An explicit solution is given below.

The Stackelberg Strategy: In a two-person game with player A, a leader, and player B a follower, a strategy $y_A{*} \in Y_A$ *is a Stackelberg equilibrium for the leader if for all* y_A, $J_A(y_A{*}, y_B^R(y_A{*})) \geq J_A(y_A, y_B^R(y_A))$, *where* $y_B = y_B^R(y_A)$ *is the best response function of the follower.*

The leader's Stackelberg solution is then: $y_A{}^* = \arg\max_{y_A \in Y_A} \{J_A(y_A, y_B^R(y_A))\}$. That is, if strategy sets are unconstrained and the payoff functions are continuously differentiable, the necessary condition for optimality of the leader is:

$$\frac{\partial J_A(y_A, y_B^R(y_A))}{\partial y_A}\Big|_{y_A = y_A{}^*} = 0.$$

To insure that the leader maximizes his objective, we check the second-order and sufficient condition:

$$\frac{\partial^2 J_A(y_A, y_B^R(y_A))}{\partial {y_A}^2}\Big|_{y_A = y_A{}^*} < 0.$$

12.5 Random Payoff and Strategic Risk Games

Random payoff games are defined as stated earlier in terms of random conditional consequences, or $\left(\tilde{\mathbf{A}} = [\tilde{a}_{ij}], \tilde{\mathbf{B}} = [\tilde{b}_{ij}]\right)$. When these matrices can be reduced to a deterministic format, the solution of the game can be reached by application of the Nash conjecture. For example, for a each pair of strategic alternatives $(i,j), i = 1, \ldots, n; j = 1, \ldots, m$, with random payoffs, $(\tilde{a}_{ij}, \tilde{b}_{ij})$, we can proceed as follows:

The Certain Equivalent bi-matrix game: Let $(u_A(.), u_B(.))$ be the utilities of two parties and $\left(E_a u_A(\tilde{a}_{ij}), E_b u_B(\tilde{b}_{ij})\right)$ be their expected utilities. Define the certain equivalents defined by $\left(u_A(c_{ij}) = E_a u_A(\tilde{a}_{ij}), u_B(d_{ij}) = E_b u_B(\tilde{b}_{ij})\right)$. The bi-matrix certain equivalent game is then: $(\mathbf{C}, \mathbf{D}) = \left(c_{ij} = {u_A}^{-1}(E_a u_A(\tilde{a}_{ij})), d_{ij} = {u_B}^{-1}(E_b u_B(\tilde{b}_{ij}))\right)$

Such a transformation reduces the game to a "standard nonzero sum game." For example, if the parties' utility functions are of the quadratic type and their certain equivalent, calculated as a function of their expected payoffs, their standard deviations and the parties index of absolute risk aversion (ρ_A, ρ_B), then the game certain equivalents payoffs are: $\left(c_{ij} = E_a(\tilde{a}_{ij}) - \rho_A \sigma_a(\tilde{a}_{ij});\ d_{ij} = E_b(\tilde{b}_{ij}) - \rho_B \sigma_b(\tilde{b}_{ij})\right)$. Since, the certain equivalents are also equal to: $\left(c_{ij} = E_a(\tilde{a}_{ij}) - \pi_{ij}^a; d_{ij} = E_b(\tilde{b}_{ij}) - \pi_{ij}^b\right)$, the game conditional consequences can be defined in terms of the risk premium associated to each conditional payoff of the bi-matrix game. The games outcomes (assuming a randomized strategy) and the game risk premium are then:

$$V_A = \sum_{i=1}^{n}\sum_{j=1}^{m}\left(E_a(\tilde{a}_{ij}) - \rho_A \sigma_a(\tilde{a}_{ij})\right)x_i y_j;\ V_B = \sum_{j=1}^{m}\sum_{i=1}^{n}\left(E_b(\tilde{b}_{ij}) - \rho_B \sigma_b(\tilde{b}_{ij})\right)x_i y_j$$

and

$$\prod_A = \sum_{i=1}^{n}\sum_{j=1}^{m} \pi_{ij}^{a} x_i y_j; \quad \prod_B = \sum_{j=1}^{m}\sum_{i=1}^{n} \pi_{ij}^{b} x_i y_j$$

where $x_i y_j$ denotes the probability that party A selects a strategy i while part B selects strategy j. In these games, a Nash rationality combined with an attitude towards risks bourn by the parties defined in terms of their utility functions can be analyzed. Of course, such an approach requires that the parties' utility functions be known which is a strong assumption.

Example: A random payoff game: Say that an exporter whose foreign income is W is faced with the decision to repatriate the income to the home country or not. If the exporter reinvests his profit out of the home country, he may lose an export incentive provided from the government. In other words, the government has at its disposal a strategic threat against exporters that do not repatriate their profits. Two situations arise, for exporters and the government alike. The exporter may repatriate profits or reinvest them externally. The government may or may not reduce its export subsidies. Four paired consequences are defined by the bi matrix (A,B) below.

Table: The bi-matrix game $\left(\tilde{A}, \tilde{B}\right)$

	Repatriate y	Do not repatriate (1-y)
Check exporter x	$\xi W(1-\tau) - C(1-\sigma);$ $\xi W\tau - C\sigma - K$	$\begin{cases} \xi W - C & wp\ 1\text{-}\delta \\ \xi W - C(1+\alpha) & wp\ \delta \end{cases};$ $\begin{cases} -K & wp\ 1\text{-}\delta \\ C\alpha - K & wp\ \delta \end{cases},\ \alpha > \sigma$
Do not check exporter 1-x	$\xi W(1-\tau) - C(1-\sigma);$ $\xi W\tau - C\sigma$	$\begin{cases} \xi W - C & wp\ 1\text{-}\delta \\ \xi W - C\sigma & wp\ \delta \end{cases};$ $\begin{cases} 0 & wp\ 1\text{-}\delta \\ -C\sigma & wp\ \delta \end{cases}$

The bi-matrix (A,B) entries define the exporter and the government's conditional payoffs and losses. If the exporter foreign revenues are W, the local profit of the exporter are $(\xi W(1-\tau) - C(1-\sigma))$ where the profit tax rate is τ and σ is a rate of subsidy on exports. As a result, the cost of export is $C(1-\sigma)$ if the exporter chooses to repatriate (and thus profit from the export subsidy). If the exporter chooses not to repatriate he can then "lie" and profit from the government subsidy, hoping not to be caught or be truthful and not profit from the subsidy. Profits in local currency are then $\xi W - C$. Let the cost of auditing the export firm be K and let the government revenues defined by the taxes it collects from the exporter repatriating his revenues less the cost of subsidies. In addition, if the exporter profits from the subsidy and does not repatriate (i.e. he lies), and is caught, a penalty is imposed which is an added revenue. This results in the bi-matrix game above. This is a "random payoff game with both the potential probability of "lying" by the exporter and the FX rate (which can be random). If we consider for simplicity expected values only, we have then the following bi-matrix game:

	Repatriate y	Do not repatriate (1-y)
Check exporter x	$\hat{\xi}W(1-\tau)-C(1-\sigma)$; $\hat{\xi}W\tau-C\sigma-K$	$\hat{\xi}W-C((1\text{-}\delta)+(1+\alpha)\delta)$; $-K+\delta C\alpha$
Do not check exporter 1-x	$\hat{\xi}W(1-\tau)-C(1-\sigma)$; $\hat{\xi}W\tau-C\sigma$	$\hat{\xi}W-C(1\text{-}\delta+\delta\sigma)$; $-C\delta\sigma$ $-C\sigma\delta$

According to this game, an exporter will always repatriate if:

$$\hat{\xi}W(1-\tau)-C(1-\sigma) \geq \hat{\xi}W-C((1\text{-}\delta)+(1+\alpha)\delta)$$

or

$$\hat{\xi}W(1-\tau)-C(1-\sigma) \geq \hat{\xi}W-C(1\text{-}\delta+\delta\sigma); -C\delta\sigma$$

$$\frac{\hat{\xi}W}{C}\tau \leq \sigma+\mathrm{Max}\{\alpha\delta,-\delta(1-\sigma))\}=\sigma+\alpha\delta$$

And therefore,

$$\frac{\hat{\xi}W}{C} \leq \frac{\sigma+\alpha\delta}{\tau}$$

The Nash equilibrium solution is defined however by the following:

$$x^*=\frac{C\sigma(1+\delta)-\hat{\xi}W\tau}{2C(\sigma+\delta(\alpha+\sigma))},\ y^*=\frac{(1+\alpha)\delta+2\delta\sigma C}{C\delta(\alpha-\sigma)}$$

Such a solution then provides a predictive model for tax reporting as a function of some of the parameters that may be accounted for when reaching such a decision and thus be use to control exporters propensity to repatriate their foreign revenues.

12.5.1 A Risk Constrained Random Payoff Games: A Heuristic Interior Solution

This case applies Neymann-Pearson types I and II risks to constraint a games' risks. The rational of such an approach is as follows. Consider the following two parties random payoffs game with matrices:

$$(\tilde{\mathbf{A}},\tilde{\mathbf{B}})=\left(\tilde{a}_{ij},\tilde{b}_{ij}\right),\sum_{i=1}^{n}x_i=1,\ 0\leq x_i\leq 1,\ \sum_{j=1}^{m}y_j=1,\ 0\leq y_j\leq 1$$

A game random payoffs are given by: $\tilde{V}_a = \mathbf{x}\tilde{\mathbf{A}}\mathbf{y}^{\mathbf{T}}, \tilde{V}_b = \mathbf{x}\tilde{\mathbf{B}}\mathbf{y}^{\mathbf{T}}$. Assume that aggregate risk constraints have been negotiated which require that for both parties:

$$P\{\tilde{\mathbf{A}}\mathbf{y} \leq \tilde{V}_a\} \geq 1 - \xi \text{ and } P\{\mathbf{x}\tilde{\mathbf{B}} \leq \tilde{V}_b\} \geq 1 - \eta$$

A risk constrained game is then defined by:

$$\begin{aligned}
&\text{MaxMax}(\hat{\mathbf{A}}, \hat{\mathbf{B}}) = (\tilde{a}_{ij}, \tilde{b}_{ij}), \text{ Subject to :} \\
&\quad P\{\tilde{\mathbf{A}}\mathbf{y} \leq \tilde{V}_a\} \geq 1 - \xi \text{ and } P\{\mathbf{x}\tilde{\mathbf{B}} \leq \tilde{V}_b\} \geq 1 - \eta \\
&\text{and } \sum_{i=1}^{n} x_i = 1, \ 0 \leq x_i \leq 1, \ \sum_{j=1}^{m} y_j = 1, \ 0 \leq y_j \leq 1
\end{aligned}$$

For example, consider the bi matrix game

$$\tilde{V}_a = \mathbf{x}\tilde{\mathbf{A}}\mathbf{y}^{\mathbf{T}} \sim N(\mu_a, \sigma_a^2), \quad \tilde{V}_b = \mathbf{x}\tilde{\mathbf{B}}\mathbf{y}^{\mathbf{T}} \sim N(\mu_b, \sigma_b^2)$$

In which case:

$$\begin{aligned}
\tilde{V}_a &= (\tilde{a}_{11} - \tilde{a}_{12} - \tilde{a}_{21} + \tilde{a}_{22})xy + (\tilde{a}_{12} - \tilde{a}_{22})x + (\tilde{a}_{21} - \tilde{a}_{22})y + \tilde{a}_{22} \\
\tilde{V}_b &= (\tilde{b}_{11} - \tilde{b}_{12} - \tilde{b}_{21} + \tilde{b}_{22})xy + (\tilde{b}_{12} - \tilde{b}_{22})x + (\tilde{b}_{21} - \tilde{b}_{22})y + \tilde{b}_{22}
\end{aligned}$$

Expected outcomes and variance can then be calculated. For example,

$$\begin{aligned}
E(\tilde{V}_a) &= (\hat{a}_{11} - \hat{a}_{12} - \hat{a}_{21} + \hat{a}_{22})xy + (\hat{a}_{12} - \hat{a}_{22})x + (\hat{a}_{21} - \hat{a}_{22})y + \hat{a}_{22} \\
E(\tilde{V}_b) &= (\hat{b}_{11} - \hat{b}_{12} - \hat{b}_{21} + \hat{b}_{22})xy + (\hat{b}_{12} - \hat{b}_{22})x + (\hat{b}_{21} - \hat{b}_{22})y + \hat{b}_{22}
\end{aligned}$$

While for statistically independent outcomes, we have a variance:

$$\begin{aligned}
\text{var}(\tilde{V}_a) &= \text{var}(\tilde{a}_{11})x^2y^2 + \text{var}(\tilde{a}_{12})x^2(1-y)^2 + \text{var}(\tilde{a}_{21})(1-x)^2y^2 \\
&\quad + \text{var}(\tilde{a}_{22})(1-x)^2(1-y)^2 \\
\text{var}(\tilde{V}_b) &= \text{var}(\tilde{b}_{11})x^2y^2 + \text{var}(\tilde{b}_{12})x^2(1-y)^2 + \text{var}(\tilde{b}_{21})y^2(1-x)^2 \\
&\quad + \text{var}(\tilde{b}_{22})(1-x)^2(1-y)^2
\end{aligned}$$

Certain equivalents for both players are then (see Chap. 7):

$$\Lambda_a = E(\tilde{V}_a) - \lambda_a\sqrt{\text{var}(\tilde{V}_a)} \text{ and } \Lambda_b = E(\tilde{V}_b) - \lambda_b\sqrt{\text{var}(\tilde{V}_b)}$$

Note that the solution of this game is then given by:

$$\frac{\partial \Lambda_a}{\partial x}=0 \text{ or} \frac{\partial E(\tilde{V}_a)}{\partial x}\sqrt{\operatorname{var}(\tilde{V}_a)}=\frac{1}{2}\frac{\partial \operatorname{var}(\tilde{V}_a)}{\partial x}$$
$$\frac{\partial \Lambda_b}{\partial y}=0 \text{ or} \frac{\partial E(\tilde{V}_b)}{\partial y}\sqrt{\operatorname{var}(\tilde{V}_b)}=\frac{1}{2}\frac{\partial \operatorname{var}(\tilde{V}_b)}{\partial y}$$

With derivatives:

$$\frac{\partial E(\tilde{V}_a)}{\partial x}=(\hat{a}_{11}-\hat{a}_{12}-\hat{a}_{21}+\hat{a}_{22})y+(\hat{a}_{12}-\hat{a}_{22}) \text{ and}$$
$$\frac{\partial E(\tilde{V}_b)}{\partial y}=(\hat{b}_{11}-\hat{b}_{12}-\hat{b}_{21}+\hat{b}_{22})x+(\hat{b}_{21}-\hat{b}_{22})$$

$$\frac{1}{2}\frac{\partial \operatorname{var}(\tilde{V}_a)}{\partial x}=var(\tilde{a}_{11})xy^2+\operatorname{var}(\tilde{a}_{12})x(1-y)^2-\operatorname{var}(\tilde{a}_{21})(1-x)y^2$$
$$-\operatorname{var}(\tilde{a}_{22})(1-x)(1-y)^2$$
$$\frac{1}{2}\frac{\partial \operatorname{var}(\tilde{V}_b)}{\partial y}=\operatorname{var}(\tilde{b}_{11})x^2y-\operatorname{var}(\tilde{b}_{12})x^2(1-y)+\operatorname{var}(\tilde{b}_{21})y(1-x)^2$$
$$-\operatorname{var}(\tilde{b}_{22})(1-x)^2(1-y)$$

Given the derivatives above, and assuming for simplicity an interior solution, a solution is found by the solution of the two nonlinear equations in x and y as a function of the two parameters (λ_a, λ_b) expressing the aversion to each of the parties. Note that this solution differs from the solution obtained by replacing all random payoffs by their expected utility.

12.6 Bayesian Theory and Bayesian Games

Should we invest in a domestic market or in a foreign one? Should we collaborate with a foreign firm or not? Should we increase the domestic interest rate as a protective measure against an exchange rate downfall and face retaliation from another country? Should we be rational and reach a decision on the basis of what we know our competition might do, or should we combine our prior and subjective assessment with an unfolding evidence-intelligence? This occurs in many instances. For example, when information is so overbearing or so limited that one must take an educated or at best an intuitive guess about competitors' potential actions. In most cases, steps are to be taken to limit and "bound" our decision procedures for otherwise no decision can be reached in a reasonable time. These "bounds" are varied and underlie theories of "bounded rationality" based on the premise that we can only do the best we can and no better! However, when problems are well defined, when they are formulated properly, meaning that the alternatives are well-stated, the potential events well-established, and their conditional consequences are determined, we can presume that a rational procedure based on the model we have built can be followed. If, in addition, the risks inherent in the problem are explicitly stated, a rational decision can be reached.

To proceed rationally it is necessary for an individual decision maker (an investor for example) to reach a judgment about: the alternatives available, the sources of uncertainties, the conditional outcomes and preferences needed to order and price alternatives. In a game theoretical framework, it is equally important to assess what the "other party" alternatives, information, estimates and intents are. Then, combine them without contradicting oneself, (i.e. by being rational) in selecting the best course of action to follow. Further, to be rational it is necessary to be self-consistent in stating what we believe or are prepared to accept and then accept the consequences of our actions. Of course, it is possible to be "too rational." To be a practical investor, one must accept that there is a "bounded rationality" and that an investment will in the end bear some risk we did not plan to assume. This understanding is an essential motivation for applying the Bayes (subjective risk) approach. Underlying its rationality is a number of relations that assume (Rubinstein 2000): Knowledge of the problem; Clear preferences; an ability to optimize; indifference to equivalent logical descriptions of alternative and choice sets. In addition, knowing equally ones "competition" is as important (and in some cases, it is more important). Psychologists and economists have doubted these assumptions. The fact that decisions are not always rational does not mean that there are no underlying procedures to the decision making process. A systematic approach to departures from rationality has been the topic of intense economic and psychological interest of particular importance in finance, emphasizing "what is" rather than "what ought to be." Both Behavioral Finance and Behavioral Games have addressed these issues as well.

Rationality is also a "bounding" qualitative and de-facto unavoidable concept that is based on essentially four dimensions: Stealth alternatives; The analysis of information; The perception of risk and: Decision-making. It may be defined and used in different manners. "Classical rationality" as defined by Muth (1961) and Lucas (1978), underlies Rational Expectations that assumes that decision makers use all the available information, perceive risk without bias and make the best possible decisions (given an ability to compute) with the information they possess at the time the decision is made. By contrast, a "Bayesian Rationality," has a philosophically different approach. Whereas Muth and Lucas suppose that an investor extrapolates from the available information to a true distribution of payoffs, Bayesian rationality supposes that we have a prior subjective distribution of payoffs that is updated through a learning mechanism as new information unfolds. Further, Muth and Lucas suppose that this prior or subjective distribution is the true one, imbedding future realizations while the Bayes' approach supposes that the investor's belief or prior is indeed subjective but evolving through learning to the true distribution. In other words, it summarizes past experiences to predict a future rather than observe and interpret it as a manifestation of a forthcoming and certain future (although framed in a probability language). These "differences of opinion" have substantive impact on how we develop our approach to decision making. In a game situation, learning what is the likelihood of a party selecting one or the other strategy has similar implications. For Muth and Lucas, the present is "the present of the future" while for Bayes it incorporates learning from one's bias (prejudice or misconception) into a measurement and hence a decision reached gradually as learning sets in. When a subjective rationality is used in games, it results in "games with learning" with information acquired during the conduct of the game to better estimate both parties' strategic probabilities and their consequences. Below, we review first some principles of Bayesian decision making.

12.6.1 Bayes Decision Making

Bayesian rational decision making involves as stated in Chap. 3, the following elements:

1. A decision to be taken from a set of known alternatives.
2. Risk defined in terms of events with associated known (subjective) probabilities.
3. Conditional consequences resulting from the selection of a decision and the occurrence of a specific event (once uncertainty, ex-post, is resolved).
4. A preference over consequences, i.e. there is a well specified preference function or procedure for selecting a specific alternative among a set of given alternatives.

In a game theoretic context, a player may know his alternatives, but the counter party's alternatives may be unknown to him or known partially. As a result, the consequences of the game may not be estimated easily. For an indifferent decision-maker there is no problem. A problem arises when certain outcomes are preferred over others (such as making more money over less) and when preferences are sensitive to the risks associated to such outcomes. What are these preferences? There are several possibilities, each based on the information available—what is known and not known and how we balance the two, our attitude toward risk (or put simply, how we relate to the probabilities of uncertain outcomes, their magnitude and their adverse consequences).

12.6.2 Examples: Bayesian Calculus

First, consider a decision problem with n states of nature $s, j = 1, 2, \ldots n$ denoting stock movements, each assumed to have a prior probability $p(s_j)$ expressing our current belief. In this case, the mean and the spread (variance) are, $\mu = \sum_{i=1}^{n} s_i p(s_i)$,

$\sigma^2 = \sum_{i=1}^{n} (s_i - \mu)^2 p(s_i)$. Now let $p(x_1|s_j)$ be the probability of observing x_1 conditional on state of nature s_j occurring (also called the likelihood of the datum x_1). Then, $p(x_1, s_j) = p(x_1|s_j)p(s_j)$. By symmetry, we have $p(x_1, s_j) = p(s_j, x_1)$ and therefore, $p(x_1, s_j) = p(s_j|x_1)p(x_1)$ leading to:

$$p(x_1) = \sum_{j=1}^{n} p(x_1|s_j)p(s_j) \quad \text{and as a result} \quad p(s_j|x_1) = \frac{p(x_1|s_j)p(s_j)}{\sum_{j=1}^{n} p(x_1|s_j)p(s_j)}$$

which is Bayes Theorem (see also Chap. 3). This problem can be extended further to say two parties (one willing to sell, the other willing to buy), each having a prior estimate of forthcoming stock movements, say that they are given for both as follows:

$$p_1^1(x_1) = \sum_{j=1}^{n} p_1^1(x_1|s_j^1) \text{ and } p_1^2(x_1) = \sum_{j=1}^{n} p_1^2(x_1|s_j^2)$$

And therefore each updating his own estimate the probabilities of future price changes using Bayes theorem, or:

$$p_1^1(s_j^1|x_1) = \frac{p_1^1(x_1|s_j^1)p^1(s_j)}{\sum_{j=1}^{n} p_1^1(x_1|s_j^1)p^1(s_j)} \text{ and } p_1^2(s_j^1|x_1) = \frac{p_1^2(x_1|s_j^1)p^2(s_j)}{\sum_{j=1}^{n} p_1^2(x_1|s_j^1)p^2(s_j)}$$

On the basis of public information, treated privately, estimates of future market states are then reached that can lead to different decisions by the parties. For example, both buying (and the market price increases), both selling (and the market price decreases) or one selling and the other buying (and the price remains unchanged).

Example. Say that the probability of making money is an "all X or nothing."

$$\tilde{x} = \begin{cases} X & w.p.\ p \\ 0 & w.p.\ 1-p \end{cases}$$

where "p" is the probability of success and "X" is the money collected. The probability is assumed to be distributed by a Beta probability distribution with parameters "r" and "s," or: $\tilde{p} \sim B(r, s)$. As information is revealed, the probability distribution is revised. If success occurs, then, the Bernoulli-beta process yields a posterior beta distribution with parameters $1 + r$, and s, $B(1 + r, s)$. However, if no success occurs, the posterior distribution is beta but given by the following

$B(r, 1+s)$. In this sense, an update scheme is given by a Beta distribution $B(r+q, 1-q+s), q = 1, 0$.

A mathematical note: Let m be the number of successes in a Bernoulli experiment and n be the total number of Bernoulli experiments. For a Beta prior probability distribution, the posterior has also a Beta distribution with:

$$\tilde{PD} \sim B(a,b), \hat{PD} = \frac{a}{a+b} = p_{0,0} : \text{Prior Distribution}$$

$$p_{0,0} \Rightarrow \begin{cases} p_{1,1} & Ep_{1,1} = \frac{a+1}{a+b+2} \\ p_{0,1} & Ep_{0,1} = \frac{a}{a+b+1} \end{cases}$$

$$p_{m,n} = B(a_{m,n}, b_{m,n}), a_{m,n} = a + m, b_{m,n} = b + n$$

Example: Reproducing Probability Models and Bayesian Learning Assume that a player 1 estimates that the prior probability of selecting one of two strategies by the counter party is $q_0 \in [0,1]$ and let $f(q_0)$ be the probability of this estimate which we assume to be given by a Beta probability distribution $B(r_0, s_0)$ with parameters (r_0, s_0). The probability distribution is:

$$f(q_0) = \frac{\Gamma(r_0)\Gamma(s_0)}{\Gamma(r_0+s_0)} {q_0}^{r_0-1}(1-q_0)^{s_0-1}, 0 < q_0 \le 1$$

$$E(q_0) = \frac{r_0}{r_0+s_0}, \quad \text{var}(q_0) = \frac{r_0 s_0}{(r_0+s_0)^2(r_0+s_0+1)}$$

Similarly, the second player has a prior estimate of the first player selecting his first strategy with probability $p_0 \in [0,1]$, with a Beta probability distribution $B(u_0, v_0)$ with mean and variance:

$$E(p_0) = \frac{u_0}{u_0+v_0} \quad \text{and} \quad \text{var}(p_0) = \frac{u_0 v_0}{(u_0+v_0)^2(u_0+v_0+1)}$$

These prior estimates can be calculated based on the initial estimate of the game matrix of outcomes. Assuming that the game solution provides a randomize strategy for each of player, we then set:

$$\frac{u_0}{u_0+v_0} = \text{Prior Probability of selecting the first strategy by player 1}$$
$$\frac{r_0}{r_0+s_0} = \text{Prior Probability of selecting the first strategy by player 2}$$

Following the play of the game, both are observed and both select the first strategy. Bayesian learning-adaptation would then lead to updating these estimates to: $B(r_0+1, s_0+1)$ and $B(u_0+1, v_0+1)$. If say the first player selects the second strategy and the second the first strategy then Bayes learning indicates that: $B(r_0+1, s_0+1)$ and $B(u_0, v_0+1)$. Generally, let $x_0 = 1, 0$ and $y_0 = 1, 0$ be indicators expressing what the first and second players actually do. Then, following the first game, the following probability distributions are obtained:

$$B(r_1, s_1),\ r_1 = r_0 + y_0, s_1 = s_0 + 1 \quad \text{and} \quad B(u_1, v_1), u_1 = u_0 + x_0, v_1 = v_0 + 1$$

In the following period, say 2, the distributions are:

$$B(r_2, s_2), r_2 = r_0 + \sum_{i=0}^{1} y_i,\ s_2 = s_0 + 2 \quad \text{and}$$

$$B(u_2, v_2), u_2 = u_0 + \sum_{i=0}^{1} x_i, v_2 = v_0 + 2$$

And generally at period t, we have:

$$B(r_t, s_t),\ r_t = r_0 + \sum_{i=0}^{t} y_i,\ s_t = s_0 + t \quad \text{and} \quad B(u_t, v_t), u_t = u_0 + \sum_{i=0}^{t} x_i, v_t = v_0 + t$$

where $(x_i, y_i), i = 0, 1, \ldots, t-1$ (a Filtration) denotes the history of game. A naïve prediction of one or the other players probability distributions at time t for time t + 1 are thus:

$$E_I(+1) = \left(\frac{u_0 + \sum_{i=0}^{t-1} x_i}{u_0 + \sum_{i=0}^{t-1} x_i + v_0 + t + 1} \right),\ E_{II}(+1) = \left(\frac{r_0 + \sum_{i=0}^{t-1} y_i}{r_0 + \sum_{i=0}^{t-1} y_i + 0, s_0 + t + 1} \right)$$

On the basis of these probability estimates, an expectation over each of the future outcomes for each player can then be calculated.

The usefulness of this example is in its simplicity. Issues the pertain to Bayesian games are however for more intricate as references at the end of this chapter indicate.

12.7 Mean Field Games and Finance

Mean Field Games (MFG, see Lasry and Lions 2006a, b, 2007a, b; Lasry et al. 2010a, b) is based on a concept of Mean Field Theory profusely used in Physics, also known as self-consistent field theory. Explicitly, consider the interactions of molecules in a medium. Each molecule cannot affect individually the medium but all the molecules together, define the medium and its characteristics. Mean Field Theory suggests that we represent the interaction of the molecule with a representative statistical behavior of the medium in a given set of conditions. MFG provides an extension by considering each "molecule" with "a mind of its own" and studying the reciprocal effects of the aggregate statistical behavior of the medium and that of "the molecule." At the limit, when the number of molecules is "infinite," the "ensemble" of all molecules, each pursuing its "preferences" defines the characteristics of the statistical medium. The MFG finance framework may then be stated in terms of four hypotheses:

- Rational expectations
- A continuum of agents
- Agents anonymity
- An interaction between agents is of the mean field type

For example, each individual pursues his preferences by selecting an investment strategy, yet his selection of stocks or fixed income assets do not affect by themselves the market. In other words the market price is the result of an infinite number of economic agents, investors, speculators with choices that define an equilibrium price. The very important results of Lasry and Lions show how such a problem can be transformed into an optimization problem consisting of two coupled equations. One equation defines the condition for the party's optimum in terms of his preferences and a statistical representation of the market mechanism (such as prices) and a second equation is a definition of this statistical (price) process. Such problems are unfortunately difficult to solve, although a number of specific problems have been solved and numerous articles appear to apply and profit from the MFG models. Chapters 8 and 9 have elaborated an approach for a multi-agents CCAPM price model based on the MFG principle that "individuals" actions (and thus their preferences), define the "medium".

Additional Reading and References

Game theory underlies numerous and fundamental results in economics (Dresher 1961; Harsanyi 1967–1968; Harsanyi and Selten 1977; Owen 1982; Shubik 1955, 2002; Rubinstein 1998 on rationality and games, Gintis 2009; Maschler 1966; Von Stengel 1991 on recursive games, Kogan and Tapiero 2007 on Supply Chain Games, Reyniers and Tapiero 1995a, b; Bassan 2003 on options and game theory Bassan et al. 2003). Combining behavioral experiments and psychology a trend to defined a solution to games based on such principles have attracted a great deal of attention (Camerer and Weber 1992; Camerer 2003, Behavioral Game Theory). These approaches are in fact challenging the basic precepts of rationality used by decision makers. There are therefore, numerous approaches one may use to define alternative solutions to the celebrated Nash solution. Satisficing games for examples express a bounded rationality in formulating an alternative solution to Nash' equilibrium includes (see also Neyman and Okada 1999, on satisficing as well as Stirling 2002; Stirling and Goodrich 1999; Reesor and McLeish 2001).

Recently, a great deal of attention is devoted to Mean Field Games, cursorily described in this chapter. This approach provides an avenue to reconcile in economics and finance both micro economic decisions and macro-economic ones (resulting from an aggregate of economic factors). Essential contributions are due to Lasry and Lion as indicated in the text. There is however a large number of publications and an interest of this approach in a large number of domains. Some additional references include, Achdou and Dolcetta (2010), Gomes et al. (2010), Gueant's (2009) thesis, Lachapelle (2010) on human crowds and groups interactions using a MFG approach, Lachapelle et al. (2010) on the computation of mean field equilibria in economics, Lachapelle and Wolfram (2011), on MFG modeling of congestion and aversion in pedestrian crowds, etc. This approach provides under specific cases, a solution to N-persons infinite games addressed to by Bensoussan and Frehse (2000, 2009) and Bensoussan et al. (2010).

References

Aas, K., Czado, C., Frigessi, A., & Bakken, H. (2009). Pair copula constructions of multiple dependence. *Insurance Mathematics and Economics, 44*, 182–198.

Abdellaoui, M., Luce, R. D., Machina, M. J., & Munier, B. (Eds.). (2007). *Uncertainty and risk: Mental, formal and experimental representations*. Berlin: Springer.

Abdellaoui, M., & Munier, B. (1997). Experimental determination of preferences under risk: The case of very low probability radiation. *Ciencia and Tecnologia dos Materiais, 9*, 25–31.

Abdellaoui, M., & Munier, B. (1999). How consistent are probability tradeoffs in individual preferences under risk? In M. J. Machina & B. Munier (Eds.), *Beliefs, interactions and preferences in decision making* (pp. 285–295). Dordrecht/Boston: Kluwer Academic.

Abe, S. (1997). *Physics Letters A, 224*, 326.

Abramowitz, M., & Stegun, I. A. (1965). *Handbook of mathematical functions*. New York: Dover.

Acerbi, C. (2002). Spectral measures of risk: A coherent representation of subjective risk aversion. In G. Szego (Ed.), Beyond VaR (special issue). *Journal of Banking and Finance, 26*. http://www.gloriamundi.org/var/wps.html.

Acerbi, C., & Tasche, D. (2002). On the coherence of expected shortfall. In G. Szego (Ed.), Beyond VaR (special issue). *Journal of Banking and Finance, 26*. http://www.gloriamundi.org/var/wps.html.

Acharya, V. V., & Johnson, T. C. (2007). Insider trading in credit derivatives. *Journal of Financial Economics, 77*, 110–141.

Achdou, Y., & Dolcetta, I. C. (2010). *Mean field games: Numerical methods*. Preprint.

Adler, M., & Dumas, B. (1980). Foreign exchange risk management. In B. Antl (Ed.), *Currency risk management* (pp. 145–158). London: Euromoney.

Adrian, T., & Brunnermeir, M. (2009). CoVaR, FRB of New York, Staff report No. 348, ssrn.com/abstract 1269446.

Agrawal, V., & Seshadri, S. (2000). Risk intermediation in supply chains. *IIE Transactions, 32*, 819–831.

Agrell, P. J., Lindroth, R., & Norrman, A. (2004). Risk, information and incentives in telecom supply chain. *International Journal of Production Economics, 90*(1), 1–16.

Ahlbrecht, M., & Weber, M. (1997). Preference for gradual resolution of uncertainty. *Theory and Decision, 43*, 167–185.

Ait-Sahalia, Y. (1996). Nonparametric pricing of interest rate derivative securities. *Econometrica, 64*, 527–560.

Ait-Sahalia, Y., & Lo, A. W. (2000a). Nonparametric estimation of state price densities implicit in financial asset prices. *Journal of Finance, 53*(1998), 499–548.

Ait-Sahalia, Y., & Lo, A. W. (2000b). Nonparametric risk management and implied risk aversion. *Journal of Econometrics, 94*, 9–51.

C.S. Tapiero, *Engineering Risk and Finance*, International Series in Operations Research & Management Science 188, DOI 10.1007/978-1-4614-6234-7,

Akerlof, G. (1970). The market for lemons: Quality uncertainty and the market mechanism. *Quarterly Journal of Economics, 84*, 488–500.

Akturk, E., Bagci, G. B., & Sever, R. (2007). Is Sharma-Mittal entropy really a step beyond Tsallis and Renyi entropies? asXiv:cond-mat/07032771v1, gbb0002@unt.edu.

Ale, B. (2002). Risk assessment practices in the Netherlands. *Safety Science, 40*, 105–126.

Ale, B., Smith, E., & Pitblado, R. (2000). *Safety around airport—Developments in 1990s and future directions*. London: Det Norske Veritas.

Aleksiejuk, A., & Holyst, J. A. (2001). A simple model of bank bankruptcies. *Physica A: Statistical Mechanics and Applications, 299*, 198–204.

Alexander, J. (1993). The lawsuit avoidance theory of why initial public offerings are underpriced. *UCLA Law Review, 41*, 17–73.

Alexander, C. (1998). *Risk management and analysis* (Volume 1 & 2). Wiley.

Allais, M. (1953). Le comportement de l'homme rationnel devant le risque, Critique des postulats et axiomes de l'ecole Americaine. *Econometrica, 21*, 503–546.

Allais, M. (1979a). The foundations of a positive theory of choice involving risk and criticism of the postulates of the American school. In M. Allais & O. Hagen (Eds.), *Expected utility hypotheses and the Allais paradox* (pp. 27–145). Dordrecht: Holland.

Allais, M. (1979b). The so-called Allais paradox and rational decisions under uncertainty. In M. Allais & O. Hagen (Eds.), *Expected utility hypotheses and the Allais paradox*. Dordrecht/Boston: Reidel.

Allen, F., & Faulhaber, G. (1989). Signaling by underpricing in the IPO market. *Journal of Financial Economics, 23*, 303–324.

Allen, F., & Gale, D. (2000). Financial contagion. *Journal of Political Economy, 108*, 1–33.

Alliance for Nuclear Responsibility. Factsheet #2 Security Just the Facts: The Five Fatal Flaws of Nuclear Power, San Luis Obispo, Public Citizen. http://a4nr.org/articles/fatalFlaws/2.security.pdf/download.

Almeida C., & Garcia, R. (2008). *Empirical likelihood estimators for stochastic discount factors*. Working Paper, Getulio Vargas Foundation and EDHEC Business School (SSRN Listing).

Altman, E. I., & Riken, H. A. (2004). How rating agencies achieve rating stability. *Journal of Banking and Finance, 28*, 2679–2714.

Amaral, L. A. N., Bulkdyrev, S. V., Havlin, S. V., Leschron, H., Mass, P., Salinger, M. A., Stanley, H. E., & Stanley, M. H. R. (1997). *Journal of Physics I,* France, 621.

Amato, J. D., & Remolona, E. M. (2005). *The pricing of unexpected credit losses*. BIS, Working Paper No. 190.

Anderlini, L., & Felli, L. (1994). Incomplete written contracts: Undescribable states of nature. *Quarterly Journal of Economics, 109*, 1086–1124.

Anderson, E. (1957). On the collective theory of risk in the case of contagion between the claims. In: *Transactions XVth International Congress of Actuaries, II* (pp. 219–222). New York.

Anderson, L., & Sidenius, J. (2004). Extensions to the Normal Copula: Random recovery and random factor loading. *Journal of Credit Risk, 1*, 29–70.

Ang, B. W. (1999). Is the energy intensity a less useful indicator than the carbon factor in the study of climate change? *Energy Policy, 27*, 943–946.

Ang, B. W., & Pandiyan, G. (1997). Decomposition of energy-induced CO2 emissions in manufacturing. *Energy Economics, 19*, 363–374.

Angell, L. C., & Klassen, R. D. (1999). Integrating environmental issues into the mainstream: An agenda for research in operations management. *Journal of Operations Management, 17*, 575–598.

Anscombe, F. J., & Aumann, R. J. (1963). A definition of subjective probability. *Annals of Mathematical Statistics, 34*, 199–205.

Antelman, G. R., & Savage, I. R. (1965). Surveillance problems: Wiener processes. *Naval Research Logistics Quarterly, 12*, 35–55.

Argonne National Laboratory, & Greenhouse Gases. (2005). Regulated emissions, and energy use in transportation (GREET 1.6) model. Transportation Technology Research and Development Center. http://www.transportation.anl.gov/ttrdc/greet. Accessed on March 10.

Arrow, K. J. (1951a). *Social choice and individual values*. New York: Wiley.

Arrow, K. J. (1951b). Alternative approaches to the theory of choice in risk-taking situations. *Econometrica, 19*, 404–437.

Arrow, K. J. (1963a). Aspects of the theory of risk bearing, YRJO Jahnsson lectures, also in 1971. *Essays in the theory of risk bearing*. Chicago: Markham Publishing Company.

Arrow, K. J. (1963b). Uncertainty and the welfare economics of medical care. *American Economic Review, 53*(5), 941–973.

Arrow, K. J. (1982). Risk perception in psychology and in economics. *Economics Inquiry, 20*, 1–9.

Arrow, K. J., Colombatto, E., Perlman, M., & Schmidt, C. (Eds.). (1996). *The rational foundations of economic behavior*. London: Macmillan.

Artzner, P. (1999). Application of coherent measures to capital requirements in insurance. *North American Actuarial Journal, 3*(2), 11–25.

Artzner, P., Delbaen, F., Eber, J. M., & Heath, D. (1999). Coherent risk measures. *Mathematical Finance, 9*, 203–228.

Artzner, P., Delbaen, F., Eber, J.-M., & Heath, D. (2000). Risk management and capital allocation with coherent measures of risk. Retrieved October, from http://www.math.ethz.ch/finance.

Artzner, P., Delbaen, F., Eber, J.-M., Heath, D., & Ku, H. (2001). Coherent multiperiod risk adjusted values. RetrievedOctober, from http://www.math.ethz.ch/finance.

Artzner, P., Delbaen, F., Eberand, J. M., & Heath, D. (1997). Thinking coherently. *Risk, 10*, 68–71.

Aschauer, D. A. (1989a). Is public expenditure productive. *Journal of Monetary Economics, 23*, 177–200.

Aschauer, D. A. (1989b). Does public capital crowd out private capital. *Journal of Monetary Economics, 24*, 171–188.

Ashby Ross, W. (1956). *An introduction to cybernetics*. London: Chapman & Hall.

Aumann, R. J. (1974). Subjectivity and correlation in randomized strategies. *Journal of Mathematical Economics, 1*, 67–96.

Aumann, R. (1976). Agreeing to disagree. *The Annals of Statistics, 4*, 1236–1239.

Avallenada, M., & Zhu, J. (2001). Distance to default. *Risk, 14*(12), 125–129.

Avenhaus, R. (1994). Decision theoretic analysis of pollutant emission monitoring procedures. *Annals of Operations Research, 54*, 23–28.

Avenhaus, R., Canty, M. D., Kilgour, D. M., von Stengel, B., & Zamir, S. (1996). Inspection games in arms control. *European Journal of Operational Research, 90*(3), 383–394.

Avenhaus, R., & Kilgour, D. M. (2004). Efficient distributions of arms-control inspection effort. *Naval Research Logistics, 51*, 1–27.

Avenhaus, R., Stengel, B. V., & Zamir, S. (1995). Inspection game. In R. J. Aumann & S. Hart (Eds.) *Handbook of game theory,* Elsevier vol. 3.

Bahr, N. (1997). *System safety engineering and risk assessment: A practical approach*. London: Taylor and Francis.

Bai, X., Wang, Y., Dai, G., Tsai, W. T., & Chen Y. (2007). *A framework for contract-based collaborative verification and validation of web services*. ACM SIGSOFT CBSE.

Bairamov, I., & Elmmastas Gultekin, O. (2010). Discrete distributions connected with the bivariate binomial, Hacettepe. *Journal of Mathematics and Statistics, 39*(1), 109–120.

Bakker, G. J., & Blom, H. A. P. (1993). Air traffic collision risk modeling. In: *Proceedings of the 32nd IEEE Conference on Decision and Control*, San Antonio.

Banks, D., Petralia, F., & Wang, S. (2011). Adversarial risk analysis, Borel games. *Applied Stochastic Models, 27*(2), 72–86.

Barlow, R., & Proschan, F. (1965). *Mathematical theory of reliability*. New York: Wiley.

Barnett, A. (2000). Free-flight and en route air safety: A first-order analysis. *Operations Research, 48*, 833–845.

Barnett, V., & Toby Lewis, L. (1994). *Outliers in statistical data* (3rd ed.). New York: Wiley.

Barrois, T. (1834). Essai sur l'application du calcul des probabilités aux assurances contre l'incendie. *Mem. Soc. Sci. De Lille,* 85–282.

Barry, C. B., & Jennings R. H. (1991). *The opening price performance of initial public offerings of common stock.* Presented at the 1991 Financial Management Association Meetings in Chicago.

Bartram, S. M., Brown, G. W., & Hund, J. E. (2007). Estimating systemic risk in the international financial system. *Journal of Financial Economics, 86*(3), 835–869. doi:10.1016/j.jfineco.2006.10.001ISSN0304-405X.

Barzel, Y. (1982a). *Knight's moral hazard theory of organization.* Mimeo, Department of Economics, University of Washington.

Barzel, Y. (1982b). Measurement cost and the organization of markets. *Journal of Law and Economics, 25,* 27–47.

Basak, S., & Shapiro, A. (2001). Value-at-risk-based risk management: Optimal policies and asset prices. *The Review of Financial Studies, 14,* 371–405.

Basawa, I. V., & Prabhu, N. U. (1981). Estimation in the single server queues. *Naval Research Logistics Quarterly, 28,* 475–487.

Basawa, I. V., & Rao, P. (1980). *Statistical inference for stochastic processes.* New York: Academic Press.

Basel Committee on Banking Supervision. (2004). *Working paper on the regulatory treatment of operational risks.* Available at http://www.bis.org.

Basel Committee on Banking Supervision. (2008). *Principles of sound liquidity risk management and supervision.* Basel, Switzerland: Bank of International Settlements (BIS).

Basi, F., Embrechts, P., & Kafetzaki, M. (1998). Risk management and quantile estimation. In R. Adler, R. Feldman, & M. Taqqu (Eds.), *Practical guide to heavy tails* (pp. 111–130). Boston: Birkhauser.

Bassan, B. (2003, November 20). *Pricing stocks and options with Bayesian games.* Dipartimento di Matematica, Universita "La Sapienza", Piazzale Aldo Moro 5, I–00185 Roma, Italy, bassan@mat.uniroma1.it.

Bassan, B., Gossner, O., Scarsini, M., & Zamir, S. (2003). Positive value of information in games. *International Journal of Game Theory, 32,* 17–31.

Basseville, M. (1988). Detecting changes in signals and systems—A survey. *Automatica, 24*(3), 309–326.

Baston, V. J., & Bostock, F. A. (1991). A generalized inspection game. *Naval Research Logistics, 38,* 171–182.

Batten, D. F., & Karlsson, C. (Eds.). (1996). *Infrastructure and the complexity of economic development* (Advances in spatial science, pp. 49–60). Heidelberg and New York: Springer.

Bauer, R. A. (1960). Consumer behavior as risk taking. In R. S. Hancock (Ed.), *Dynamic marketing for a changing world.* Chicago: American marketing Association.

Bauer, C. (2000). Value at risk using hyperbolic distributions. *Journal of Economics and Business, 52,* 455–467.

Baxter, M. (2007). Gamma process dynamic modeling of credit. *Risk, 20*(10), 98–101.

Bay, S. D., & Pazzani, M. J. (1999). *Detecting change in categorical data: Mining contrast sets.* Fifth ACM SIGKDD International Conference on Knowledge Discovery and Data Mining. New York: The Association for Computing Machinery.

Beard, R. E., Pentikainen, T., & Pesonen, E. (1979). *Risk theory* (2nd ed.). London: Methuen and Co.

Beatty, R., & Ritter, J. (1986). Investment banking, reputation, and the underpricing of initial public offerings. *Journal of Financial Economics, 15,* 213–232.

Beck, U. (1992). *Risk society.* London: Sage.

Beck, C. (2001). Dynamical foundations of nonextensive statistical mechanics. *Physics Review Letters, 87,* 180601. 68

Beckers, S. (1996). A survey of risk measurement theory and practice. In C. Alexander (Ed.), *Handbook of risk management and analysis.*

Bell, D. E. (1982). Regret in decision making under uncertainty. *Operations Research, 30*, 961–981.

Bell, D. E. (1983). Risk premiums for decision regrets. *Management Science, 29*, 1156–1166.

Bell, D. E. (1985). Disappointment in decision making under uncertainty. *Operation Research, 33*, 1–27.

Bell, D. (1995). Risk, return and utility. *Management Science, 41*, 23–30.

Bendor, J., & Kumar, S. (2003, October 4). *Satisficing and optimality*. Working paper, Graduate School of Business, Stanford University.

Bensoussan, A. (1982). *Stochastic control by functional analysis methods*. Amsterdam: North-Holland.

Bensoussan, A., & Frehse, J. (2000). Stochastic games for N players. *Journal of Optimization Theory and Applications, 105*(3), 543–565.

Bensoussan, A., & Frehse, J. (2009). On diagonal elliptic and parabolic systems with super-quadratic Hamiltonians. *Communications on Pure and Applied Analysis, 8*, 83–94.

Bensoussan, A., Frehse, J., & Vogelgesang, J. (2010). Systems of Bellman equations to stochastic differential games with non-compact coupling. *Discrete and Continuous Dynamical Systems, 27*(4), 1375–1389.

Bensoussan, A., Sung, K. C. J., Yam, S. C. P., & Yung, S. P. (2012). Linear-Quadratic Mean Field Games Workshop Ajou University, 2012, and NYU-POLY September.

Bensoussan, A., & Tapiero, C. S. (1982). Impulsive control in management: Prospects and applications. *Journal of Optimization Theory and Applications, 37*, 419–442.

Benston, G. (1996). *The origins of and justification for the Glass–Steagall Act. Universal banking: Financial system design reconsidered* (pp. 31–69). Illinois: Howewood

Ben-Tal, A., & Teboulle, M. (1987). Penalty functions and duality in stochastic programming via-divergence functionals. *Mathematics of Operations Research, 12*, 224–240.

Ben-Tal, A., & Teboulle, M. (2007). An old-new concept of convex risk measures: The optimized certainty equivalent. *Mathematical Finance, 17*(3), 449–476.

Benveniste, A., Basseville, M., & Moustakides, G. (1987). The asymptotic local approach to change detection and model validation. *IEEE Transactions on Automatic Control, AC-32*(7), 583–592.

Benveniste, L., & Spindt, P. (1989). How investment bankers determine the offer price and allocation of new issues. *Journal of Financial Economics, 24*, 343–362.

Beran, J. (1994). *Statistics for long-memory processes*. London: Chapman and Hall.

Berger, E., (1992). *Etude d'un Marché agricole non soutenu*. Mimeo, Note de Recherche GRID nr. 92-02, E.N.S. de Cachan.

Berndt, E. R., & Harrison, B. (1991). *Measuring the contribution of public infrastructure in Sweden*. NBER, Working Paper no. 3842.

Bernstein, P. (1998). *Against the Gods: The remarkable history of risk*. New York: Wiley.

Bhargava, R., & Fraser, D. (1998). On the wealth and risk effects of commercial bank expansion into securities underwriting: An analysis of Section 20 subsidiaries. *Journal of Banking and Finance, 22*, 447–465.

Bird, C. G., & Kortanek, K. O. (1974). Game theoretic approaches to some air pollution regulation problems. *Socio-Economic Planning Sciences, 8*, 141–147.

Bisawa, A., & Wang, J. S. (2002). A new bivariate binomial distribution. *Statistics and Probability Letters, 60*, 231–240.

Bisias, D., Flood, M., Lo, A. W., & Vlvanis, S. (2012). *A survey of systemic risk analytics*. Office of Financial Research. Working Paper #0001, January 5.

Bismut, J. M. (1975). Growth and intertemporal allocation of risks. *Journal of Economic Theory, 10*, 239–257.

Bismut, J. M. (1976). Theorie Probabiliste du Controle des Diffusions. *Memoirs of the American Mathematical Society, 4*(167).

Bismut, J. M. (1978). An introductory approach to duality in optimal stochastic control. *SIAM Review, 20*, 62–78.

Bjerve, S., & Doksum, K. (1993). Correlation curves: Measures of association as function of covariate values. *The Annals of Statistics, 21*, 890–902.

Black, F. (1972). Capital market equilibrium with restricted borrowing. *Journal of Business, 45*, 444–454.

Black, F., & Scholes, M. (1973). The pricing of options and corporate liabilities. *Journal of Political Economy, 81*, 637–659.

Bloemena, A. R. (1964). *Sampling from a graph*. Amsterdam, The Netherlands: Mathematics Centrum.

Blom, H. A. P., Bakker, G. J., Blanker, P. J. G., Daams, J., Everdij, M. H. C., & Klompstra, M. B. (1998). Accident risk assessment for advanced ATM. In *Proceedings of the Second USA/Europe Air Traffic Management R&D Seminar*, Orlando.

Blom, H. A. P., Corker, K. M., & Stroeve, S. H. (2005). Study on the integration of human performance and accident risk assessment models: AIR-MIDAS & TOPAZ. In *Proceedings of the Sixth USA/Europe Air Traffic Management R&D Seminar*, Baltimore.

Blom, H. A. P., Klompstra, M. B., & Bakker, G. J. (2003). *Accident risk assessment of simultaneous converging instrument approaches*. National Airspace Laboratory (Report NLR-TP-2003-557), Amsterdam.

Blom, H. A. P., Stroeve, S. H., & de Jong, H. H. (2006). Safety risk assessment by monte carlo simulation of complex safety critical operations. In *Proceedings of the 14th Safety Critical Systems Symposium*, Bristol.

Boeing Commercial Airplanes. (2006). *Statistical summary of commercial jet airplane accidents: Worldwide operations 1959–2005*. Seattle: Boeing Commercial Airplanes.

Bohn, J. R. (2000). A survey of contingent-claims approaches to risky debt valuation. *Journal of Risk Finance, 1*(3), 53–78.

Bohn, J. R., & Choi, S. (1996). Fraud in the new-issue market: Empirical evidence on securities class actions. *University of Pennsylvania Law Review, 144*, 903–982.

Bolland, P. J., & Proschan, F. (1994). Stochastic order in system reliability theory. In M. Shaked & J. G. Shantikumar (Eds.), *Stochastic orders and their applications* (pp. 485–508). San Diego, CA: Academic Press.

Bollerslev, T. (1986). Generalized autoregresive conditional heteroskedasticity. *Journal of Econometrics, 31*, 307–327.

Bollerslev, T. (1990). Modeling the coherence in short run nominal exchange rates: A multivariate generalized ARCH model. *The Review of Economics and Statistics, 72*, 498–505.

Bollerslev, T., Chu, R. Y., & Kroner, K. F. (1992). ARCH modeling in finance. *Journal of econometrics, 52*, 5–59.

Bollerslev, T., Engle, R. F., & Nelson, D. B. (1994). ARCH models. In R. F. Engle & D. L. McFadden (Eds.), *Handbook of econometrics* (Vol. 4, pp. 2959–3038). Amsterdam: North Holland.

Booth, G., Kaen, F., & Koveos, P. (1982). R/S analysis of foreign exchange rates under two international monetary regimes. *Journal of Monetary Economics, 10*, 407–415.

Boots, B., & Tapiero, C. S. (1973). Structural inference in transportation networks. *Environment and Planning, 6*(1974), 411–418.

Borch, K. H. (1968). *The economics of uncertainty*. Princeton, NJ: Princeton University Press.

Borch, K. (1974). *The mathematical theory of insurance*. Lexington, MA: Lexington Books.

Borge, D. (2001). *The book of risk*. New York: Wiley.

Borges, E. P., & Roditi, I. (1998). A family of nonextensive entropies. *Physics Letters A, 246*, 399–402.

Borland, L., & Bouchaud, J. P. (2004). A non Gaussian option pricing model with skew. *Quantitative Finance, 4*(5), 499–514.

Born, M. (1954). *Nobel lecture*. Published in "Les Prix Nobel". Stockholm: Nobel Foundation.

Bouchaud, J. P., & Potters, M. (2003). *Theory of financial risks and derivatives pricing, from statistical physics to risk management* (2nd ed.). London: Cambridge University Press.

Box, G. E. P. (1988). Signal to noise ratios, performance criteria and transformation (with discussion). *Technometrics, 30*, 1–40.

Box, G. (1993). Quality improvement – The new industrial revolution. *International Statistical Review, 61*, 3–19.

Box, G. E. P., & Draper, N. R. (1987). *Empirical model building and response surfaces.* New York: Wiley.

Box, G. E. P., Hunter, J. S., & Hunter, W. G. (1978). *Statistics for experiments.* New York: Wiley.

Boyd, J., & Graham, S. (1993). Bank holding company mergers with nonbank firms: Effects on the risk of failure. *Journal of Banking and Finance, 17*(1), 43–63.

Brandt, M., & Diebold, F. (2006). A no arbitrage approach to range based estimation of return covariances and correlations. *Journal of Business, 79*, 61–74.

Breeden, D. (1979). An inter-temporal asset pricing model with stochastic consumption and investment opportunities. *Journal of Financial Economics, 7*(3), 265–296.

Breeden, D., & Litzenberger, R. (1978). Prices of state-contingent claims implicit in option prices. *Journal of Business, 51*, 621–651.

Brock, W. (1986). Distinguishing random and deterministic systems. *Journal of Economic Theory, 40*(1), 168–195.

Brock, W. A., & Dechert, W. D. (1988). Theorems on distinguishing deterministic systems. In W. Barnett, E. Berndt, & H. White (Eds.), *Dynamic econometric modeling* (pp. 247–265). Cambridge: Cambridge University Press.

Brockett, P. L., & Xia, X. (1995). Operations research in insurance, a review. *Transaction of the Society of Actuaries, 47*, 7–82.

Brody, D. C., Buckley, I. R. C., & Constantinou, I. C. (2007). Option price calibration from Renyi entropy. *Physics Letters A, 366*, 298–3007.

Bryis, E., Eeckoudt, L., & Loubergé, H. (1989). Endogenous risk and the risk premium. *Theory and Decision, 26*, 37–46.

Buhlmann, H. (1970). *Mathematical methods in risk theory.* Berlin: Springer.

Buhlmann, H. (1980). An economic premium principle. *ASTIN Bulletin, 11*, 52–60.

Buzacott, J. A., & Yao, D. D. (1986). On queuing network models of flexible manufacturing systems. *Queueing Systems, 1*(1986), 5–27.

Caballe, J., & Pomanski, A. (1996). Mixed risk aversion. *Journal of Economic Theory, 71*, 485–513.

Camerer, C. (2003). *Behavioral game theory, experiments in strategic interaction.* Priceton, NJ: Princeton University Press.

Camerer, C. F., & Weber, M. (1992). Recent developments in modeling preferences: Uncertainty and ambiguity. *Journal of Risk and Uncertainty, 5*, 325–370.

Campbell, J. Y. (1999). Asset prices, consumption, and the business cycle. Part 3 of *Handbook of macroeconomics* (Vol. 1, pp. 1231–1303).

Campbell, J. Y. (2000). Assets pricing at the millennium. *Journal of Finance, 55*, 1515–1567.

Campbell, J. Y. (2003). Consumption-based asset pricing. *Handbook of the economics of finance* (Chap. 13, pp. 801–885). Amsterdam: Elsevier.

Campbell, J. Y., & Cochrane, J. H. (1999). By force of habit: A consumption-based explanation of aggregate stock market behavior. *Journal of Political Economy, 107*, 205–251.

Canty, M. J., Rothenstein, D., & Avenhaus, R. (2001). Timely inspection and deterrence. *European Journal of Operational Research, 131*, 208–223.

Capobianco, M. F. (1973). Recent advances in statgraphics. *Annals of the New York Academy of Sciences.*

Capocelli, R. M., & Ricciardi, L. M. (1972). On the inverse of the first passage time probability problem. *Journal of Applied Probability, 9*, 270–287.

Cariou, P. M., Mejia, Q., & Wolff, F. C. (2008). On the effectiveness of port state control inspections. *Transportation Research, E44*, 491–503.

Carreira-Perpinan, M. A., & Reinals, S. (2000). Practical identifiability of finite mixtures of multivariate Bernoulli distributions. *Neural Computations, 12*(1), 141–152.

Cartea, A., & Howison, S. (2003) *Distinguished limits of Levy-Stable processes, and applications to option pricing*. University of Oxford, Mathematical Institute. http://www.finance.ox.ac.uk/file_links/mf_papers/2002mf04.pdf.

Casualty Actuarial Society. (2001, November). Report of the Advisory Committee on Enterprise Risk Management, 1100 N. Glebe Rd, #600, Arlington, VA, 2201, office@casact.org.

Caves, R. E., & Murphy, W. E. (1976). Franchising firms, markets and intangible assets. *Southern Economic Journal, 42*, 572–586.

Chankong, V., & Haimes, Y. Y. (1983). *Multi objective decision making*. New York: North Holland.

Chapman, D. (1997). Approximating the asset pricing kernel. *The Journal of Finance, 1*(4), 1383–1410.

Chappe, R., & Semmler, W. (2012). Financial reform in the US: A critical survey of Dodd-Frank and what is needed for Europe, IMK (Macroeconomic Policy Institute). *Hans Bockler Stiftung*. Germany: Dusseldorf.

Chemmanur, T. J., & Fulghieri, P. (1999). A theory of the going-public decision. *Review of Financial Studies, 12*, 249–279.

Chemmanur, T. J., & He, J. (2011). IPO waves, product market competition, and the going public decision: Theory and evidence. *Journal of Financial Economics, 101*, 382–412.

Chemmanur, T. J., & Yan, A. (2009). Product market advertising and new equity issues. *Journal of Financial Economics, 92*, 40–65.

Chen, X., & Fan, Y. (2006). Estimation and model selection of semiparametric copula-based multivariate dynamic models under copula misspecification. *Journal of Econometrics, 135*, 125–154.

Cheng, S., Liu, Y., & Wang, S. (2004). Progress in risk measurement. *Advanced Modelling and Optimization, 6*, 1–20.

Cherubini, U., Luciano, E., & Vecchiato, W. (2004). *Copulas methods in Finance*. New York: Wiley.

Cheung, Y. W. (1983). Long memory in foreign exchange rates. *Journal of Business and Economic Statistics, 11*, 93–101.

Chib, S., & Greenberg, E. (1998). Analysis of multivariate probit models. *Biometrika, 85*, 347–361.

Cho, B.-H., & Hooker, N. H. (2009). Comparing food safety standards. *Food Control, 20*, 40–47.

Chow, Y. S., Robbins, H., & Siegmund, D. (1971a). *The theory of optimal stopping*. New York: Dover.

Chow, Y. S., Robbins, H., & Siegmund, D. (1971b). *Great expectations: The theory of optimal stopping*. Boston: Hougton Mifflin Co.

Christoffersen, P., Jacobs, K., & Vaiberg, G. (2008). *Forward looking betas*. Working Paper, Mc Gill University.

Claes, I., & Van den Broeck, C. (1987). Random walk with persistence. *Journal of Statistical Physics, 49*(1–2), 383–392.

Coase, R. H. (1937). The nature of the firm. *Economica, 4*(16), 386–405.

Coase, R. H. (1960). The problem of social cost. *Journal of Law and Economics, 3*(2), 1–44.

Cochran, W. G., & Cox, G. M. (1957). *Experimental designs*. New York: Wiley.

Cochrane, J. H. (1989). The sensitivity of test of the intertemporal allocation of consumption. *American Economic Review, 79*(3), 319–337.

Cochrane, J. H. (1991). A simple test of consumption insurance. *Journal of Political Economy, 99*, 957–976.

Cochrane, J. H. (2000). *A resurrection of the stochastic discount factor model*. Mimeo, University of Chicago.

Cochrane, J. (2005). *Asset pricing* (Rev. Version). Princeton, NJ: Princeton University Press.

Cochrane, J. H. (2007). The dog that did not bark: A defense of return predictability. *Review of Financial Studies, 21*, 1533–1575.

Cogger, K. O., & Yu, P. L. (1985). Eigenweight vectors and least distance approximation for revealed preference in pairwise weight ratios. *Journal of Optimization Theory and Applications, 46*, 483–491.

Coile, R. C., Jr. (1990). *The new medicine: Reshaping medical practice and health care management*. Rockville, MD: Aspen Publishers.

Coleman, D. E., & Montgomery, D. C. (1993). A systematic approach to planning for a designed industrial experiment. *Technometrics, 35*, 1–12.

Coles, M. G. (1993). The price of worker reliability. *Economic Letters, 41*(2), 149–155.

Colombo, S., & Lavoine, J. (1972). Transformations de Laplace et de Mellin, Formularires, Mode d'Utilisation, *Memorial des Sciences Mathematique*, Gauthiers-Villars.

Comaro, A. (2003). *America's best hospitals*. US News World Report. 28:46.

Condorcet, M. J., De Caritat, A. N., & Marquis de. (1785). Essai sur l' application de l'analyse a la probabilite de decisions rendues a la pluralite des voix (Essays on the application of analysis to the probability of majority decisions) Imprimerie Royale, Paris, France. Reprinted be the American Mathematical Society, Providence, RI, 1972.

Corrado, C., & Su, T. (1996). Skewness and kurtosis in S&P 500 index returns implied by S&P 500 index option prices. *Journal of Derivatives, 4*, 8–19.

Cox, L. A., Jr. (2009). *Risk analysis and complex uncertain systems*. New York: Springer.

Cox, D. R., & Miller, H. D. (1965). *The theory of stochastic processes*. New York: Wiley.

Cox, J. C., & Ross, S. A. (1978). A survey of some new results in financial option pricing theory. *Journal of Finance, 31*, 383–402.

Cox, J. J., & Tait, N. (1991). *Reliability, safety and risk management*. Oxford: Butterworth-Heinemann.

Cramer, H. (1930). *On the mathematical theory of risk*. Skandia Jubilee Volume, Stockholm.

Cramer, H. (1955). *Collective risk theory*. Jubilee Volume, Skandia Insurance Company.

Crosbie, P. (1999). *Modeling default risk*. KMV Corporation. (http://www.kmv.com).

Crosby, P. B. (1984). *Quality without tears*. New York: Mc Graw Hill Book Co.

Culp, C. L. (2006). *Structured finance and insurance*. New York: Wiley-Finance.

Cumperayot, P. J., Dan_ɀelsson, J., Jorgensen, B. N., & de Vries, C. G. (2001). On the (ir)relevance of value at risk regulation. In F. Franke, W. Haerde, & G. Stahl (Eds.), *Measuring risk in complex stochastic systems* (pp. 99–117). Berlin: Springer.

Curado, E. M. F., & Tsallis, C. (1991, 1992). *Journal of Physics*, L69, E-24, 3187, E-25, 1992, 1019.

Dachraoui, K., Dionne, G., Eeckoudt, L., & Goodfroid, P. (1999). *Comparative risk aversion*. HEC, Montreal Working Paper.

Dacorogna, M. M. (2004). Insurance and Finance Convergence and Divergence, First Bonzenfreies Colloquium on Market Dynamics and Quantitative Economics, Alessandria, Italy, Sept. 9–10.

Darooneh, A. H., Nacimi, G., Mehri, A., & Sadeghi, P. (2010). Tsallis entropy, Escort probability and incomplete information theory, entropy, *12*, 2497–2503. http://www.mdpi.com/journal/entropy.

Das, S. R. (1995). Credit risk derivatives. *Journal of Derivatives, 2*, 7–21.

Daudin, J. J., & Tapiero, C. S. (1996). Les Outils et la Gestion de la Qualité, *Economica*.

Dayananda, P. A. W. (1970). Optimal reinsurance. *Journal of Applied Probability, 7*, 134–156.

Daykin, C., Pentkanen, T., & Pesonen, T. (1994). *Practical risk theory for actuaries*. London: Chapman and Hall.

De Bondt, W. F. M. (2005). *The psychology of world equity markets*. Cheltenham: Edward Elgar.

De Bondt, W., Muradoglu, G., Shefrin, H., & Staikouras, S. (2008). Behavioral Finance: Quo Vadis? *Formal of Applied Finance*, *Fall/Winter*.

de Finetti, B. (1937). La Prevision: Ses Lois Logiques, ses Sources Subjectives. *Annales de l'Institut Henri Poincaré, 7*, 1–68.

De la Selva, S. M. T., Lindenberg, K., & West, B. J. (1988). Correlated random walks. *Journal of Statistical Physics, 53*(½), 203–219.

Debreu, G. (1952). A social equilibrium existence theorem. *Proceedings of the National Academy of Science, 38*, 886–893.

Debreu, G. (1953). *Une economie de l'incertain*. Working Paper, Electricite de France, Paris, France (Published in 1960 in *Economie Appliquee*, 13, no.1, pp. 111–116).

Debreu, G. (1959). *Theory of value: An axiomatic analysis of economic equilibrium* (Cowles Foundation, Monograph 17). New Haven, CT: Yale University Press.

DeCanio, S. J. (1997). Economic modeling and the false tradeoff between environmental protection and economic growth. *Contemporary Economic Policy, 15*, 10–27.

Dekker, R., Wildeman, R. E., & van der Duyn Schouten, F. A. (1997). A review of multi-component maintenance models with economic dependence. *Mathematical Methods of Operations Research, 45*(3), 411–435.

DeLoach, J. (1998). *Enterprise-wide risk management*. Englewood Cliffs, NJ: Prentice Hall.

Dembo, R. S. (1989). *Scenario optimization. Algorithmics Inc*. Research Paper 89.01.

Dembo, R. S. (1993). Scenario immunization. In S. A. Zenios (Ed.), *Financial optimization*. London: Cambridge University Press.

Deming, E. W. (1982). *Quality, productivity and competitive position*. Cambridge, MA: MIT Press.

Dempster, M. A. H., Medova, E. A., & Yang, S. W. (2007). *Empirical copulas for CDO tranche pricing using relative entropy*. Center for Financial Research, Judge Business School, University of Cambridge, UK.

Denuit, M. (2001). Laplace transform ordering of actuarial quantities. *Insurance Mathematics and Economics, 29*, 83–102.

Denuit, M., Dhaene, J., Goovaerts, M., & Kaas, R., (2006, June). *Actuarial theory for dependent risks: Measures, orders and models*. ISBN: 978-0-470-01644-2, Adobe E-Book, 458 pages.

Denuit, M., & Lamber, P. (2005). Constraints on concordance measures in bivariate discrete data. *Journal of Multivariante Analysis, 82*, 1–16.

DesHarnais, S. I. (1990). Current uses of large data sets to assess the quality of providers: Construction of risk-adjusted indexes of hospital performance. *International Journal of Technology Assessment in Health Care, 6*, 229–238.

DesHarnais, S. I., Chesney, J. D., Wroblewski, R. T., Fleming, S. T., & McMahon, L. F., Jr. (1988). The risk-adjusted mortality index: A new measure of hospital performance. *Medical Care, 26*(12), 1129–1148.

DesHarnais, S. I., McMahon, L. F., Jr., & Wroblewski, R. T. (1991). Measuring outcomes of hospital care using multiple risk-adjusted indexes. *Health Services Research, 26*(4), 425–445.

DesHarnais, S. I., McMahon, L. F., Jr., Wroblewski, R. T., & Hogan, A. J. (1990). Measuring hospital performance: The development and validation of risk-adjusted indexes of mortality, readmissions, and complications. *Medical Care, 28*(12), 1127–1141.

Dhaene, J., Denuit, M., Goovaerts, M. J., Kaas, R., & Vyncke, D. (2002a). The concept of comonotonicity in actuarial science and finance: Theory. *Insurance Mathematics and Economics, 31*, 3–33.

Dhaene, J., Denuit, M., Goovaerts, M. J., Kaas, R., & Vyncke, D. (2002b). The concept of comonotonicity in actuarial science and finance: Applications. *Insurance Mathematics and Economics, 31*, 133–161.

Diebold, F. X., Doherty, N. A., & Herring, R. J. (Eds.). (2010). *The known, the unknown and the unknowable in financial risk management*. Princeton, NJ: Princeton University Press.

Diebold, F. X., & Lopez, N. A. (1995). *Modeling volatility dynamics*. Federal Reserve Bank of New York.

Diebold, F. X., & Rudenbusch, G. D. (1989). Long memory and persistence in aggregate outputs. *Journal of Monetary Economics, 24*, 189–209.

Dimukes, K. (2005). The limits of expertise: The misunderstanding role of pilot error in airline accidents. *ASPA/ICAO Regional Seminar*.

Dionne, G., & Eeckouldt, L. (1985). Self insurance, self protection and increased risk aversion. *Economic Letters, 17*, 19–42.

Dittmar, R. (2002). Nonlinear pricing kernels, kurtosis preference and evidence from the cross section equity returns. *Journal of Finance, 57*(1), 369–403.

Dodge, H. F. (1943). Sampling inspection plan for a continuous production. *Annals of Mathematical Statistics, 14*, 264–279.

Doherty, N. A. (2000). *Integrated risk management: Techniques and strategies for managing corporate risk*. New York: McGraw Hill.

Domokos, S., & Thoth, B. (1984). Persistent random walks in a one dimensional random environment. *Journal of Statistical Physics, 37*, 27–38.

Drake, P., & Vetsuypens, M. (1993). IPO underpricing and insurance against legal liability. *Financial Management, 22*, 64–73.

Dresher, M. (1961). *Games of strategy: Theory and applications*. Englewood Cliffs, NJ: Prentice Hall.

Dreze, J., & Modigliani, F. (1966). Epargne et Consommation en Avenir Aleatoire. *Cahiers du Seminaire d' Econometrie*.

Dreze, J., & Modigliani, A. (1972). Consumption decisions under uncertainty. *Journal of Economic Theory, 5*, 308–385.

Duffie, D. (1988). *Security markets: Stochastic models*. New York: Academic Press.

Duffie, D. (1992). *Dynamic asset pricing theory*. Princeton, NJ: Princeton University Press.

Duffie, D. (1999). Credit swap valuation. *Financial Analysts Journal, 55*, 73–87.

Duffie, D., & Garleanu, N. (2001). Risk and valuation of collateralized debt valuation. *Financial Analyst Journal, 57*(1), 41–62.

Duffie, D., & Zame, W. (1989). The consumption based capital asset pricing model. *Econometrica, 57*, 1279–1297.

Duncan, A. J. (1974). *Quality control and industrial statistics* (4th ed.). Homewood, IL: Irwin.

Dyer, J. S. (1990). Remarks on the analytic hierarchy process. *Management Science, 36*(1990), 249–258.

Dyer, J. S., & Jia, J. (1997). Relative risk-value model. *European Journal of Operational Research, 103*, 170–185.

Eberlein, E., & Keller, U. (1995). Hyperbolic distributions in finance. *Bernoulli, 1*, 261–299. Available from http://www.neyman.mathematik.uni-freiburg.de/homepages/eberlein/papers/eberlein.keller.

Economides, N. (1996). The economics of networks. *International Journal of Industrial Organization, 14*(2).

Eeckhoudt, L., Gollier, C., & Schlesinger, H. (1996). Changes in background risk and risk taking behavior. *Econometrica, 64*, 683–689.

Eeckoudt, L., & Kimball, M. (1991). Background risk prudence and the demand for insurance. In G. Dionne (Ed.), *Contributions to insurance economics*. Boston: Kluwer Academic Press.

Ehrenberg, A. C. S. (1972). *Repeat-buying*. Amsterdam: North Holland.

Eichner, T., & Wagener, A. (2005). Measures of risk attitude and correspondence between mean variance and expected utility. *Decisions in Economics and Finance, 28*, 53–67.

Einstein, A. (1906). On the theory of the Brownian movement. *Annalen der Physik, 19*, 371–381.

Einstein, A. (1956). *Investigations on the theory of Brownian movement*. New York: Dover.

Ellsberg, D. (1961). *Risk, ambiguity and the savage axioms* (pp. 643–669). 75: Quarterly Journal of Economics.

Embrechts, P. (2000a). Actuarial versus financial pricing insurance. *Risk Finance, 1*, 17–2.

Embrechts, P. (Ed.). (2000b). *Extremes and integrated risk management*. London: Risk Books.

Embrechts, P., Furrer, H., & Kaufmann, R. (2009). Different kinds of risk. In T. G. Andersen, R. A. Davis, J.-P. Kreiss, & T. Mikosch (Eds.), *Handbook of financial time series*. New York: Springer.

Embrechts, P., Höing, A., & Juri, A. (2003). Using copulae to bound the value-at-risk for functions of dependent risks. *Finance and Stochastics, 7*, 145–167.

Embrechts, P., Klupperberg, C., & Mikosch, T. (1997). *Modeling extremal events in insurance and finance*. Berlin: Springer.

Embrechts, P., Lindskog, F., McNeil, A. (2001). *Modeling dependence with copulas and applications to risk management*. Working Paper, Department of Mathematics, ETHZ CH-8092 Zurich.

Embrechts, P., McNeil, A., & Straumann, D. (2002). Correlation and dependence properties in risk management: Properties and pitfalls. In M. Dempster (Ed.), *Risk management: Value at risk and beyond*. Cambridge: Cambridge University Press.

Engle, R. (1987). Autoregressive conditional heteroskedasticity with estimates of the variance of U.K. inflation. *Econometrica, 50*, 987–1008.

Engle, R. F. (1995). *ARCH selected reading*. Oxford: Oxford University Press.

Epstein, L. G., & Zin, S. E. (1989). Substitution, risk aversion and the temporal behavior of consumption and asset returns: A theoretical framework. *Econometrica, 57*, 937–969.

Epstein, L. G., & Zin, S. E. (1991a). Substitution, risk aversion and the temporal behavior of consumption and asset returns: An empirical analysis. *Journal of Political Economy, 99*, 263–286.

Esscher, F. (1932). On the probability function of the collective theory of risk. *Scandinavian Actuarial Journal, 15*, 175–195.

Evans, J. S. B. T., & Over, D. E. (1996). *Rationality and reasoning*. Hove: Psychology Press.

Everitt, B. S., & Hand, D. J. (1981). *Finite mixtures distributions* (Monographs on statistics and applied probability). London: Chapman and Hall.

Fama, E. F. (1970a). Efficient capital markets: A review of theory and empirical work. *Journal of Finance, 25*, 383–417.

Fama, E. F. (1970b). Multiperiod consumption-investment decisions. *American Economic Review, 60*, 163–174.

Fama, E. F. (1996). The CAPM is wanted, dead or alive. *The Journal of Finance, 51*, 1947.

Fama, E. F., & French K. R. (1997). *The equity premium*. Mimeo, University of Chicago.

Fang, K. T., Kotz, S., & Ng, K. W. (1987). *Symmetric multivariate and related distributions*. London: Chapman & Hall.

Federal Aviation Administration. (2005). European Organization for Safety of Air Navigation, 2005. ATM Safety Techniques and Toolbox, Safety Action Plan–15. Federal Aviation Administration, European Organization for Safety of Air Navigation, Washington, DC.

Federal Aviation Administration, & Air Traffic Organization. (2006). *Aircraft Accident and Incident Notification, Investigation, and Reporting*. Order 8020.16. http://www.faa.gov/airports_airtraffic/air_traffic/publications/at_orders/media/AAI.pdf.

Feigenbaum, A. V. (1983). *Total quality control* (3rd ed.). New York: Mc Graw Hill.

Feller, W. (1951). The asymptotic distribution of the range of sums of independent random variables. *Annals of Mathematical Statistics, 22*, 427–432.

Feller, W. (1971). *An introduction to probability theory and its applications* (2nd ed., Vol. 2). New York: Wiley.

Fellner, W. (1961). Distortion of subjective probabilities as a reaction to uncertainty. *Quarterly Journal of Economics, 75*, 670–690.

Fields, L. P., & Fraser, D. R. (2004). Effects of IPO mispricing on the risk and reputational capital of commercial banks. *Review of Financial Economics, 13*, 65–77.

Fishburn, P. C. (1970). *Utility theory for decision making*. New York: Wiley.

Fishburn, P. C. (1981). Foundations of risk measurement: Effects of gains on risk. *Journal of Mathematical Psychology, 25*(226), 242.

Fishburn, P. C. (1988). *Nonlinear preference and utility theory*. Baltimore, MD: The Johns Hopkins Press.

Florens, J. P., & Foucher, C. (1999). Pollution monitoring: Optimal design of inspection an economic analysis of the use of satellite information to deter oil pollution. *Journal of Environmental Economics and Management, 38*, 81–96.

Focacci, A. (2003). Empirical evidence in the analysis of the environmental and energy policies of a series of industrialised nations, during the period 1960–1997, using widely employed macroeconomic indicators. *Energy Policy, 31*, 333–352.

Follmer, H., & Schied, A. (2008, October 8) (pdf). *Convex and coherent risk measures*. http://wws.mathematik.hu-berlin.de/~foellmer/papers/CCRM.pdf. Retrieved July 22, 2010.

Forthman, M. T., Gold, R. S., Dove, H. G., & Henderson, R. D. (2010). Risk-adjusted indices for measuring the quality of inpatient care. *Qualtiy Management in Health Care, 19*(3), 265–277.

Fou, J. C., & Koutras, M. V. (1994). Distribution theory of runs: A Markov chain approach. *Journal of the American Statistical Association, 89*, 1050–1058.

Fraisse, P. (1957). *Psychologie du Temps*. Paris: P.U.F. (Presses Universitaires de France).

Franke, F., Harde, W., & Stahl, G. (2000). *Measuring risk in complex stochastic systems*. Berlin: Springer.

Fraundorf, P. (2007). Thermal roots of correlation-based complexity. *Complexity, 13*(3), 18–26.

Frees, E., & Valdez, E. (1998). Understanding relationships using copulas. *North American Actuarial Journal, 2*, 1–25.

Frenkel, M., Karmann, A., & Scholtens, B. (Eds.). (2010). *Sovereign risks and financial crises*. Heidelberg: Springer.

Frey, R., & McNeil, A. J. (2001). *Modeling dependent defaults*. Presented at the Conference on "Statistical and Computational Problems in Risk Management", University of Rome "La Sapienza", June 14–16. Available from http://www.math.ethz.ch/_frey/credit-paper.pdf.

Friedman, M. (1976). *Price theory*, Aldine de Gruyler.

Friedman, M., & Savage, L. J. (1948). The utility analysis of choices involving risk. *Journal of Political Economy, 56*, 279–304.

Friedman, M., & Savage, L. J. (1952). The expected utility hypothesis and the measurability of utility. *Journal of Political Economy, 60*, 463–474.

Friend, G. (1998). EcoMetrics: Integrating direct and indirect environmental costs and benefits into management information systems. *Environmental Quality Management* (Spring), 19–29.

Frittelli, M., & Rosazza Gianin, E. (2002). Putting order in risk measures. In G. Szego (Ed.), "Beyond VaR" (special issue). *Journal of Banking and Finance*, 26, July.

Frittelli, M., & Rosazza Gianin, E. (2005). Law-invariant convex risk measures. *Advances in Mathematical Economics, 7*, 33–46.

FSA. (2005). Formal Safety Assessment. Passenger ship safety: effective voyage planning for passenger ships, Formal Safety Assessment—large passenger ships navigation, Sub Committee on Safety of Navigation, 50th session, NAV 50/11/1/http://research.dnv.com/skj/FSALPS/FSA-LPS-NAV.htmS.

FSA. (2006a). Formal Safety Assessment. Consideration on utilization of Bayesian network at step 3 of FSA, Maritime Safety Committee, 81st session, MSC 81/18/1.

FSA. (2006b). Formal Safety Assessment. FSA Study on ECDIS/ENCs. Maritime Safety Committee, 81st session, MSC81/24/5/http://research.dnv.com/skj/FSA-ECDIS/ECDIS.htmS.

Fujiwara, Y. (2004). Zipf law in firms bankruptcy. *Physica A: Statistical Mechanics and Applications, 337*, 219–230.

Fung, H.-G., & Lo, W.-C. (1995). Examining the long range dependence in exchange rates. *Journal of International Financial Markets, Institutions and Money, 5*, 19–29.

GAIN. (2003). Guide to methods & tools for safety analysis in air traffic management. *Global Aviation Information Network*. http://www.gainweb.orgS.

Galambos, J. (1978). *The asymptotic theory of extreme order statistics*. New York: Wiley.

Garlaschelli, S., Battiston, M. C., Servedio, V. D. P., & Caldarelli, G. (2005). The scale free nature of market investment network. *Physica A: Statistical Mechanics and Applications, 350*(2005), 491–499.

Gattorna, J. (Ed.) (1988). *Strategic supply chain alignment*, Chapter 27. Gower: Aldershot.

Gawande, K., & Bohara, A. K. (2005). Agency problems in law enforcement: Theory and application to the U.S. coastguard. *Management Science, 51*, 1593–1609.

Gerber, H. U. (1974). On additive premium calculation principles. *ASTIN Bulletin, 7*, 215–222.

Gerber, H. U. (1979). An introduction to mathematical risk theory, Monograph No. 8. Huebner Foundation. Philadelphia: University of Pennsylvania.

Gerber, H. U. (1985). On additive principles of zero utility. *Insurance: Mathematics and Economics, 4*(4), 249–251.

Gerber, H. U., & Goovaerts, M. J. (1981). On the representation of additive principles of premium calculation. *Scandinavian Actuarial Journal, 4*, 221–227.

Gerber, H. U., & Shiu, E. S. W. (1994). Options pricing by Esscher transforms. *Transactions of the Society of Actuaries, 46*, 99–191.

Gibbs, J. W. (1961). *The scientific papers of J.W. Gibbs* (Vol. 1). New York: Dover.

Gilboa, I., & Schmeidler, D. (1995). Case-based decision theory. *Quarterly Journal of Economics, 99*, Août 1995, 605–639.

Gintis, H. (2009). *The bounds of reason: Game theory and the unification of the behavioral sciences*. Princeton, NJ: Princeton University Press.

Gleick, J. (1987). *Chaos: Making a new science*. New York: Viking Press.

Godfrey, A. B., & Kenett, R. S. (2007). Joseph M. Juran, a perspective on past contributions and future impact. *Quality and Reliability Engineering International, 23*, 653–663.

Gollier, C. (1995). The comparative statics of changes in risk revisited. *Journal of Economic Theory, 66*, 522–536.

Gollier, C. (2000). *The economics of risk and time*. Cambridge, MA: MIT Press.

Gollier, C., & Pratt, J. W. (1996). Risk vulnerability and the tempering effect of background risk. *Econometrica, 64*, 1109–1123.

Golub, B., & Crum, C. C. (2010). *Risk management lessons worth remembering from the credit crisis of 2007 – 2009*. Working Paper, Blackrock Co.

Gomes, D., Mohr, J., & Souza, R. R. (2010). Discrete time, finite state space mean field games. *Journal de Mathematiques Pures et Appliquees, 93*(2), 308–328.

Good, I. J. (1950). *Probability and the weighing of evidence*. London: Griffin.

Good, I. J. (1965). *The estimation of probabilities: An essay on modern Bayesian methods* (Research Monograph no 30). Cambridge, MA: MIT Press.

Good, I. J. (1968). Utility of a distribution. *Nature, 219*, 139.

Goovaerts, M. J., De Vijlder, F., & Haezendonck, J. (1984). *Insurance premiums*. Amsterdam: North-Holland.

Goovaerts, M., Kaas, R., & Dhaene, J. (2003a). Economic capital allocation derived from risk measures. *North American Actuarial Journal, 7*(2), 44–59.

Goovaerts, M., Kaas, R., Dhaene, J., & Tang, Q. (2003b). A unified approach to generate risk measures. *ASTIN Bulletin, 33*(2), 173–192.

Goovaerts, M., Kaas, R., Dhaene, J., & Tang, Q. (2003c). *Some new classes of consistent risk measures*. Mimeo.

Gordon, W. J., & Newell, G. F. (1967). Closed queuing networks with exponential servers. *Operations Research, 15*, 252.

Goss, R. (1989). *Rational approaches to maritime safety*. The 23rd Blackadder Lecture, Library of the University of Wales, Cardiff.

Gourieroux, C., Laurent, J. P., & Scaillet, O. (2000). Sensitivity analysis of values at risk. *Journal of Empirical Finance, 7*, 225–245.

Goyal, S. (2007). *Connections: An introduction to the economics of networks*. Princeton: Princeton University Press.

Gradshtein, I. S., & Ryzhik, M. (1965). *Tables of integrals, series and products*. New York: Academic Press.

Gramlich, E. M. (1994). Infrastructure: A review essay. *Journal of Economic Literature, 32*, 1176–1196 (see important list of references at the end of the paper).

Grandmont, J., & Malgrange, P. (1986). Nonlinear economic dynamics: Introduction. *Journal of Economic Theory, 40*, 3–12.

Granger, C. (1980). Long memory relationships and the aggregation of dynamic models. *Journal of Econometrics, 14*, 227–238.

Granovskii, A., Dincer, I., & Rosen, M. A. (2006). Environmental and economic aspects of hydrogen production and utilization in fuel cell vehicles. *Journal of Power Sources, 157*(1), 411–421.

Grant, E. L., & Leavenworth, R. S. (1988). *Statistical quality control* (6th ed.). New York: McGraw Hill.

Grau-Viladrich, M. (2003). Monitoring policies to prevent oil spills: Lessons from the theoretical literature. *Marine Policy, 27*, 249–263.

Greaker, M. (2003a). Strategic environmental policy; eco-dumping or a green strategy? *Journal of Environmental Economics and Management, 45*, 692–707.

Greaker, M. (2003b). Strategic environmental policy when the governments are threatened by relocation. *Resource and Energy Economics, 25*, 141–154.

Greaker, M. (2006). Spillovers in the development of new pollution abatement technology: A new look at the Porter-hypothesis. *Journal of Environmental Economics and Management, 52*, 411–420.

Green, M. T., & Fielitz, B. (1977). Long term dependence in common stock returns. *Journal of Financial Economics, 4*, 339–349.

Green, M. T., & Fielitz, B. (1980). Long term dependence and least squares regression in investment analysis. *Management Science, 26*(10), 1031–1038.

Greening, L. A., Greene, D. L., & Difiglio, C. (2000). Energy efficiency and consumption—The rebound effect—A survey. *Energy Policy, 28*, 389–401.

Gross, D., & Harris, C. M. (1985). *Fundamentals of queueing theory* (2nd ed.). New York: Wiley.

Grossman, S. J., & Oliver, D. H. (1983). An analysis of the Principal-Agent problem, *Econometrica, 51*(1), 7–45.

Gueant, O. (2009). *Mean field games and applications to economics*. Ph.D. Thesis. Universite Paris Dauphine, Paris.

Gul, F. (1991). A theory of disappointment aversion. *Econometrica, 59*, 667–686.

Gumbel, E. J. (1958). *Statistics of extremes*. New York: Columbia University Press.

Ha, J. S., & Seong, P. H. (2003). *A method for risk-informed safety significance categorization using the analytic hierarchy process and Bayesian belief networks*. Department of Nuclear and Quantum Engineering, Korea Advanced Institute of Science and Technology, 373-1, Guseong-Dong, Yuseong-Gu, Daejeon 305-701, South Korea.

Hacking, I. (2006). *The emergence of probability* (2nd ed.). Cambridge: Cambridge University Press.

Hadar, J., & Russell, W. R. (1969). Rules for ordering uncertain prospects. *American Economic Review, 59*, 25–34.

Haimes, Y. Y. (1998). *Risk modeling, assessment and management*. New York: Wiley-Interscience.

Hale, A. (2002). Risk contours and risk management criteria for safety at major airports, with particular reference to the case of Schiphol. *Safety Science, 40*, 299–323.

Hallikas, J., Karvonen, I., Pulkkinen, U., Virolainen, V.-M., & Tuominen, M. (2004). Risk management processes in supplier networks. *International Journal of Production Economics, 90*(1), 47–58.

Hamada, K., Fujimoto, Y., & Shintaku, E. (2002). Ship inspection support system using a product model. *Journal of Marine Science and Technology, 6*, 205–215.

Hamel, A., & Heyde, F. (2008, December 11). *Duality for set-valued risk measures*. http://www.princeton.edu/~ahamel/SetRiskHamHey.pdf.

Hansen, M., & Zhang, Y. (2004). Safety efficiency: Link between operational performance and operation errors in the national airspace system. Transportation Research Record, *Journal of Transportation Research Board*, (1888), 15.

Hansson, S. O. (1996). What is philosophy of risk? *Theoria, 62*, 169–186.

Hansson, S. O. (2001). *The structure of values and norms*. New York: Cambridge University Press.

Hansson, S. O., & Peterson, M. (2001). Rights, risks and residual obligations. *Risk Decision and Policy, 6*, 157–166.

Harary, F. (1969). *Graph theory*. Reading, MA: Addison-Wesley.

Hardle, W., & Stahl, G. (Eds.), (2012). *Measuring risk in complex stochastic systems* (Vol. 147). Lecture notes on statistics. Berlin: Springer.

Hare, J. (1997). Port state control: Strong medicine to cure a sick industry. *Georgia Journal of International and Comparative Law, 26*, 571–594.

Harford, J. D. (1987). Self-reporting of pollution and the firm's behavior under imperfectly enforceable regulations. *Journal of Environmental Economics and Management, 14*, 293–303.

Harland, C., Brencheley, H., & Walker, H. (2003). Risk in supply networks. *Journal of Purchasing and Supply Management, 9*(2), 51–62.

Harsanyi, J. (1967–1968). Games with incomplete information played by Bayesian players: I–III, *Management Science*, 14, 159–182, 320–334, 486–502.

Harsanyi, J. C., & Selten, R. (1987). *A general theory of equilibrium selection in games*. Cambridge, MA: MIT Press.

Haubrich, J., & Lo, A. (1989). *The sources and nature of long term dependence in the business cycle*. NBER Working Paper no. 2951.

Helms, B., Kaen, F., & Rosenman, R. (1984). Memory in commodity futures contracts. *Journal of Futures Markets, 4*, 559–567.

Henry, C. (1974a). Option value in the economics of irreplaceable assets. *Review of Economic Studies, 41*, 89–104.

Henry, C. (1974b). Investment decisions under uncertainty: The irreversibility effect. *American Economic Review, 64*, 1006–1012.

Hermann, S., & Vallois, P. (2010a). *Probabilistic tools for safety management of the geological disposal*. Working paper, Ecole des Mines de Nancy.

Hermann, S., & Vallois, P. (2010b). From persistent random walk to the telegraph equation. *Stochastic and Dynamics, 10*(2), 161–196.

Hicks, J. (1979). *Causality in economics*. New York: Basic Books (first published in 1939).

Hillier, F., & Lieberman, G. (2010). *An introduction to operations research* (9th ed.). Mc Graw Hill: New York.

Hirshleifer, J., & Riley, J. G. (1979). The analysis of uncertainty and information: An expository survey. *Journal of Economic Literature, 17*, 1375–1421.

Hirshleifer, J., & Riley, J. G. (1992). *The analytics of uncertainty and information*. Cambridge: Cambridge University Press.

Ho, Y. C., & Cao, X. R. (1983). Perturbation analysis and optimization of queueing networks. *Journal of Optimization Theory and Applications, 40*, 559.

Holmstrom, B. (1979). Moral hazard and observability. *Bell Journal of Economics, 10*, 74–91.

Holmstrom, B. (1982). Moral hazard in teams. *Bell Journal of Economics, 13*, 324–340.

Holmstrom, B., & Myerson, R. B. (1983). Efficient and durable decision rules with incomplete information. *Econometrica, 51*, 1799–1819.

Holtz-Eakin, D., & Schwartz, A. E. (1995). Infrastructure in a structural model of economic growth. *Regional Science and Urban-Economics, 25*(2), 131–151.

Holtz-Eakin, D., & Schwartz, A. E. *Infrastructure in a structural model of economic growth*. National Bureau of Economic Research Working Paper: 4824, August 1994 (Copies available from: *National Bureau of Economic Research,* 1050 Massachusetts Avenue, Cambridge, MA 02138).

Hsieh, D. A. (1991). Chaos and nonlinear dynamics application to financial markets. *Journal of Finance, 46*, 1839–1877.

Hsu, L. F., Lin, C., & Tapiero, C. S. (1993). Network of queues modeling in flexible manufacturing systems: A survey. *RAIRO, 27*, 1–45.

Hsu, L. F., & Tapiero, C. S. (1987a). Quality control of the M/M/1 queue. *International Journal of Production Research, 25*, 447–455.

Hsu, L. F., & Tapiero, C. S. (1987b). Maintenance of the M/G/1 queue like job shop. *Queueing Systems: Theory and Applications*.

Hsu, L. F., & Tapiero, C. S. (1988a). Quality control of an unreliable random FMS with Bernoulli and CSP sampling. *International Journal of Production Research, 26*, 1125–1135.

Hsu, L. F., & Tapiero, C. S. (1988b). A Bayes approach to the quality control of an M/G/1 job shop queue. *Naval Research Logistics Quarterly, 35*, 327–343.

Hsu, L. F., & Tapiero, C. S. (1992). Integrated quality control and process maintenance. *International Journal of Production Research, 30.*

Hsu, L. F., & Tapiero, C. S. (1994). Inspection of an unreliable flexible manufacturing system: With repairable parts and non negligible inspection times. *Production Planning and Control.*

Hull, J., & White, A. (2000). Valuing credit default swaps I: No counterparty default risk. *Journal of Derivatives, 8*(1), 29–40.

Hull, J., & White, A. (2001). Valuing credit default swaps II: Modeling default correlations. *The Journal of Derivatives, Spring*, 12–21.

Hull, J., & White, A. (2003a). The valuation of credit default swap options. *Journal of Derivatives, 10*(3), 40–50.

Hull, J., & White, A. (2006). Valuing credit derivatives using an implied copula approach. *Journal of Derivatives, 14*(2), 8–28.

Hurst, H. E. (1951). Long terms storage of reservoirs. *Transaction of the American Society of Civil Engineers, 116.*

Iezzoni, L. I. (Ed.) (1994). *Risk adjustment for measuring health care outcomes.* Ann Arbor, MI: Health Administration Press, 30–32, 200.

Iezzoni, L. I. (1997). The risk of risk adjustment. *Journal of the American Medical Association, 278*(19), 1600.

Ijiri, Y., & Simon, H. A. (1977). *Skew distributions and the size of business firms.* New York: North Holland.

Imhof, J. P. (1985). On the range of Brownian motion and its inverse process. *The Annals of Probability, 13*(3), 1011–1017.

Ingene, C. A., & Hughes, M. A. (1985). Risk management by consumers. *Research in Consumer Behavior, 1*, 103–158.

Innes, R. (1996). Regulating automobile pollution under certainty, competition, and imperfect information. *Journal of Environmental Economics and Management, 31*, 219–239.

Innes, R. (1999). Remediation and self-reporting in optimal law enforcement. *Journal of Public Economics, 72*, 379–393.

International Maritime Organization (IMO). (2002). Guidelines for the application of Formal Safety Assessment (FSA) for use in the IMO rule-making process, 2002 /http://www.imo.orgS.

Irwin, J. O. (1925). On a criterion for the rejection of outlying observations. *Biometrika, 17*, 237–250.

Jackson, F. H. (1909). Mess. Math. 38, 1909, 57; *Quarterly Journal of Pure and Applied Mathematics, 41*, 193.

Jackson, J. R. (1957). Networks of waiting lines. *Operations Research, 5*, 518.

Jackson, J. R. (1963). Jobshop-like queueing systems. *Management Science, 10*, 131.

Jackwerth, J. C. (1999). Option implied risk neutral distributions and implied binomial trees: A literature review. *Journal of Derivatives, 7*, 66–82.

Jackwerth, J. C. (2000). Recovering risk aversion from option prices and realized returns. *The Review of Financial Studies, 13*(2), 433–451.

Jacoby, J., & Kaplan, L. (1972). The components of perceived risk. In M. Venkatesan, (Ed.), *Proceedings: Third Annual Conference, Atlanta Association for Consumer Research* (pp. 382–393).

Jacque, L., & Tapiero, C. S. (1986). The expected cost of ruin and insurance premiums in mutual insurance. *The Journal of Risk and Insurance, LIV*(3), 594–602.

Jacque, L., & Tapiero, C. S. (1987). Premium valuation in international insurance. *Scandinavian Actuarial Journal*, 50–61.

Jacque, L., & Tapiero, C. S. (1988a). Insurance premium allocation and loss prevention in a large firm: A principal agent analysis. *Studies in Banking and Finance, 16*, 231–245.

Jacque, L., & Tapiero, C. S. (1988b) Joint insurance and capitalization costs. *Insurance: Economics and Mathematics.*

Jacque, L., & Tapiero, C. S. (1990). Premium allocation and risk avoidance in a large firm: A continuous model. *Insurance: Economics and Mathematics.*

Janic, M. (2000). An assessment of risk and safety in civil aviation. *Journal of Air Transport Management, 6*, 43–50.

Jarrow, A., Lando, D., & Turnbull, M. (1997). A Markov model for the term structure of credit risk spreads. *The Review of Financial Studies, 10*(2), 481–523.

Jarrow, R., & Turnbull, S. (1995). Pricing derivatives on financial securities subject to credit risk. *Journal of Finance, 50*, 53–86.

Jaynes, E. T. (1957a). Information theory and statistical mechanics. *Physical Review, 106*, 620.

Jaynes, E. T. (1957b). Information theory and statistical mechanics II. *Physical Review, 108*, 171.

Jensen, F. V. (1996). *An introduction to Bayesian networks*. London: UCL Press.

Jewell, W. S. (1974). Operations research in the insurance industry: A survey of applications. *Operations Research, 22*, 918–928.

Jia, J., & Dyer, J. S. (1996). A standard measure of risk and risk-value models. *Management Science, 42*(12), 1691–1705.

Jia, J., Dyer, J. S., & Butler, J. C. (2001). Generalized disappointment models. *Journal of Risk and Uncertainty, 22*(1), 59–78.

Joe, H. (1997). *Multivariate models and dependence concepts* (Monographs on statistics and applied probability, Vol. 37). London: Chapman and Hall.

Joe, H., & Hu, T. (1996). Multivariate distributions from mixtures of max-infinitely divisible distributions. *Journal of Multivariate Analysis, 57*, 240–265.

John, P. M. (1962). *A tolerance region for multivariate distributions*, Sankhia, Series A, pp. 363–368.

John, P. (1990). *Statistical methods in engineering and quality assurance*. New York: Wiley-Interscience.

Johnson, N. L., & Kotz, S. (1969). *Discrete distributions*. New York: Houghton Mifflin.

Johnson, N. L., & Kotz, S. (1970a). *Continuous univariate distributions-1*. New York: Houghton Mifflin.

Johnson, N. L., & Kotz, S. (1970b). *Continuous univariate distributions-2*. New York: Houghton Mifflin.

Johnson, N. L., Kotz, S., & Balakrishnan, N. (1997). *Discrete multivariate distributions*. New York: Wiley.

Jorion, P. (2000). *VaR: The new benchmark for managing financial risk*. New York: McGraw Hill.

Jouini, E., Meddeb, M., & Touzi, N. (2004). Vector-valued coherent risk measures. *Finance and Stochastics, 8*, 531–552.

Juran, J. M. (1974). *Quality control handbook* (3rd ed.). New York: Mc Graw Hill.

Juran, J. M. (1980). *Quality planning and analysis*. New York: Mc Graw Hill.

Kaas, R., Goovaerts, M., Dhaene, J., & Denuit, M. (2001). *Modern actuarial risk theory*. Dordrecht: Kluwer Academic.

Kadane, J. B., & Larkey, P. D. (1982). Subjective probability and the theory of games. *Management Science, 28*, 113–120.

Kahane, Y., Neumann, S., & Tapiero, C. S. (1988). Computer backup pools, disaster recovery and default risk. *Communications of the ACM, 31*, 78–83.

Kahneman, D., & Lovallo, D. (1993). Timid choices and bold forecasts: A cognitive perspective on risk taking. *Management Science, 39*, 17–31.

Kahneman, D., & Riepe, W. W. (1998). Aspects of investor psychology. *Journal of Portfolio Management, 24*(4).

Kahneman, D., Tversky, A. (1979). Prospect theory: An analysis of decision under risk, *Econometrica*, March, 263–291.

Kahneman, D., & Tversky, A. (Eds.). (2010). *Choices, values and frames*. Cambridge: Cambridge University Press.

Kalman, R. E. (1994). Randomness reexamined. *Modeling, Identification and Control, 15*(3), 141–151.

Kaniadakis, G. (2009). Maximum entropy principles and power law tailed distributions. *The European Physical Journal, B, 70*, 3–13.

Kapur, J. N., & Kesavan, H. K. (1992). *Entropy optimization principles with applications.* San Diego, CA: Academic Press.

Karni, E. (1979). On multivariate risk aversion. *Econometrica, 47*, 1391–1401.

Katz, M., & Shapiro, C. (1985). Network externalities, competition, and compatibility. *American Economic Review, 75*(3), 424–440.

Katz, M., & Shapiro, C. (1994). Systems competition and network effects. *Journal of Economic Perspectives, 8*(2), 93–115.

Katzman, M. T. (1988). Pollution liability insurance and catastrophic environmental risks. *The Journal of Risk and Insurance, 55*, 75–100.

Kenett, R. S., & Zacks, S. (2004). *Modern industrial statistics: Design and control of quality and reliability*, Duxbury Press, San Francisco, 1998, Spanish edition, 2000, 2nd edition 2003, Chinese edition, 2004.

Keynes, J. M. (1921). *A treatise on probability*. London: MacMillan (reprinted, 1948).

Keynes, J. M. (1935). The General Theory of Employment, Interest and Money. Fellow of the King's College, Cambridge, published by Harcourt, Brace and Company, First Published: Macmillan Cambridge University Press, for Royal Economic Society in 1936.

Kihlstrom, R. E., & Mirman, L. J. (1974). Risk aversion with many commodities. *Journal of Economic Theory, 8*, 337–360.

Kimball, M. (1990). Precautionary saving in the small and in the large. *Econometrica, 58*, 53–78.

Kindleberger, C. (1978). *Manias, panics, and crashes: A history of financial crises.* New York: Basic Books.

Klein, B., & Saft, L. F. (1985). The law and economics of franchise tying contracts. *Journal of Law and Economics*, 345–349.

Klugman, S. A., Panier, H. H., & Willmot, G. E. (2000). *Loss models: From data to decisions* (3rd ed.). Wiley, Wiley Series in probability and Statistics.

KMV-Corporation. (1997). Modeling default risk, *Technical Document*, http://www.kmv.com.

Knapp, S., & Franses, P. H. (2007). Econometric analysis on the effects of port state control inspections on the probability of casualty. Can targeting of subsequent ships for inspections be improved? *Marine Policy, 31*, 550–563.

Knight, F. (1921). *Risk, uncertainty and profits.* Boston, MA: Houghton, Mifflin & Co.

Kogan, K., & Tapiero, C. S. (2007). In F. Hillier (Series Ed.), *Supply chain games: Operations management and risk valuation.* Series in Operations Research and Management Science. Springer.

Kotowitz, Y., & Mathewson, F. (1979). Advertising, consumer information and product quality. *Bell Journal of Economics, 10*, 566–588.

Krantz, D., Luce, R. D., Suppes, P., & Tversky, A. (1971). *Foundations of measurement* (Vol. 1). New York: Academic Press.

Kuhre, W. L. (1998). *ISO 14031: Environmental performance evaluation.* Englewood Cliffs, NJ: Prentice-Hall.

Kullback, S. (1959). *Information theory and statistics.* New York: Wiley.

Kullback, S. (1987). The Kullback-Leibler distance. *The American Statistician, 41*, 340–341.

Kullback, S., & Leibler, R. A. (1951). On Information and sufficiency. *Annals of Mathematical Statistics, 22*(1), 79–86. doi:10.1214/aoms/1177729694 DOI:10.1214%2Faoms%2F1177729694. MR39968.

Kunitomo, N. (1995). Improving the Parkinson method of estimating security price volatilities. *Journal of Business, 65*(2), 295–302.

Kusuoka, S. (2001). On law invariant coherent risk measures. *Advanced Mathematical Economics, 3*, 83–95.

Kutler, J., (2009). The years of the Swans. *Risk Professional, April*, 12–16.

La Londe, B., & Cooper, M. (1989). *Partnership in providing customer service: A third part perspective*. Oak Brook, IL: Council of Logistic Management.

Lachapelle, A. (2010). *Human crowds and groups interactions: A mean field games approach*. Preprint.

Lachapelle, A., Salomon, J., & Turinici, G. (2010). Computation of mean field equilibria in economics. *Mathematical Models and Methods in Applied Sciences, 20*(4), 567–588.

Lachapelle, A., Salomon, J., & Turinici, G. (2012). *Computation of mean field equilibria in economics*. DRAFT CEREMADE, UMR CNRS 7534, Université de Paris-Dauphine, Place du Maréchal De Lattre De Tassigny, 75775 Paris cedex 16, France.

Lachapelle, A., Wolfram, M.-T. (2011). *On a mean field game approach modeling congestion and aversion in pedestrian crowds*. Preprint

Laffont, J. J. (1989). *The economics of uncertainty and information*. Cambridge, MA: MIT Press.

Laffont, J. J. (1995). Regulation, moral hazard and insurance of environmental risks. *Journal of Public Economics, 58*, 319–336.

Landis, J. R. (1995). Comparing hospital mortality in adult patients with pneumonia: A case study of statistical methods in a managed care program. *Annals of Internal Medicine, 122*(2), 125–131.

Langlois, N., & Cosgel, M. M. (1993). Frank Knight on risk, uncertainty and the firm: A new reinterpretations. *Economic Inquiry, 31*(3), 456–465.

Laplace, P. S. (1774). Memoire sur la Probabilite des Causes Par les Evenements, *Memoires de l'Academie Royale des Sciences*, Paris, 6, 621–656

Larkin, J. (2003). *Strategic reputation risk management*. London: Palgrave MacMillan.

Lasry, J. M., & Lions, P.-L. (2006a). Jeux a champ moyen. I. Le cas stationnaire. *C. R. Math. Acad. Sci. Paris, 343*(9), 619–625.

Lasry, J. M., & Lions, P.-L. (2006b). Jeux a champ moyen. II. Horizon fini et controle optimal. *C. R. Math. Acad. Sci. Paris, 343*(10), 679–684.

Lasry, J. M., & Lions, P.-L. (2007a). Mean field games. *Japanese Journal of Mathematics, 2*(1), 229–260.

Lasry, J. M., & Lions, P.-L. (2007b). Mean field games. *Cahiers de la Chaire Finance et Developpement Durable*.

Lasry, J. M., Lions, P.-L., & Gueant, O. (2010a). *Application of mean field games to growth theory*. Preprint.

Lasry, J. M., & Lions, P.-L., & Gueant, O. (2010b). Mean field games and applications. *Paris-Princeton lectures on Mathematical Finance*.

Lax, M., Cai, W., & Xu, M. (2006). *Random processes in physics and finance*. London, England: Oxford University Press.

Leadbetter, M. R., Lindgren, G., & Rootzen, H. (1983). *Extremes and related properties of random sequences and processes*. New York: Springer.

Lee, J. (1993). Generating random binary deviates having fixed marginal distributions and specified degrees of association. *Statistical Computing, 47*, 209–215.

Lee, H., & Tapiero, C. S. (1986). Quality control and the sales process. *Naval Research Logistics Quarterly, 33*, 569–587.

Lee, H., & Tapiero, C. S. (1989). Quality control and product servicing. *European Journal of Operation Research, 39*, 261–273.

Lehar, A. (2005). Measuring systemic risk: A risk management approach. *Journal of Banking Finance, 29*(10), 2577–2603. doi:10.1016/j.jbankfin.2004.09.007 ISSN0378-4266.

Leland, H., & Pyle, D. (1977). Information asymmetries, financial structure and financial intermediation. *Journal of Finance, 32*, 371–387.

Lemke, C. E., & Howson, J. T. (1964). Equilibrium points of bi-matrix games. *SIAM Journal, 12* (2), 413–423.

Leroy, S. F., & Singell, L. D., Jr. (1987). Knight on risk and uncertainty. *Journal of Political Economy, 95*(21), 394–406.

Levy, P. (1937). *Théorie de l'addition des variables aléatoires*. Paris: Gauthier-Villars.

Levy, H., & Levy, L. (1991). Arrow-Pratt measures of risk aversion: the multivariate case. *International Economic Review, 32*, 891–898.

Li, D. X. (1999). The valuation of basket credit derivatives. *CreditMetrics Monitor*, April.

Li, D. X. (2000a). On default correlation: A Copula approach. *Journal of Fixed Income, 9*, 43–54.

Li, D. X. (2000b). On default correlation: A Copula function approach. *The Journal of Fixed Income, 6*, 43–54.

Li, K. X., & Cullinane, K. (2003). An economic approach to maritime risk management and safety regulation. *Maritime Economics and Logistics, 5*, 268–284.

Li, K. X., & Zheng, H. (2008). Enforcement of law by the Port State Control (PSC). *Maritime Policy & Management, 35*(1), 61–71.

Liebowitz, S., & Margolis, S. (1994). Network externality: An uncommon tragedy. *Journal of Economic Perspectives, 8*(2), 133–150.

Lim, W. (2001). Producer-supplier contracts with incomplete information. *Management Science, 47*(5), 709–715.

Lindskog, F. (2000). *Modeling dependence with copulas*. ETH, Zurich: RiskLab.

Lintner, J. (1965a). The valuation of risky assets and the selection of risky investments in stock portfolios and capital budgets. *Review of Economic and Statistics, 47*, 13–37.

Lintner, J. (1965b). Security prices, risk and maximum gain from diversification. *Journal of Finance, 20*, 587–615.

Liu, L., Kashyap, B. R. K., & Templeton, J. G. C. (1990). On the $GI^X/G/\infty$ system. *Journal of Applied Probability, 27*, 671–683.

Ljungqvist, L., & Sargent, T. J. (2000). *Recursive macroeconomic theory*. Cambridge, MA: MIT Press.

Lo, A. W. (1991). Long term memory in stock market prices. *Econometrica, 59*, 1279–1313.

Lo, A. W. (1997). Fat tails, long memory and the stock market since 1960's. *Economic Notes, 26*, 213–245.

Loomes, G., & Sugden, R. (1982). Regret theory: An alternative to rational choice under uncertainty. *The Economic Journal, 92*, 805–824.

Loomes, G., & Sugden, R. (1987). Some implications of a more general form of regret theory. *Journal of Economic Theory, 41*, 270–287.

Lootsma, F. A. (1988). Numerical scaling of human judgement in pairwise comparison methods for fuzzy multi-criteria decision analysis. In G. Mitra (Ed.), *Mathematical models for decision support* (pp. 57–88). Berlin: Springer.

Lorenz, E. (1966). Large-scale motions of the atmosphere: Circulation. In P. M. Hurley (Ed.), *Advances in earth science*. Cambridge, MA: MIT Press.

Losq, E., & Chateau, J. P. D. (1982). A generalization of the CAPM based on a property of the covariance operator. *Journal of Financial and Quantitative Analysis, 17*, 783–797.

Lowenstein, R. (2000). *When genius failed: The rise and fall of long-term capital management*. New York: Random House.

Lucas, R. (1978). Asset prices in an exchange economy. *Econometrica, 46*, 1429–1446.

Luce, R. D. (1980). Several possible measures of risk. *Theory and Decision, 12*, 217–228.

Luce, R. D. (1981). Correction to several possible measures of risk. *Theory and Decision, 13*, 381.

Luce, R. D., & Raiffa, H. (1958). *Games and decisions*. New York: Wiley.

Luce, R. D., & Raiffa, H. (1967). *Games and decisions: Introduction and critical survey*. New York: Wiley.

Luce, R. D., & Weber, E. (1986). An axiomatic theory of conjoint expected risk. *Journal of Mathematical Psychology, 30*, 188–205.

Lundberg, F. (1909). *Zur Theorie der Ruckversicherung Verdandlungskongress fur Versicherungsmathematik*. Wien.

Lundberg, O. (1940). *On random processes and their applications to sickness and accident statistics*. Upsala: Almquist and Wiksells.

Machina, M. J. (1982). Expected utility analysis without the independence axiom. *Econometrica, 50*, 277–323.

Machina, M. J. (1987). Choice under uncertainty: Problems solved and unsolved. *Journal of Economic Perspectives, 1*, 121–154.

Machina, M. J., & Munier, B. (Eds.). (1999). *Beliefs, interactions and preferences in decision making*. Boston: Kluwer Academic.

Magill, M., & Quinzii, M. (1996). *Theory of incomplete markets*. Cambridge, MA: MIT Press.

Majumdar, A., Dupuy, M. D., & Ochieng, W. O. (2006). A framework for the development of safety indicators for New Zealand airspace: The categorical analysis of factors affecting loss of separation incidents. Transportation Research Board (TRB) annual conference.

Malvergne, Y., & Sornette, D. (2002). Multivariate Weibull Distributions for Asset Retturns-I, ISFA, Graduate School of Actuarial Science, University of Lyon 1,France malevergne@univ-lyon1.fr and sornette@unice.fr.

Mandelbrot, B. B. (1963). *The variation of certain speculative prices* (pp. 392–417). XXXVI: Journal of Business.

Mandelbrot, B. (1972). Statistical methodology for non-periodic cycles: From the covariance to R/S analysis. *Annals of Economic and Social Measurement, 1*, 259–290.

Mandelbrot, B. (1997). Three fractal models in finance: Discontinuity, concentration, risk. *Economic Notes, 26*, 171–212.

Mandelbrot, B. (2004). *The misbehavior of markets*. New York: Basic Books.

Mandelbrot, B., & Hudson, R. (2006). *The misbehavior of markets: A fractal view of financial turbulence*. New York: Basic Books.

Mandelbrot, B., & Taqqu, M. (1979). Robust R/S analysis of long run serial correlation. *Bulletin of the International Statistical Institute, 48*(Book 2), 59–104.

Mandelbrot, B., & Van Ness, J. (1968). Fractional brownian motion, fractional noises and applications. *SIAM Review, 10*, 422–437.

Mandelbrot, B., & Wallis, J. (1968). Noah, Joseph and operational hydrology. *Water Resources Research, 4*, 909–918.

Mandelbrot, B., & Wallis, J. R. (1969). Computer experiments with fractional noises. *Water Resources Research, 5*, 228–267.

Marcus, P. A., & Willig, J. T. (Eds.). (1997). *Moving ahead with ISO 14000: Improving environmental management and advancing sustainable development*. New York: Wiley.

Markowitz, H. (1952). Portfolio selection. *Journal of Finance, 7*, 77–99.

Markowitz, H. (1959). *Portfolio selection – Efficient diversification of investments*. New York: Wiley.

Marshall, A., & Olkin, I. (1988). Families of multivariate distributions. *Journal of the American Statistical Association, 83*, 834–841.

Martιnez-Jaramillo, S., Perez Perez, O., Avila Embriz, F., & Lopez Gallo Dey, F. (2010). Systemic risk, financial contagion and financial fragility. *Journal of Economic Dynamics and Control, 34*, 2358–2374.

Marvel, H. (1982). Exclusive dealing. *Journal of Law and Economics, 25*, 1–26.

Maschler, M. (1966). A price leadership method for solving the inspector's non-constant sum game. *Naval Research Logistics Quarterly, 13*, 11–33.

Mathewson, G. F., & Winter, R. A. (1986). The economics of franchise contracts. *Journal of Law and Economics, 28*, 503–526.

May, R. (1974). Biological populations with non-overlapping generations: Stable points, stable cycles, and chaos. *Science, 186*, 645–647.

May, R. (1976). Simple mathematical models with very complicated dynamics. *Nature, 261*, 459–467.

Mazzoleni, P. (2004). Risk measures and return performance: A critical approach. *European Journal of Operations Research, 155*, 268–275.

McCarthy, D. (1997). Rights, explanation, and risks. *Ethics, 107*, 205–225.

McKerlie, D. (1986). Rights and risk. *Canadian Journal of Philosophy, 16*, 239–251.

McNeil, A. J., Frey, R., & Embrechts, R. (2005). *Quantitative risk management: Concepts, techniques and tools*. Princeton, NJ: Princeton University Press.

Medio, A. (1992). *Chaotic dynamics*. Cambridge: Cambridge University Press.

Merton, R. (1969). Lifetime portfolio selection under uncertainty: The continuous time case. *The Review of Economics and Statistics, 50*, 247–257.

Merton, R. C. (1973a). Theory of rational option pricing. *Bell Journal of Economics and Management Science, 4*, 141–183.

Merton, R. C. (1977). Optimum consumption and portfolio rules in a continuous time model. *Journal of Economic Theory, 3*, 373–413.

Merton, R. C. (1990). *Continuous time finance*. Cambridge, MA: Blackwell.

Mikosch, T., & Resnick, S. (Eds.), (2001). *Levy Processes: Theory and Applications*. Basel: Birkhauser-Verlag.

Mikosch, T., & Resnick, S. (Eds.), (2012). *Levy processes: Theory and Applications*. Basel: Birkhauser-Verlag.

Minsky, H. (1993). The financial instability hypothesis. *Handbook of Radical Political Economy* (Working paper no. 74, 1992). Aldershot.

Miyahara, Y. (2012). *Option pricing in incomplete markets*. London: Imperial College Press.

Montgomery, D. C. (1991). *Design and analysis experiments* (3rd ed.). New York: Wiley.

Mood, A. M. (1940). The distribution theory of runs. *Annals of Mathematical Statistics, 11*, 367–392.

Morgan, J. P. (1994). *Riskmetrics*. New York, October and November 1995.

Morgan, J. P. (1997). *Credit metrics*. New York.

Morgenstern, O. (1967). *Game theory: A new paradigm of social science*. Paper in "New Methods of Thought and Procedure". Springer.

Morrison, C., & Schwartz, A. (1996). State infrastructure and productive performance. *American Economic Review, 86*, 1095–1111.

Mosteller, F. (1941). Note on the application of runs to quality control charts. *Annals of Mathematical Statistics, 12*, 228–232.

Moustakides, G. V. (1986). Optimal stopping times for detecting changes in distributions. *The Annals of Statistics, 13*, 1379–1387.

Mulvey, J. M., Vanderbei, R. J., & Zenios, S. A. (1991). *Robust optimization of large scale systems*. Report SOR 13, Princeton University.

Munier, B. (1986). Complexité et Décision stratégique dans l'Incertain: Que peut-on retenir de la Théorie?. In M. Boiteux, Th. de Montbrial, & B. Munier (Eds.), *Marchés, Capital et Incertitude*. Paris: Economica.

Munier, B. (1991). Market uncertainty and the process of belief formation. *Journal of Risk and Uncertainty, 4*, 233–250.

Munier, B. (1995). From instrumental to cognitive rationality: Contributions of the last decade to risk modeling. *Revue d'Economie Politique, 105*, 5–70 (French, with abstract in English).

Munier, B. (Ed.). (2012). *Global uncertainty and the volatility of agricultural commodities*. Amsterdam, The Netherlands: IOS Press.

Munier, B., Selten, R., et al. (1999). Bounded rationality modeling. *Marketing Letters, 10*(3), 233–248.

Munier, B., & Tapiero, C. S. (2008). Risk attitudes. In B. Everitt & E. Melnick (Eds.), *Encyclopedia of quantitative risk assessment*. Hoboken, NJ: Wiley.

Munnell, A. H. (1992). Infrastructure investment and economic growth. *Journal of Economic Perspectives, 6*, 189–198.

Muth, J. (1961). Rational expectations and the theory of price movements. *Econometrica, 29*, 315–335.

Mzasoliver, J., Porra, J. M., & Weiss, G. H. (1992). The continuum limit of a two dimensional persistent random walk. *Physica A: Statistical Mechanics and Applications, 182*, 593–598.

Nash, F. (1950). Equilibrium points in N-person games. *Proceedings of the National Academy of Sciences, 36*, 48–49.

Nasr, M. (2007). Quality by Design (QbD) – A Modern System Approach to Pharmaceutical Development and Manufacturing – FDA Perspective", *FDA Quality Initiatives Workshop*, Maryland, USA.

Nau, R. (1999). Arbitrage, incomplete models, and other people's brains. In M. J. Machina & B. Munier (Eds.), *Beliefs, interactions and preferences in decision making*. Dordrecht/Boston: Kluwer Academic.

Nau, R. (2011). Risk, ambiguity and state-preference theory. *Economic Theory, 48*(2–3), 437–467.

Naudts, J. (2007). *Generalized thermostatistics*. Belgium: University of Antwerpen.

Nelsen, R. B. (1999). *An introduction to copulas*. Berlin: Springer.

Nelson, P. (1970). Information and consumer behavior. *Journal of Political Economy, 78*, 311–29.

Nelson, D. B., & Foster, D. P. (1994). Asymptotic filtering theory for univariate ARCH model. *Econometrica, 62*, 1–41.

Nerlove, M., & Arrow, K. J. (1962). Optimal advertising policy under dynamic conditions. *Economica, 42*, 129–142.

Netjasov, F., & Janic, M. (2008). A review of research on risk and safety modeling in civil aviation. *Journal of Air Transport Management, 14*, 213–220.

Neyman, A., & Okada, D. (1999). Strategic entropy and complexity in repeated games. *Games and Economic Behavior, 29*, 191–223.

Neyman, J., & Scott, E. L. (1971). Outliers proneness of phenomena and of related distributions. In J. Rustagi (Ed.), *Optimizing methods in statistics*. New York: Academic Press.

Nijkamp, P., & Blaas, E. (1993). *Impact assessment and evaluation in transportation planning*. Amsterdam: Kluwer Academic.

Nijkamp, P., & Ubbels, B. (1999). How reliable are estimates of infrastructure costs? A comparative analysis. *International Journal of Transport Economics, 26*(1), 23–53.

Nikoloupoulos, A., & Karlis, D. (2008). Multivariate logit copula model with application to dental data. *Statistics in Medicine, 27*, 6393–6406.

Noah, T. (2012). *The great divergence: America's growing inequality crisis and what we can do*. New York: Bloomsbury Press.

Oakes, D. (1994). Multivariate survival distributions. *Journal of Nonparametric Statistics, 3*, 343–354.

Otway, T. H. (1995). Records of the Florentine proveditori degli cambiatori: An example of an antipersistent time series in economics. *Chaos, Solitons and Fractals, 5*, 103–107.

Owen, G. (1982). *Game theory*. New York: Academic Press.

Page, E. S. (1954). Continuous inspection schemes. *Biometrika, 41*, 100–114.

Pareto, V. (1896). *Le cours d'Economie Politique*. London: Macmillan.

Parkinson, M. (1980). The extreme value method for estimating the variance of the rate of return. *Journal of Business, 53*, 61–65.

Paroush, J., & Tapiero, C. S. (1975). Optimal location of a polluting plant on a line under uncertainty. *Journal of Regional Science, 16*(3), 367–374.

Patient Safety Indicators Overview. (2006). *AHRQ quality indicators*. Rockville, MD: Agency for Healthcare Research and Quality.

Patlak, C. (1953). Random walk with persistence and external bias. *Bulletin of Mathematical Biology, 15*(3), 311–338.

Patton, A. (2006). Modeling asymmetric exchange rate dependence. *International Economic Review, 47*, 527–556.

Patton Andrew, J. (2007). Copula-based models for financial time series. In T. G. Andersen, R. A. Davis, J. -P. Kreiss, & T. Mikosch (Eds.), *Handbook of financial time series*.Springer, forthcoming. Available: June 3, 2009

Pedersen, C., & Satchell, S. E. (1998). An extended family of financial-risk measures. *The Geneva Papers on Risk and Insurance Theory, 23*, 89–117.

Peizer, B., & Pratt, J. W. (1977). A normal approximation for binomial, F, Beta and other common, related tail probabilities. *Journal of the American Statistical Association, 1457–1494, Part II*

Penner, I. (2007). *Dynamic convex risk measures: Time consistency, prudence, and sustainability.* Ph.D. thesis, Humboldt-Universitat zu Berlin.

Peter, E. E. (1995). *Chaos and order in capital markets.* New York: Wiley.

Phadke, M. S. (1986). *Quality engineering using robust design.* Englewood Cliffs, NJ: Prentice Hall.

Pierskalla, W. P., & Voelker, J. A. (1976). A survey of maintenance models: The control and surveillance of deteriorating systems. *Naval Research Logistics Quarterly, 23*(3), 353–388.

Pollatsek, A., & Tversky, A. (1970). A theory of risk. *Journal of Mathematical Psychology, 7*, 540–553.

Posner, M., & Tapiero, C. S., (1988). The manufacturing quality density function and quality control. *International Journal of Production research, 26.*

Post, T., van der Assem, N. J., Baltussen, G., & Thaler, R. H. (2008). Deal or no deal? Decision making under risk in a large-payoff game show. *American Economic Review, 98*(1), 38–71.

Pottier, N. (1996). Analytic study of the effect of persistence on a one-dimensional biased random walk. *Physica A: Statistical Mechanics and Applications, 230*, 563–576.

Prabhu, N. U. (1965). *Stochastic processes.* New York: MacMillan.

Prabhu, N. U. (1980). *Stochastic storage processes.* New York: Springer.

Pratt, J. W. (1964). Risk aversion in the small and in the large. *Econometrica, 32*, 122–136.

Pratt, J. W. (1977). A normal approximation for binomial, F, Beta and other common, related tail probabilities. *Journal of the American Statistical Association, 63*, 1416–1456. Part I.

Pratt, J. W. (1990). The logic of partial-risk aversion: Paradox lost. *Journal of Risk and Uncertainty, 3*, 105–113.

Pratt, J. W., & Zeckhauser, R. (1987). Proper risk aversion. *Econometrica, 55*, 143–154.

Prigogine, I., & Stengers, I. (1988). *Entre le Temps et l'Eternité.* Paris: Fayard.

Pykhtin, M. (Ed.). (2005). *Counterparty credit risk modeling.* London: Risk Books.

Pyndick, R. (1991). Irreversibility, uncertainty and investment. *Journal of the Economic Literature, 29*, 1110–1148.

Quiggin, J. (1982). A theory of anticipated utility. *Journal of Economic Behavior and Organization, 3*(4), 323–343.

Quiggin, J. (1993). *Generalized expected utility theory-the rank-dependent expected utility model.* Dordrecht: Kluwer Academic.

Rabin, M. (1998). Psychology and economics. *Journal of Economic Literature, 36*, 11–46.

Ramsey, F. P. (1931). Truth and probability. In *The foundations of mathematics and other logical essays* (pp.156–198). London: Routledge and Kegan Paul

Rau, J. G. (1970a). *Optimization and probability in systems engineering.* New York: Van Nostrand Reinhold Co.

Reason, J. (1997). *Managing the risks of organisational accidents.* Aldershot: Ashgate.

Reesor, R. M., & McLeish, D. L. (2001). *Risk, entropy and the transformation of distributions.* Working paper, Department of Statistics and Actuarial Science, University of Waterloo, Waterloo, Ontario, Canada.

Reisman, A., Ritchken, P., & Tapiero, C. S. (1987a). Reliability, pricing and quality control. *European Journal of Operations Research, 31*, 37–45.

Reiss, R. D., & Thomas, M. (2001). *Statistical analysis of extreme values.* Basel: Birkhauser.

Rényi, A. (1961). On measures of information and entropy. In *Proceedings of the 4th Berkeley Symposium on Mathematics, Statistics and Probability* (pp. 547–561), 1960.

Renyi, A. (1970). Foundations of Probability, volume 9, Holden Day, San Francisco, 1970, 23, 64–67.

Rey, P. (1992, February). *The economics of franchising.* ENSAE Paper.

Rey, P., & Tirole, J. (1986). The logic of vertical restraints. *American Economic Review, 76*, 921–939.

Reyniers, D. J. (1992). Supplier-customer interactions in quality control. *Annals of Operations Research, 34*, 307–330.

Reyniers, D. J., & Tapiero, C. S. (1995a). The delivery and control of quality in supplier-producer contracts. *Management Science, 41*(10), 1581–1589.

Reyniers, D. J., & Tapiero, C. S. (1995b). Contract design and the control of quality in a conflictual environment. *European Journal of Operations Research, 82*(2), 373–382.

Riedel, F. (2004). Dynamic coherent risk measures. *Stochastic Processes and their Applications, 112*(2), 185–200.

Riordan, M. (1984). Uncertainty, asymmetric information and bilateral contracts. *Review of Economic Studies, 51*, 83–93.

Ritchken, P., & Tapiero, C. S. (1986). Contingent claim contracts and inventory control. *Operation Research, 34*, 864–870.

Ritter, J. R. (1991). The long-run performance of initial public offerings. *Journal of Finance, 46*, 3–27.

Rockafellar, T., Uryasev, S., & Zabarankin, M. (2002). *Deviation measures in risk analysis and optimization*. http://www.ise.ufl.edu/uryasev/Deviation_measures_wp.pdf. Retrieved October 13, 2011.

Rockafellar, R. T., & Wets, R. J.-B. (1992). A dual strategy for the implementation of the aggregation principle in decision making under uncertainty. *Applied Stochastic Models and Data Analysis, 8*, 245–255.

Rodriguez, R. N. (1977). A guide to the Burr type XII distributions. *Biometrika, 64*, 129–134.

Rogers, L. C. G., & Satchell, S. E. (1991). Estimating variance from high, low and closing prices. *The Annals of Applied Probability, 1*, 504–512.

Ross, S. A. (1981). Some stronger measures of risk aversion in the small and in the large with applications. *Econometrica, 49*, 621–638.

Ross, S. A. (1999). Adding risks: Samuelson's fallacy of large numbers revisited. *Journal of Financial and Quantitative Analysis, 34*, 323–340.

Rothenstein, D., & Zamir, S. (2002). Imperfect inspection games over time. *Annals of Operations Research, 109*, 175–192.

Rothschild, M., & Stiglitz, J. (1970). Increasing risk I: A definition. *Journal of Economic Theory, 2*, 225–243.

Rubenstein, R. Y. (1981). *Simulation and the Monte Carlo methods*. New York: Wiley.

Rubinstein, A. (1982). Perfect equilibrium in a bargaining model. *Econometrica, 50*, 97–109.

Rubinstein, M. (1994). Implied binomial trees. *Journal of Finance, 64*, 771–818.

Rubinstein, A. (1998). *Modeling bounded rationality*. Cambridge, MA: MIT Press.

Ruckle, W. H. (1992). The upper risk of an inspection agreement. *Operations Research, 40*, 877–884.

Russell, C. S. (1990). Game models for structuring monitoring and enforcement systems. *Natural Resource Modeling, 4*, 143–176.

Saaty, T. L. (1961). *Elements of queuing theory*. New York: McGraw Hill Book Co.

Saaty, T. L. (1980). *The analytic hierarchy process*. New York: McGraw Hill.

Saito, Y. U., Watanabe, T., & Iwamura, M. (2007). Do larger firms have more interfirm relationships? *Physica A: Statistical Mechanics and Applications, 383*, 158–163.

Sakai, Y. (1985). The value of information in a simple duopoly model. *Journal of Economic Theory, 36*, 36–54.

Samuelson, P. A. (1963). Risk and uncertainty: A fallacy of large numbers. *Scientia, 98*, 108–163.

Samuelson, P. A. (1965). Proof that properly anticipated prices fluctuate randomly. *Industrial Management Review, 6*, 41–49.

Sanchez-Robles, B. (1998a). The role of infrastructure investment in development: Some macroeconomic considerations. *International Journal of Transport Economics, 25*(2), June, 113–136. http://wwwdte.spol.uniroma1.it/trasporti/journal.html.

Sanchez-Robles, B. (1998b). Infrastructure investment and growth: Some empirical evidence. *Contemporary Economic Policy, 16*(1), 98–108.

Sandmann, K., & Sondermann, D. (1993). A term structure model and the pricing of interest rates derivatives. *The Review of Future Markets, 12*(2), 391–423.

Sato, A.-H. (2010). q-Gaussian distributions and multiplicative stochastic processes for analysis of multiple financial time series, Mathematical Analysis of Generalized Entropies and their Applications. *Journal of Physics*, Conference Series 201, 012008.

Savage, L. J. (1954). *The foundations of statistics*. New York: Wiley.

Savov, A. (2011). Asset pricing with garbage. *The Journal of Finance, 66*(1), 177–201.

Schied, A. (2006). Risk measures and robust optimization problems. *Stochastic Models, 22*, 753–831.

Schied, A. (2007). Optimal investments for risk- and ambiguity-averse preferences: A duality approach. *Finance and Stochastics, 11*(1), 107–129.

Schmeidler, D. (1989). Subjective probability and expected utility without additivity. *Econometrica, 57*, 571–587.

Schonbucher, P. (2003). *Credit derivatives pricing models, pricing, implementation*. London: Wiley-Finance.

Schweizer, B. (1991). Thirty years of copulas. In *Advances in probability distributions with given marginals* (pp. 13–50; 879–885). Dordrecht: Kluwer Academic.

Seal, H. L. (1969). *Stochastic theory of a risk business*. New York: Wiley.

Seal, H. (1978). *Survival probabilities: The goals of risk theory*. New York: Wiley, Interscience.

Sebenius, J. K., & Geanakopolos, J. (1983). Don't bet on it: Contingent agreements with asymmetric information. *Journal of the American Statistical Association, 78*, 424–426.

Sen, A. (1973). *On economic inequality*. Oxford: Clarendon Press.

Sethi, S. P. (1977). Dynamic optimal control models in advertising: A survey. *SIAM Review, 19*(4), 685–725.

Shackle, G. L. S. (1958). *Time in economics*. Amsterdam: North Holland.

Shaked, M., & Shantikumar, J. G. (Eds.). (1994). *Stochastic orders and their applications*. San Diego: Academic Press.

Shalit, H., & Yitzhaki, S. (1984). Mean-Gini, portfolio theory and the pricing of risky assets. *Journal of Finance, 29*(5), 1449–1468.

Sharpe, W. (1964). Capital asset prices: A theory of market equilibrium under conditions of risk. *Journal of Finance, 19*, 425.

Sheffrin, H. (2008a). *A behavioral approach to asset pricing* (2nd ed.). Boston: Elsevier Academic Press.

Sheffrin, H. (2008b). Risk and return in behavioral SDF-based asset pricing models. *Journal of Investment Management, 6*(3), 1–19.

Shewart, W. A. (1931). *The economic control of manufactured product*. New York: Van Nostrand.

Shubik, M. (1955). The uses of game theory in management science. *Management Science, 2*(1), 40–55.

Shubik, M. (2002). Game theory and operations research: Some musings 50 years later. *Operations Research, 50*, 192–196.

Siebenaler, Y. (1997). *Etude de l'amplitude pour certains processus de Markov*. Thesis, Department of Mathematics, University of Nancy I, France (under supervision of Professor Pierre Vallois).

Siebenaler, Y., Tapiero, C. S., & Vallois, P. (1997). *The average and the variance of the range run length in birth and death random walks*. Mimeo.

Siegmund, D. (1985). *Sequential analysis, tests and confidence intervals*. Berlin: Springer.

Simon, H. A. (1955). A behavioral model of rational choice. *Quarterly Journal of Economics, 69*, 99–118.

Simon, H. A. (1976). From substantive to procedural rationality. In S. J. Latsis (Ed.), *Methods and appraisal in economics*. Cambridge, MA: MIT Press.

Simon, H. A. (1979). *Models of man*. New Haven, CT: Yale University Press.

Simon, H. A. (1982). *Models of bounded rationality* (Vol. 2). Cambridge, MA: MIT Press.

Sivazlian, B. D., & Stanfel, L. E. (1975). *Analysis of systems in operations research*. Englewood Cliffs, NJ: Prentice Hall.

Sklar, A. (1959). Fonctions de répartition à n dimensions et leurs marges. *Publications de l'Institut de Statistique de l'Université de Paris, 8*, 229–231.

Sklar, A. (1973). Random variables, joint distributions, and copulas. *Kybernetica, 9*, 449–460.

Slovic, P. (1972). Psychological study of human judgment: Implications for investment decision making. *Journal of Finance, 27*(4), 779–799.

Slutsky, E. (1937). The summation of random causes as the source of cyclic processes. *Econometrica, 5*, 105–146.

Smith, A. (1776). *The wealth of nations.* (Reprinted in 1994 by Modern Library). New York, NY.

Solberg, J. J. (1977). A mathematical model of computerized manufacturing systems, In *Proceedings of the 4th International Conference on Production Research*, Tokyo, Japan.

Soros, G. (2008). *The new paradigm for financial markets: The credit crisis of 2008 and what it means.* New York: Public Affairs.

Spence, A. M. (1977). Consumer misperceptions, product failure and product liability. *Review of Economic Studies, 44*, 561–572.

Srivastava, L. (1997). Energy and CO2 emissions in India: Increasing trends and alarming portents. *Energy Policy, 25*, 941–949.

Stackleberg, H. V. (1952). *The theory of the market economy*. Translated by Peacock AT, London, William Hodge and Co.

Stanley, M. H. R., Amaral, L. A. N., Bulkdyrev, S. V., Havlin, S. V., Leschron, H., Mass, P., Salinger, M. A., & Stanley, H. E. (1996). *Nature, 397*, 804.

Stein, C. (1973, September). Estimation of the mean of a multivariate normal distribution. In *Proceedings of the Prague Symposium on Asymptotic Statistics.*

Stiglitz, J. E. (1987). The causes and consequences of the dependence of quality on price. *Journal of Economic Literature, 5*, 1–48.

Stiglitz, J. E., et al. (1989). *The economic role of the state*. Oxford: Basil Blackwell.

Stirling, W. C. (2002). *Satisficing games and decision making: With applications to engineering and computer science*. London: Cambridge University Press.

Stirling, W. C., & Goodrich, M. A. (1999). Satisficing games. *Information Sciences, 114*, 255–280.

Stone, B. (1973). A general class of three-parameter risk measures. *Journal of Finance, 28*, 675–685.

Stratonovich, R. L. (1963). *Topics in the theory of random noise* (Vol. 1). New York: Gordon and Breach.

Stulz, R. M. (2009). Credit default swap and the credit crisis. http//ssrn.com/abstract=1475323.

Sugden, R. (1993). An axiomatic foundation of regret theory. *Journal of Economic Theory, 60*, 150–180.

Szego, G. (2002). (Special Issue Ed.) *Journal of Banking and Finance*, *26*, 1253–1272

Tadikamalla, P. R. (1980). A look at the Burr and related distributions. *International Statistical Review, 48*, 337–344.

Taguchi, G. (1985). Quality engineering in Japan. *Communications in Statistics – Theory and Methods, 14*(11), 2785–2801.

Taguchi, G., & Clausing, D. (1990). Robust quality. *Harvard Business Review*, *January*.

Taguchi, G., & Wu, Y. (1979). *Introduction to off-line quality control*. Nagoya, Japan: Central Japan Quality Control Association.

Taleb, N. N. (2007). *The Black Swan: The impact of the highly improbable*. New York: Random House.

Taleb, N. N. (2008). The fourth quadrant: A map of the limits of statistics. *Edge*. http://www.edge.org/3rd_culture/taleb08/taleb08_index.html.

Taleb, N. N. (2009). Errors, robustness, and the fourth quadrant. *International Journal of Forecasting, 25*(4), 744–759.

Taleb, N. N., & Tapiero, C.S. (2010). Risk externalities and too big to fail. *Physica A: Statistical Mechanics and Applications.*

Tang, L.-C., Lam, S.-W., & Goh, T.-N. A task-based six sigma roadmap for healthcare services.

Tapiero, C. S. (1975a). Random walk models of advertising, their diffusion approximations and hypothesis testing. *Annals of Economics and Social Measurement, 4*, 293–309.

Tapiero, C. S. (1975b). On line and adaptive optimum advertising, control by a diffusion approximation. *Operations Research, 23*, 890–907.

Tapiero, C. S. (1976). Accounting for the probable spatial impact of local activities. *Environment and Planning, 8*, 917–926.

Tapiero, C. S. (1977a). A stochastic model of sales response to advertising. *Metroeconomica, 29*, 159–167.

Tapiero, C. S. (1977b). *Managerial planning: An optimum and stochastic control approach.* New York: Gordon Breach (2 volumes).

Tapiero, C. (1978a) Time, Dynamics and the Process of Management Modeling. *TIMS Studies in Management Science, Special Issue on Applied Optimum Control.*

Tapiero, C. S. (1978b). Optimal advertising and goodwill under uncertainty. *Operations Research, 26*(3), 450–463.

Tapiero, C. S. (1979). A generalization of the nerlove-arrow model to multi firms advertising under uncertainty. *Management Science, 25*, 907–915.

Tapiero, C. S. (1980). A probability model for the effects of distance and the demand for multiple facilities. *Environment and Planning, 12.*

Tapiero, C. S. (1982a). A stochastic diffusion model with advertising and word-of-mouth effects. *European Journal of Operations Research, 12*, 348–356.

Tapiero, C. S. (1982b). A stochastic model of consumer behavior and optimal advertising. *Management Science, 28*, 1054–1064.

Tapiero, C. S. (1984). Mutual insurance, a diffusion stochastic control problem. *Journal of Economic Dynamics and Control, 7*, 241–260.

Tapiero, C., Reisman, A., & Ritchken, P. (1987). Product failures, manufacturing reliability and quality control: A dynamic framework. *INFOR (Canadian Journal of Operations Research), 25*(2), 1986, 152–163.

Tapiero, C. S. (1987). Learning and Quality Control. *IIE Transactions (American Institute of Industrial Engineers Journal), 19*, 362–370.

Tapiero, C. S. (1988). *Applied stochastic models and control in management.* Amsterdam: North-Holland.

Tapiero, C. S. (1990). An economic model for determining the optimal quality and process control policy in a queue like production system. *International Journal of Production Research, 28*, 1447–1457.

Tapiero, C. S. (1994a). The qualities of manufacturing and economic quality. *OR Insight.*

Tapiero, C. S. (1994b). Complexity and the industrial management. *OR Insight*, 12–18.

Tapiero, C. S. (1994c). Stochastic modeling: Art and science. *Internatinal Journal of Continuing Engineering Education.*

Tapiero, C. (1995a). Complexity and industrial systems. Special Issue Ed., *RAIRO.*

Tapiero, C. S. (1995b). Acceptance sampling in a producer-supplier conflicting environment: Risk neutral case. *Applied Stochastic Models and Data Analysis, 11*, 3–12.

Tapiero, C. S. (1996). *The management and the control of quality.* England: Chapman and Hall.

Tapiero, C. S. (1998a). The economic effects of reliable and unreliable testing technologies. *International Journal of Computer Integrated Manufacturing, 11*(3), 232–240.

Tapiero, C. S. (1998b). *Applied stochastic models and control in finance and insurance.* Boston: Kluwer Academic Press.

Tapiero, C. S. (2000a). The NBD repeat purchase process and M/G/Infinity queues. *International Journal of Production Economics, 63*, 141–145.

Tapiero, C. S. (2000b). Ex-post inventory control. *International Journal of Production Research, 38*(6), 1397–1406.

Tapiero, C. S. (2004a). Environmental quality control and environmental games. *Environmental Modeling and Assessment, 9*(4), 201–206.

Tapiero, C. S. (2004b). *Risk and financial management: Mathematical and computational concepts*. London: Wiley.

Tapiero, C. S. (2005a). *Risk management, encyclopedia on actuarial and risk management*. New York and London: Wiley.

Tapiero, C. S. (2005b). Modeling environmental queue control: A game model. *Stochastic Environmental Research and Risk Assessment, 19*(1), 59–70.

Tapiero, C. S. (2005c). Reliability design and RVaR. *International Journal of Reliability Quality and Safety Engineering (IJRQSE), 12*(4), 347–53.

Tapiero, C. S. (2005d). Environmental quality and satisficing games. *Journal of Science and Engineering B, 2*(1–2), 7–30.

Tapiero, C. S. (2005e). Advertising and advertising claims over time. In C. Deissenberg & R. F. Hartl (Eds.), *Optimal control and dynamic games, applications in finance, management science and economics*. Dordrecht: Springer.

Tapiero, C. S. (2005f). Value at risk and inventory control. *European Journal of Operations Research, 163*(3), 769–775.

Tapiero, C. S. (2005g). Reliability design and RVaR. *International Journal of Reliability, Quality and Safety Engineering (IJRQSE), 12*(4).

Tapiero, C. S. (2006). Risk and assets pricing. In H. Pham (Ed.), *Handbook of engineering statistics*. Berlin and New York: Springer.

Tapiero, C. S. (2007). Consumers risk and quality control in a collaborative supply chain. *European Journal of Operations Research, 182*, 683–694.

Tapiero, C. S. (2008). Orders and inventory commodities with price and demand uncertainty in complete markets. *International Journal of Production Economics, 115*, 12–18.

Tapiero, C. S. (2010a). *Risk finance and assets pricing*. Hoboken, NJ: Wiley.

Tapiero, C. S. (2010b). The future of financial risk management. *The Journal of Financial Transformation, 29*, 17–25.

Tapiero, C. S. (2012a). Insurance and finance: Convergence or completion. *Risk and Decision Analysis, 3*, 19–35.

Tapiero, C. S. (2012b). The price of safety and economic reliability. In H. Pham (Ed.) New York: Springer.

Tapiero, C. S., Capobianco, M. F., & Lewin, A. Y. (1975). Structural inference in organizations. *Journal of Mathematical Sociology, 4*, 121–130.

Tapiero, C. S., & Hsu, L. F. (1988). Quality control of an unreliable random FMS Bernoulli and CSP sampling. *International Journal of Production Research, 26*(6), 1125–1135.

Tapiero, C. S., & Kogan, K. (2009). Risk-averse order policies with random prices in complete markets and retailers' private information. *European Journal of Operational Research, 196*, 594–599.

Tapiero, O. (2012). *Implied risk neutral distribution: The non extensive entropy approach*, Ph.D Dissertation, Bar Ilan University, Israel.

Tapiero, C. S, & Totoutuom-Tangho, D. (2012). CDO: A modeling perspective. *Risk and Decision Analysis, 3*(1–2), 75–88.

Tapiero, C. S., & Vallois, P. (1996). Run length statistics and the Hurst exponent in random and birth-death random walks. *Chaos, Solitons and Fractals, September*.

Tapiero, C. S., & Vallois, P. (2000). The inter-event range process and testing for chaos in time series. *Neural Network World, 10*(1–2), 89–99.

Tetens, J. N. (1786). *Einleitung zur Berchnung der Leibrenten und Antwartschaften*, Leipzig.

Thaler, R. H., et al. (1997). The effect of myopia and loss aversion on risk taking: An experimental test. *Quarterly Journal of Economics, 112*, 647–661.

Tietenberg, T. (1998). *Environmental economics and policy*. Reading, MA: Addison-Wesely.

Tintner, G., & Sengupta, S. K. (1975). *Stochastic economic: Stochastic processes, control and programming*. New York: Academic Press.

Tirole, J. (1988). *The theory of industrial organization*. Cambridge, MA: MIT Press.

Tobin, J. (1958). Liquidity preferences as behavior towards risk. *Review of Economic Studies, 25*, 65–86.

Tolley, H. D., Bahr, M. D., & Dotson, P. K. (1989). A statistical method for monitoring social insurance claims. *Journal of Risk and Insurance, LVI*(4), 670–687.

Toth, B. (1986). Persistent random walks in random environments. *Probability Theory and Related Fields, 71*, 615–662.

Tsallis, C. (1994). *Physics Letter A, 195*, 329.

Tsallis, C., Anteneodo, L., Borland, L., & Osorio, R. (2003). Nonextensive statistical mechanics and economics. *Physica A: Statistical Mechanics and Applications, 324*(1), 89–100.

Tversky, A., & Kahneman, D. (1974). Judgement under uncertainty: Heuristics and biases. *Science, 185*(4157), 1124–1131.

Tversky, A., & Kahneman, D. (1992). Advances is prospect theory: Cumulative representation of uncertainty. *Journal of Risk and Uncertainty, 5*, 297–323.

Vallois, P. (1995). On the range process of a Bernoulli random walk. In J. Janssen & C. H. Skiadas (Eds.), *Proceedings of the Sixth International Symposium on Applied Stochastic Models and Data Analysis* (Vol. II, pp. 1020–1031). Singapore: World Scientific.

Vallois, P. (1996). The range of a simple random walk on Z. *Advances in Applied Probability, 28*, 1014–1033.

Vallois, P., & Tapiero, C. S. (1995a). Moments of an amplitude process in a random walk and approximations: Computations and applications. *Recherche Operationnelle/Operation Research (RAIRO), 29*(1), 1–17.

Vallois, P., & Tapiero, C. S. (1995b). The average run length of the range in birth and death random walks. In *Proceedings of the Conference on Applied Stochastic Models and Data Analysis*, Dublin.

Vallois, P., & Tapiero, C. S. (1996). The range process in random walks: Theoretical results and applications. In H. Ammans, B. Rustem, & A. Whinston (Eds.), *Advances in computational economics*. Dordrecht: Kluwer.

Vallois, P., & Tapiero, C. S. (1997a). Range reliability in random walks. *Mathematical Methods of Operational Research, 45*, 325–345.

Vallois, P., & Tapiero, C. S. (1997b). The range process in random walks: Theoretical results and applications. In H. Ammans, B. Rustem, & A. Whinston (Eds.), *Advances in computational economics* (pp. 291–307). Dordrecht: Kluwer Publications.

Vallois, P., & Tapiero, C. S. (1998). R/S analysis and the birth-death random walk. In *Proceedings IFAC, and Computational Economics Meeting in Cambridge (England)*, June 29, 1998.

Vallois, P., & Tapiero, C. S. (2001). The range inter-event process in a symmetric birth-death random walk. *Applied Stochastic Models in Business and Industry, Sept. 17*(3), 293–306.

Vallois, P., & Tapiero, C. S. (2007). Memory-based persistence in a counting random walk process. *Physica A: Statistical Mechanics and Applications, 386*(1), 303–317.

Vallois, P., & Tapiero, C. S. (2009). A claims persistence process and insurance. *Insurance Economics and Mathematics, 44*(3), 367–373.

Vander Beken, T., Dorn, N., & Van Daele, S. (2010). Security risks in nuclear waste management: Exceptionalism, opaqueness and vulnerability. *Journal of Environmental Management, 91*, 940–948.

Vives, X. (1984). Duopoly information equilibrium: Cournot and Bertrand. *Journal of Economic Theory, 24*, 71–94.

Von Neumann, J., & Morgenstern, O. (1944). *Theory of games and economic behavior*. Princeton, NJ: Princeton University Press.

von Stackleberg, H. (1934). *Marktform and Gleichgweicht*. Vienna: Springer.

Von Stengel, B. (1991). *Recursive inspection games*. Technical Report S-9106. Munich: University of the Federal Armed Forces.

Wakker, P. P. (1994). Separating marginal utility and probabilistic risk aversion. *Theory and Decision, 36*, 1–44.

Wakker, P. P. (2001). On the composition of risk preference and belief. *Psychological Review, 111*, 236–241.

Wakker, P. P. (2010). *Prospect theory for risk and ambiguity*. Cambridge: Cambridge University Press.

Wakker, P., & Tversky, A. (1993). An axiomatization of cumulative prospect theory. *Journal of Risk and Uncertainty, 7*, 147–176.

Wald, A. (1947). *Sequential analysis*. New York: Dover Publications.

Wald, A. (1961). *Statistical decision functions*. New York: Wiley.

Wald, A., & Wolfowitz, J. (1940). On a test whether two populations are from the same population. *Annals of Mathematical Statistics, 11*, 147–162.

Walras, L. (1874/1877). *Elements d'Economie Politique Pure*, Corbaz, Lausanne (English Translation in Allen and Unwin, London, UK, 1954).

Weber, I. (2002). Literature review, summary of recent articles on the convergence of insurance and financial markets and services. http://katie.cob.ilstu.edu/faculty_staff/literature_review_full.shtml.

Weber, T. A. (2003). An exact relation between willingness to accept and willingness to pay. *Economics Letters, 80*(3), 311–315.

Weber, S. (2006). Distribution-invariant risk measures, information, and dynamic consistency. *Mathematical Finance, 16*, 419–442.

Weber, T. A. (2008). *Price theory in economics*. Department of Management Science and Engineering, 442 Terman Engineering Center, Stanford University, Stanford, CA 94305-4026, webert@stanford.edu.

Wei John, K. C., & Lee, C. F. (1988). The generalized Stein/Rubinstein covariance formula and its application to estimate real systematic risk. *Management Science, 34*(10), 1266–1270.

Weil, P. (1990). Nonexpected utility in macroeconomics. *Quarterly Journal of Economics, 105*, 29–42.

Weiss, G. H. (1994). *Aspects and applications of the random walk*. Amsterdam: North-Holland.

Weiss, G. H. (2002). Some applications of persistent random walks and the telegrapher's equation. *Physica A: Statistical Mechanics and Applications, 311*, 381–410.

Weiss, G. H., & Rubin, R. J. (1983). Random walks: Theory and selected applications. *Advances in Chemical Physics, 52*, 363–505.

Welch, I. (1989). Seasoned offerings, imitation costs, and the underpricing of initial public offerings. *Journal of Finance, 44*, 421–449.

Wetherhill, G. B., & Brown, D. W. (1991). *Statistical process control*. London: Chapman and Hall.

Whitt, W. (1983). The queuing network analyzer. *Bell Systems Technical Journal, 62*, 2779.

Whittle, P. (1990). *Risk sensitive optimal control*. New York: Wiley.

Wiener, N. (1950). *The human use of human beings: Cybernetics and society*. New York: Avon Books.

Wiggins, J. B. (1992). Estimating the volatility of S&P 500 futures prices using the extreme-value method. *The Journal of Futures Markets, 12*(3), 265–273.

Willasen, Y. (1981). Expected utility, Chebychev bounds, mean variance analysis. *Economic Journal, 83*, 419–428.

Willasen, Y. (1990). Best upper and lower Tchebycheff bounds on expected utility. *Review of Economic Studies, 57*, 513–520.

Willsky, A. S. (1976). A survey of design methods for failure detection in dynamic systems. *Automatica, 12*(1976), 601–611.

Willsky, A. S. (1986). Detection of abrupt changes in dynamic systems, in detection of abrupt changes in signals and dynamical systems. In M. Basseville & A. Benveniste (Eds.), *Lecture notes in control and information sciences*, LNCIS 77. Springer, 27–49.

Willsky, A. S., & Jones, H. L. (1976). A generalized likelihood ratio approach to the detection and estimation in linear systems of jumps in linear systems. *IEEE Transactions on Automatic Control, 21*, 108–112.

Wolfowitz, J. (1943). On the theory of runs with some applications to quality control. *Annals of Mathematical Statistics, 14*, 280–288.

Xepapadeas, A. P. (1991). Environmental policy under imperfect information: Incentives and moral hazard. *Journal of Environmental Economics and Management, 20*, 113–126.

Xepapadeas, A. P. (1992). Environmental policy design and dynamic nonpoint-source pollution. *Journal of Environmental Economics and Management, 23*, 22–39.

Xepapadeas, A. P. (1994). Controlling environmental externalities: Observability and optimal policy rules. In C. Dosi & T. Tomasi (Eds.), *Nonpoint source pollution regulation: Issues and analysis*. Dordrecht: Kluwer Academic.

Xepapadeas, A. P. (1995). Observability and choice of instrument mix in the control of externalities. *Journal of Public Economics, 56*, 485–498.

Yaari, M. E. (1987). The dual theory of choice under risk. *Econometrica, 55*(1), 95–115.

Yang, D., & Zhang, Q. (2000). Drift independent volatility estimation based on high, low, open, and close prices. *Journal of Business, 73*, 477–492.

Yaniv, G. (1991). Absenteeism and the risk of involuntary unemployment. *Journal of Socio Economics, 20*(4), 359–372.

Yaron, D., & Tapiero, C. S. (Eds.). (1980). *Operations research in agriculture and water resources*. Amsterdam: North Holland Book Co.

Yip, T. L. (2007). Port traffic risks – A study of accidents in Hong Kong waters. *Transportation Research E44*.

Yitzhaki, S. (1983). On an extension of the Gini inequality index. *International Economic Review, 24*(3), 617–628.

Ziegler, A. (2004). *A game theory analysis of options*. Heidelberg: Springer.

Zsidisin, G. A. (2003). A grounded definition of supply risk. *Journal of Purchasing and Supply Management, 9*, 217–224.

Zsidisin, G. A., Panelli, A., & Upton, R. (2001). Purchasing organization involvement in risk assessments, contingency plans, and risk management: An exploratory study. *Supply Chain Management: An International Journal, 5*, 187–197.

Additional References

Abe, S. (2000). Remark on the escort distribution representation of non-extensive statistical mechanics. *Physics Letters A, 275*, 250–253.

Abel, A. (1990). Asset prices under habit formation and catching up with the Joneses. *American Economic Review, 80*, 38–42.

Achaya, V. V., Engle, R. F., Figlewski, S., Lynch, A. W., & Subrahmanyam, G. M. (2009). Centralized clearing for credit derivatives. In V. V. Acharya & M. Richardson (Eds.), *Restoring financial stability*. Hoboken NJ: Wiley.

Albanese C., Chen, O., & Dalessandro, A. (2005). *Dynamic credit correlation model*. Working paper.

Anand, K., Gai, P., & Marsili, M. (2011). Rollover risk, network structure and systemic financial crises. http://papers.ssrn.com/sol3/papers.cfm?abstract_id=1507196.

Arnsdorf, M., & Halperin, I. (2007). BSLP: Markovian bivariate spread-loss model for portfolio credit derivatives. http://www.defaultrisk.com.

Asian Financial Crisis. (1997). Wikipedia. http://en.wikipedia.org/wiki/1997, Asian Financial Crisis.

Ayres, R. U. (1979). *Uncertain futures: Challenges for decision-makers*. New York: Wiley.

Ayres, R. U, & Ayres, L. W. (Eds.) (2002). *A handbook of industrial ecology*. Cheltenham, UK and Northampton, MA: Edward Elgar Publishing. ISBN 1-84064-506-7. Retrieved November 22, 2010.

Ayres, R. U., & Simonis, U. E. (Eds.) (1994). *Industrial metabolism: Restructuring for sustainable development*. Tokyo and New York: United Nations University Press. ISBN 92-808-0841-9. Retrieved November 23, 2010.

Ayres, R. U., Weaver, P. M. (Eds.) (1998). *Eco-restructuring: Implications for sustainable development*. Tokyo, New York and Paris: United Nations University Press. ISBN 92-808-0984-9. Retrieved November 22, 2010.

Bachelier, L. J. B. (1900). *Theorie de la Speculation*. Paris: Gauthier-Villars (Reprinted in 1995, Editions Jaques Gabay, Paris, 1995).

Balinth, T. (1986). Persistent random walks in random environments. *Probability Theory and Related Fields, 71*, 615–625.

Barndorff-Nielsen, O. E., Mikosh, T., & Resnick, S. (Eds.). (2001). *Levy process – Theory and applications*. Birkauser: Basal.

Bawa, V. S. (1975). Optimal rules for ordering uncertain prospects. *Journal of Financial Economics, 2*, 95–121.

Ben-Avraham, D., & Havlin, S. (2000). *Diffusion and reactions in fractals and disordered systems*. Cambridge: Cambridge University Press.

Benhabib, J. (1992). *Cycles and chaos in economic equilibruim*. Princeton, NJ: Princeton University Press.

Brock, W. A., Hsieh, D. A., & LeBaron, B. (1992). *Nonlinear dynamics, chaos and instability*. Cambridge, MA: MIT Press.

Campos, D. (2010). Renyi and Tsallis entropies for incomplete or overcomplete systems of events. *Physica A: Statistical Mechanics and Applications, 389*, 981–992.

Cecchetti, S., Lam, P., & Mark, N. (1990). Evaluating empirical tests of asset pricing models. *American Economic Review, 80*(2), 48–51.

Chassin, M. R., Loeb, J. M., Schmaltz, S. P., & Wachter, R. M. (2010). Accountability measures – Using measurement to promote quality improvement. *The New England Journal of Medicine, 363*(7), 683–88.

Cohan, W. D. (2009). *House of cards*. New York, NY: Doubleday.

Cootner, P. H. (Ed.). (1964). *The random character of stock market prices*. Cambridge, MA: MIT Press.

Cox, D. (1967). *Renewal theory*. London: Methuen (First ed. in 1962).

Cox, D., & Isham, V. (1979). *Point processes*. London: Chapman & Hall.

Culp, C. L. (2002). *The art of risk management: Alternative risk transfer, capital structure and convergence between insurance and capital markets*. New York: Wiley.

Dacorogna, M. M., Gencay, R., Muller, U. A., Olsen, R. B., & Pictet, O. V. (2001). *An introduction to high frequency finance*. San Diego: Academic Press.

Darooneh, A. H., & Dadashinia, C. (2008). Analysis of the spatial and temporal distribution between successive earthquakes: Nonextensive statistical mechanics viewpoint. *Physica A: Statistical Mechanics and Applications, 387*, 3647–3654.

Darrel, I. (1991). *Software quality and reliability*. London: Chapman and Hall.

Dreze, J. H. (1977). *Demand theory under quantity rationing: A note*. Mimeo, CORE.

Duffie, D., & Pan, J. (1997). An overview of value-at-risk. *Journal of Derivatives, 4*, 27–49.

Dufresne, F., Gerber, H. U., & Shiu, E. (1991). Risk theory and the gamma process. *ASTIN Bulletin, 22*, 177–192.

Eberlein, E. (2001). Application of generalized hyperbolic levy motions to finance. In *Levy process: Theory and applications*. Birkhauser: Basel.

Eberlein, E., & Keller, U. (1995b). Hyperbolic distributions in finance. *Bernouilli, 1*, 281–299.

Eberlein, E., Keller, U., & Praue, K. (1998). New insights into smile, mispricing and value at risk: The hyperbolic model. *Journal of Business, 71*(3), 371–406.

Eberlein, E., Prause, K. (2000). The generalized hyperbolic model: Financial derivatives and risk measures. In *Mathematical Finance-Bachelier Congress 2000*. Berlin: Springer.

Epstein, L., & Zin, S. (1991b). Substitution, risk aversion, and temporal behavior of consumption and asset returns: A theoretical framework. *Econometrica, 57*(4): 937–969.

Failed bank list, Federal Deposit Insurance Corporation. http://www.fdic.gov/bank/individual/failed/banklist.html.

Fishburn, P. C. (1997). Mean-risk analysis with risk associated with below-target returns. *The American Economics Review, 67*, 116–126.

Frittelli, M., & Gianin, E. R. (2002). Putting order in risk measures. *Journal of Banking and Finance, 26*, 1473–1486.

Fung, H. G., & Lo, W. C. (1993). Memory in interest rate futures. *The Journal of Futures Markets, 13*, 865–873.

Fung, H.-G., Lo, W.-C., & Peterson, J. E. (1994). Examining the dependency in intra-day stock index futures. *The Journal of Futures Markets, 14*, 405–419.

Gibbs, G. W. (1928). *Elementary principle in statistical mechanics*. New York, NY: Longmans Green and Company.

Goodhart, C., & O'Hara, M. (1997). High-frequency data in financial markets: Issues and applications. *Journal of Empirical Finance, 4*, 73–114.

Grandell, J. (1991). *Aspects of risk theory*. Berlin: Springer.

Grandmont, J.-M. (1985). On endogenous competitive business cycles. *Econometrica, 53*(5), 995–1045.

Gregory, J., & Laurent, J. (2003). *Basket default swaps, CDO's and factor copulas*. Working paper, BNP Paribas, France.

Gregory, J., & Laurent, J. -P. (2003). I will survive. *Risk*, 103–107.

Helbing, D. (2010, October). Systemic risks in society and economics. International Risk Governance Council. http://irgc.org/IMG/pdf/Systemic_Risks_in_Society_and_Economics_Helbing.pdf.

Hilfer, R. (1995). Exact solution for a class of fractal time random walks. *Fractals, 3*, 211–216.

Hirsch, M. W., Smale, S., & Devaney, R. L. (2004). *Differential equations, dynamical systems, an introduction to chaos* (2nd ed.). San Diego, CA: Elsevier Academic Press.

Ho, T.-S., Stapleton, R. C., & Subrahmanyam, M. G. (1995). Multivariate binomial: Approximations for asset prices with nonstationary variance and covariance characteristics. *The Review of Financial Studies, Winter, 8*(4), 1125–1152.

Hommes, C. (2008). Interacting agents in finance. In *The new Palgrave dictionary of economics*. Palgrave Macmillan.

Huillet, T. (2002). Renewal processes and the Hurst effect. *Journal of Physics A: Mathematical and Theoretical, 35*, 4395–4413.

Hull, J., & White, A. (2001b). Valuing credit default swaps II: Modeling default correlations. *The Journal of Derivatives, 8*, 12–21.

Hull, J., & White, A. (2003b). The valuation of credit default swap options. *Journal of Derivatives, 10*(3), 40–50.

Inui, K., & Kijima, M. (2005). On the significance of expected shortfall as a coherent risk measure. In Risk Measurement. *Journal of Banking & Finance, 29*, 853–864.

Jacoby, J., & Kaplan, L. (1972). The components of perceived risk. In M. Venkatesan (Ed.), *Proceedings: Third Annual Conference, Atlanta Association for Consumer Research* (pp. 382–393).

Jarrow, R., & Yu, F. (2001). Counterparty risk and the pricing of defaultable securities. *Journal of Finance, 56*, 1765–1800.

Jouiniy, E., & Napp, C. (2009, December 2). *A class of models satisfying a dynamical version of the CAPM*. Université Paris IX-Dauphine and CREST.

Kaniadakis, G., Lavagno, A., & Quarati, P. (1996). *Physics Letters B, 369*, 308.

Khintchine, A. Y. (1955). *Mathematical methods in the theory of queuing*. London: Charles Griffin, 1960 (Translated from the 1955 Russian Edition).

Kopoci, B. (1999). Multivariate negative binomial distribution generated by multivariate exponential distributions. *Applicationes Mathematicae, 25*(4), 463–472.

Krokhmal, P. (2007). Higher moment coherent risk measures. *Quantitative Finance, 7*, 373–387.

Krokhmal, P., Uryasev, S., & Zrazhevsky, G. (2002). Risk management for hedge fund portfolios: A comparative analysis of linear rebalancing strategies. *Journal of Alternative Investments, 5*, 10–29.

Lancaster, K. (1971). *Consumer demand, a new approach*. New York: Columbia University Press.

Lucas, D., Goodman, L., & Fabozzi, F. (2006). *Collateralized debt obligations structures and analysis* (2nd ed.). Hoboken, NJ: Wiley.

Madan, D. B., & Seneta, E. (1990). The VG model for share market returns. *Journal of Business, 63*, 511–524.

Mankiw, N. G., & Zeldes, S. P. (1991). The consumption of stockholders and non stockholders. *Journal of Financial Economics, 29*, 97–112.

Martens, M., & van Dijk, D. (2007). Measuring volatility with the realized range. *Journal of Econometrics, 138*, 181–207.

Masoliver, J., Montero, M., Perello, J., & Weiss, G. H. The CTRW in finance: direct and inverse problem. http://xxx.lanl.gov/abs/cond-mat/0308017.

McKibbin, W. J., & Stoeckel, A. (2009). Modelling the global financial crisis. *Oxford Review of Economic Policy, 25*(4), 581–607.

Merton, R. C. (1973b). An intertemporal capital asset pricing model. *Econometrica, 41*, 867–887.

Minsky, H. P. (1992). *The financial instability hypothesis*. The Jerome Levy Economics Institute Working Paper No. 74.

Nau, R. F., & McCardle, K. F. (1990). Coherent behaviour in non-cooperative games. *Journal of Economic Theory, 50*, 424–444.

Naudts, J. (2006). Parameter estimation in non-extensive thermostatistics. *Physica A: Statistical Mechanics and Applications, 365*, 42–49.

Naudts, J. (2008). Generalised exponential families and associated entropy functions. *Entropy, 10*, 131–149.

Neri, U., & Venturi, B. (2007). Stability and bifurcations in IS-LM economic models. *International Review of Economics, 54*(1), 53–65.

Orazio, A. P., & Weber, G. (1995). Is consumption growth consistent with intertemporal optimization? Evidence from the consumer expenditure survey. *Journal of Political Economy, 103*, 1121–1157.

Rau, J. G. (1970b). *Optimization and probability in systems engineering*. New York: Van Nostrand.

Renyi, A. (1961). On measures of entropy and information. In *Proceedings of the 4th Berkeley Symposium on Mathematics, Statistics and Probability* (Vol. 1, pp. 547–561). Statistical Laboratory of the University of California, Berkeley, CA, USA, 20 June–30 June, 1960; University of California Press: Berkeley, California, 1961.

Renyi, A. (1970). *Probability theory*. Amsterdam, The Netherlands: North-Holland.

Rockfellar, R. T., & Uryasev, S. (2002). Conditional value-at-risk for general loss distributions. *Journal of Banking and Finance, 26*, 1443–1471.

Ross, S. A. (1976). The arbitrage theory of capital asset pricing. *Journal of Monetary Economics, 13*(3), 341–360.

Ross, S. M. (1997). *Introduction to probability models*. New York: Academic Press.

Sato, K. I. (1999). *Levy process and infinitely divisible distributions*. Cambridge: Cambridge University Press.

Scalas S. (2004). Five years of continuous-time random walks in econophysics. In A. Namatame (Ed.), *Proceedings of WEHIA 2004*, Kyoto. Downloadable from http://ideas.repec.org/p/wpa/wuwpfi/0501005.html.

Shannon, C. E. (1948). A mathematical theory of communication. *Bell System Technical Journal, 27*(379–423), 623–656.

Shooman, M. L. (1968). *Probabilistic reliability: An engineering approach*. New York: McGraw Hill.

Stulz, R. M. (2009, September). Credit default swaps and the credit crisis. Electronic copy available at http://ssrn.com/abstract=1475323.

Taguchi, G., Elsayed, E. A., & Hsiang, T. (1989). *Quality engineering in production systems.* New York: McGraw Hill.

Tapiero, C. S. (1975c). Random walk models of advertising, their diffusion approximations and hypothesis testing. *Journal of Economic and Social Measurement, 4*, 293–309.

Tapiero, C. S. (2004c). Risk management. In J. Teugels & B. Sundt (Eds.), *Encyclopedia on actuarial*. New York: Wiley.

Tapiero, C. S. (2011). Insurance and finance: Convergence and/or competition. *Risk and Decision Analysis*. (also SSRN, NYU-POLY, 2010).

Tapiero, C. S., & Farley, J. V. (1975). Optimal control of sales force effort in time. *Management Science, 21*, 976–985.

Tasche, D. (2002). Expected shortfall and beyond. *Journal of Banking and Finance, 26*, 1519–1553.

Thurner, S., Farmer, J. D., & Geanakoplos, J. (2010). *Leverage causes fat tails and clustered volatility*. Cowles Foundation Discussion Paper No. 1745. http://cowles.econ.yale.edu/P/cd/d17a/d1745.pdf.

Totoutuom-Tangho, D. (2007, November). *Dynamic COPULAS: Applications to finance and economics*. Doctoral Thesis, Mines Paris/ParisTech, Ed. No.396.

Truccoa, P., Cagnoa, E., Ruggerib, F., & Grandea, O. (2008). A bayesian belief network modelling of organizational factors in risk analysis: A case study in maritime transportation. *Reliability Engineering and System Safety, 93*, 823–834.

Tsallis, C. (1988). Possible generalization of Boltzmann-Gibbs statistics. *Journal of Statistical Physics, 52*, 479–488.

Tsallis, C. (1998). The role of constraints within generalized nonextensive statistics. *Physica A: Statistical Mechanics and Applications, 261*, 534–554.

Tsallis, C. (2001). *Non extensive statistical mechanics and its applications*. In S. Abe, Y. Okamoto (Eds.), Springer: Berlin.

Tsallis, C. (2004). *Nonextensive entropy-interdisciplinary applications*. In M. Gell-Mann, C. Tsallis (Eds.), New York, NY: Oxford University Press.

Tsallis, C. (2009). *Introduction to nonextensive statistical mechanics*. Berlin: Springer.

Tsallis, C., Gell-Mann, M., & Sato, Y. (2005). Special issue on the nonextensive statistical mechanics. *Europhysics News, 36*, 186–189.

Tsallis, C., Plastino, A. R., & Alvarez-Estrada, R. F. (2009). Escort mean values and the characterization of power-law-decaying probability densities. *Journal of Mathematics and Physics, 50*, 043303.

Wang, Q. A. (2001). Incomplete statistics: Nonextensive generalization of statistical mechanics. *Chaos, Solitons and Fractals, 12*, 1431–1437.

Wang, Q. A. (2002). Nonextensive statistics and incomplete information. *The European Physical Journal B, 26*, 357–368.

Index

A
Acceptance sampling, 377–387
Accountable (events), 2
Actuarial risk, 36–37, 288
Adverse selection, 27, 303, 400
Advertising risk, 45–46, 240
AHP. *See* Analytic hierarch process (AHP)
Allais's paradox, 234
Ambiguity, 196, 373
Analytic hierarch process (AHP), 211, 212
Anti-fragility, 33
Arbitrage, 21, 107, 214, 224, 226, 227, 256, 265, 266, 286, 305, 315, 323, 325, 334, 335
Arrow Pratt index of risk aversion, 234, 242, 243, 245
Assets pricing, 21, 30, 31, 65, 140, 229, 230, 281, 301, 334, 336–338
Audits, 29, 377, 378, 403

B
Background risks, 87, 208, 242–245, 373
Basel Committee, 31, 435
Bayesian analysis, 208
Bayesian controls, 383–387
Bayes theorem, 182, 459
Bernoulli, 64, 70, 79, 80, 83–87, 89, 90, 92, 117–119, 122–127, 129, 132, 133, 142–144, 150, 153, 159, 168, 191, 384, 385, 426, 459, 460
Beta distribution, 353, 459, 460
Big-data, 5, 199–201
Binomial distribution, 81, 84, 85, 87, 91–92, 117, 125, 154, 155, 165, 185, 191, 217, 241, 379, 414
Bi-variate Bernoulli, 117, 119, 124–126, 129, 133
Bi-variate beta distribution, 128–130
Bi-variate binomial distribution, 120, 126–127, 138
Bi-variate Lomax distribution, 130
Bi-variate Poisson distribution, 127
Boltzmann–Gibbs entropy, 333, 337, 338, 340, 346, 347
Bonus-malus, 37, 244
Booms and bust, 3, 248
Bounded rationality, 4, 167, 202, 209, 213, 255, 356–357, 373, 456, 457, 462
Branding risk, 45–46
Burr distribution, 101–104

C
Capital adequacy ratio (CAR), 36, 178, 324
Capital assets pricing model (CAPM), 38, 206, 265–269, 280, 319, 320
CAR. *See* Capital adequacy ratio (CAR)
Catastrophic, 5–8, 111, 415
Causal dependence, 110–112, 139–141, 198
CCAPM. *See* Consumption capital assets pricing model (CCAPM)
CDS. *See* Credit default swaps (CDS)
Certain equivalent, 233, 235, 236, 238, 241, 452
Chance constraints, 69, 75
Characteristic function, 76, 128, 180, 181
Coherent risk measurement
  homogeneity, 213–214
  monotonicity, 213, 214
  sub-additivity, 213
  translation invariance, 213–215
Collateral, 9, 39, 284–287
Collective risk, 159–161, 192, 280
Competing default risks, 114–115

C.S. Tapiero, *Engineering Risk and Finance*, International Series in Operations Research & Management Science 188, DOI 10.1007/978-1-4614-6234-7,

Complete markets, 18, 225, 231, 250, 256, 272, 275, 280, 300, 325, 334, 335, 349
Complexity, 3, 11, 13–15, 17–19, 50, 52, 53, 71, 85, 87, 167, 190, 201, 202, 206, 253, 255, 259, 297–299, 323, 324, 331, 367, 368, 373, 374, 378, 383, 393, 412, 413, 421, 439
Complexity risks, 13–14, 19, 53
Compound Poisson process, 82–83, 159–161
Concavity, 233
Concurrence, 15
Conditional Value at Risk (CVaR), 36, 67, 219
Consumption capital assets pricing model (CCAPM), 38, 215, 243, 251–281, 283, 290–297, 300, 327, 336, 342, 348, 462
Contagion, 1, 12, 45, 48, 50, 51, 53, 111, 149, 168–171, 178, 182–183, 187–190, 192, 214, 299, 368, 372, 382, 413, 415, 417
Contagious risks, 16
Control charts, 11, 41, 57, 65, 66, 173, 178, 387–392
Controls, 2, 33, 57, 154, 196, 246, 253, 308, 361, 375, 439
Convexity, 213
Copula, 109, 127, 130–138, 168, 374
Corporate risks, 23–25
Countable (events), 2, 244, 363
Counter-cyclicality, 413
Counter-party, 8–10, 39, 47, 285, 312, 330, 376, 438, 441, 444, 448, 458, 460
Credit
  derivatives, 10, 35, 38, 126, 138, 238, 250, 323, 325, 330, 331, 363, 378
  risk, 8–10, 38, 39, 125, 130, 250, 284–285, 287–288, 299, 303, 310–312, 330, 335
Credit default swaps (CDS), 10, 250, 289, 331
CVaR. *See* Conditional Value at Risk (CVaR)
Cyber risks, 1, 13, 299

D
Data envelopment analysis (DEA), 211, 212
Data management, 34
DEA. *See* Data envelopment analysis (DEA)
Debt, 1, 8–10, 20, 38–40, 64, 65, 251, 256, 266, 272, 279, 283–298, 303, 310–317, 330, 439
Default, 5, 39, 62, 109, 153, 202, 236, 283, 335, 377, 439
Desertification, 5, 6, 299, 303
Diagnosis, 16, 167, 206
Disappointment, 8, 45, 204, 240, 363
Distribution tail, 60, 130, 169, 248
Divergence, 25, 216–218, 370
Diversification, 10, 114, 213, 324
Dodd–Frank, 13

E
Economic inequalities, 20, 215, 218, 226, 272, 280, 300
Endogenous risks, 28, 33, 438
Entropy, 12, 23, 107, 216–218, 232, 330, 337–356, 374
Environmental pollution, 49, 366–367
Ex-ante, 2, 34, 48, 161, 204, 240, 334, 362, 363, 375, 441, 443
Ex-ante risk management, 33, 34
Excess-loss, 244
Exchange, 2, 5, 8, 10, 12, 18, 21, 22, 27–30, 35–38, 41, 47, 53, 65, 116, 138, 141, 152, 223, 224, 228, 229, 232, 236, 252–256, 265, 272, 289, 297, 298, 300–312, 319, 321, 322, 326, 328, 335, 336, 363, 374, 403, 440–442, 448, 449, 456
Experimental risks, 16
Exponential distribution, 84, 95, 151, 154
Ex-post, 2, 4, 23, 33, 34, 46, 204, 209, 218, 240, 247, 357, 360, 362–364, 375, 377, 382, 383, 441, 458
Ex-post risk management, 33, 333
Extended CCAPM, 215, 243, 256, 266
External risks, 24, 25, 28–31, 39, 40, 49, 197, 331, 333, 364–368, 370–373, 379, 412, 415, 417, 422–424, 431, 432, 434
Extreme probability distribution, 100
Extreme statistics, 60–61, 191
Extreme winds, 6, 7

F
Failure mode evaluation and critical analysis (FMECA), 207–208, 345
Fault tolerance, 41
Fault tolerant system, 393
Fault tree analysis, 208
Filtration, 59–60, 143, 226, 227, 229, 265, 287, 307, 461
Financial price, 335
Financial risks, 14, 18, 19, 30, 35, 38, 68, 138, 191, 206, 213, 214, 339, 366
Financial valuation, 203, 226, 249
FMECA. *See* Failure mode evaluation and critical analysis (FMECA)
Food safety, 1
Fractal, 110, 142, 168–172, 192, 374

G

Game
  and risk, 439–443
  theory, 29, 259, 403–404, 437–440, 444, 462
Gamma probability distribution, 81, 93, 95–96, 101, 344, 350
Generalized Pareto distribution (GPD), 100, 178–180
Genetics, 206
Geometric distribution, 84, 87, 147, 276, 397, 419, 427
Globalization, 4–5, 10, 11, 18, 20, 24, 266, 297–299, 365, 378, 412
GPD. *See* Generalized Pareto distribution (GPD)
Group technology, 52
Gumbel, 100, 134

H

Hazard, 8, 17, 26–28, 40, 63–65, 89, 101, 103, 114, 115, 179, 220, 221, 287, 288, 290, 300, 324, 325, 331, 403, 441
Hazard rate, 63–65, 89, 101, 103, 115, 179, 220, 221, 287, 288, 290
Health care, 1, 8, 14, 16, 17, 27, 31, 35, 40, 41, 55, 57, 168, 203–206, 253, 393, 398, 400, 438
Hurst index, 142, 169–172, 192
Hyperbolic absolute risk aversion (HARA) utility function, 236, 243, 329–330
Hyper geometric distribution, 414

I

Implied distribution, 340
Implied preferences, 199
Incomplete markets, 18, 21–22, 226, 256, 280, 334, 370
Incompleteness, 22, 226, 325, 334, 338, 342, 344, 347, 348, 373, 438
Incomplete state preferences, 344–356, 373, 441
Index of criticality, 207
Industrial risk, 41, 55, 66
Information asymmetry, 17, 22, 27–30, 37, 204, 226, 325, 374, 403, 434, 440
Infrastructure, 1, 8, 20, 196, 310, 331, 365, 369–370
Initial Public Offering (IPO), 1, 46–49, 55, 280, 331, 368
Integrity risks, 52, 53
International Standardization Organization (ISO) certification, 15
Inverse problem, 325
IPO. *See* Initial Public Offering (IPO)

J

Jensen's inequality, 233, 246

K

Kendall's tau, 112, 116, 117
Kernel, 229–232, 267, 292, 293, 295, 296, 302, 328–329, 348, 349, 356
Kernel pricing, 230, 265–267, 271, 275, 277, 278, 281, 290, 295, 296, 302, 306, 313, 314, 316, 325, 342
Kurtosis, 58–61, 67, 82, 85, 93, 94, 135, 184, 233, 281

L

Laplace, 30, 203, 218, 235, 337, 338, 340
Laplace Stiljes, 76
Laplace transform (LT), 76, 188, 220, 221
Law of requisite variety, 14
Legal risk, 9
Lexis probability distribution, 85
Liability risks, 16, 196
Linear correlation, 110, 112, 115, 116
Linear regression, 72, 115, 268
Logistic(s)
  distribution, 97–98, 156
  risks, 65, 70
LOGIT, 71–73, 138, 206, 353
Lognormal distribution, 94–95, 340

M

Macro-prudential, 4
Macro risks, 20
Management of quality, 14–15
Managing uncertainty, 4
Marketing, 1, 36, 44–49, 55, 191, 327, 335, 435
Market price, 37, 38, 42, 48, 110, 126, 150, 196, 224, 226, 228, 229, 231, 245, 252–254, 256, 272, 306, 315, 325, 327, 343, 348, 349, 437, 459, 462
Markov chains, 51, 142, 156–158, 183
Martingale, 224–232, 325, 336
Maximax, 203
Measures of deviation, 199
Medial correlation, 112, 117, 135

Memory, 61, 84, 90, 95, 109, 110, 120, 139–143, 146–148, 151–153, 168–171, 182–190, 192, 225–227
Memory processes, 142, 168, 189
Micro-financial, 18, 21, 262
Micro–macro mismatch, 19–21
Micro-prudential, 4
Minimax, 204, 360, 362, 443, 444, 450
Minimin, 203
Moment generating function, 76–79, 81, 82, 94, 96, 343, 344
Moments, 58–61, 66, 68, 75–81, 83, 84, 90, 93–96, 99, 107, 111, 118, 119, 125–127, 129, 135, 145, 174, 181, 184–186, 189, 206, 216, 218, 231–233, 264, 268, 270, 289, 295, 296, 339, 343, 344, 351, 355
Monte Carlo, 104–106, 135, 167, 174, 224, 400
Moral hazard, 17, 27, 28, 40, 300, 324, 325, 331, 403, 441
Mortgage Backed securities (MBS), 1, 10, 20, 35, 126, 198, 284, 325, 378
Multinomial distribution, 117
Multiple objectives, 198, 202, 203, 212
Multivariate Bernoulli distribution, 117, 125
Multivariate discrete distributions, 117–127
Multivariate Poisson distribution, 127
Multivariate probability distributions, 109–138

N

Negative binomial distribution, 84, 85, 92–93, 154
Networking, 11–13, 15, 24, 299, 368, 374
Networks, 12, 13, 17, 26, 31, 40, 50–54, 89, 191, 208, 265, 325, 365, 368–369, 372, 374, 376, 378, 394, 395, 400–403, 435
Normal probability distribution, 60, 61, 68, 70, 74, 75, 81, 85, 87, 93–98, 128, 138, 144, 162, 167, 168, 180, 233, 246–248, 340, 342
Numeraire, 226, 254, 255

O

Operations risks, 52, 376
Outliers risks, 172–173
Outsourcing, 11, 12, 15, 24, 298, 326, 370, 413, 441

P

Poisson distribution, 84, 85, 87, 90–92, 95, 127, 149, 155, 159, 160, 386
Poisson process, 82–83, 90, 141, 146–149, 154, 155, 159–161
Power asymmetries, 11, 26, 29, 253, 255, 364, 403, 413
Preferences, 11, 14, 27, 30, 36, 38, 140, 195, 198–204, 211, 212, 218, 219, 223, 224, 226, 227, 232–235, 242, 243, 246, 249, 251, 256, 265–267, 272, 275, 281, 288, 294, 300, 301, 308, 325, 334–336, 344–357, 373, 438–441, 443, 457, 458, 462
Privatization, 370
Probability generating function, 76, 77, 79–82, 85, 91, 118, 120, 125, 127, 129, 130, 153, 160, 183, 185, 187, 188
Probability measure, 207, 225, 228, 229, 267, 271, 272, 279, 280, 302, 305, 307, 308, 313, 314, 316, 319–321, 323, 342, 343, 348, 374
Process capability, 389, 392–394
Prospect theory, 23, 345, 373
Prudence, 242–245, 329

Q

Quality, 1, 11, 14–17, 27, 29–31, 39, 41, 42, 48, 50, 55, 57, 66, 70–71, 75, 131, 140, 168, 191, 203–206, 208, 209, 213–215, 225, 233, 240, 246, 250, 253, 254, 299, 335, 361, 368, 375, 393, 434, 435, 439, 442
Quality assurance, 11, 70–71
Queue control, 54, 394–403
Queues, 50, 51, 53, 54, 191, 394–403, 435
Queuing networks, 394, 401–403
Queuing theory, 51, 191

R

Random increment processes, 143
Randomness, 3, 53, 74, 83, 87, 104, 105, 166–168, 182, 217, 218, 389, 439, 441
Random payoff games, 403, 405, 438, 440, 452, 454–456
Random process, 150, 190, 191
Random walk, 85, 139, 141–156, 163–165, 168–175, 178, 189–193
Range, 58, 60–61, 99, 107, 130, 142, 169–175, 177, 178, 190, 192, 193, 206, 213, 216, 361, 367, 388, 392
Range processes, 142, 192
Rare events, 3, 7, 18, 28, 111, 166–181, 208, 373

Rationality, 4, 21, 22, 167, 199, 201, 202, 204, 209, 211, 213, 224, 232, 234, 255, 256, 288, 325, 334, 356–364, 373, 430, 438, 441, 443, 450, 453, 456–458, 462
Rational valuation, 34
Recovery, 9, 10, 25, 26, 33, 34, 49, 105, 250, 284, 289, 357, 362
Re-engineering, 15
Regime-unstable, 3
Regret, 23, 33, 34, 45, 202, 204, 250, 357, 360, 362, 363, 373
Regulation, 1, 5, 8–14, 18, 19, 25, 27–29, 31, 36–40, 43, 46, 47, 49, 50, 111, 119, 120, 153, 154, 178, 196–198, 205, 214, 224, 232, 239, 240, 244, 248, 249, 251, 284, 298–300, 324, 331, 333, 335, 358, 364–373, 375–435, 440, 442, 443
Regulatory distortion, 10
Reliability, 30, 51, 52, 57, 62–65, 70, 88–90, 96, 101, 102, 107, 121, 125, 130, 141, 190, 193, 207, 208, 220, 224, 248, 250, 258, 331, 368, 398, 428, 429
Reliability Availability Maintainability (RAM), 43, 57, 207–208
Reputation risk, 1, 16, 44–45, 47, 48
Risk analysis, 33, 34, 55, 57, 105, 106, 403, 413
Risk assessment, 23, 25, 90, 167, 303
Risk attitudes, 2, 22, 23, 40, 203, 223, 233, 235, 239, 242–244, 288, 308, 325, 328–330, 403, 437
Risk averse, 36, 48, 196, 233, 236, 239, 242, 244–246, 250, 257
Risk avoidance, 233
Risk bearing, 232
Risk consequences, 2, 13, 27, 39, 64, 69, 70, 98, 169, 196, 198, 205, 206, 211, 215, 216, 266, 365, 376, 434
Risk convergence, 36
Risk derivatives, 197, 205, 335
Risk design, 2, 33, 34, 68, 379
Risk engineering, 1–31, 190, 191, 204, 298
Risk exposure, 9, 18, 25, 33, 34, 47, 68, 110, 161–163, 173, 174, 178, 189, 208, 213, 214, 246, 248, 301
Risk externality, 24, 25, 28–31, 39, 40, 49, 197, 331, 364–373, 379, 412, 415–417, 422–424, 431
Risk factors, 48, 64, 66, 110, 112–114, 125, 126, 131, 136, 167, 205, 206, 360, 375
Risk finance, 18, 30, 31
Risk incentive, 8, 326
Risk management, 3, 4, 9, 11, 15, 30, 33–55, 57, 66, 110, 178, 199, 201, 213, 250, 324, 438
Risk measurement, 34, 37, 69, 110, 195–221, 244, 250
Risk minimization, 38
Risk of certainty, 18–19, 196
Risk of complexity, 18, 19, 255
Risk of incomplete markets, 21–22
Risk premium, 8, 21, 22, 34, 36–38, 66, 213, 233–239, 241, 243, 244, 247, 255, 268, 271, 272, 288, 314, 320, 322, 324, 335, 452
Risk pricing, 20, 33, 34, 57, 65, 203, 205, 224, 229–232, 256, 259–260, 266, 283–331, 342
Risk science (RS), 192
Risk sharing, 2, 33–35, 244, 288, 326, 441
Risks of regulation, 18, 19
Risk technology, 11–12
Risk tooling, 248
Risk valuation, 2, 33, 34, 195, 203, 219, 223–250
Robust design, 15, 33, 34, 361
Robustness, 15, 41, 208, 357–362, 393
RS. *See* Risk science (RS)
Run time stochastic models, 153–154

S

Sabotage, 11
Safety, 1, 13, 16–17, 30, 31, 45, 55, 63, 196, 197, 207, 208, 218, 223, 250, 334, 358, 369, 372, 375, 403, 438
Sampling errors, 71
Scarcity, 252, 253, 255–265
Scenario, 26, 361–364, 373
SDF. *See* Stochastic discount factor (SDF)
Securitization, 10, 34, 39, 238, 284, 325, 335
Security, 1, 8, 10, 13, 20, 24, 31, 35, 36, 45, 46, 54, 61, 63, 66, 137, 197, 200, 268, 302, 303, 334, 340, 363, 378, 438
Sensitivity analysis, 250, 358, 405, 411
Severity, 206, 376, 426
Short term memory, 109, 182–190, 226
Simplicity, 13–15, 58, 63, 96, 153, 161, 207, 230, 241, 247, 257, 259, 260, 303, 309, 311, 329, 360, 384, 387, 399, 407, 423, 425, 428, 445, 454, 456, 461
Simulation, 28, 40, 51, 64, 100, 104–106, 127, 167, 177, 191, 357, 359, 360, 363, 377, 400
Six sigma (6 sigma), 15, 16, 57, 388, 392–393

Skewness, 58–62, 67, 82, 83, 85, 94, 95, 135, 169, 181, 233, 268, 342
Social risks, 12
Social security, 8
Space, 5, 7, 54, 140, 159, 229, 365, 394
SPC. *See* Statistical process control (SPC)
Spearman's rho, 112, 116, 135
Spread, 12, 13, 35, 115, 142, 172, 255, 258, 267, 280, 289–290, 323, 338, 458
Stake-holders, 11, 15, 16, 25, 26, 203, 204, 285, 378
Standard deviation, 41, 59, 61, 62, 65, 66, 68, 170–172, 192, 214, 216, 218, 238, 247, 279, 302, 361, 388, 391–393, 452
Statistical controls, 14–15, 41, 43, 53, 71, 215, 378, 382, 396, 398, 412, 414–428, 434, 443, 447
Statistical functional dependence, 109
Statistical process control (SPC), 14, 41, 57, 178
Statistical risks, 27, 41, 69–70, 372, 377–387
Stealth firms, 12, 16, 43
Stochastic discount factor (SDF), 265, 281
Stochastic dominance, 220, 221, 345
Stop-loss, 68, 244
Strategic risks, 11, 18, 24, 28, 29, 48, 197, 208, 227, 259, 299, 375–435, 439, 441, 444, 452–456
Subjective valuation, 34, 224
Super and hyper Poisson distribution, 92
Supply chains, 1, 15, 25, 26, 31, 40, 50, 51, 55, 111, 368, 378, 401, 440, 462
Sustainability, 24, 25, 324, 365, 366, 413
Syndication, 10
Systemic risks, 1, 12, 22, 29, 39, 192, 248, 283, 324, 365, 366, 377, 382, 395, 412, 417, 432, 434

T

TBTB. *See* Too Big to Bear (TBTB)
TBTF. *See* Too Big To Fail (TBTF)
Technology, 1, 6, 7, 10–15, 20, 24, 27, 49, 50, 52, 111, 167, 212, 257, 259, 298, 299, 324, 326, 327, 367, 368, 393, 404, 405
Time VaR, 161–164, 177, 178, 248
Too Big to Bear (TBTB), 1, 39, 214
Too Big To Fail (TBTF), 30, 35, 39, 40, 214, 253, 323, 324, 364, 376, 412, 413, 415
Total Quality Control (TQC), 15
Total Quality Management (TQM), 15, 41, 393
TQC. *See* Total Quality Control (TQC)
TQM. *See* Total Quality Management (TQM)
Transportation risks, 52
Trinomial process, 163–166
Trinomial random walk, 145, 174
Tsallis entropy, 337, 338, 342, 345–356, 374
Type I error (alpha), 71, 379, 423, 429, 431
Type II error (beta), 71, 379, 387, 403, 430

U

Uncertainty economics, 216, 220, 333–374
Uncertainty models, 22–23, 231, 374
Underpricing, 48, 331
Unstable systems, 3
Utility valuation, 213, 224, 232, 247, 437

V

Value at Risk (VaR), 36, 67–68, 153, 161–164, 173, 177, 178, 196, 215, 216, 244, 246–248, 250, 299, 335, 358, 372, 435, 440
Variance, 38, 58, 59, 61, 62, 65–68, 80–83, 85–87, 90, 92–96, 101, 102, 110, 112–115, 120, 121, 124, 125, 128, 135, 143–145, 147, 149, 150, 153, 160, 162, 167–171, 174, 175, 180, 184, 188–190, 192, 203, 214, 216, 233–235, 238, 241, 246, 249, 264, 268, 280, 339, 340, 342, 356, 361, 374, 379, 388, 419, 455, 458, 460
Volatility, 1, 18, 47, 59, 61, 65–67, 70, 81, 142, 163, 167–178, 191, 192, 279, 280, 297, 299, 322, 373, 374
Volatility at risk, 163, 173–178

W

Warranty, 27, 35, 46, 224, 229, 240, 254, 272, 405
Weibull, 76, 95, 100–103, 134, 175
Work sampling, 11

9781461462330